MW01012067

QUANTUM MECHANICS
An Introduction

Springer
Berlin
Heidelberg
New York
Barcelona
Hong Kong
London
Milan
Paris
Singapore
Tokyo

Greiner
Quantum Mechanics
An Introduction 4th Edition

Greiner
Quantum Mechanics
Special Chapters

Greiner · Müller
Quantum Mechanics
Symmetries 2nd Edition

Greiner
Relativistic Quantum Mechanics
Wave Equations 3rd Edition

Greiner · Reinhardt
Field Quantization

Greiner · Reinhardt
Quantum Electrodynamics
2nd Edition

Greiner · Schramm · Stein
Quantum Chromodynamics
2nd Edition

Greiner · Maruhn
Nuclear Models

Greiner · Müller
Gauge Theory of Weak Interactions
3rd Edition

Greiner
Mechanics I
(in preparation)

Greiner
Mechanics II
(in preparation)

Greiner
Classical Electrodynamics

Greiner · Neise · Stöcker
Thermodynamics and Statistical Mechanics

Walter Greiner

QUANTUM MECHANICS

An Introduction

With a Foreword by
D. A. Bromley

Fourth Edition
With 57 Figures
and 88 Worked Examples and Problems

Springer

Professor Dr. Walter Greiner

Institut für Theoretische Physik der
Johann Wolfgang Goethe-Universität Frankfurt
Postfach 11 19 32
60054 Frankfurt am Man
Germany

Street address:

Robert-Mayer-Strasse 8–10
60325 Frankfurt am Main
Germany

e-mail: greiner@th.physik.uni-frankfurt.de

Title of the original German edition: *Theoretische Physik,* Band 4: *Quantenmechanik Teil I, Eine Einführung* © Verlag Harri Deutsch, Thun, 1989

Library of Congress Cataloging-in-Publication Data applied for.

Die Deutsche Bibliothek – CIP-Einheitsaufnahme

Quantum mechanics / Walter Greiner; Berndt Müller. With a foreword by D. A. Bromley. – Berlin; Heidelberg; New York; Barcelona; Hong Kong; London; Milan; Paris; Singapore; Tokyo: Springer (Theoretical physics; ...)
Einheitssacht.: Quantenmechanik <engl.>
An introduction: with 88 worked examples and problems / with a foreword by D. A. Bromley. – 4. ed. – 2000
(Theoretical physics; Vol. 1) (Physics and astronomy online library) ISBN 3-540-67458-6

ISBN 3-540-67458-6 4th Edition Springer-Verlag Berlin Heidelberg New York

ISBN 3-540-58079-4 3rd Edition Springer-Verlag Berlin Heidelberg New York

Springer-Verlag Berlin Heidelberg New York
a member of BertelsmannSpringer Science+Business Media GmbH

Typesetting: Data conversion by LE-TEX, Leipzig
Cover design: Design Concept, Emil Smejkal, Heidelberg
Copy Editor: V. Wicks
Production Editor: P. Treiber
Printed on acid-free paper SPIN 10767248 56/3141/tr - 5 4 3 2 1 0

Foreword to Earlier Series Editions

More than a generation of German-speaking students around the world have worked their way to an understanding and appreciation of the power and beauty of modern theoretical physics – with mathematics, the most fundamental of sciences – using Walter Greiner's textbooks as their guide.

The idea of developing a coherent, complete presentation of an entire field of science in a series of closely related textbooks is not a new one. Many older physicists remember with real pleasure their sense of adventure and discovery as they worked their ways through the classic series by Sommerfeld, by Planck and by Landau and Lifshitz. From the students' viewpoint, there are a great many obvious advantages to be gained through use of consistent notation, logical ordering of topics and coherence of presentation; beyond this, the complete coverage of the science provides a unique opportunity for the author to convey his personal enthusiasm and love for his subject.

The present five volume set, *Theoretical Physics*, is in fact only that part of the complete set of textbooks developed by Greiner and his students that presents the quantum theory. I have long urged him to make the remaining volumes on classical mechanics and dynamics, on electromagnetism, on nuclear and particle physics, and on special topics available to an English-speaking audience as well, and we can hope for these companion volumes covering all of theoretical physics some time in the future.

What makes Greiner's volumes of particular value to the student and professor alike is their completeness. Greiner avoids the all too common "it follows that ... " which conceals several pages of mathematical manipulation and confounds the student. He does not hesitate to include experimental data to illuminate or illustrate a theoretical point and these data, like the theoretical content, have been kept up to date and topical through frequent revision and expansion of the lecture notes upon which these volumes are based.

Moreover, Greiner greatly increases the value of his presentation by including something like one hundred completely worked examples in each volume. Nothing is of greater importance to the student than seeing, in detail, how the theoretical concepts and tools under study are applied to actual problems of interest to a working physicist. And, finally, Greiner adds brief biographical sketches to each chapter covering the people responsible for the development of the theoretical ideas and/or the experimental data presented. It was Auguste Comte (1798–1857) in his *Positive Philosophy* who noted, "To understand a science it is necessary to know its history". This is all too often forgotten in modern physics teaching and the bridges that Greiner builds to the pioneering figures of our science upon whose work we build are welcome ones.

Greiner's lectures, which underlie these volumes, are internationally noted for their clarity, their completeness and for the effort that he has devoted to making physics an integral whole; his enthusiasm for his science is contagious and shines through almost every page.

These volumes represent only a part of a unique and Herculean effort to make all of theoretical physics accessible to the interested student. Beyond that, they are of enormous value to the professional physicist and to all others working with quantum phenomena. Again and again the reader will find that, after dipping into a particular volume to review a specific topic, he will end up browsing, caught up by often fascinating new insights and developments with which he had not previously been familiar.

Having used a number of Greiner's volumes in their original German in my teaching and research at Yale, I welcome these new and revised English translations and would recommend them enthusiastically to anyone searching for a coherent overview of physics.

Yale University
New Haven, CT, USA
1989

D. Allan Bromley
Henry Ford II Professor of Physics

Preface to the Fourth Edition

We are pleased once again that the need has arisen to produce a new edition of *Quantum Mechanics: Introduction*. We have taken this opportunity to make several amendments and improvements to the text. A number of misprints and minor errors have been corrected and explanatory remarks have been given at various places. Also, several new examples and exercises have been added.

We thank several colleagues and students for helpful comments, particularly Dr. Stefan Hofmann who supervised the preparation of this fourth edition of the book. Finallly, we acknowledge the agreeable collaboration with Dr. H. J. Kölsch and his team at Springer-Verlag, Heidelberg.

Frankfurt am Main
July 2000

Walter Greiner

Preface to the Third Edition

The text *Quantum Mechanics – An Introduction* has found many friends among physics students and researchers so that the need for a third edition has arisen. There was no need for a major revision of the text but I have taken the opportunity to make several amendments and improvements. A number of misprints and minor errors have been corrected and a few clarifying remarks have been added at various places. A few figures have been added or revised, in particular the three-dimensional density plots in Chap. 9.

I am grateful to several colleagues for helpful comments, in particular to Prof. R.A. King (Calgary) who supplied a comprehensive list of corrections. I also thank Dr. A. Scherdin for help with the figures and Dr. R. Mattiello who has supervised the preparation of the third edition of the book. Furthermore I acknowledge the agreeable collaboration with Dr. H. J. Kölsch and his team at Springer-Verlag, Heidelberg.

Frankfurt am Main
July 1994

Walter Greiner

Preface to the Second Edition

Like its German companion, the English edition of our textbook series has also found many friends, so that it has become necessary to prepare a second edition of this volume. There was no need for a major revision of the text. However, I have taken the opportunity to make several minor changes and to correct a number of misprints. Thanks are due to those colleagues and students who made suggestions to improve the text. I am confident that this textbook will continue to serve as a useful introduction to the fascinating topic of quantum mechanics.

Frankfurt am Main,
November 1992

Walter Greiner

Preface to the First Edition

Quantum Mechanics – An Introduction contains the lectures that form part of the course of study in theoretical physics at the Johann Wolfgang Goethe University in Frankfurt. There they are given for students in physics and mathematics in their fourth semester. They are preceded by Theoretical Mechanics I (in the first semester). Theoretical Mechanics II (in the second semester), and Classical Electrodynamics (in the third semester). Quantum Mechanics I – An Introduction then concludes the foundations laid for our students of the mathematical and physical methods of theoretical physics. Graduate work begins with the courses Thermodynamics and Statistical Mechanics, Quantum Mechanics II – Symmetries, Relativistic Quantum Mechanics, Quantum Electrodynamics, the Gauge Theory of Weak Interactions, Quantum Chromodynamics, and other, more specialized lectures.

As in all the other fields mentioned, we present quantum mechanics according to the inductive method, which comes closest to the methodology of the research physicist: starting with some key experiments, which are idealized, the basic ideas of the new science are introduced step by step. In this book, for example, we present the concepts of "state of a system" and "eigenstate", which then straightforwardly lead to the basic equation of motion, i.e. to the Schrödinger equation; and, by way of a number of classic, historically important observations concerning the quantization of physical systems and the various radiation laws, we infer the duality of waves and particles, which we understand with Max Horn's conception of a "guiding field".

Quantum mechanics is then further developed with respect to fundamental problems (uncertainty relations; many-body systems; quantization of classical systems; spin within the phenomenological Pauli theory and through linearization of wave equations; etc.), applications (harmonic oscillator; hydrogen atom; Stern–Gerlach, Einstein–de Haas, Frank–Hertz, and Rabi experiments), and its mathematical structure (elements of representation theory; introduction of the *S* Matrix, of Heisenberg, Schrödinger, and interaction pictures; eigendifferentials and the normalization of continuum wave functions; perturbation theory; etc.). Also, the elements of angular-momentum algebra are explained, which are so essential in many applications of atomic and nuclear physics. These will be presented in a much broader theoretical context in *Quantum Mechanics – Symmetries.*

Obviously an introductory course on quantum theory cannot (and should not) cover the whole field. Our selection of problems was carried out according to their physical importance, their pedagogical value, and their historical impact on the development of the field.

Students profit in the fourth semester at Frankfurt from the solid mathematical education of the first two years of studies. Nevertheless, in these lectures new mathematical tools and methods and their use have also to be discussed. Within this category belong the solution of special differential equations (especially of the hypergeometrical and confluent hypergeometrical differential equations), a reminder of the elements of matrix calculus, the formulation of eigenvalue problems, and the explanation of (simple) perturbation methods. As in all the lectures, this is done in close connection with the physical problems encountered. In this way the student gets a feeling for the practical usefulness of the mathematical methods. Very many worked examples and exercises illustrate and round off the new physics and mathematics.

Furthermore, biographical and historical footnotes anchor the scientific development to the general side of human progress and evolution. In this context I thank the publishers Harri Deutsch and F.A. Brockhaus (*Brockhaus Enzyklopädie*, F.A. Brockhaus, Wiesbaden – marked by BR) for giving permission to extract the biographical data of physicists and mathematicians from their publications.

The lectures are now in their 5th German edition. Over the years many students and collaborators have helped to work out exercises and illustrative examples. For the first English edition I enjoyed the help of Maria Berenguer, Snježana Butorac, Christian Derreth, Dr. Klaus Geiger, Dr. Matthias Grabiak, Carsten Greiner, Christoph Hartnack, Dr. Richard Hermann, Raffaele Mattiello, Dieter Neubauer, Jochen Rau, Wolfgang Renner, Dirk Rischke, Thomas Schönfeld, and Dr. Stefan Schramm. Miss Astrid Steidl drew the graphs and pictures. To all of them I express my sincere thanks.

I would especially like to thank Mr. Béla Waldhauser, Dipl.-Phys., for his overall assistance. His organizational talent and his advice in technical matters are very much appreciated.

Finally, I wish to thank Springer-Verlag; in particular. Dr. H.-U. Daniel, for his encouragement and patience, and Mr. Mark Seymour, for his expertise in copy-editing the English edition.

Frankfurt am Main
July 1989

Walter Greiner

Contents

Contents of Examples and Exercises

1. The Quantization of Physical Quantities

1.1 Light Quanta

In order to explain physical phenomena caused by light, two points of view have emerged, each of which has its place in the history of physics. Almost simultaneously in the second half of the seventeenth century the corpuscle theory was developed by Newton and the wave theory of light was created by Huygens. Some basic properties like the rectilinear propagation and reflection of light can be explained by both theories, but other phenomena, such as interference, the fact that light plus light may cause darkness, can be explained only by the wave theory.

The success of Maxwell's electrodynamics in the nineteenth century, which interprets light as electromagnetic waves, seemed finally to confirm the wave theory. Then, with the discovery of the photoelectric effect by ***Heinrich Hertz*** in 1887, a development began which led ultimately to the view that light has to be described by particles or waves, depending on the specific problem or kind of experiment considered. The "particles" of light are called *quanta of light* or *photons*, the co-existence of waves and particles wave–particle duality.

In the following we shall discuss some experiments which can be explained only by the existence of light quanta.

1.2 The Photoelectric Effect

The ejection of electrons from a metal surface by light is called the *photoelectric effect*. An experiment by ***Philipp Lenard*** showed that the energy of the detached electrons is given by the frequency of the irradiating light (Fig. 1.1).

Monochromatic light yields electrons of a definite energy. An increase in light intensity leads to the emission of more electrons, but does not change their

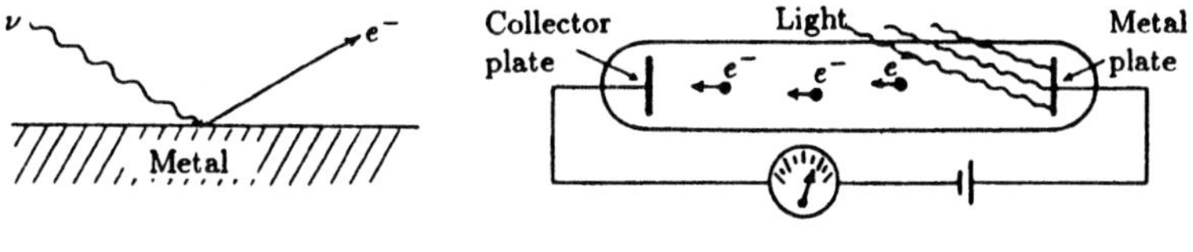

Fig. 1.1. Measurement of the photoelectric effect: light (→) shines on a metal thereby liberating electrons (e^-)

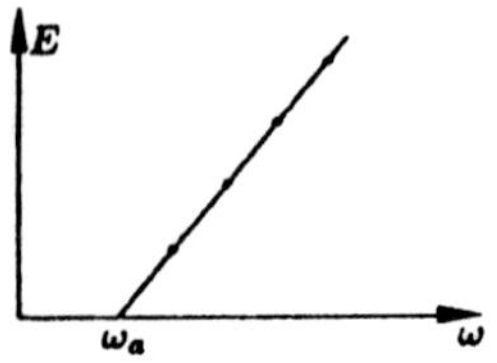

Fig. 1.2. The linear increase of photoelectron energy with frequency ω of the incident light

energy. This is in clear contradiction to classical wave theory, where the energy of a wave is given by its intensity. If we carry out the experiment with monochromatic light of different frequencies, a linear dependence between energy and frequency is obtained, as shown in Fig. 1.2:

$$E \propto (a + b\omega) \ . \tag{1.1}$$

The proportionality factor, i.e. the slope of the straight line, is found to be Planck's constant h divided by 2π, so that

$$E = \hbar(\omega - \omega_a) = h(\nu - \nu_a) \tag{1.2}$$

with $h = 2\pi\hbar = 6.6 \times 10^{-34}\,\mathrm{Ws}^2$.

Einstein interpreted this effect by postulating discrete quanta of light (photons) with energy $\hbar\omega$. Increasing the intensity of the light beam also increases the number of photons, which can break off correspondingly more electrons from the metal.

In these experiments, a frequency limit ω_a appears, which depends on the kind of metal. Below this frequency limit, no electrons can leave the metal. This means that a definite *escape energy* $\hbar\omega_a$ is needed to raise electrons from the surface of the metal.

The light quantum that has to be postulated to understand the photoelectric effect moves with the velocity of light. Hence it follows from Einstein's *Theory of Relativity* that the rest mass of the photon is equal to zero.

If we set the rest mass equal to zero in the general relation for the total energy

$$E^2 = (m_0c^2)^2 + p^2c^2 = \hbar^2\omega^2 \tag{1.3}$$

and express the frequency by the wave number $k = \omega/c$, the momentum of the photon follows as

$$p = \hbar k = \hbar\omega/c \ , \tag{1.4}$$

or written as a vector identity, assuming that the direction of the momentum of the photon should correspond to the propagation direction of the light wave,

$$\boldsymbol{p} = \hbar\boldsymbol{k} \ . \tag{1.5}$$

1.3 The Compton Effect

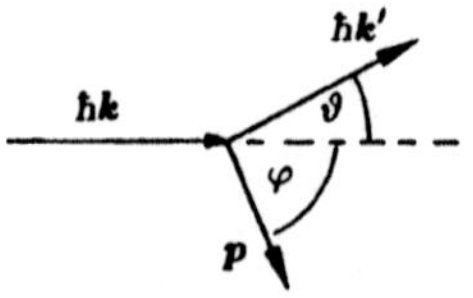

Fig. 1.3. Conservation of momentum in Compton scattering

When X-rays are scattered by electrons, a shift in frequency can be observed, the amount of this shift depending on the scattering angle. This effect was discovered by ***Compton*** in 1923 and explained on the basis of the photon picture simultaneously by Compton himself and ***Debye***.

Figure 1.3 illustrates the kinematical situation. We assume the electron is unbound and at rest before the collision. Then the conservation of energy and momentum reads:

$$\hbar\omega = \hbar\omega' + \frac{m_0c^2}{\sqrt{1-\beta^2}} - m_0c^2 , \qquad (1.6)$$

$$\hbar \boldsymbol{k} = \hbar \boldsymbol{k}' + \frac{m_0 \boldsymbol{v}}{\sqrt{1-\beta^2}} . \qquad (1.7)$$

To obtain a relation between the scattering angle ϑ and the frequency shift, we divide (1.7) into components parallel and vertical to the direction of incidence. This yields, with $k = \omega/c$,

$$\frac{\hbar\omega}{c} = \frac{\hbar\omega'}{c}\cos\vartheta + \frac{m_0 v}{\sqrt{1-\beta^2}}\cos\varphi \quad \text{and} \qquad (1.8)$$

$$\frac{\hbar\omega'}{c}\sin\vartheta = \frac{m_0 v}{\sqrt{1-\beta^2}}\sin\varphi . \qquad (1.9)$$

From these two component equations, we can first eliminate φ and then, by (1.6), the electron velocity v ($\beta = v/c$). Hence for the frequency difference we have

$$\omega - \omega' = \frac{2\hbar}{m_0c^2}\omega\omega' \sin^2\frac{\vartheta}{2} . \qquad (1.10)$$

If we put $\omega = 2\pi c/\lambda$, we obtain the *Compton scattering formula* in the usual form with the difference in wavelength as a function of the scattering angle ϑ:

$$\lambda' - \lambda = 4\pi\frac{\hbar}{m_0c}\sin^2\frac{\vartheta}{2} . \qquad (1.11)$$

The scattering formula shows that the change in wavelength depends only on the scattering angle ϑ. During the collision the photon loses a part of its energy and the wavelength increases ($\lambda' > \lambda$).

The factor $2\pi\hbar/m_0c$ is called the *Compton wavelength* λ_c of a particle with rest mass m_0 (here, an electron). The Compton wavelength can be used as a measure of the size of a particle. The kinetic energy of the scattered electron is then

$$T = \hbar\omega - \hbar\omega' = \hbar c 2\pi\left(\frac{1}{\lambda} - \frac{1}{\lambda'}\right) , \qquad (1.12)$$

or (see Fig. 1.4)

$$T = \hbar\omega\frac{2\lambda_c \sin^2\vartheta/2}{\lambda + 2\lambda_c\sin^2\vartheta/2} . \qquad (1.13)$$

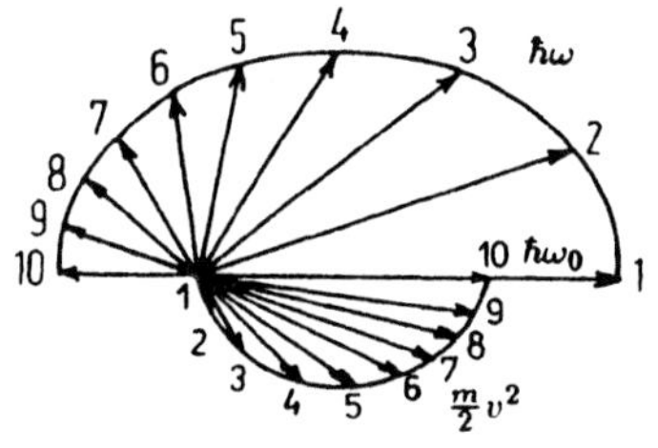

Fig. 1.4. The Compton effect energy distribution of photons and electrons, showing dependence on the scattering angle

Thus the energy of the scattered electron is directly proportional to the energy of the photon. Therefore the Compton effect can only be observed in the domain of short wavelengths (X-rays and γ-rays). To appreciate this observation fully, we have to remember that in classical electrodynamics, no alteration in frequency is permitted in the scattering of electromagnetic waves; only light quanta with momentum $\hbar \boldsymbol{k}$ and energy $\hbar\omega$ make this possible. Thus the idea of light quanta has been experimentally confirmed by the Compton effect. A relatively broad Compton line appears in the experiment, due to certain momentum distributions of the electrons and because the electrons are bound in atoms.

The Compton effect is a further proof for the concept of photons and for the validity of momentum and energy conservation in interaction processes between light and matter.

1.4 The Ritz Combination Principle

In the course of investigations of the radiation emitted by atoms, it appeared that characteristic spectral lines belong to each atom and that these lines can be formally arranged into certain spectral series (for example, the Balmer series in the hydrogen atom). The ***Ritz*** combination principle (1908) states that new spectral lines can be found by additive and subtractive combination of two known spectral lines. The existence of spectral lines means that transitions (of the electrons) between discrete energy levels take place within the atom.

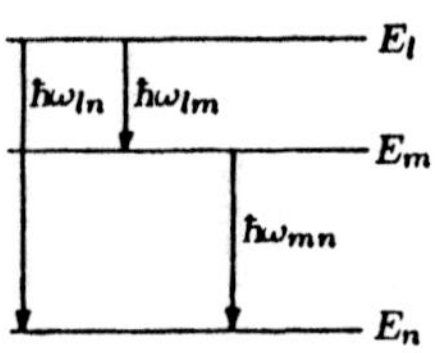

Fig. 1.5. The Ritz combination principle

The frequency condition $E = \hbar\omega$ yields an explanation for the Ritz combination principle. Considering the transition of an atom from a state with energy E_l to a state with energy E_n (Fig. 1.5), we have

$$\hbar\omega_{ln} = E_l - E_n = E_l - E_m + E_m - E_n \tag{1.14}$$

or for the frequencies,

$$\omega_{ln} = \omega_{lm} + \omega_{mn} \quad . \tag{1.15}$$

In Fig. 1.5, the energy levels and the corresponding transitions are represented schematically. The spectral series result from transitions from different higher energy levels into a common "ground state" E_n. Thus, the spectral analysis of atoms suggests quite clearly that only discrete energy levels exist in an atom and that energy can be transferred only by light quanta with a definite energy.

1.5 The Franck–Hertz Experiment

Another experiment demonstrating the quantization of energy was performed by ***Franck*** and ***Hertz*** in 1913, using a triode filled with mercury vapour. The triode consists of an axial cathode K in a cylindrical grid A closely surrounded by a third electrode Z. The electrons are accelerated between K and A to reach Z through the anode grid. A small countervoltage prevents very slow electrons from reaching Z. The experiment yields a current–voltage characteristic between K and Z as shown in Fig. 1.6.

As long as the energy of the electron in the field does not exceed 4.9 eV, the electrons can cross the tube without a loss of energy. The exchange of energy due to elastic collisions between electrons and mercury atoms can be neglected. The current increases steadily, but as soon as the energy of the electrons has reached 4.9 eV, the current drops drastically. A mercury atom obviously can take up exactly this much energy from the electrons in a collision. Thereafter an electron has insufficient kinetic energy to reach the second anode Z and the atom emits this energy with the characteristic wavelength of $\lambda = 2537$ Å. On increasing the voltage further, the electrons can regain kinetic energy and the process repeats itself.

The Franck–Hertz experiment shows the existence of discrete energy levels (quantization of energy) in the mercury atom.

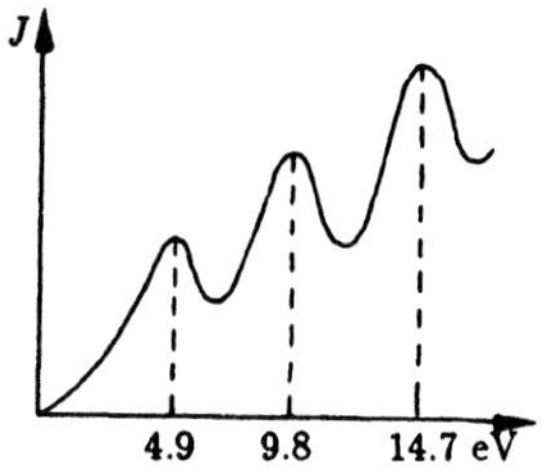

Fig. 1.6. Current (J)–voltage (eV) characteristics for the Franck–Hertz experiment showing regular maxima and minima

1.6 The Stern–Gerlach Experiment

In their experiment performed in 1921, ***Stern*** and ***Gerlach*** observed the splitting of an atomic beam in an inhomogeneous magnetic field. If an atom possesses a magnetic moment $\boldsymbol{m}$, it will be affected not only by a torque, but also by a force $\boldsymbol{F}$ when in an inhomogeneous magnetic field $\boldsymbol{H}$. The potential energy in the magnetic field is given by $V = -\boldsymbol{m} \cdot \boldsymbol{H}$; the force is given by the gradient, i.e. $\boldsymbol{F} = -\operatorname{grad} V = \operatorname{grad} \boldsymbol{m} \cdot \boldsymbol{H}$.

In the experiment, a beam of neutral silver atoms was sent through an inhomogeneous magnetic field and the distribution of the atoms after passing the field was measured (for a detailed discussion see Chap. 12). Classically one would expect a broadening of the beam, due to the varying strength of the magnetic field. In practice, however, the beam is split into two distinct partial beams. The intensity distribution on the screen is shown qualitatively in Fig. 1.7.

This doubly peaked distribution obviously means that the magnetic moment of the silver atoms cannot orient itself arbitrarily with respect to the magnetic field; rather, only two opposing orientations of the magnetic moment in the field are possible. This cannot be understood classically. Obviously the phenomena of quantization appearing in the atomic domain are not restricted to energy and momentum only, but are also found in other physical quantities. This particular quantization is called directional quantization (or quantization of the angular momentum, see Sect. 4.8).

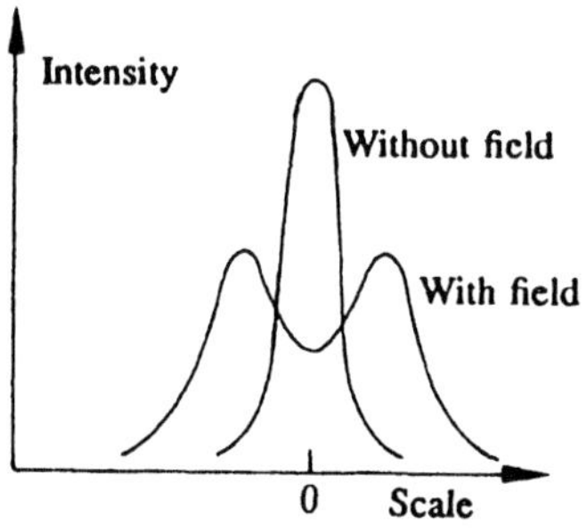

Fig. 1.7. Intensity distributions of the Ag atoms after their transition through an inhomogeneous magnetic field

1.7 Biographical Notes

HERTZ, Heinrich Rudolf, German physicist, *Hamburg 22.2.1857, †Bonn 1.1.1894, a professor of physics in Karlsruhe and Bonn, confirmed with his experiments concerning the propagation of electromagnetic waves the predictions of Maxwell's electromagnetic theory of light in 1887/88. He discovered the so-called *Hertz waves*, which are the physical fundamentals of modern radio engineering. He proved the influence of ultraviolet light on electrical discharge (1887), which led to the discovery of the photoelectric effect by W. Hallwachs. In 1892 H. observed the transmission of cathode rays through thin metal plates and gave P. Lenard the task of explaining their nature. H. also gave an exact definition of hardness.

LENARD, Philipp, German physicist, *Preßburg 7.6.1862, †Messelhausen (Baden-Württ.) 20.5.1947, student of H. Hertz, was a professor in Breslau, Aachen, Kiel and Heidelberg. Using the window tube suggested by Hertz, L. was the first to investigate cathode rays as free electrons independent of the way they were generated and made a major contribution to the explanation of the nature of these rays. Among other things he showed that the rate of absorption of cathode rays is nearly proportional to the mass of the radiated substance. Furthermore he demonstrated that the velocity of electrons emitted due to the photoelectric effect is independent of the intensity of light, but depends on its frequency. Thus he created the experimental foundation for the fundamental photoelectric law formulated by Einstein. Of equal importance was his verification that the active centre of an atom is concentrated in a nucleus, which is tiny in comparison with the radial dimension of the whole atom. Later this fact was also experimentally proved by E. Rutherford. The explanation of the mechanism of phosphorescence and the proof that an electron must have a definite minimum energy to ionize an atom are further achievements by L. He also introduced the "electron-volt" (eV) as a unit of measurement. In 1905 he received the Nobel Prize in Physics. L. was as renowned an experimental physicist as his contemporaries J.J. Thomson and E. Rutherford, but was sceptical of Einstein's Special Theory of Relativity. He rejected the Weimar Republic and gradually developed into a fanatical anti-semite and national socialist. [BR]

EINSTEIN, Albert, German physicist, *Ulm 31.4.1879, †Princeton (N.J.) 18.4.1955. Having grown up in Munich, he moved to Switzerland at the age of 15. As a "technical expert third class" at the patent office in Bern, he published in 1905 in Vol. 17 of *Annalen der Physik* three most important papers. In his "On the theory of Brownian motion" he published a direct and conclusive proof, based on a purely classical picture, of the atomistic structure of matter. In his paper, "On the electrodynamics of moving bodies", he set up with his profound analysis of the terms "space" and "time" the *Special Theory of Relativity.* From this he concluded a few months later the general equivalence of mass and energy, expressed by the famous formula $E = mc^2$. In his third article, E. extended the quantum approach of M. Planck (1900) in "On a heuristic viewpoint concerning the production and transformation of light" and made the second decisive step towards the development of quantum theory, directly leading to the idea of the duality of particles and waves. The concept of light quanta was considered too radical by most physicists and was very sceptically received. A change in the opinion of physicists did not take place until Niels Bohr proposed his theory of atoms (1913). E., who became a professor at the University of Zürich in 1909, went to Prague in 1911, and returned to Zürich a year later where he joined the Eidgenössische Technische Hochschule. In 1913 he was called to Berlin as a full-time member of the Preussische Akademie der Wissenschaften and director of the Kaiser-Wilhelm-Institut für Physik. In 1914/15 he developed the *General Theory of Relativity*, starting from the strict proportionality of gravitational and inertial mass. As a result of the successful testing of his theory by a British solar eclipse expedition, E. became well known to the general public. His political and scientific opponents tried unsuccessfully to start a campaign against him and his theory of relativity. The Nobel Prize Committee therefore considered it advisable to award E. the 1921 Nobel Prize in Physics not for his theory of relativity, but for his contributions to quantum theory. Beginning in 1921 E. tried to set up his unified theory of matter which aimed to incorporate electrodynamics as well as gravitation. Even after it had been shown by H. Yukawa that other forces exist besides gravitation and electrodynamics, he continued with his efforts which, however, remained unsuccessful. Although he published a paper in 1917 which was instrumental to the statistical interpretation of quantum theory, he later raised severe objections, based on his philosophical point of view to the "Copenhagen Interpretation" proposed by N. Bohr and W. Heisenberg. Several attacks because of his Jewish background caused E. in 1933 to relinguish all the academic positions he held in Germany; at the Institute for Advanced Study in Princeton, in the U.S.A., he found somewhere new to continue his studies. The final stage of E.'s life was overshadowed by the fact that although a life-long pacifist, fearing German aggression he initiated the development of the American atomic bomb by writing, together with others, to President Roosevelt on 8.2.1939. [BR]

COMPTON, Arthur Holly, American physicist, *Wooster (Ohio) 10.9.1892, † Berkeley (CA) 15.3.1962, became a professor at Washington University, St. Louis, in 1920 and at the University of Chicago in 1923. In 1945 he became chancellor of Washington University. In the course of his investigations on X-rays he discovered the Compton effect in 1922. He and Debye simultaneously gave the quantum-theoretical explanation for this effect. C. was also the first to prove the total reflection of X-rays. Together with R.L. Doan, he achieved the diffraction of X-rays from a diffraction grating. Jointly with C.T.R. Wilson he was awarded the Nobel Prize in Physics in 1927. In cooperation with his students C. carried out extensive investigations on cosmic rays. During the Second World War he participated in the development of the atomic bomb and radar as director of the plutonium research project of the American Government. [BR]

DEBYE, Petrus Josephus Wilhelmus, Dutch physicist, naturalized in America in 1946, *Maastricht (Netherlands) 24.3.1884, † Ithaca (N.J.) 2.11.1966, was called "the Master of the Molecule". In 1911 he became a professor at the University of Zürich as successor to A. Einstein, then in Utrecht (1912–1914), Göttingen (1914–1920), at the Eidgenössische Technische Hochschule in Zürich (1920–1927), in Leipzig (1927–1935), and was director of the Kaiser-Wilhelm-Institut für Physik in Berlin, 1935–1939. In 1940 he emigrated to the United States and became a professor of chemistry at Cornell University (Ithaca) in 1948. There he directed the chemistry department from 1940 until his retirement in 1952. D. was famous both as a theoretical and as an experimental physicist. He formulated the T^3 law for the decrease of the specific heat of solids at low temperatures. He developed the Debye–Scherrer method (1917 independently of A.W. Hull) and, jointly with E. Hückel, formulated a theory of dissociation and conductivity of strong electrolytes. Independently of F.W. Glaugue and almost at the same time D. pointed out the possibility of reaching low temperatures by adiabatic demagnetization of ferromagnetic substances. During extensive research, he determined the dipole moments of molecules. This research together with results of the diffraction experiments of X-rays and electron rays from gases and liquids enabled him to establish their molecular structure; for this he was awarded the Nobel Prize in Chemistry in 1936. After his retirement, he developed methods to determine the molecular weight and the molecular expansion of giant molecules of highly polymerized substances. [BR]

RITZ, Walter, Swiss physicist, *Sitten 22.2.1878, † Göttingen 7.7.1909, formulated the combination principle for spectral lines in 1908.

FRANCK, James, German physicist, *Hamburg 20.8.1882, † Göttingen (during a journey through Germany) 21.5.1964. Franck was a member of the Kaiser-Wilhelm-

Institut für Physikalische Chemie, and, beginning in 1920, a professor in Göttingen; he left Germany in 1933. From 1935, F. was a professor of physics at Johns Hopkins University in Baltimore; 1938–1947, professor of physical chemistry in Chicago; from 1941 on, he was also active at the University of California. Jointly with G. Hertz, at the Physikalisches Institut in Berlin, F. investigated the energy transfer of electrons colliding with gas atoms. His results sustained Planck's quantum hypothesis as well as the theory of spectral lines postulated by Bohr in 1913. For this work F. and Hertz were awarded the Nobel Prize in Physics in 1926. Extending these investigations, F. measured for the first time the dissociation energy of chemical compounds by optical means and determined the lifetime of metastable states of atoms. In addition he developed the law for the intensity distribution within a band structure, which is known today as the *Franck–Condon principle*. In the U.S.A. he devoted himself primarily to the investigation of photochemical processes within the living plant cell. During the Second World War F. worked on a project involving the technical utilization of nuclear energy. In 1945 he warned of the political and economic consequences of the use of atomic bombs in a petition which has become well known as the *Franck Report*. [BR]

HERTZ, Gustav, German physicist, nephew of Heinrich Hertz, *Hamburg 22.7.1887, † Berlin 30.10.1975, first was a professor in Halle and Berlin and head of the research laboratory of the Siemens factories. From 1945–1954, H. built up an institute at Suchumi on the Black Sea together with former students and collaborators; in 1954 he directed a university institute in Leipzig. From 1911 on, together with J. Franck, he investigated the excitation of atoms by collisions with electrons; they shared the Nobel Prize in Physics in 1926. In 1932, H. developed the technique of isotope separation with a diffusion cascade consisting of many single steps. He applied this method to the extraction of uranium 235 on a large technical scale in the Soviet Union. [BR]

STERN, Otto, German physicist, *Sorau (Niederlausitz) 17.2.1888, † Berkeley (CA) 17.8.1969, became a professor in Rostock in 1921, and in Hamburg in 1923, where he also acted as director of the Physikalisch-Chemisches Institut. Beginning in 1915, S. developed the method of using molecular rays for the determination of atomic and nuclear properties. He had particular success in discovering the quantization of the magnetic moment (the Stern–Gerlach experiment), in his diffraction experiments with molecular hydrogen and helium rays (1929), and in determining the magnetic moment of the proton (begun in 1933). S. emigrated to the United States in 1933 and worked at the Carnegie Institute of Technology in Pittsburgh. In 1943, he was awarded the Nobel Prize in Physics. [BR]

GERLACH, Walther, German physicist, *Biebrich a. Rh. 1.8.1889, † Munich 1979, professor in Frankfurt, Tübingen and from 1929 in Munich, determined the value of the Stefan–Boltzmann constant by precision measurements (1916). Together with Otto Stern he showed the quantization of the magnetic moment by deflection of atomic rays in an inhomogeneous magnetic field (1921). At that time G. was a lecturer at Physikalisches Institut der Universität Frankfurt a.M.; Otto Stern was visiting lecturer at the Institute für Theoretische Physik in Frankfurt, which was directed by Max Born (as successor to Max v. Laue) at that time. Futhermore, G. worked on quantitative spectral analysis and the coherence between atomic structure and magnetism. In extensive historical analyses of science, G. tried to point out the "humanistic value of physics". [BR]

2. The Radiation Laws

The energy density $\varrho(\omega, T)$ of the radiation emitted by a black body was described by two separate, contradictory theorems in classical physics. The Rayleigh–Jeans radiation law accounted for experiments in the region of long-wave radiation; Wien's law for those in the region of short-wave radiation. By introducing a new constant h, Planck was successful in finding an interpolation between the two laws.

Planck's radiation law covered the whole range of frequencies and contained the two other radiation laws as specific extreme cases (Fig. 2.1). In the beginning, Planck's law was only an interpolation formula, but later he was able to show that this radiation law could be deduced under the assumption that the energy exchange between radiation and black body was discontinuous. The quanta of energy transfer are given by the relation $E = h\nu$, ν being the frequency. From the historical point of view, this was the beginning of quantum mechanics.

In the following we will explicitly discuss the various radiation laws.

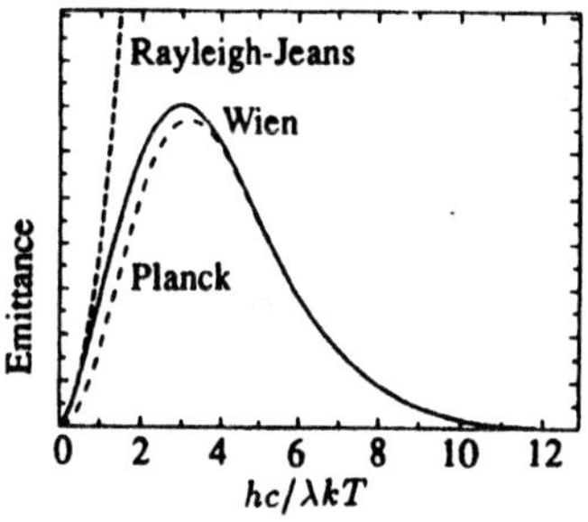

Fig. 2.1. Qualitative behaviour of the emittance according to the laws of Rayleigh–Jeans and Wien: the spectrum of a black body as a function of $\hbar\omega/kT = hc/kT\lambda$

2.1 A Preview of the Radiation of Bodies

If radiation hits a body, it can enter its interior or be reflected by its surface. The reflection is *regular* if the angle between the incident ray and the normal to the surface is the same as between reflected ray and normal and if all of these together, i.e. incident, reflected ray and normal, are situated in the same plane. On the other hand, if the rays are also reflected in other directions, we then speak of *diffuse* reflection. If the reflected part is the same in all directions, independent of the direction of incidence and of the colour of the light, then one calls the reflecting surface *gray*. If all the incident light is reflected without any loss in this way, the surface is *white*.

A white surface element $\mathrm{d}F$ reflects (emits) the radiant flux

$$J(\omega, T) \cos\theta \, \mathrm{d}F \, \mathrm{d}\Omega \, \mathrm{d}\omega$$

into the solid angle $\mathrm{d}\Omega$ at an angle θ relative to the vertical of $\mathrm{d}F$. Its radiance $J(\omega, T)$ is the same for all directions. The radiant flux is proportional to the cosine of the angle θ between the direction of reflection and the normal of the surface (*Lambert's cosine law*). If a gray or white surface of arbitrary shape is illuminated from an arbitrary direction, it will reflect with the same apparent

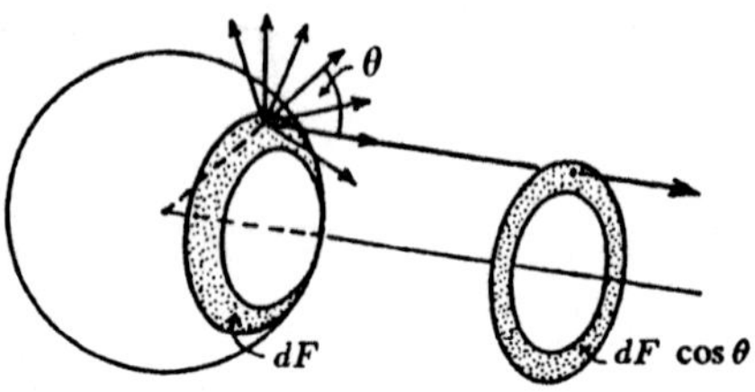

Fig. 2.2. Radiation emitted by surface dF at angle θ to the normal of the surface appearing to arise from the projected surface $dF \cos\theta$

radiance in all directions. The quantity of light reflected by every surface element dF is proportional to the projection of $dF \cos\theta$ onto a plane perpendicular to the direction of reflection (Fig. 2.2). Therefore a white or gray surface seems to have the same brightness, seen from any direction, but a different size.

The radiation which is not reflected at the surface of a body penetrates it, either passing through it or being absorbed by it. A body that absorbs all radiation that hits it without letting any part through or reflecting it, is called *black*.

2.2 What is Cavity Radiation?

Now we consider the radiation field that exists inside a cavity with walls formed by absolute black bodies with temperature T. If the black walls emitted no radiation, then none could exist inside the cavity since it would very quickly be absorbed. However, it is an experimental fact that black bodies do emit light at high temperatures. Without any exact knowledge concerning this emission by black bodies, we can nevertheless draw various conclusions from its existence:

(1) After a short period of time, the radiation inside the cavity will reach a thermal equilibrium caused by the emission and absorption by the walls. If this equilibrium is reached, the radiation field will no longer vary.

(2) Everywhere in the cavity, the radiance $J(\omega, T)$ is independent of the direction of the light rays. The radiation field is isotropic and independent of the shape of the cavity or of the material of its walls. If this were not true, we could place a black body in the form of a small disc having the same temperature as the walls into the cavity, and it would heat up if the plane of the disc were perpendicular to the direction in which $J(\omega, T)$ is largest. This, however, would contradict the 2nd law of thermodynamics.

(3) The radiation field has the same properties at each point of the cavity. $J(\omega, T)$ is independent of spatial coordinates. If this were not the case, little carbon sticks could be set at two different points that have the temperature of the wall, a stick would absorb more of the radiation at a point where the radiation field is stronger than at a point where it is weaker. As a result, the two sticks in the cavity radiation would reach different temperatures. Again, this is not possible according to the 2nd law of thermodynamics.

(4) The cavity radiation hits all surface elements of the wall with radiance $J(\omega, T)$. The surface has to emit as much radiation as it absorbs, therefore the radiance of a black body is $J(\omega, T)$. Thus the emission of all black bodies of the same temperature is identical and depends only on the temperature. Their radiance is independent of direction. The unpolarized light flux

$$2J(\omega, T) \cos\theta \, d\omega \, dF \, d\Omega$$

is sent into a cone of aperture $d\Omega$, whose axis is inclined to the normal vector of the black surface element dF by an angle θ (Fig. 2.3). The factor 2 is caused by the two possible polarizations of each ray. $J(\omega, T)$ depends only on temperature and frequency. For emission by black bodies, Lambert's cosine theorem is valid, as it is for reflection by white surfaces. A glowing black body appears bright all over, seen from any direction.

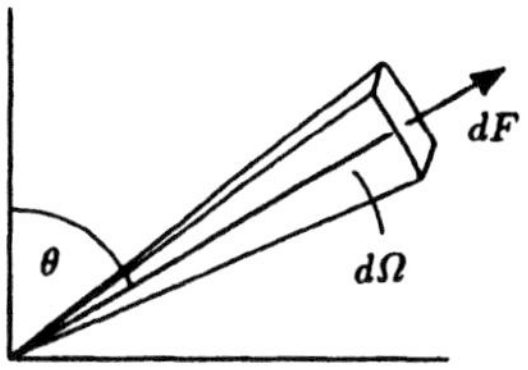

Fig. 2.3. Cone of aperture $d\Omega$, whose axis is inclined to the normal vector of the surface element dF by an angle θ

(5) In a cavity enclosed by walls impenetrable to radiation, the same radiation exists as in a cavity with black walls. If we put a little carbon stick inside, it has to be in thermal equilibrium with the walls and the radiation. This will only be the case if the radiation is the same as for black walls. The radiation inside a cavity enclosed by impervious or black walls is called *black body radiation.*

(6) A single electromagnetic wave of polarization α causes an energy density ϱ_α which is related to the current density $|S_\alpha|$ by

$$\varrho_\alpha = \frac{|S_\alpha|}{c} \; . \tag{2.1}$$

Here, $\boldsymbol{S}_\alpha$ denotes the Poynting vector of the electromagnetic wave with polarization α. Linearly polarized radiation with a frequency between ω and $\omega + d\omega$ and a direction of propagation within a certain solid angle Ω and $\Omega + d\Omega$ yields the following contribution to the energy density:

$$\frac{J(\omega, T) \, d\omega d\Omega}{c} \; , \tag{2.2}$$

and twice as much, if we take into account the two possible directions of polarization. By integrating over $d\Omega$, we get for the energy density of the total radiation

$$\varrho(\omega, T) \, d\omega = \frac{8\pi J(\omega, T) \, d\omega}{c} \tag{2.3}$$

in the interval $d\omega$, since, for cavity radiation, $J(\omega, T)$ is independent of the direction Ω. Thus we obtain a general relation between the radiance $J(\omega, T)$ and the energy density $\varrho(\omega, T)$ of the cavity radiation, namely,

$$J(\omega, T) = \frac{c\varrho(\omega, T)}{8\pi} \; . \tag{2.4}$$

The following exercise will help in further understanding this relation.

EXERCISE

2.1 On Cavity Radiation

Problem. Clarify once more the relation between the radiance (intensity of emitted energy per unit solid angle) and the energy density $\varrho(\omega, T)$ of cavity radiation.

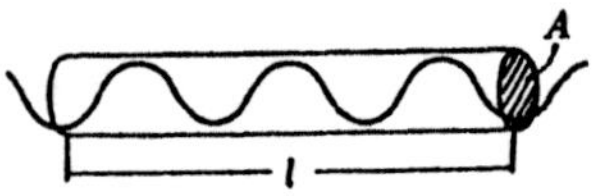

Solution. Concerning energy and intensity:
For a plane wave we have

$$t = \frac{l}{c}\,, \quad E = \varrho_0 V = \varrho_0 l A\,,$$

$$J_0 = \frac{P}{A} = \frac{E}{At} = \frac{\varrho_0 l A}{A(l/c)} = c\varrho_0\,, \quad \text{where} \qquad (1)$$

$\varrho_0 =$ energy density of radiation for a single plane wave,

$J_0 =$ intensity $=$ radiant power per unit area for a single plane wave,

$E = \varrho_0 V =$ radiant energy of the volume V for a single plane wave,

$V = lA =$ volume,

$P = E/t =$ radiant power.

Let a cavity contain an isotropic electromagnetic radiation field. We would like to know the radiant power per area, i.e. the intensity of the radiation that emerges from the aperture of area A.

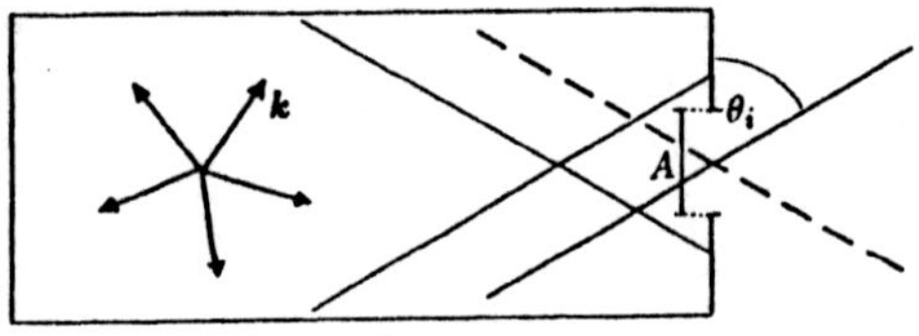

A = area of emergence

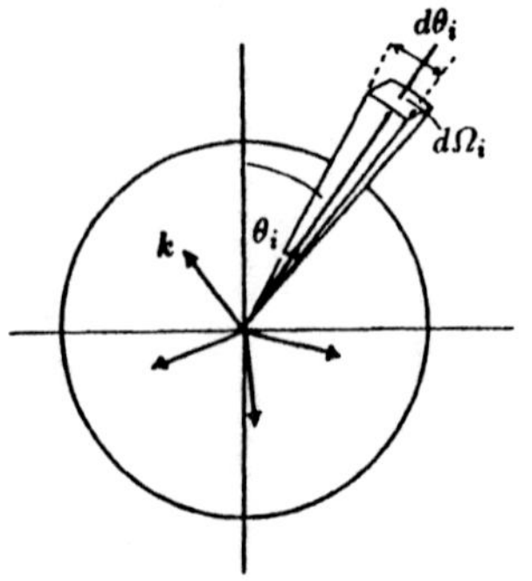

We construct the isotropic radiation by N plane waves with $\boldsymbol{k}$ vectors pointing equally in all directions of the space. Then we have for any wave:

$$J_0 = c\varrho_0\,.$$

n_i is the number of those $\boldsymbol{k}$ that point into the solid angle $\mathrm{d}\Omega_i$ for which $\theta_i \le \theta < \theta_i + \mathrm{d}\theta_i$ is valid.
Now we have

$$\frac{n_i}{N} = \frac{\mathrm{d}\Omega_i}{\Omega} = \frac{1}{2}\sin\theta_i\,\mathrm{d}\theta_i\,. \qquad (2)$$

Exercise 2.1

Because of the isotropy we have integrated over φ. Through A flows

$$P_{0i} = 2J_0 A \cos\theta_i \tag{3}$$

(P = radiant power) per plane wave. The factor 2 appears because there are two degrees of freedom of polarization. Hence

$$P_{\text{tot}} = \sum_i n_i P_{0i} = \sum_i N\tfrac{1}{2}\sin\theta_i\,\mathrm{d}\theta_i\,2J_0 A\cos\theta_i \;, \tag{4}$$

$$P_{\text{tot}} = ANJ_0 \sum_i \sin\theta_i \cos\theta_i\,\mathrm{d}\theta_i \;; \tag{5}$$

θ_i runs from 0 to $\frac{\pi}{2}$. The sum can be replaced by an integral.

$$\begin{aligned} J_{\text{tot}} = \frac{P_{\text{tot}}}{A} &= NJ_0 \int\limits_0^{\pi/2} \sin\theta\cos\theta\;\mathrm{d}\theta \\ &= \frac{1}{2}NJ_0 \int\limits_0^{\pi/2} \sin 2\theta\;\mathrm{d}\theta = \frac{1}{2}NJ_0\left[-\frac{1}{2}\cos 2\theta\right]_0^{\pi/2} \\ &= \frac{1}{2}NJ_0\left[+\frac{1}{2}+\frac{1}{2}\right] = \frac{1}{2}NJ_0 \;. \end{aligned} \tag{6}$$

For the total energy density ϱ, we have

$$\varrho = 2\sum_N \varrho_0 = 2N\varrho_0 \;.$$

Again, the factor 2 is due to the two possible polarizations per plane wave.
Together with $J_0 = c\varrho_0$ we get

$$J_{\text{tot}} = \frac{1}{2}NJ_0 = \frac{1}{2}Nc\varrho_0 = \frac{c}{4}\varrho \;.$$

Therefore the total intensity is

$$J_{\text{tot}} = \frac{c}{4}\varrho \;. \tag{7}$$

It is emitted into a half-space with the solid angle $\Omega_{\text{H}} = 2\pi$. The intensity per unit solid angle (the radiance) is therefore

$$J = \frac{1}{2\pi}J_{\text{tot}} = \frac{c}{8\pi}\varrho \;. \tag{8}$$

2.3 The Rayleigh–Jeans Radiation Law: The Electromagnetic Eigenmodes of a Cavity

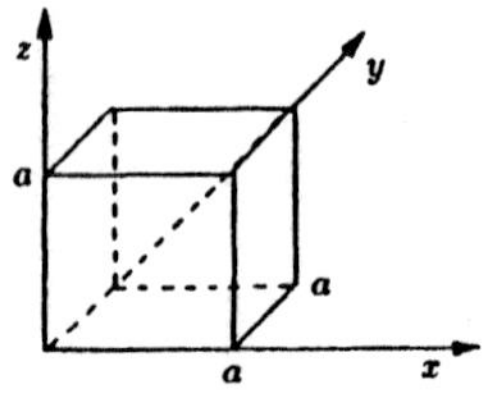

Fig. 2.4. Radiation field assumed to be enclosed in a box

First of all we calculate the radiation density of a radiation field in thermodynamic equilibrium. The average energy per degree of freedom is then given by $\frac{1}{2}k_B T$. To determine the number of degrees of freedom of the radiation field given by the vector potential $\boldsymbol{A}$, we consider a cubical volume of edge length a. We now assume that the volume contains neither charges nor currents and has perfectly reflecting (mirrored) inner surfaces (hence the often-used name, *cavity radiation*). The vector potential obeys *d'Alembert's equation*:

$$\Box \boldsymbol{A}(\boldsymbol{r},t) = \left(\Delta - \frac{1}{c^2}\frac{\partial^2}{\partial t^2}\right)\boldsymbol{A} = 0 \ . \tag{2.5}$$

By separating off the time dependence, i.e. $\boldsymbol{A}(\boldsymbol{r},t) = \boldsymbol{A}(\boldsymbol{r})\exp(\mathrm{i}\omega t)$, we get for the space-dependent part of the function the *Helmholtz equation*:

$$\left(\Delta + \frac{\omega^2}{c^2}\right)\boldsymbol{A}(\boldsymbol{r}) = (\Delta + k^2)\boldsymbol{A}(\boldsymbol{r}) = 0 \ . \tag{2.6}$$

Here,

$$\boldsymbol{A}(\boldsymbol{r},t) = \boldsymbol{A}(\boldsymbol{r})\sin(\omega t) \quad \text{or} \quad \boldsymbol{A}(\boldsymbol{r},t) = \boldsymbol{A}(\boldsymbol{r})\cos(\omega t) \tag{2.7}$$

and $k = \omega/c$ is the wave number. We do not want to solve the two wave equations explicitly, but use the boundary conditions of the problem to determine the number of degrees of freedom.

The vector potential is free of sources (Coulomb gauge); therefore

$$\operatorname{div} \boldsymbol{A} = 0 \ .$$

This condition is equivalent to the transversality of the plane waves in the box. For every wave vector $\boldsymbol{k}$, there exist two independent amplitudes $\boldsymbol{A}$ (polarizations) of the vector potential, each of them of the form:

$$\boldsymbol{A}(\boldsymbol{r}) = \boldsymbol{A}\sin(\boldsymbol{k}\cdot\boldsymbol{r}) \quad \text{or} \quad \boldsymbol{A}(\boldsymbol{r}) = \boldsymbol{A}\cos(\boldsymbol{k}\cdot\boldsymbol{r}) \quad \text{with} \tag{2.8}$$

$$\boldsymbol{A}\cdot\boldsymbol{k} = 0 \quad \text{or} \quad A_x k_x + A_y k_y + A_z k_z = 0 \ , \quad \text{and} \tag{2.9}$$

$$k_x^2 + k_y^2 + k_z^2 = k^2 = \omega^2/c^2 \ . \tag{2.10}$$

For the components of the wave vector k_x, k_y, k_z, we deduce conditions by demanding that the tangential components of $\boldsymbol{A}$ vanish on the reflecting inner surfaces (mirrors) of the cube. These conditions exclude the second solution (2.8) and yield for the first type

$$\sin(k_x a) = \sin(k_y a) = \sin(k_z a) = 0 \ ,$$

from which the wave numbers

$$k_x = \frac{n_x\pi}{a} \ , \quad k_y = \frac{n_y\pi}{a} \ , \quad k_z = \frac{n_z\pi}{a} \ , \quad n_x,\, n_y,\, n_z = 1,\, 2,\, 3,\, \ldots \tag{2.11}$$

follow. The numbers n_x, n_y, n_z are restricted to positive integer values only, because we are looking for stationary waves in the volume. The number of linearly independent functions $\boldsymbol{A}(\boldsymbol{r}, t)$ in the $(\mathrm{d}n_x, \mathrm{d}n_y, \mathrm{d}n_z)$ of number space is the volume element in number space itself, i.e.

$$\mathrm{d}n_x\,\mathrm{d}n_y\,\mathrm{d}n_z\ . \tag{2.12}$$

Together with $n^2 = n_x^2 + n_y^2 + n_z^2$, we adopt spherical coordinates and get for the number of lattice points in the first octant of the spherical shell (see Fig. 2.5)

$$\tfrac{1}{8}4\pi n^2\,\mathrm{d}n = \tfrac{1}{2}\pi n^2\,\mathrm{d}n\ . \tag{2.13}$$

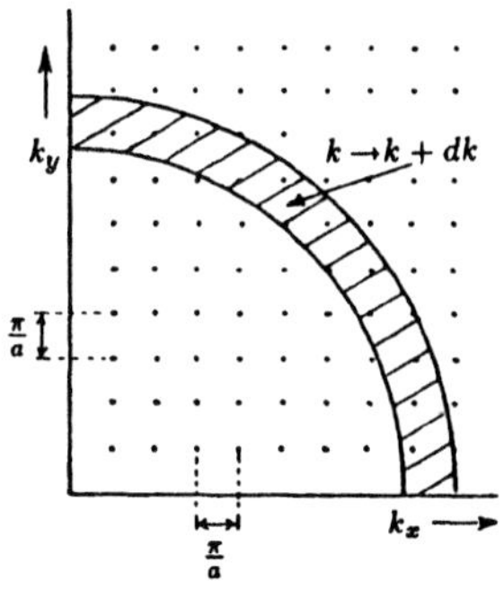

Fig. 2.5. Two-dimensional illustration of the counting method. Every eigenfrequency is represented by a point with the coordinates $n_x\pi/a$, $n_y\pi/a$. The distance between the origin and this point is the wave number of the eigenmode. The two circles have radii k and $k + \mathrm{d}k$, respectively

With (2.9) and (2.10) we get the number $\mathrm{d}N'(\omega)$ of independent solutions for the $\boldsymbol{A}$ field, situated in the frequency interval between ω and $\omega + \mathrm{d}\omega$ such that

$$\frac{1}{2}\pi n^2\,\mathrm{d}n = \frac{\pi}{2}\left(\frac{a}{\pi}\right)^3 k^2\,\mathrm{d}k = \frac{V}{2\pi^2 c^3}\omega^2\,\mathrm{d}\omega = \mathrm{d}N'(\omega)\ , \tag{2.14}$$

where $V = a^3$ is the volume of the cube. Now, if we also take into account the two directions of polarization for every electromagnetic normal mode, we finally get the density $\mathrm{d}N/\mathrm{d}\omega$ of the possible electromagnetic states in a cavity,

$$\frac{\mathrm{d}N(\omega)}{\mathrm{d}\omega} = \frac{V}{\pi^2 c^3}\omega^2\ . \tag{2.15}$$

According to statistical thermodynamics, the average kinetic energy per degree of freedom is given by $\frac{1}{2}k_{\mathrm{B}}T$.[1] The energy per volume and frequency interval $\mathrm{d}\omega$ is given by taking into account the two possible directions of polarization:

$$\frac{\mathrm{d}E}{V\,\mathrm{d}\omega} \equiv \varrho(\omega, T) = \frac{1}{V}\frac{\mathrm{d}N}{\mathrm{d}\omega}k_{\mathrm{B}}T = k_{\mathrm{B}}T\frac{\omega^2}{\pi^2 c^3}\ . \tag{2.16}$$

The spectral energy density is therefore

$$\varrho(\omega, T) = \frac{k_{\mathrm{B}}T}{\pi^2 c^3}\omega^2\ . \tag{2.17}$$

Using the relation between the emittance and the energy density of the radiation already deduced in (2.4), we can immediately write down the emittance:

$$J(\omega, T)\,\mathrm{d}\omega = \frac{c}{8\pi}\varrho(\omega, T)\,\mathrm{d}\omega = \frac{k_{\mathrm{B}}T}{8\pi^3 c^2}\omega^2\,\mathrm{d}\omega\ . \tag{2.18}$$

This is the ***Rayleigh–Jeans** radiation law.*

The equations deduced here coincide with experiment only for low frequencies ω. We can already see from (2.15) and (2.17) that they cannot be valid for high frequencies, since the energy density becomes infinite as $\omega \to \infty$.

---

[1] The average kinetic energy for an oscillator is also $\frac{1}{2}k_{\mathrm{B}}T$, but its mean potential energy is of the same magnitude (virial theorem). Therefore the average energy per frequency is $k_{\mathrm{B}}T$.

2.4 Planck's Radiation Law

In contrast with the classical derivation of the radiation density according to Rayleigh–Jeans, we now want to determine the density of photons. Photons are emitted or absorbed by the transition of an atom from one energy state to another. The main problem here is the quantum mechanical calculation of the transition probabilities. However, it is possible to determine the proportions of the transition probabilities and to ascertain the energy density of the photon field by comparison with the Rayleigh–Jeans law, without explicit calculation of the transition rates. This derivation of ***Planck***'s radiation law originates from Albert Einstein.

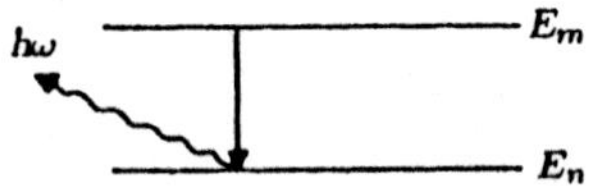

Fig. 2.6. Photon of frequency ω being emitted in the transition from a state of energy E_m to a state of energy E_n

Two of the energy eigenstates (energy levels) of an atom are represented in Fig. 2.6. The atom jumps spontaneously from the energy level E_m to the state of lower energy E_n and emits a photon of frequency $\omega = (E_m - E_n)/\hbar$.

The photon has two different polarizations as a transverse electromagnetic wave. We identify them by the index α ($\alpha = 1, 2$). The probability for the *spontaneous transition* is denoted by $a^n_{m\alpha}$. The probability for the emission of a photon into a solid-angle element $d\Omega$ is

$$dW'_e = a^n_{m\alpha}\, d\Omega \; . \tag{2.19}$$

The presence of photons in the radiation field in the vicinity of an atom in the state E_m stimulates it to an induced emission with the probability dW''_e. This has to be added to the spontaneous emission probability. The total probability is therefore

$$dW_e = dW'_e + dW''_e \; .$$

The probability that an atom in the state E_n is able to absorb a photon of energy ω and move to a state E_m is denoted by dW_a.

Einstein, who first made these considerations, set the probability of absorption and induced emission proportional to the number of photons contained in the radiation field. The *energy per element of solid angle and interval of frequency* for photons of polarization α is again denoted by the spectral energy density $\varrho_\alpha(\omega, T, \Omega)$.

The number of photons in the frequency range between ω and $\omega + d\omega$ and in the direction between the solid angle Ω and $\Omega + d\Omega$ is given by

$$\frac{\varrho_\alpha(\omega, T, \Omega)\, d\omega\, d\Omega}{\hbar\omega} \; .$$

For the probabilities of absorption and induced emission, we write analogously to (2.19)

$$dW''_e = b^n_{m\alpha}\varrho_\alpha(\omega, T, \Omega)\, d\Omega \; , \qquad dW_a = b^m_{n\alpha}\, \varrho_\alpha(\omega, T, \Omega)\, d\Omega \; . \tag{2.20}$$

The coefficients b are the transition probabilities per unit spectral energy density. Therefore they have a different dimension from the transition probability $a^n_{m\alpha}$.

Now we denote by $N_n(N_m)$ the number of atoms in the state of energy $E_n(E_m)$. Then radiative equilibrium is characterized by

$$N_m(\mathrm{d}W'_\mathrm{e} + \mathrm{d}W''_\mathrm{e}) = N_n \,\mathrm{d}W_\mathrm{a} \;, \tag{2.21}$$

that is, the number of emission and absorption processes is equal.

We insert (2.19) and (2.20) and get

$$N_m\big(b^n_{m\alpha}\, \varrho_\alpha(\omega, T, \Omega) + a^n_{m\alpha}\big) = N_n b^m_{n\alpha}\, \varrho_\alpha(\omega, T, \Omega) \;. \tag{2.22}$$

The coefficients $a^n_{m\alpha}$, $b^m_{n\alpha}$, $b^n_{m\alpha}$ characterize properties of the atom and are related to each other. It is permissible, however, to choose special conditions (like radiative equilibrium) to determine the relations between the coefficients, since the latter are not changed by doing so.

The number of atoms in the state E_n is given by the Boltzmann distribution, i.e. we have the relation

$$N_n : N_m = \exp(-E_n/k_\mathrm{B}T) : \exp(-E_m/k_\mathrm{B}T) \;.$$

Here, as before in Sect. 2.3, we have added the index B that distinguishes the Boltzmann constant k_B from the wave number, so that (2.22) becomes

$$\begin{aligned} &\exp(-E_m/k_\mathrm{B}T)\big(b^n_{m\alpha}\, \varrho_\alpha(\omega, T, \Omega) + a^n_{m\alpha}\big) \\ &\quad = \exp(-E_n/k_\mathrm{B}T) b^m_{n\alpha}\, \varrho_\alpha(\omega, T, \Omega) \;. \end{aligned} \tag{2.23}$$

For very high temperatures $T \to \infty$, the exponential function approaches 1 and the spectral energy density also becomes very large. Then we can neglect the term $a^n_{m\alpha}$ and obtain the relation

$$b^n_{m\alpha} = b^m_{n\alpha} \;.$$

We rearrange (2.23) and get for the spectral energy density, by making use of $\hbar\omega_{mn} = E_m - E_n$:

$$\varrho_\alpha(\omega, T, \Omega) = \frac{a^n_{m\alpha}}{b^n_{m\alpha}} \frac{1}{\exp(\hbar\omega/k_\mathrm{B}T) - 1} \;.$$

A comparison with the Rayleigh–Jeans law for the spectral energy density, which is valid for low frequencies, gives the proportion of the transition coefficients. By expanding the denominator for $\hbar\omega/k_\mathrm{B}T \ll 1$, we find

$$\varrho_\alpha(\omega, T, \Omega) = \frac{a^n_{m\alpha}}{b^n_{m\alpha}} \frac{k_\mathrm{B}T}{\hbar\omega} \;. \tag{2.24}$$

Comparing now (2.24) and (2.17), we have to pay attention to the fact that $\varrho_\alpha(\omega, T, \Omega)$ is the spectral energy density per polarization and solid angle:

$$\varrho_\alpha(\omega, T, \Omega) = \frac{1}{8\pi} \varrho(\omega, T) \;.$$

Hence we obtain

$$\frac{a^n_{m\alpha}}{b^n_{m\alpha}} = \frac{1}{8\pi^3}\frac{\hbar\omega^3}{c^3} ,$$

and therefore

$$\varrho_\alpha(\omega, T, \Omega) = \frac{\hbar\omega^3}{8\pi^3c^3}\frac{1}{\exp(\hbar\omega/k_BT)-1} . \tag{2.25}$$

By integrating over the solid angle and adding the two directions of polarization, it follows that

$$\varrho(\omega, T) = \frac{\hbar\omega^3}{\pi^2c^3}\frac{1}{\exp(\hbar\omega/k_BT)-1} . \tag{2.26}$$

Corresponding to the derivation of (2.18), we get for the radiance:

$$J(\omega, T)\,\mathrm{d}\omega = \frac{\hbar\omega^3}{8\pi^3c^2}\frac{1}{\exp(\hbar\omega/k_BT)-1}\,\mathrm{d}\omega . \tag{2.27}$$

For high frequencies $\hbar\omega \gg k_BT$, Planck's radiation formula turns into ***Wien's radiation law***. Indeed, (2.27) yields

$$\varrho(\omega, T) = \frac{\hbar\omega^3}{\pi^2c^3}\exp(-\hbar\omega/k_BT) \tag{2.28}$$

for $\hbar\omega \gg k_BT$.

Let us consider the number density of the photons in the two limits of the ***Rayleigh–Jeans*** and the ***Wien*** radiation laws. With the frequencies ω_1 and ω_2, where $\omega_2 \gg \omega_1$, we obtain from (2.17)

$$\mathrm{d}N_{\mathrm{RJ}} = \frac{\varrho_{\mathrm{RJ}}(\omega_1, T)\,\mathrm{d}\omega}{\hbar\omega_1} = \frac{k_BT}{\pi^2\hbar c^3}\omega_1\,\mathrm{d}\omega , \tag{2.29}$$

and from (2.28)

$$\mathrm{d}N_{\mathrm{W}} = \frac{\varrho_{\mathrm{W}}(\omega_2, T)\,\mathrm{d}\omega}{\hbar\omega_2} = \frac{\omega_2^2}{\pi^2c^3}\exp(-\hbar\omega_2/k_BT)\,\mathrm{d}\omega . \tag{2.30}$$

The relation between the photon number densities follows as

$$\frac{\mathrm{d}N_{\mathrm{W}}}{\mathrm{d}N_{\mathrm{RJ}}} = \frac{\exp(-\hbar\omega_2/k_BT)\hbar\omega_2^2}{k_BT\omega_1} \ll 1 . \tag{2.31}$$

Since $\hbar\omega_2/k_BT \gg 1$, the exponential function makes the ratio of the number densities small. We can conclude from this that the wave character of light always becomes evident when many photons of low energy are present. The particulate character of light becomes noticeable in the case of photons of high energy and low number density.

EXAMPLE

2.2 The Derivation of Planck's Radiation Law According to Planck

It is not only of historical but also of physical interest to see on which arguments Planck based his radiation law. This derivation is different from Einstein's (done some years later) about which we have just learned.

The radiation in a cavity whose walls are kept at a temperature T is homogeneous, isotropic and unpolarized. Its energy density E/V and frequency distribution of the energy density

$$E/V = \int \varrho(\omega, T)\,\mathrm{d}\omega \tag{1}$$

are determined by the temperature alone. This is clear, but can also be verified in detail (Kirchhoff's theorem); we will not do so here.

The energy of the radiation in a cavity can be interpreted as the energy of the electromagnetic field; the frequencies that occur are the eigenoscillations (resonances) of this field. Let V be the volume of the cavity and

$$\frac{\mathrm{d}N(\omega)}{\mathrm{d}\omega}\mathrm{d}\omega \tag{2}$$

be the number of eigenoscillations of the field in the frequency range $\mathrm{d}\omega$; furthermore, let $\bar{\varepsilon}(\omega, T)$ be the mean energy of an eigenoscillation of frequency ω at temperature T. Then the energy content of the cavity in the frequency interval $\mathrm{d}\omega$ is

$$V\varrho(\omega, T)\,\mathrm{d}\omega = \frac{\mathrm{d}N(\omega)}{\mathrm{d}\omega}\bar{\varepsilon}(\omega, T)\,\mathrm{d}\omega \ ; \quad \text{thus}$$

$$\varrho(\omega, T) = \frac{1}{V}\frac{\mathrm{d}N(\omega)}{\mathrm{d}\omega}\bar{\varepsilon}(\omega, T) \ . \tag{3}$$

The number of eigenoscillations is given [see (2.15)] by

$$\frac{\mathrm{d}N(\omega)}{\mathrm{d}\omega} = \frac{V}{\pi^2 c^3}\omega^2 \tag{4}$$

and therefore

$$\varrho(\omega, T) = \frac{1}{\pi^2 c^3}\omega^2\bar{\varepsilon}(\omega, T) \ . \tag{5}$$

Because the eigenoscillations of the electromagnetic field are harmonic, each eigenoscillation has two degrees of freedom, corresponding to the two polarization states of the radiation. Applying the theorem of equipartition of energy from statistical thermodynamics, the mean energy $\bar{\varepsilon}$ of each eigenoscillation is

$$\bar{\varepsilon} = k_\mathrm{B}T \ . \tag{6}$$

Example 2.2

Hence the Rayleigh–Jeans radiation formula follows:

$$\varrho(\omega, T) = \frac{k_B T}{\pi^2 c^3}\omega^2 \ . \tag{7}$$

For small values of $\hbar\omega/k_B T$, this is in agreement with experience, but it fails completely for higher frequencies. In particular, it does not show the characteristic decrease in energy density towards higher frequencies (see Fig. 2.1). That reality is at variance with the theoretical radiation formula (7) means that eigenoscillations with larger values of $\hbar\omega/k_B T$ contain less energy than is expected from the law of equipartition.

Planck replaced this rule (6) with a completely different one: The energy of a harmonic oscillation is an integer multiple of an energy step proportional to the frequency:

$$\varepsilon_n = n\hbar\omega \ ; \quad n = 0,\ 1,\ 2,\ \ldots \ . \tag{8}$$

Every energy state defined here is to be treated as distinct from every other one. Evaluating the mean energy in thermal equilibrium gives

$$\bar{\varepsilon} = \frac{\sum\limits_n \varepsilon_n \, \mathrm{e}^{-\varepsilon_n/k_B T}}{\sum\limits_n \mathrm{e}^{-\varepsilon_n/k_B T}} \ . \tag{9}$$

Introducing the equipartition sum

$$Z = \sum_n \mathrm{e}^{-\beta\varepsilon_n} \ , \quad \beta = 1/k_B T \ , \tag{10}$$

as an abbreviation, (9) becomes

$$\bar{\varepsilon} = -\frac{\mathrm{d}}{\mathrm{d}\beta}\ln Z \ . \tag{11}$$

From the evaluation of

$$Z = \sum_n \left[\exp(-\beta\hbar\omega)n\right] = \frac{1}{1-\exp(-\beta\hbar\omega)} \tag{12}$$

we get

$$\bar{\varepsilon} = \frac{\hbar\omega}{\exp(\hbar\omega/k_B T) - 1} \ , \tag{13}$$

and with (5) we have Planck's radiation law:

$$\varrho(\omega, T) = \frac{1}{\pi^2 c^3}\frac{\hbar\omega^3}{\exp(\hbar\omega/k_B T) - 1} \ . \tag{14}$$

This is (of course) identical to (2.27) given before.

The figure besides illustrates the dependence of the functions (13) and (14) on temperature (abscissa $k_B T/\hbar\omega$, ordinate $\bar{\varepsilon}/\hbar\omega$). Function (13) [or (14)] can be approximated for small $\hbar\omega/k_B T$ by a power series of the exponential. Then the mean energy becomes

$$\bar{\varepsilon} = k_B T \ .$$

For small $\hbar\omega/k_B T$ it makes no difference whether the energies of the eigenoscillations have the discrete values given by (8) or have continuously distributed values, as in classical theory.

The approximation in the limit of large $\hbar\omega/k_B T$

$$\bar{\varepsilon} = \hbar\omega \exp(-\hbar\omega/k_B T)$$

leads to *Wien's radiation formula*:

$$\varrho(\omega, T) = \frac{\hbar\omega^3}{\pi^2 c^3} \exp(-\hbar\omega/k_B T) \ .$$

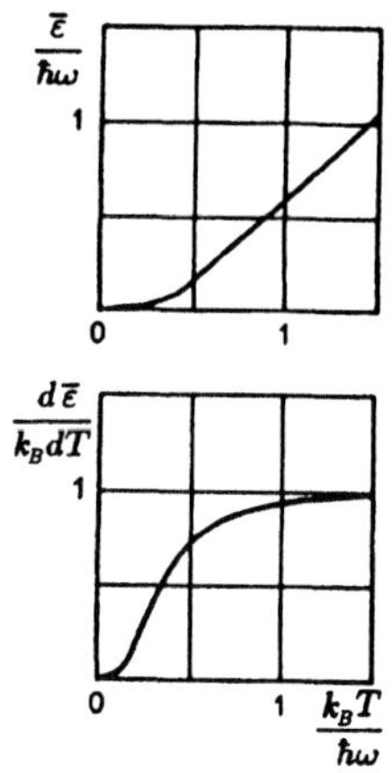

Mean energy and specific heat of an oscillator

The factor $[\exp(\hbar\omega/k_B T) - 1]^{-1}$ in the mean energy (13) can be interpreted as the number of photons in a state characterized by the photon energy $\hbar\omega$. This is important in Exercise 2.3.

Planck's hypothesis, that for harmonic oscillations only the energies $E = n\hbar\omega$ are legitimate, is in contradiction with the intuitive idea of an oscillation. His ingenious guess of the correct quantized states of the harmonic oscillator led to the breakthrough; that the introduction of the quantum of action h coincides in general with the nonclassical description of microsystems became clear only in the course of later developments.

EXERCISE

2.3 Black Body Radiation

From the foregoing, we know the density per frequency interval $\mathrm{d}N(\omega)/\mathrm{d}\omega$ of the electromagnetic field in a cavity resonator with volume V (2.15), i.e.

$$\frac{\mathrm{d}N(\omega)}{\mathrm{d}\omega} = \frac{V}{\pi^2 c^3}\omega^2 \ .$$

Here, the two directions of polarization are included.

Problem. (a) Consider the cavity as a container of photons and calculate the spectral distribution $\frac{1}{V}(\mathrm{d}E(\omega)/\mathrm{d}\omega)$ of the black body radiation that escapes from a tiny hole in the cavity. Consider that photons are spin-1 particles (bosons) and that the number of photons in a state of energy E at temperature T is given by

$$f_{\mathrm{BE}} = (\mathrm{e}^{E/k_B T} - 1)^{-1} \ \text{(the Bose–Einstein distribution)} \ .$$

Compare the result with Planck's law.

Exercise 2.3

(b) Show that the total electromagnetic energy in the cavity with walls kept at temperature T is proportional to T^4, and evaluate the factor of proportionality.

Hint:

$$\int_0^\infty \frac{x^3}{e^x-1}\,dx = \int_0^\infty dx\, x^3 e^{-x}\frac{1}{1-e^{-x}} = \int_0^\infty dx\, x^3 e^{-x}\sum_{n=0}^\infty e^{-nx}$$
$$= \sum_{n=0}^\infty \int_0^\infty dx\, x^3 e^{-(n+1)x} = \sum_{n=0}^\infty \frac{1}{(n+1)^4}\int_0^\infty dy\, y^3 e^{-y}$$
$$= 6\sum_{n=0}^\infty \frac{1}{(n+1)^4} = \frac{\pi^4}{15}\,.$$

Solution. (a) In a state with energy E at temperature T, we find

$$f_{BE} = [\exp(\hbar\omega/k_B T) - 1]^{-1} \tag{1}$$

photons with the energy $E = \hbar\omega$. The density of states is $dN/d\omega$; therefore, the density of photons with energy $\hbar\omega$ in the resonator is given by

$$\frac{dn}{d\omega} = f_{BE}\frac{dN}{d\omega} = \frac{V}{\pi^2 c^3}\omega^2 \frac{1}{\exp(\hbar\omega/k_B T) - 1} \tag{2}$$

and the energy density for the volume V and frequency interval $d\omega$ is

$$\frac{1}{V}\frac{dE(\omega)}{d\omega} = \frac{1}{V}\frac{dn(\omega)}{d\omega}\hbar\omega = \frac{\hbar\omega^3}{\pi^2 c^3}\frac{1}{\exp(\hbar\omega/k_B T) - 1}\,. \tag{3}$$

This is exactly Planck's radiation formula [see (2.26)].

(b) The total energy in the cavity is given by

$$E = \int_0^\infty \frac{dE(\omega)}{d\omega}\,d\omega = \frac{V\hbar}{\pi^2 c^3}\int_0^\infty \frac{\omega^3\,d\omega}{\exp(\hbar\omega/k_B T) - 1}$$
$$= \frac{V k_B^4}{\hbar^3 \pi^2 c^3}T^4 \int_0^\infty \frac{q^3\,dq}{e^q - 1}$$
$$= \frac{V k_B^4 \pi^2}{15\hbar^3 c^3}T^4 \quad \text{(the Stefan–Boltzmann law)}\,. \tag{4}$$

Then the energy density dE/dV in the cavity is

$$\frac{dE}{dV} = \frac{E}{V} = aT^4\,,$$
$$a = \frac{\pi^2 k_B^4}{15\hbar^3 c^3} = 7.56\times 10^{-15}\,\mathrm{erg\,cm^{-3}\,K^{-4}}\,. \tag{5}$$

Exercise 2.3

This gives rise to homogeneous, isotropic radiation of density K, where K is given by

$$K = \frac{c}{4\pi}\frac{\mathrm{d}E}{\mathrm{d}V}\,\mathrm{erg\,cm^{-2}} \ .$$

The emittance is then

$$\varepsilon(T) = \frac{c}{4}\frac{E(T)}{V} = \sigma T^4 \ , \tag{6}$$

where $\sigma = 5.42 \times 10^{-5}$ ($\mathrm{erg\,cm^{-2}\,s^{-1}\,K^{-4}}$). Indeed, the total energy emitted from a surface element $\mathrm{d}f$ in the time interval $\mathrm{d}t$ in the *forward* direction (see Figure) turns out to be

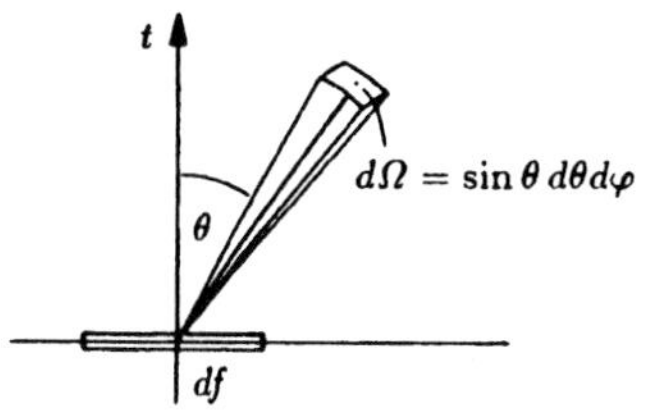

Emittance in time interval $\mathrm{d}t$ from a surface element $\mathrm{d}f$

$$\begin{aligned}\mathrm{d}E &= \int_0^{\pi/2}\int_0^{2\pi} \sin\theta\,\mathrm{d}\theta\,\mathrm{d}\varphi K(\cos\theta\,\mathrm{d}f)\,\mathrm{d}t \\ &= K\,\mathrm{d}f\,\mathrm{d}t \int_0^{\pi/2}\int_0^{2\pi} \mathrm{d}\theta\,\mathrm{d}\varphi\cos\theta\sin\theta = \pi K\,\mathrm{d}f\,\mathrm{d}t \ .\end{aligned}$$

The total power per surface area, the intensity, therefore is given by

$$\varepsilon = \frac{\mathrm{d}E}{\mathrm{d}f\,\mathrm{d}t} = \pi K = \frac{c}{4}\frac{\mathrm{d}E}{\mathrm{d}V} \ . \tag{7}$$

The radiation of stars is approximately described by black-body radiation. Therefore the Stefan–Boltzmann law can be used to estimate the stellar surface temperature if we measure the radiation energy per $\mathrm{cm^2}$ perpendicular to the direction of emission.

For example, consider the Sun. Its radius R is 0.7×10^{11} cm. Therefore the total emitted radiation energy is given by

$$4\pi(0.7 \times 10^{11})^2(5.42 \times 10^{-5})\,T^4\,3.34 \times 10^{18}\,T^4\,\mathrm{erg\,s^{-1}} \ , \ T \text{ in Kelvin} \ . \tag{8}$$

The radiation energy hitting 1 $\mathrm{cm^2}$ of the Earth's surface in one second, taking the mean Sun–Earth distance to be 1.5×10^{13} cm, is calculated thus:

$$\frac{3.34 \times 10^{18}\,T^4}{4\pi(1.5 \times 10^{13})^2\,\mathrm{cm^2}}\frac{\mathrm{erg}}{\mathrm{sec}} = 0.96 \times 10^{-9}\,T^4\frac{\mathrm{erg}}{\mathrm{cm^2}} \ . \tag{9}$$

This quantity is called the *solar constant* and is determined experimentally. Its measured value is $1.94\,\mathrm{cal\,cm^{-2}\,min^{-1}}$. Transformation of units

$$1.94\frac{\mathrm{cal}}{\mathrm{cm^2\,min}} = \frac{1.94 \times 4.2 \times 10^7}{60}\frac{\mathrm{erg}}{\mathrm{cm^2\,s}} = 1.36 \times 10^6\,\mathrm{erg\,cm^{-2}\,s^{-1}} \tag{10}$$

easily gives

$$T^4 = 1.52 \times 10^{15}\,(\mathrm{K})^4 \text{ or } T \approx 6000\,\mathrm{K} \ . \tag{11}$$

EXERCISE

2.4 Wien's Displacement Law

Problem. Derive Wien's displacement law, i.e.

$$\lambda_{\max} T = \text{const.}$$

from Planck's spectral energy density $\frac{1}{V}\mathrm{d}E/\mathrm{d}\omega$. Here $\lambda_{\max}$ is that wavelength where $\frac{1}{V}\mathrm{d}E/\mathrm{d}\omega$ achieves its maximum. Interpret the results.

Solution. We are looking for the maximum of Planck's spectral distribution:

$$\begin{aligned}\frac{\mathrm{d}}{\mathrm{d}\omega}\left[\frac{1}{V}\frac{\mathrm{d}E}{\mathrm{d}\omega}\right] &= \frac{\mathrm{d}}{\mathrm{d}\omega}\left[\frac{\hbar\omega^3}{\pi^2c^3}\left(\exp\left(\frac{\hbar\omega}{k_\mathrm{B}T}\right)-1\right)^{-1}\right] \\ &= \frac{3\hbar\omega^2}{\pi^2c^3}\left[\exp\left(\frac{\hbar\omega}{k_\mathrm{B}T}\right)-1\right]^{-1} \\ &\quad - \frac{\hbar\omega^3}{\pi^2c^3}\frac{\hbar}{k_\mathrm{B}T}\frac{\exp(\hbar\omega/k_\mathrm{B}T)}{[\exp(\hbar\omega/k_\mathrm{B}T)-1]^2} = 0 \\ &\Rightarrow 3-\frac{\hbar\omega}{k_\mathrm{B}T}\exp\left(\frac{\hbar\omega}{k_\mathrm{B}T}\right)\left[\exp\left(\frac{\hbar\omega}{k_\mathrm{B}T}\right)-1\right]^{-1} = 0\ . \qquad (1)\end{aligned}$$

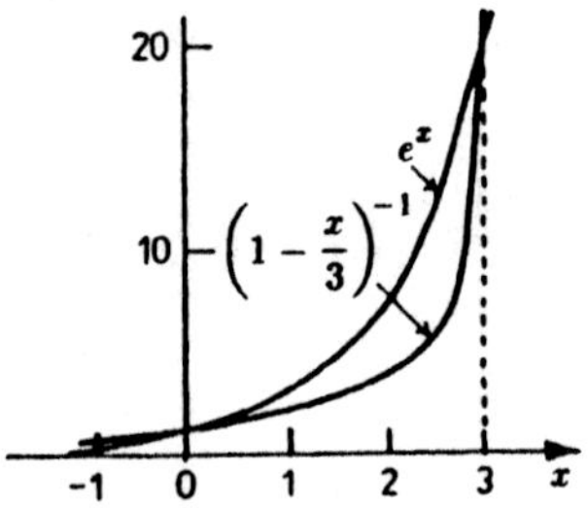

The crossing points of the two curves yield the solution to (2)

With the shorthand notation $x = \hbar\omega/k_\mathrm{B}T$, we get the transcendental equation

$$e^x = \left(1-\tfrac{x}{3}\right)^{-1}\ , \qquad (2)$$

which must be solved graphically or numerically. Besides the trivial solution $x = 0$ (minimum), a positive solution exists (see Figure). Therefore

$$x_{\max} = \frac{\hbar\omega_{\max}}{k_\mathrm{B}T}\ , \qquad (3)$$

and because $\omega_{\max} = 2\pi\nu_{\max} = 2\pi c/\lambda_{\max}$ we have

$$\lambda_{\max}T = \text{const.} = 0.29\,\text{cm K}\ . \qquad (4)$$

This means the wavelength emitted most intensely by a black body is inversely proportional to the temperature of the body. $\lambda_{\max}$ shifts ("displaces") with the temperature, this is why it is called a "displacement law". This law may be used to determine the temperature of bodies (stars), too.

Inserting the solar surface temperature $T \approx 6000$ K (see Exercise 2.3) into Wien's displacement law, we find

$$\lambda_{\max} = \frac{0.29}{6000}\,\text{cm} = 4.8\times10^{-5}\,\text{cm} = 4800\,\text{Å}\ , \qquad (5)$$

where 1 Å $= 10^{-8}$ cm. This is approximately the wavelength of yellow light. These estimates are within 20% of the exact values.

EXERCISE

2.5 Emitted Energies of a Black Body

Problem. Calculate the proportion of energy emitted by a black body radiator at $T = 2000$ K in two bands of width 100 Å, one centred at 5000 Å (visible light) and the other at 50 000 Å (infrared).

Solution. We define $\lambda_1 = 5000$ Å, $\lambda_2 = 50\,000$ Å, $\Delta\lambda = 50$ Å, and calculate

$$W = \frac{\Delta E_{\lambda_2}}{\Delta E_{\lambda_1}} \tag{1}$$

$$= \frac{1}{V}\int\limits_{\lambda_2-\Delta\lambda}^{\lambda_2+\Delta\lambda}\left|\frac{\mathrm{d}E}{\mathrm{d}\lambda}\right|\mathrm{d}\lambda \Bigg/ \frac{1}{V}\int\limits_{\lambda_1-\Delta\lambda}^{\lambda_1+\Delta\lambda}\left|\frac{\mathrm{d}E}{\mathrm{d}\lambda}\right|\mathrm{d}\lambda \approx \left.\frac{\mathrm{d}E}{\mathrm{d}\lambda}\right|_{\lambda=\lambda_2} \Bigg/ \left.\frac{\mathrm{d}E}{\mathrm{d}\lambda}\right|_{\lambda=\lambda_1} .$$

Because $\omega = 2\pi c/\lambda$, we get

$$\frac{\mathrm{d}E}{\mathrm{d}\lambda} = \frac{\mathrm{d}E}{\mathrm{d}\omega}\left|\frac{\mathrm{d}\omega}{\mathrm{d}\lambda}\right| = \frac{\hbar(2\pi c/\lambda)^3}{\pi^2 c^3}\left[\exp\frac{\hbar 2\pi c}{k_\mathrm{B}T\lambda} - 1\right]^{-1}\frac{2\pi c}{\lambda^2}$$

$$= \frac{8\pi hc}{\lambda^5}\left[\exp\frac{hc}{k_\mathrm{B}T\lambda} - 1\right]^{-1} , \tag{2}$$

and with $hc = 12\,400$ eV Å, $k = 8.62 \times 10^{-5}$ eV K^{-1}, it follows that $W = 5.50$. Thus, only a small fraction of energy is emitted as visible light.

EXAMPLE

2.6 Cosmic Black Body Radiation

During the last decade, black body radiation has gained special importance. In the late 1940s, George Gamov, first by himself, but later with R. Alpher and H. Bethe, investigated some consequences of the "Big Bang Model" of the creation of the Universe. One of those consequences was that the remnants of the intense radiation field created in the beginning should be present as a black body radiation field. Calculations predicting such a radiation field at a temperature of 25 K proved unreliable. Until 1964, no attempts had been made to measure this radiation. Then A.A. Penzias and R.W. Wilson discovered strong thermal noise with their radio-astronomical detector, and renewed interest in this problem arose. A group under the leadership of R.H. Dicke, consisting of P.J. Peebles, P.G. Roll and D.T. Wilkinson, performed measurements of cosmic background radiation and understood at once the meaning of the thermal noise. It corresponds to black body radiation which today we believe to be of 2.65 ± 0.09 K. The measurements were not easy to make, because any antenna is deluged by signals

Example 2.6

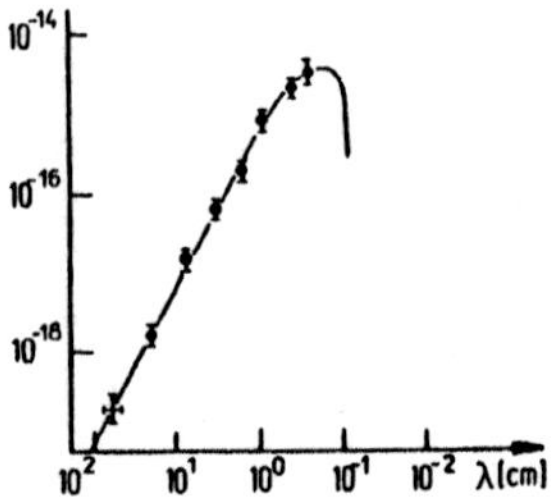

Measurements of the background radiation in erg s^{-2} cm^{-1}steradian^{-1}Hz^{-1} as a function of wavelength in cm. The drawn curve is the predicted spectrum for $T = 2.7$ K

from the Earth's surface, from the atmosphere, from several cosmic point sources and also by the noise generated by the electrical circuits in the measuring apparatus. In 1945, Dicke had constructed an instrument to measure radiation that could be used for these experiments. His idea was to construct a radio receiver that switches back and forth between the sky and a bath of liquid helium, 100 times a second. The receiver's output signal is filtered: only signals varying with a frequency of 100 Hz are measured. These represent the difference between radiation from space and liquid helium. By varying the measuring apparatus, the atmospheric component could be separated out.

The experimental verification of radiation at a temperature corresponding to the calculations made by Dicke and his collaborators is one of the strongest arguments in support of the "Big Bang Model" (see Figure). More precise measurements will determine how fast we (Earth, solar system, local group of galaxies) move relative to the background radiation. Currently, our velocity relative to the radiation field is less than 300 km s^{-1}, roughly corresponding to the velocity of the solar system with respect to the local group of galaxies, caused by the rotation of our own galaxy.

2.5 Biographical Notes

RAYLEIGH, John Williams Strutt, 3rd baron R., English physicist, *Langford Grove (Essex) 12.11.1842, † Terling Place (Chelmsford) 30.6.1919, was a professor at the Cavendish Laboratory in Cambridge from 1879–84, at the Royal Society in London from 1884 to 1905. R. investigated, among other things, the intensity of sound by measuring sound pressure exerted on an easily movable plate (*Rayleigh disc*), deduced the blue colour of the sky was caused by the scattering of light by the molecules of the air (*Rayleigh scattering*) and postulated a radiation law in 1900, known as the *R.–Jeans law*, representing a special case of Planck's law. Discrepancies in the measurements of the sound velocity of nitrogen led him and W. Ramsay to the discovery of argon in 1894, which was rewarded with the Nobel Prize in Physics and Chemistry in 1904 [BR].

JEANS, James Hopwood, Sir, English mathematician, physicist and astronomer, *Southport 11.9.1877, † Dorking 16.9.1946, was a professor of astronomy at the Royal Society from 1912–1946. J. did pioneering work mainly in the fields of thermodynamics, stellar dynamics and cosmogony. He wrote natural philosophical works and also fascinating popular astronomical books [BR].

PLANCK, Max, German physicist, *Kiel 23.4.1858, † Göttingen 4.10.1947, received his doctor's degree at the age of 21 after submitting a thesis on thermodynamics. In 1885 he became a professor in Kiel; in 1889 he was a professor of theoretical physics and continued to work long after his retirement. During his studies on entropy in 1894 P. devoted himself to thermal radiation. In doing so, he discovered (not later than May, 1899), while still of the opinion that Wien's radiation formula was correct, a new constant of nature,

Planck's quantum of action. In mid-October, 1900, he deduced his radiation formula by an ingenious interpolation, which turned out to be the correct *law of black-body radiation.* The 14th of December, 1900, when P. reported on the derivation of this formula from the principles of physics at the meeting of the Deutsche Physikalische Gesellschaft in Berlin, is considered as the "birthday of quantum theory". While P. remained sceptical concerning Einstein's light quantum hypothesis, he immediately recognized the importance of the theory of special relativity established by Einstein in 1905; it is due mainly to P. that it was so quickly accepted in Germany. In 1918 he was awarded the Nobel Prize in Physics. Because of his scientific works and straightforward and uncompromising character, and because of his gentlemanly behaviour, he occupied a unique position among German physicists. As one of the four permanent secretaries, he directed the Preußische Akademie der Wissenschaften for more than twenty-five years. For many years he was president of the Deutsche Physikalische Gesellschaft and copublisher of the "Annalen der Physik". The Deutsche Physikalische Gesellschaft founded the Max Planck medal on his 70th birthday; P. was the first to win this award. After the end of the Second World War, the Kaiser Wilhelm Gesellschaft zur Förderung der Wissenschaften, of which P. had been president for seven years, was renamed the Max Planck Gesellschaft zur Förderung der Wissenschaften e.V. [BR].

WIEN, Wilhelm, German physicist, *Gaffken (Ostpreußen) 13.1.1864, †München 30.8.1928, was a professor in Aachen, Gießen, Würzburg, München; in 1893, still assistant of H.v. Helmholtz, he discovered his *displacement law*; in 1896 he published his important (though only approximately valid) *radiation law* which had already been found. For these works (papers) W. was awarded the Nobel Prize (1911); their continuation by M. Planck led to quantum theory. In 1896, W. turned to writing about particle beams: among other things, he identified the cathode rays as negatively charged particles; he noticed that the channel rays consist of a mixture of predominantly positive ions and determined their specific charge and velocity. He worked on charge transfer and glow processes, determined the mean free path of the particles and the glow time of atoms glowing unperturbed in high vacuum. As editor of "Annalen der Physik" (beginning in 1906) he greatly influenced the development of science [BR].

3. Wave Aspects of Matter

3.1 De Broglie Waves

Investigations on the nature of light showed that, depending on the kind of experiment performed, light must be described by electromagnetic *waves* or by *particles* (photons). Thus the wave aspect appears in the context of diffraction and interference phenomena, whereas the particulate aspect shows up most distinctly in the photoelectric effect. So for light, the relations describing wave–particle duality are already known. But what about material particles? Their particulate nature is rather obvious; do they also possess a wave aspect?

In addition to their corpuscular properties, ***de Broglie*** assigned wave properties to particles, thus transferring the relations known from light to matter. What is true for photons should be valid for any type of particle. Hence, according to the particulate picture, we assign to *a particle*, for example an electron with mass m, *propagating uniformly with velocity* $\boldsymbol{v}$ *through field-free space*, an energy E and a momentum $\boldsymbol{p}$. In the wave picture, the particle is described by a frequency ν and a wave vector $\boldsymbol{k}$. Following de Broglie, we now speculate: since these descriptions should only designate two different aspects of the same object, the following relations between the characteristic quantities should be valid:

$$E = h\nu = \hbar\omega \quad \text{and} \tag{3.1}$$

$$\boldsymbol{p} = \hbar\boldsymbol{k} = \frac{h}{\lambda}\frac{\boldsymbol{k}}{|\boldsymbol{k}|} \ . \tag{3.2}$$

We have seen in the preceding chapters that these equations are true for the photon (electromagnetic field); now they are postulated to be valid for *all* particles. Then to every free particle, understood in the above sense, a plane wave determined up to an amplitude factor A is assigned:

$$\psi(\boldsymbol{r}, t) = A \exp[\mathrm{i}(\boldsymbol{k}\cdot\boldsymbol{r} - \omega t)] \ , \tag{3.3}$$

or, using the above relations,

$$\psi(\boldsymbol{r}, t) = A \exp[\mathrm{i}(\boldsymbol{p}\cdot\boldsymbol{r} - Et)/\hbar] \ . \tag{3.3a}$$

Following de Broglie, the plane wave connected to the particle has a wavelength

$$\lambda = \frac{2\pi}{k} = \frac{h}{p} = \frac{h}{mv} \ , \tag{3.4}$$

where the second relation is valid only for particles with nonvanishing rest mass. Because of the small value of the quantum of action, the particle mass must be sufficiently small to generate a measurable wavelength. For this reason, the wave character of matter first appears in atomic regions. The phase

$$\alpha = \boldsymbol{k} \cdot \boldsymbol{r} - \omega t \tag{3.5}$$

of the wave $\psi(\boldsymbol{r}, t)$ (3.3) propagates with velocity $\boldsymbol{u} = \dot{\boldsymbol{r}}$ according to the relation

$$\frac{\mathrm{d}\alpha}{\mathrm{d}t} = \boldsymbol{k} \cdot \dot{\boldsymbol{r}} - \omega = \boldsymbol{k} \cdot \boldsymbol{u} - \omega = 0 \ . \tag{3.6}$$

Hence, we get for the magnitude of the phase velocity $\boldsymbol{u}$ ($\boldsymbol{k}$ and $\boldsymbol{u}$ have the same direction):

$$|\boldsymbol{u}| = \frac{\omega}{k} \ . \tag{3.7}$$

In the following, we will show that matter waves – in contrast with electromagnetic waves – even show dispersion in a vacuum. We must therefore calculate $\omega(k)$. The relativistic energy theorem for free particles

$$E^2 = m_0^2 c^4 + p^2 c^2$$

can, for $v \ll c$, be put into the form

$$E = mc^2 = \sqrt{m_0^2 c^4 + p^2 c^2} = m_0 c^2 + \frac{p^2}{2m_0} + \dots \ . \tag{3.8}$$

With (3.1) and (3.2) we can give the frequency as a function of the wave number:

$$\omega(k) = \frac{m_0 c^2}{\hbar} + \frac{\hbar k^2}{2m_0} + \dots \ . \tag{3.9}$$

Therefore, the phase velocity $u = \omega/k$ is a function of k in a vacuum, i.e.

$$u = \frac{m_0 c^2}{\hbar k} + \frac{\hbar k}{2m_0} + \dots \ , \tag{3.10}$$

so that the matter waves show dispersion even there, i.e. waves with a different wave number (wavelength) have different phase velocities. On the other hand, for the *phase velocity* u, the following relation holds:

$$u = \frac{\omega}{k} = \frac{\hbar\omega}{\hbar k} = \frac{E}{p} = \frac{mc^2}{mv} = \frac{c^2}{v} \ . \tag{3.11}$$

Because $c > v$, the phase velocity of matter waves is always larger than the velocity of light in a vacuum. Hence, it cannot be identified with the velocity of the assigned particles. Because these are massive, they can only propagate more slowly than light does.

The group velocity is calculated using

$$v_g = \frac{d\omega}{dk} = \frac{d(\hbar\omega)}{d(\hbar k)} = \frac{dE}{dp} \; ; \tag{3.12}$$

(we shall prove this below). The variation of energy dE of the particle moving under the influence of a force $\boldsymbol{F}$ along a path $d\boldsymbol{s}$ is $dE = \boldsymbol{F} \cdot d\boldsymbol{s}$ and because $\boldsymbol{F} = d\boldsymbol{p}/dt$ we therefore have

$$dE = \frac{d\boldsymbol{p}}{dt} \cdot d\boldsymbol{s} = d\boldsymbol{p} \cdot \boldsymbol{v} \; . \tag{3.13}$$

Since $\boldsymbol{v}$ and $p = m\boldsymbol{v}$ are parallel, the following equations hold:

$$dE = |\boldsymbol{v}| \; |d\boldsymbol{p}| = v dp \quad \text{or} \quad \frac{dE}{dp} = v \; . \tag{3.14}$$

Hence, the group velocity of a matter wave is identical with the particle velocity, i.e.

$$v_g = v \; . \tag{3.15}$$

We can also deduce this result in a different way. If we want to describe a particle as a spatially limited entity, we cannot describe it by a plane wave (3.3). Instead, we try to describe the particle by a finite wave packet, which, with the help of a Fourier integral, is written as a superposition of harmonic waves, differing in wavelength and phase velocity. For simplicity we investigate a group of waves propagating in the x direction

$$\psi(x, t) = \int_{k_0-\Delta k}^{k_0+\Delta k} c(k) \exp\{i[kx - \omega(k)t]\} \, dk \; . \tag{3.16}$$

Here, $k_0 = 2\pi/\lambda_0$ is the mean wave number of the group and Δk is the measure of the extension (frequency spread) of the wave packet, assumed to be small ($\Delta k \ll k_0$). Therefore, we can expand the frequency ω, which according to (3.9) is a function of k, in a Taylor series in the interval Δk about k_0, and neglect terms of the order $(\Delta k)^n = (k - k_0)^n$, $n \geq 2$, i.e.

$$\omega(k) = \omega(k_0) + \left(\frac{d\omega}{dk}\right)_{k=k_0} (k - k_0) + \frac{1}{2}\left(\frac{d^2\omega}{dk^2}\right)_{k=k_0} (k - k_0)^2 + \ldots \; . \tag{3.17}$$

We take $\xi = k - k_0$ as new integration variable ξ and assume the amplitude $c(k)$ to be a slowly varying function of k in the integration interval $2\Delta k$. The term $(d\omega/dk)_{k=k_0} = v_g$ is the group velocity. Thus, (3.16) becomes

$$\psi(x, t) = \exp\{i[k_0 x - \omega(k_0)t]\} \int_{-\Delta k}^{\Delta k} \exp\left[i(x - v_g t)\xi\right] c(k_0 + \xi) \, d\xi \; . \tag{3.18}$$

Integration, transformation and the approximation $c(k_0+\xi) \approx c(k_0)$ lead to the result

$$\psi(x,t) = C(x,t)\exp\{\mathrm{i}[k_0 x - \omega(k_0)t]\} \quad \text{with} \tag{3.19a}$$

$$C(x,t) = 2c(k_0)\frac{\sin[\Delta k(x - v_g t)]}{x - v_g t} \ . \tag{3.19b}$$

Since the argument of the sine contains the small quantity Δk, $C(x,t)$ varies only slowly depending on time t and coordinate x. Therefore, we can regard $C(x,t)$ as the amplitude of an approximately monochromatic wave and $k_0 x - \omega(k_0)t$ as its phase. Multiplying numerator and denominator of the amplitude by Δk and abbreviating the term

$$\Delta k(x - v_g t) = z \ ,$$

we see that variation in amplitude is determined by the factor

$$\frac{\sin z}{z} \ .$$

This has the properties

$$\lim_{z\to 0}\frac{\sin z}{z} = 1 \quad \text{for} \quad z=0 \ , \qquad \frac{\sin z}{z} = 0 \quad \text{for} \quad z = \pm\pi \ , \ \pm 2\pi \ . \tag{3.20}$$

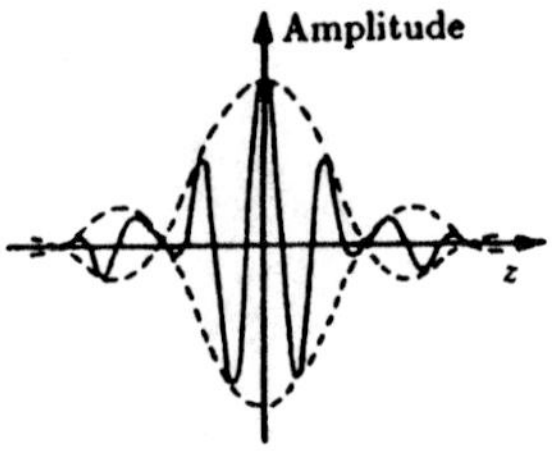

Fig. 3.1. A wave packet: several rapidly oscillating waves superpose, thus generating a group with finite extent

If we further increase the absolute value of z, the function $(\sin z)/z$ runs alternately through maxima and minima, the function values of which are small compared with the principal maximum at $z=0$, and quickly converges to zero. Therefore, we can conclude that superposition generates a wave packet whose amplitude is nonzero only in a finite region, and is described by $(\sin z)/z$. Figure 3.1 illustrates the form of such a wave packet at a certain time.

The modulating factor $(\sin z)/z$ of the amplitude assumes for $z \to 0$ the maximal value 1. Therefore, for $z = 0$

$$v_g t - x = 0 \ ,$$

which means that the maximum of the amplitude is a plane, propagating with velocity

$$\frac{\mathrm{d}x}{\mathrm{d}t} = v_g \ . \tag{3.21}$$

The propagation velocity of the plane of maximal amplitude has to be identified with the group velocity v_g, which, as we determined earlier, is the velocity of energy transport. The group velocity is the velocity of the whole wave packet ("matter-wave" group).

We can understand this in another, shorter way: if we demand $|\psi(x,t)|^2$ in (3.18) to be constant, i.e. $|\psi(x,t)|^2 = \text{const.}$, we conclude from (3.18) that

$v_g t - x = \text{const.}$, and, hence, by differentiation, $\dot{x} = v_g$. Thus, the fixed constant value of $|\psi(x,t)|^2$ moves with the group velocity v_g. Differentiating the dispersion relation (3.9) of $\omega(k)$, we get for v_g:

$$v_g = \left(\frac{d\omega}{dk}\right)_{k=k_0} = \left(\frac{\hbar k}{m_0}\right)_{k=k_0} = \frac{\hbar k_0}{m_0} = \frac{p}{m_0} . \tag{3.22}$$

From this, we must not in general conclude that the group velocity of a matter-wave group coincides with the classical particle velocity. All results derived up to now were gained under the simplification that all terms in the expansion (3.17) of $\omega(k)$ higher than first order can be neglected. This is allowed so long as the medium is free of dispersion. Since de Broglie waves show dispersion even in a vacuum, the derivative $d^2\omega/dk^2 \neq 0$, i.e. it is nonzero. This implies that the wave packet does not retain its form, but gradually spreads (each of the many monochromatic waves forming the packet has a slightly different frequency and therefore a different propagation velocity). If the dispersion is small, i.e.

$$\frac{d^2\omega}{dk^2} \approx 0 , \tag{3.23}$$

for a certain time, we can assign a particular form to the wave packet. Then we can consider the matter-wave group as moving as a whole with the group velocity v_g.

Following de Broglie, we assign to each uniformly moving particle a plane wave with wavelength λ. To determine this wavelength, we start with de Broglie's basic equations, (3.1) and (3.2). For the wavelength, the following holds:

$$\lambda = \frac{2\pi}{k} = \frac{2\pi\hbar}{p} = \frac{h}{p} . \tag{3.24}$$

If we assume the velocity of the particle to be small $v \ll c$, and use the equation

$$E = \frac{p^2}{2m_0} ,$$

we get the wavelength

$$\lambda = \frac{h}{\sqrt{2m_0 E}} , \tag{3.25}$$

meaning that we must know the rest mass of the particle in motion to determine its wavelength. If we consider, for example, an electron with kinetic energy $E = 10\,\text{keV}$ and rest mass $m_0 = 9.1 \times 10^{-28}$ g, then its matter wavelength is $\lambda_e = 0.122\,\text{Å} = 0.122 \times 10^{-8}$ cm.

As we know, the resolution of a small object under a microscope depends upon the wavelength of the light used to illuminate the object (see Examples 3.7 and 3.8 later on). The shorter the wavelength, the shorter the distance is between two points that can be seen distinctly through the microscope. The wavelength of visible light can typically be chosen as $\lambda_L \approx 5000$ Å, permitting a magnification of about 2000 times. If electrons are used instead of visible light to "scan" an object, magnification of up to 500 000 times and a resolution of about 5–10 Å can be achieved. Finally, protons and mesons in the GeV region (10^9 eV) have wavelengths so small that it is possible to use them to investigate the inner structure of elementary particles.

3.2 The Diffraction of Matter Waves

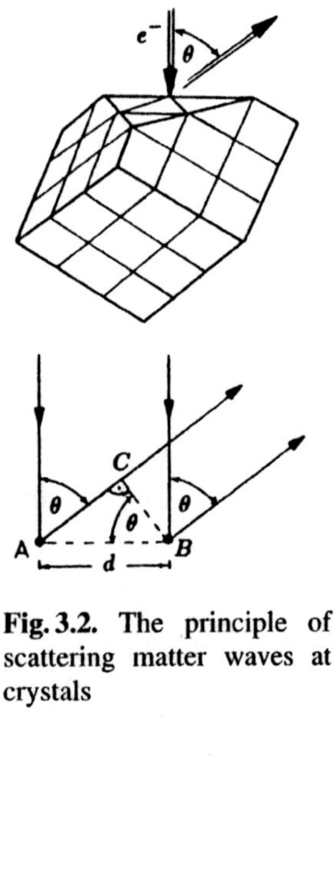

Fig. 3.2. The principle of scattering matter waves at crystals

Interference and diffraction phenomena are unique proofs of the occurrence of waves. In particular, destructive interference cannot be explained using the corpuscular picture. While the photoelectric effect and the Compton effect show the corpuscular nature of light, the diffraction of electron rays proves the existence of matter waves.

Since the wavelength of electrons is too small for diffraction by an artificial grid, crystal lattices are used for scattering. These experiments are in general a repetition of the corresponding structural investigations performed with X-rays.

Davisson and ***Germer*** applied ***Laue***'s method for X-ray diffraction. Here, the surface of a monocrystal is used as a plane diffraction grid. The electrons are scattered at the surface of the crystal, but do not penetrate it. Figure 3.2 shows the experimental setup and the path of the electron rays.

As can be seen in the figure, diffraction maxima appear if the condition

$$n\lambda = d\sin\theta \tag{3.26}$$

is fulfilled. If the electron passes through an accelerating voltage U, its energy is given by eU, and from (3.25) it follows that

$$\frac{nh}{d\sqrt{2m_0e}} = \sqrt{U}\sin\theta\ , \tag{3.27}$$

which, indeed, is confirmed by experiment.

Tartakowski and Thomson correspondingly used the *Debye–Scherrer method* of X-ray scattering (Exercise 3.1). Here, monochromatic X-radiation is diffracted by a body consisting of compressed crystal powder. The crystal powder represents a spatial diffraction grid. Because of the disordered arrangement, there are always crystals that comply with the bending condition. In Fig. 3.3 we can see the path of the rays.

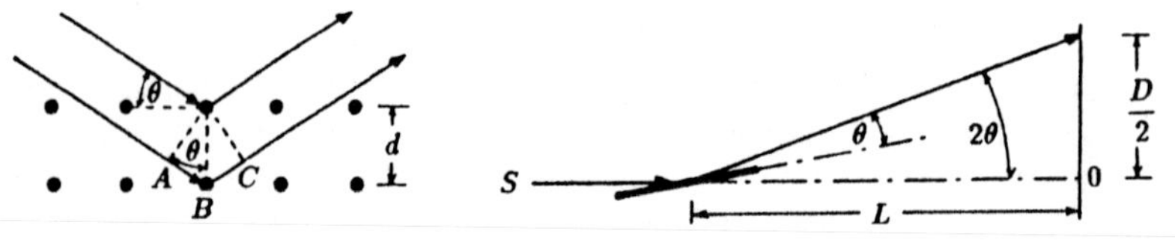

Fig. 3.3. Scattering of a matter wave at a crystal

The diffraction maxima appear under the condition (the *Wulf–Bragg relation*)

$$2d \sin\theta = n\lambda \ . \tag{3.28}$$

Owing to the statistical distribution of the minicrystals in the crystal powder, the apparatus – and correspondingly the diffraction figures – are symmetrical with respect to the $\overline{SO}$ axis. Because of this radial symmetry of the interference patterns, circles appear around O on the screen. Obviously, the relation $\tan(2\theta) = D/2L$ is valid, where L is the distance between scatterer and screen. The experimental setup is chosen in such a way that all angles are small, thus permitting the approximation $\tan(2\theta) \simeq 2\theta$. From the Wulf–Bragg equation we get

$$Dd = 2nL\lambda \ . \tag{3.29}$$

If electron rays are used, we insert the de Broglie wavelength (3.25) into the above relation and find that

$$D\sqrt{U} = \frac{2nLh}{d\sqrt{2m_0 e}} \ , \tag{3.30}$$

i.e. the square root of the accelerating voltage times the radius of the diffraction circles has to be constant for any order of diffraction.

The experimental results were in perfect agreement with this formula. Nowadays, electron rays and, in particular, neutron rays are an important tool used in solid state physics to determine crystal structures.

EXERCISE

3.1 Diffraction Patterns Generated by Monochromatic X-rays

Problem. (a) What are (schematically) the diffraction patterns generated by a monochromatic X-ray on an ideal crystal?

(b) The Debye–Scherrer method uses crystal powder rather than a crystal. What do these interference patterns look like?

Solution. (a) An ideal crystal consists of completely regularly arranged atoms. Incoming radiation of wavelength λ is reflected a little by each of these many grid planes. Macroscopic reflection occurs only if the reflected rays of several parallel net planes interfere constructively. Let d be the distance of any two parallel planes; then reflection occurs only if Bragg's condition $2d \sin\theta = n\lambda$ (n integer) is satisfied. Here θ is the angle between the ray and the grid plane. Since we assume that we have an ideal crystal, the angle θ is determined by the orientation of the crystal with regard to the ray direction. In the Bragg condition, d, θ and λ are already determined. Hence there is in general no n that fulfils the condition. Normally, no reflection occurs!

Exercise 3.1

To overcome this drawback, one avoids monochromatic X-rays for structure analysis, but uses a continuous spectrum (Laue method) instead. In this case the diffraction patterns consist of single points which are regularly arranged. Another possibility is to change θ by turning the crystal (rotating crystal method).

(b) Here, monochromatic radiation is scattered by crystal powder. Therefore (see above) for most crystals no reflection occurs. Only for those crystals oriented accidentally in one of the scattering angles θ does constructive interference occur, and the ray is deflected through 2θ. Because the crystals are uniformly distributed with respect to the azimuth, a cone of reflection results (see next figure).

Scattering by silicon: photography and diffractometer plot generated by the Debye-Scherrer powder method

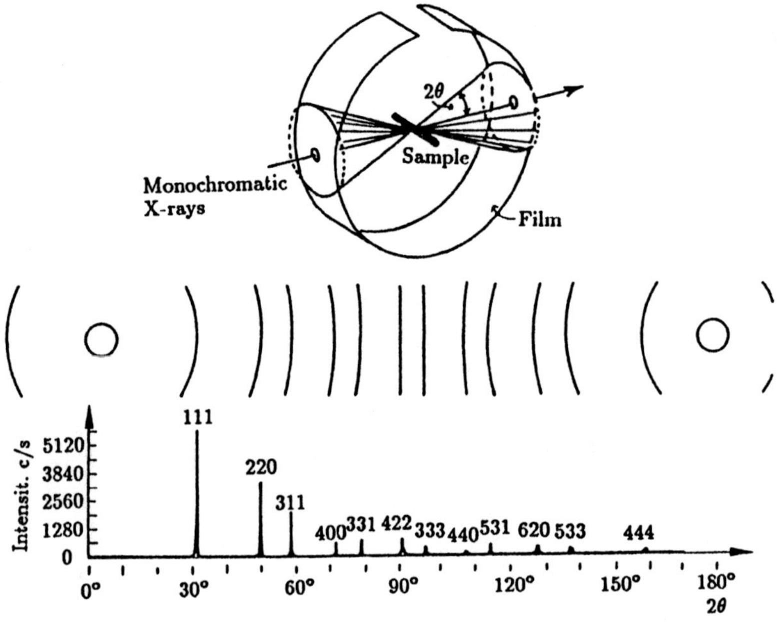

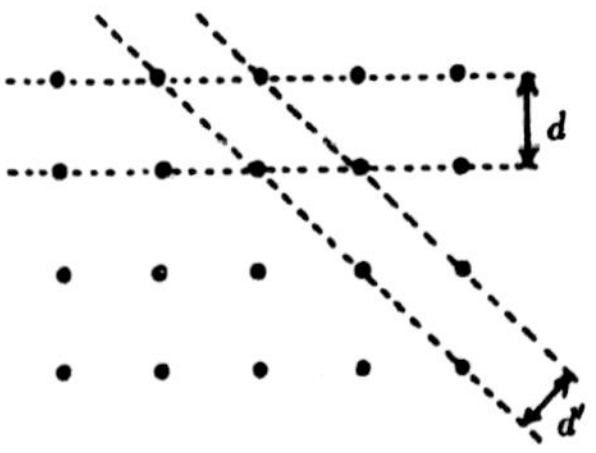

EXERCISE

3.2 Scattering of Electrons and Neutrons

Problem. (a) Calculate the wavelengths of the (electromagnetic, probability, matter) waves of 10-keV X-rays, 1-keV electrons and 5-eV neutrons.

(b) Do the interference patterns change if X-radiation is replaced by neutrons of the same wavelength?

(c) Additional question: How can "monochromatic" neutrons be created?

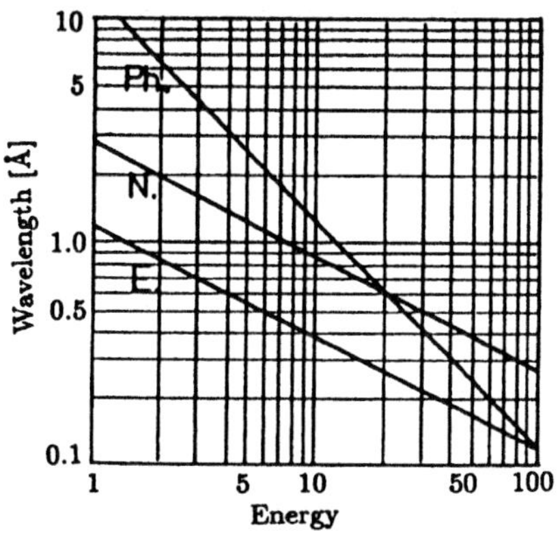

Wavelength as function of particle energy for photons (keV), neutrons (0.01 eV) and electrons (100 eV)

Solution. (a) The wavelength $\lambda = 2\pi/k$ of a particle of mass m is related to its momentum p; the latter is determined by the total energy E using

$$p = \sqrt{(E/c)^2 - m_0^2c^2} \ .$$

From the de Broglie relation $k = p/\hbar$, we therefore have

$$\lambda = 2\pi\hbar[(E/c)^2 - m_0^2c^2]^{-1/2} \ .$$

For photons $m_0 = 0$ and therefore $\lambda_{ph} = 2\pi\hbar c/E_{ph}$. In the nonrelativistic limit, we get for electrons and neutrons

$$p = \sqrt{2m_0E_{kin}} \ , \quad \text{and thus} \quad \lambda = 2\pi\hbar\sqrt{2m_0E_{kin}} \ .$$

Inserting this and using

$$m_e = 0.911 \times 10^{-27}\,\text{g} \ , \quad m_N = 1675 \times 10^{-27}\,\text{g} \ ,$$
$$1\,\text{eV} = 1.602 \times 10^{-12}\,\text{erg} \ , \quad 2\pi\hbar = 6.62 \times 10^{-27}\,\text{erg s} \ ,$$
$$1\,\text{pm} = 10^{-12}\,\text{m} = 10^{-10}\,\text{cm} = 10^{-2}\,\text{Å}$$

we have

$$\lambda_{ph}(10\,\text{keV}) = 120\,\text{pm} \ , \quad \lambda_e(1\,\text{keV}) = 39\,\text{pm} \ , \quad \lambda_N(5\,\text{eV}) = 13\,\text{pm} \ .$$

(b) Replacing the X-radiation by neutrons of the same wavelength leaves the scattering angles unaltered at first. But because the gamma rays interact with the electron distribution, whereas the neutrons interact with the atomic nuclei and, when ultimately coming into being, magnetic dipole moments, the relation of the intensities is different.

(c) To create monochromatic neutrons, Bragg reflection is again used. A polychromatic neutron ray strikes a monochromator crystal. At a certain angle 2θ, only a few wavelengths ($n\lambda = 2d\sin\theta$) are emitted (see Figure). In general one reflection ($n = 1$ and one fixed d) is much stronger than the others.

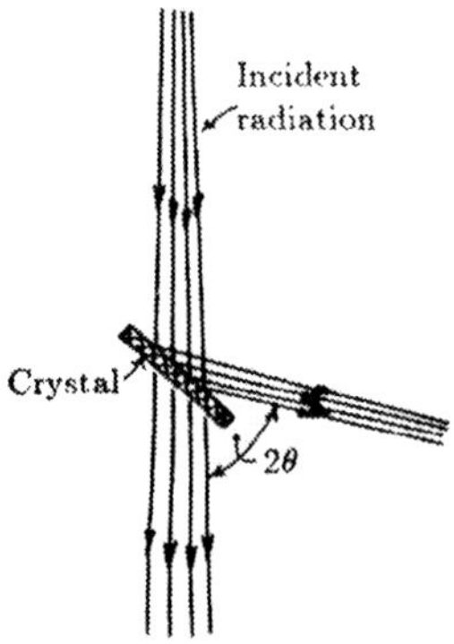

Principle of the creation of monochromatic neutrons by reflection

3.3 The Statistical Interpretation of Matter Waves

The question of how to interpret a wave describing a particle, and whether this wave should be assigned physical reality, was subject to discussion in the first years of quantum mechanics. A single electron acts as a particle, but interference patterns only arise if many electrons are scattered.

Max ***Born*** paved the way for the statistical interpretation of the wave function describing a particle. He created the term *guiding field* (in German: "Führungsfeld") as an interpretation of wave functions. The idea actually originated with Einstein, who called it a *Ghostfield* ("Gespensterfeld"). The guiding field is a scalar function ψ of the coordinates of all particles and of time. According to the basic idea, the motion of a particle is determined only by the laws of energy and momentum conservation and by the boundary conditions which depend on the particular experiment (apparatus). The particle is kept within these set limits by the guiding field. The probability that the particle will follow a particular path is given by the intensity, i.e. the absolute square of the guiding field. In the case of electron scattering, this means that the intensity of the matter wave (guiding field) determines at every point the probability of finding an electron there. We shall now further investigate this interpretation of matter waves as probability fields.

The square of the amplitude of the wave function ψ is the intensity. It ought to determine the probability of finding a particle at a certain place. Since ψ may be complex, whereas the probability is always real, we do not define ψ^2 as a measure of intensity, but instead

$$|\psi^2| = \psi\psi^* , \tag{3.31}$$

where ψ^* is the complex conjugate of ψ. In addition, the probability of finding a particle is proportional to the size of the volume under consideration. Let $\mathrm{d}W(x, y, z, t)$ be the probability of finding a particle in a certain volume element $\mathrm{d}V = \mathrm{d}x\,\mathrm{d}y\,\mathrm{d}z$ at time t. According to the statistical interpretation of matter waves, the following hypothesis is adopted:

$$\mathrm{d}W(x, y, z, t) = |\psi(x, y, z, t)|^2 \mathrm{d}V .$$

In order to get a quantity independent of volume, we introduce the *spatial probability density*

$$w(x, y, z, t) = \frac{\mathrm{d}W}{\mathrm{d}V} = |\psi(x, y, z, t)|^2 . \tag{3.32}$$

It is normalized to one, i.e. the amplitude of ψ is chosen so that

$$\int_{-\infty}^{\infty} \psi\psi^* \,\mathrm{d}V = 1 . \tag{3.33}$$

This means that the particle must be somewhere in space. The normalization integral is time independent:

$$\frac{\mathrm{d}}{\mathrm{d}t}\int_{-\infty}^{\infty} \psi\psi^* \,\mathrm{d}V = \frac{\mathrm{d}}{\mathrm{d}t}1 = 0 \;;$$

otherwise, we could not compare probabilities referring to different times. The wave function ψ can only be normalized if it is square-integrable, i.e. if the improper integral

$$\int_{-\infty}^{\infty} |\psi|^2 \mathrm{d}V \text{ converges,} \quad \text{i.e.} \quad \int_{-\infty}^{\infty} \psi\psi^* \,\mathrm{d}V < M \;,$$

M being a real constant. The probability interpretation for the ψ field expressed in (3.32) is a small step, but nevertheless it is only a hypothesis. Its validity must be proved – and *will* be proved, as we shall see – by the success of the results it predicts. A state is *bound*, if the motion of the system is restricted; if it is not restricted, we have *free* states. In the course of this book we shall establish the following facts: the wave function ψ for bound states ($E < 0$) is square-integrable, whereas $|\psi|^2$ is not integrable for free states. This is intuitively understood from Fig. 3.4; bound states are localized within the potential well and can propagate only within its interior; thus, they are confined. Free states are located above the potential well and are not bound.

A normalized wave function ψ is determined only up to a phase factor of modulus one, i.e. up to a factor $\mathrm{e}^{\mathrm{i}\alpha}$ with an arbitrarily real number α. This lack of uniqueness stems from the fact that only the quantity $\psi\psi^* = |\psi|^2$, the probability density, has physical relevance.

An example of wave functions that cannot be normalized according to the requirement (3.33) is the wave function

$$\psi(\boldsymbol{r}, t) = N\,\mathrm{e}^{\mathrm{i}(\boldsymbol{k}\cdot\boldsymbol{r}-\omega t)} \;, \tag{3.34}$$

where N is a real constant. This plane wave describes the motion of a free particle with momentum $\boldsymbol{p} = \hbar\boldsymbol{k}$ and undefined locality. But we can normalize the

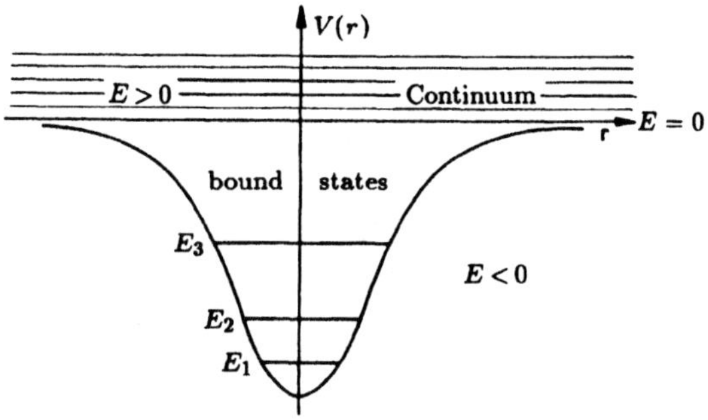

Fig. 3.4. Bound and continuum states of a particle in a potential well

function (3.34) if we define all functions within a large, finite volume considered to be a cube with an edge length L (*box normalization*):

$$\psi = \begin{cases} N\,\mathrm{e}^{\mathrm{i}(\boldsymbol{k}\cdot\boldsymbol{r}-\omega t)} & \text{for} \quad \boldsymbol{r} \text{ within } V = L^3 \\ 0 & \text{for} \quad \boldsymbol{r} \text{ outside } V = L^3 \end{cases} . \tag{3.35}$$

Another method for normalizing such "continuum wave functions" will be presented in Chap. 5.

On the surface of this volume, the wave functions must satisfy certain boundary conditions. We assume L to be large by microscopic standards ($L \gg 10^{-8}$ m). Then the influence of the boundary conditions on the motion of a particle in the volume $V = L^3$ is very small. Therefore, we can choose the boundary conditions in quite a simple, arbitrary form. Very often a periodicity with period L is selected as boundary condition; we require that

$$\psi(x,y,z) = \psi(x+L,x,z) = \psi(x,y+L,z) = \psi(x,y,z+L) \ . \tag{3.36}$$

Now we determine the normalization factor N in (3.34), keeping in mind the definition (3.35):

$$1 = \int\limits_{-\infty}^{\infty} \psi\psi^* \,\mathrm{d}V = N^2 \int\limits_{V=L^3} \mathrm{d}V = N^2 L^3 \ .$$

Hence, it follows that

$$N = \frac{1}{\sqrt{L^3}} = \frac{1}{\sqrt{V}} \ .$$

Thus, we get the normalized wave function

$$\psi_{\boldsymbol{k}}(\boldsymbol{r},t) = \frac{1}{\sqrt{V}}\mathrm{e}^{\mathrm{i}\boldsymbol{k}\cdot\boldsymbol{r}-\mathrm{i}\omega(k)t} = \psi_{\boldsymbol{k}}(\boldsymbol{r})\,\mathrm{e}^{-\mathrm{i}\omega(k)t} \ ; \qquad \psi_{\boldsymbol{k}}(\boldsymbol{r}) = \frac{1}{\sqrt{V}}\mathrm{e}^{\mathrm{i}\boldsymbol{k}\cdot\boldsymbol{r}} \ . \tag{3.37a}$$

The boundary conditions of our problem restrict the possible values of the vector $\boldsymbol{k}$:

$$\boldsymbol{k} = \frac{2\pi}{L}\boldsymbol{n} \ , \quad \boldsymbol{k} = \{k_x, k_y, k_z\} \ , \quad \boldsymbol{n} = \{n_x, n_y, n_z\} \ . \tag{3.38a}$$

Written as components, we have, respectively,

$$k_x = \frac{2\pi}{L}n_x \ , \quad k_y = \frac{2\pi}{L}n_y \ , \quad k_z = \frac{2\pi}{L}n_z \ , \tag{3.38b}$$

where n_x, n_y and n_z take on all integer values. Hence, the momentum $\boldsymbol{p} = \hbar\boldsymbol{k} = (2\pi\hbar/L)\boldsymbol{n}$ is quantized. The same is valid for the energy $E = \hbar\omega(\boldsymbol{k})$, and so for the frequency of the wave

$$\omega(\boldsymbol{k}) = \frac{E}{\hbar} = \frac{p^2}{2\hbar m} = \frac{\hbar k^2}{2m} = \left(\frac{2\pi}{L}\right)^2 \frac{\hbar}{2m}\left(n_x^2+n_y^2+n_z^2\right) \ . \tag{3.39}$$

Inserting the values for $\boldsymbol{k}$ into the normalized wave function (3.16), we get

$$\psi_{\boldsymbol{k}}(\boldsymbol{r},t) = \frac{1}{\sqrt{V}}\exp\{\mathrm{i}[(2\pi/L)\boldsymbol{n}\cdot\boldsymbol{r}-\omega(k)t]\}$$
$$= \frac{1}{\sqrt{L^3}}\exp\{\mathrm{i}[(2\pi/L)\boldsymbol{n}\cdot\boldsymbol{r}-\omega(k)t]\} = \psi_{\boldsymbol{k}}(\boldsymbol{r})\,\mathrm{e}^{-\mathrm{i}\omega(k)t} \ . \qquad (3.37\mathrm{b})$$

For these wave functions, we can explicitly check that the periodicity conditions (3.36) are fulfilled. Owing to the boundary conditions (3.36), the vector $\boldsymbol{k}$ (and therefore the momentum $\boldsymbol{p} = \hbar\boldsymbol{k}$) takes on discrete values, given by the conditions (3.38). In the limit $L \to \infty$, the difference between neighbouring values of $\boldsymbol{k}$ [and of the energy (3.39)] converges to zero, and finally we return to the motion of a free particle in infinite space.

Now we shall prove that these normalized wave functions $\psi_{\boldsymbol{k}}(\boldsymbol{r},t)$ of (3.37) constitute an orthonormal function system, so that

$$\int_V \psi^*_{\boldsymbol{k}}(\boldsymbol{r})\psi_{\boldsymbol{k}'}(\boldsymbol{r})\,\mathrm{d}V = \delta_{\boldsymbol{k}\boldsymbol{k}'} \ . \qquad (3.40)$$

Here, only the spatial part of the wave functions $\psi_{\boldsymbol{k}}$ (plane waves) from (3.37a) is considered. The factor containing the time dependence $\exp[+\mathrm{i}\omega(k)t]$ does not change anything in the orthonormality relation (3.40). We insert the wave functions and calculate:

$$\int_{V=L^3} \psi^*_{\boldsymbol{k}}(\boldsymbol{r})\psi_{\boldsymbol{k}}(\boldsymbol{r})\,\mathrm{d}V$$
$$= \frac{1}{L^3}\int_{-L/2}^{L/2}\mathrm{e}^{\mathrm{i}(k'_x-k_x)x}\,\mathrm{d}x\int_{-L/2}^{L/2}\mathrm{e}^{\mathrm{i}(k'_y-k_y)y}\int_{-L/2}^{L/2}\mathrm{e}^{\mathrm{i}(k'_z-k_z)z}$$
$$= \frac{1}{L^3}\int_{-L/2}^{L/2}\exp\left\{\mathrm{i}\left[\frac{2\pi}{L}(n'_x-n_x)x\right]\right\}\mathrm{d}x\int_{-L/2}^{L/2}\exp\left\{\mathrm{i}\left[\frac{2\pi}{L}(n'_y-n_y)y\right]\right\}\mathrm{d}y$$
$$\times\int_{-L/2}^{L/2}\exp\left\{\mathrm{i}\left[\frac{2\pi}{L}(n'_z-n_z)z\right]\right\}\mathrm{d}z \ ,$$
$$= \frac{\sin[\pi(n'_x-n_x)]}{\pi(n'_x-n_x)}\,\frac{\sin[\pi(n'_y-n_y)]}{\pi(n'_y-n_y)}\,\frac{\sin[\pi(n'_z-n_z)]}{\pi(n'_z-n_z)}$$
$$= \delta_{n'_x n_x}\delta_{n'_y n_y}\delta_{n'_z n_z} = \delta_{\boldsymbol{k}'\boldsymbol{k}} \ .$$

Obviously $\sin[\pi(n'_x-n_x)] = 0$ for $n'_x \neq n_x$. Therefore, only the cases where $n'_x = n_x$, $n'_y = n_y$ and $n'_z = n_z$ contribute. Thus, the wave functions $\psi_{\boldsymbol{k}}(\boldsymbol{r},t)$ of (3.37) indeed constitute an orthonormal function system. In addition, the $\psi_{\boldsymbol{k}}$ are a complete system, i.e. it is impossible to find an additional function ϕ that is orthogonal to all ψ_k in the sense of the relation (3.40). Then the following completeness relation is valid:

$$\int_V \psi\psi^* \,\mathrm{d}V = \int_V |\psi|^2 \mathrm{d}V = \sum_k |a_k|^2 \ , \tag{3.41}$$

where the a_k are the expansion coefficients of the arbitrary wave function ψ in terms of a complete set of the ψ_k,

$$\psi = \sum_k a_k \psi_k(x) \ . \tag{3.42}$$

If completeness (3.41) is proved (the proof is omitted here), we can always expand according to (3.42), i.e. the ψ_k constitute an orthonormal basis of a ***Hilbert*** space. A Hilbert space is a finite or infinite complete vector space on the basic field of complex numbers. In this space a scalar product is defined such that it assigns a complex number to each pair of functions $\psi(x)$ and $\phi(x)$ out of a set of linear functions. This scalar product meets four requirements:

$$\begin{aligned}
&(1) \quad \langle\psi|\phi\rangle && = \int \psi^*\phi \mathrm{d}V = \left(\int \phi^*\psi \,\mathrm{d}V\right)^* = (\langle\phi|\psi\rangle)^* \ , \\
&(2) \quad \langle\psi|a\phi_1 + b\phi_2\rangle && = a\,\langle\psi|\phi_1\rangle + b\,\langle\psi|\phi_2\rangle \quad \text{or} \\
&\qquad \int \psi^*(a\phi_1 + \phi_2)\,\mathrm{d}V = a\int \psi^*\phi_1\,\mathrm{d}V + b\int \psi^*\phi_2\,\mathrm{d}V \ , && \text{(linearity)} \\
&(3) \quad \langle\psi|\psi\rangle && = \int \psi^*\psi\,\mathrm{d}V \geq 0 \ , \\
&(4) \quad \text{from } \langle\psi|\psi\rangle && = \int \psi^*\psi\,\mathrm{d}V = 0 \ , \\
&\qquad \text{follows } \psi(x) && = 0 \ .
\end{aligned} \tag{3.43}$$

The state vectors (= wave functions) of a quantum-mechanical system constitute a Hilbert space (hence, the Hilbert space is a function space). In the following we show that (3.42) leads to the completeness relation (3.41). We multiply (3.42) on both sides by its complex conjugate, integrate over the total space and use the orthonormality condition (3.40). Then we get

$$\int_V \psi\psi^* \,\mathrm{d}V = \int_V \sum_{k,k'} a_k a_{k'}^* \psi_k \psi_{k'}^* \,\mathrm{d}V = \sum_{k,k'} a_k a_{k'}^* \int_V \psi_k \psi_{k'}^* \,\mathrm{d}V = \sum_{k,k'} a_k a_{k'}^* \delta_{k,k'}$$

i.e. we have the completeness relation (3.41).

To determine the coefficients a_k of the expansion (3.42) we multiply this equation by ψ_k^* and integrate over the volume V:

$$\int_V \psi\psi_{k'}^* \,\mathrm{d}V = \sum_k a_k \int_V \psi_k \psi_{k'}^* \,\mathrm{d}V = \sum_k a_k \delta_{kk'} = a_{k'} \ , \tag{3.44a}$$

$$\text{i.e.} \quad a_k = \int \psi\psi_k^* \,\mathrm{d}V \ .$$

With the help of the normalization integral, we obtain

$$1 = \int_{-\infty}^{\infty} \psi\psi^* \, \mathrm{d}V = \sum_{kk'} a_k a_{k'}^* \int_V \psi_k \psi_{k'}^* \, \mathrm{d}V = \sum_{kk'} a_k a_{k'}^* \delta_{kk'} = \sum_k a_k a_k^* \ .$$

Hence

$$\sum_k |a_k|^2 = 1 \ .$$

Now we interpret the quantity $|a_k|^2$ as the probability of finding a particle with momentum

$$\boldsymbol{p} = \hbar \boldsymbol{k}$$

in the state ψ. This interpretation is very reasonable in view of (3.42) and the fact that the $\psi_{\boldsymbol{k}}(\boldsymbol{r}, t)$ in (3.37) are wave functions with definite momentum $\boldsymbol{p} = \hbar \boldsymbol{k}$.

3.4 Mean (Expectation) Values in Quantum Mechanics

In the following we will calculate the mean values of position, momentum and other physical quantities in a certain state if the normalized wave function ψ is known.

1. The Mean Value of Position Coordinates. Let a quantum mechanical system be in the state ψ. The position probability density is then given by the term $\psi\psi^*$. The function of state ψ is normalized to unity. Thus the mean value of the position vector is given by

$$\langle \boldsymbol{r} \rangle = \int_V \boldsymbol{r} \, \psi^*(\boldsymbol{r}) \psi(\boldsymbol{r}) \, \mathrm{d}V = \int_V \psi^*(\boldsymbol{r}) \, \boldsymbol{r} \, \psi(\boldsymbol{r}) \, \mathrm{d}V \ .$$

Accordingly, we have for the *mean (expectation) value* of a function $f(\boldsymbol{r})$, which depends on $\boldsymbol{r}$ only,

$$\langle f(\boldsymbol{r}) \rangle = \int_V f(\boldsymbol{r}) \psi^*(\boldsymbol{r}) \psi(\boldsymbol{r}) \, \mathrm{d}V = \int_V \psi^*(\boldsymbol{r}) \, f(\boldsymbol{r}) \psi(\boldsymbol{r}) \, \mathrm{d}V \ .$$

2. The Mean Value of Linear Momentum. It has already been shown that an arbitrary wave function ψ can be expanded in terms of an orthonormal basis of the Hilbert space $\{\psi_{\boldsymbol{k}}\}$. Then the absolute square of the expansion coefficients represents the momentum probability (3.44) and the relation

$$\langle \boldsymbol{p} \rangle = \sum_{\boldsymbol{k}} a_{\boldsymbol{k}}^* (\hbar \boldsymbol{k}) a_{\boldsymbol{k}}$$

holds. By insertion of expression (3.44a) for a_k into this relation for the expectation value of the momentum, it follows that

$$\langle \boldsymbol{p}\rangle = \sum_{k}\left(\int_{V'} \psi_k(\boldsymbol{r}')\psi^*(\boldsymbol{r}')\,\mathrm{d}V'\right)\hbar\boldsymbol{k}\left(\int_{V} \psi(\boldsymbol{r})\psi_k^*(\boldsymbol{r})\,\mathrm{d}V\right) \quad ; \quad \text{or} \tag{3.45a}$$

$$\langle \boldsymbol{p}\rangle = \sum_{k}\int_{V}\int_{V'} \psi^*(\boldsymbol{r}')\psi_k(\boldsymbol{r}')\hbar\boldsymbol{k}\psi_k^*(\boldsymbol{r})\psi(\boldsymbol{r})\,\mathrm{d}V\,\mathrm{d}V' \,. \tag{3.45b}$$

It is easy to verify the relation

$$\hbar\boldsymbol{k}\psi_k^*(\boldsymbol{r}) = \mathrm{i}\hbar\nabla\psi_k^*(\boldsymbol{r}) \tag{3.46}$$

by using the wave function

$$\psi_k^*(\boldsymbol{r}) = \frac{\mathrm{e}^{-\mathrm{i}\boldsymbol{k}\cdot\boldsymbol{r}}}{\sqrt{V}} \,,$$

which has already been introduced together with the corresponding boundary condition in (3.37). Inserting (3.46) into (3.45b) yields

$$\langle \boldsymbol{p}\rangle = \sum_{k}\int_{V'} \psi^*(\boldsymbol{r}')\psi_k(\boldsymbol{r}')\,\mathrm{d}V' \int_{V} [\mathrm{i}\hbar\nabla\psi_k^*(\boldsymbol{r})]\psi(\boldsymbol{r})\,\mathrm{d}V \,.$$

Now, according to the condition of periodicity (3.36), the values of ψ and ψ_k are equal at opposite planes ($x=0$, L or $y=0$, L or $z=0$, L) of the cube with a volume $V=L^3$. Thus we get by partial integration of the x component, for example,

$$\begin{aligned}&\mathrm{i}\hbar\iint\left[\int\left(\frac{\mathrm{d}\psi_k^*}{\mathrm{d}x}\right)\psi\,\mathrm{d}x\right]\mathrm{d}y\;\mathrm{d}z\\ &\quad = \mathrm{i}\hbar\iint \psi_k^*\Big|_{x=0,L}\mathrm{d}y\;\mathrm{d}z - \mathrm{i}\hbar\int\psi_k^*\psi\left(\frac{\mathrm{d}\psi}{\mathrm{d}x}\right)\mathrm{d}V = -\mathrm{i}\hbar\int\psi_k^*\left(\frac{\mathrm{d}\psi}{\mathrm{d}x}\right)\mathrm{d}V \,.\end{aligned}$$

The expectation value of the momentum $\boldsymbol{p}$ is then given by

$$\langle \boldsymbol{p}\rangle = \int_{V}\int_{V'}\left\{\psi^*(\boldsymbol{r}')(-\mathrm{i}\hbar\nabla\psi(\boldsymbol{r}))\sum_{k}\psi_k(\boldsymbol{r}')\psi_k^*(\boldsymbol{r})\right\}\mathrm{d}V\;\mathrm{d}V' \,. \tag{3.47}$$

Now we make use of the relation

$$\sum_{k}\psi_k(\boldsymbol{r}')\psi_k^*(\boldsymbol{r}) = \delta(\boldsymbol{r}'-\boldsymbol{r}) \,, \tag{3.48}$$

which can easily be proved by expanding the delta function[1] $\delta(\boldsymbol{r}'-\boldsymbol{r})$ in terms of the complete set of functions $\psi_k(\boldsymbol{r}) = V^{-1/2}\mathrm{e}^{\mathrm{i}\boldsymbol{k}\cdot\boldsymbol{r}}$

$$\delta(\boldsymbol{r}'-\boldsymbol{r}) = \sum_{k} b_k(\boldsymbol{r}')\psi_k(\boldsymbol{r}) \,, \tag{3.49}$$

[1] The definition and properties of the delta function $\delta(x)$ are discussed in Chap. 5.

and by calculating the expansion coefficients $b_k(\boldsymbol{r}')$. Multiplying both sides of (3.49) by $\psi^*_{k'}(\boldsymbol{r})$ and integrating over $\boldsymbol{r}$ gives

$$\int\limits_V \psi^*_{k'}(\boldsymbol{r})\delta(\boldsymbol{r}'-\boldsymbol{r})\,\mathrm{d}V = \sum_{\boldsymbol{k}} b_{\boldsymbol{k}}(\boldsymbol{r}') \int\limits_V \psi_{\boldsymbol{k}}(\boldsymbol{r})\psi^*_{\boldsymbol{k}'}(\boldsymbol{r})\,\mathrm{d}V \ ,$$

so that with the help of the orthonormality relation (3.40),

$$\psi^*_{\boldsymbol{k}'}(\boldsymbol{r}') = \sum_{\boldsymbol{k}} b_{\boldsymbol{k}}\delta_{\boldsymbol{k}\boldsymbol{k}'} = b_{\boldsymbol{k}'} \ , \quad \text{i.e. } b_{\boldsymbol{k}'}(\boldsymbol{r}') = \psi^*_{\boldsymbol{k}'}(\boldsymbol{r}')$$

results, which, in turn, immediately yields (3.48).

Applying (3.48) and (3.49) to (3.47), we obtain the final form, whose structure is similar to the formula for the mean value of the position vector, namely

$$\langle \boldsymbol{p}\rangle = \int\limits_{V=L^3} \psi^*(\boldsymbol{r})(-\mathrm{i}\hbar\nabla)\psi(\boldsymbol{r})\,\mathrm{d}V \ . \tag{3.50}$$

This relation directly expresses the mean value of the momentum with the help of the wave function $\psi(\boldsymbol{r})$ for the corresponding state. The structure of (3.50) remains valid even in the limit $L \to \infty$. Thus, in the general case of an unlimited space, this formula is also valid for calculating the mean values of the momentum. In a similar way we can deduce that the mean value for an arbitrary power of the momentum can be calculated as

$$\langle \boldsymbol{p}^n\rangle = \int\limits_V \psi^*(\boldsymbol{r})(-\mathrm{i}\hbar\nabla)^n\psi(\boldsymbol{r})\,\mathrm{d}V \ . \tag{3.51}$$

We can immediately generalize this result for an arbitrary integral rational function $F(\boldsymbol{p})$ with $F(\boldsymbol{p}) = \sum_\nu a_\nu \boldsymbol{p}^\nu$ of momentum

$$\langle F(\boldsymbol{p})\rangle = \int\limits_V \psi^*(\boldsymbol{r})\hat{F}(-\mathrm{i}\hbar\nabla)\psi(\boldsymbol{r})\,\mathrm{d}V \ . \tag{3.52}$$

Here, $\hat{F}$ is an operator. The momentum $\boldsymbol{p}$ is related to the *differential operator* by

$$\hat{\boldsymbol{p}} = -\mathrm{i}\hbar\nabla \quad \text{and} \quad \hat{F}(\boldsymbol{p}) = \sum a_\nu \hat{\boldsymbol{p}}^\nu \ . \tag{3.53}$$

The importance of this relation lies in the fact that if we want to calculate the expectation value of the quantity $F(\boldsymbol{p})$, we need not go through a Fourier decomposition of the wave function $\psi(\boldsymbol{r})$ and then calculate $\langle F(\boldsymbol{p})\rangle = \sum_k F(\hbar\boldsymbol{k})a^*_{k'}a_k$, as was done in (3.45a) for the momentum $\boldsymbol{p} = \hbar\boldsymbol{k}$. Instead, the entire calculation can be abbreviated by introducing the operator $\hat{F}(\hat{\boldsymbol{p}})$ instead of the function $F(\boldsymbol{p})$, and performing the integral (3.52) directly. In the following we shall apply these relations and calculate three operators which are of particular importance in quantum mechanics.

3.5 Three Quantum-Mechanical Operators

1. The Kinetic Energy Operator. In the nonrelativistic case we have for the kinetic energy $T = p^2/2m$. With $\nabla^2 = \Delta$, we obtain the operator $\hat{T}$ as

$$\hat{T} = \frac{\hat{p}^2}{2m} = \frac{(-\mathrm{i}\hbar\nabla)^2}{2m} = -\frac{\hbar^2}{2m}\Delta \ ; \tag{3.54}$$

this is a special case of (3.51) or (3.52).

2. The Angular Momentum Operator. With the classical angular momentum of a particle $\boldsymbol{L} = \boldsymbol{r} \times \boldsymbol{p}$, we obtain the quantum-mechanical angular momentum operator

$$\hat{\boldsymbol{L}} = \boldsymbol{r} \times (-\mathrm{i}\hbar\nabla) = -\mathrm{i}\hbar\boldsymbol{r} \times \nabla \ . \tag{3.55a}$$

The individual components of this operator are

$$\hat{L}_x = -\mathrm{i}\hbar\left(y\frac{\partial}{\partial z} - z\frac{\partial}{\partial y}\right) \ , \quad \hat{L}_y = -\mathrm{i}\hbar\left(z\frac{\partial}{\partial x} - x\frac{\partial}{\partial z}\right) \ ,$$
$$\hat{L}_z = -\mathrm{i}\hbar\left(x\frac{\partial}{\partial y} - y\frac{\partial}{\partial x}\right) \tag{3.55b}$$

(A detailed discussion of the angular momentum operator will be given in Chap. 4.)

3. The Hamiltonian Operator. The total energy of time-independent physical systems is described classically by the Hamiltonian function,

$$H = T + V(\boldsymbol{r}) \ .$$

Here T denotes the kinetic energy and $V(\boldsymbol{r})$ the potential energy. This yields the Hamiltonian operator (Hamiltonian),

$$\hat{H} = -\frac{\hbar^2}{2m}\Delta + \hat{V}(\boldsymbol{r}) \ . \tag{3.56}$$

In quantum mechanics, an operator is assigned to any observable quantity (in signs: $A \to \hat{A}$). Let $A(\boldsymbol{r}, \boldsymbol{p})$ be a function of $\boldsymbol{r}$ and $\boldsymbol{p}$. We construct the corresponding operator by replacing the quantities $\boldsymbol{r}$ and $\boldsymbol{p}$ in the expression for A by the assigned operators $\hat{\boldsymbol{r}} = \boldsymbol{r}$ and $\hat{\boldsymbol{p}} = -\mathrm{i}\hbar\nabla$. Here the position operator is identical with the position vector. (But we must be careful: this is not generally valid; it is only true in the Cartesian coordinate representation chosen here! See our discussion of quantization in curvilinear coordinates in Chap. 8).

EXERCISE

3.3 The Expectation Value of Kinetic Energy

Problem. Calculate the expectation value (mean value) of the kinetic energy $\hat{T} = \hat{p}^2/2m$ with $\hat{p} = -\mathrm{i}\hbar\nabla$ and of the potential $\hat{V} = -e^2/r$ for the $1s$ electron in the ground state of hydrogen with the wave function

$$\psi_{1s} = \frac{1}{\sqrt{\pi a^3}}\mathrm{e}^{-r/a} \;, \qquad a = \frac{\hbar^2}{me^2} \;.$$

Solution. The expectation value is defined by

$$\langle \hat{T} \rangle = \int \mathrm{d}^3 r \psi_{1s}^*(\boldsymbol{r}) \hat{T} \psi_{1s}(\boldsymbol{r}) \;,$$
$$\langle \hat{V} \rangle = \int \mathrm{d}^3 r \psi_{1s}^*(\boldsymbol{r}) \hat{V} \psi_{1s}(\boldsymbol{r}) \;.$$

Using spherical coordinates we get

$$\begin{aligned}
\langle \hat{T} \rangle &= \int \mathrm{d}^3 r \psi_{1s}^* \frac{\hat{p}^2}{2m} \psi_{1s} \\
&= \frac{1}{\pi a^3} 4\pi \int_0^\infty r^2 \,\mathrm{d}r \,\mathrm{e}^{-r/a} \left(\frac{-\hbar^2}{2m} \frac{1}{r^2} \frac{\partial}{\partial r} r^2 \frac{\partial}{\partial r} \right) \mathrm{e}^{-r/a} \\
&= -\frac{2\hbar^2}{ma^3} \int_0^\infty \mathrm{d}r \,\mathrm{e}^{-r/a} \left(-\frac{1}{a} \left[2r - \frac{r^2}{a} \right] \right) \mathrm{e}^{-r/a} \\
&= \frac{2\hbar^2}{ma^4} \int_0^\infty \mathrm{d}r \left(2r - \frac{r^2}{a} \right) \mathrm{e}^{-2r/a} = \frac{1}{2} \frac{me^4}{\hbar^2} \;, \\
\langle \hat{V} \rangle &= \frac{1}{\pi a^3} 4\pi \int_0^\infty r^2 \,\mathrm{d}r \,\mathrm{e}^{-r/a} \left(\frac{-e^2}{r} \right) \mathrm{e}^{-r/a} \\
&= \frac{-4e^2}{a^3} \int_0^\infty \mathrm{d}r \, r \,\mathrm{e}^{-2r/a} = -\frac{me^4}{\hbar^2} \;.
\end{aligned}$$

The total energy is $E = \langle \hat{T} + \hat{V} \rangle = -\frac{1}{2}(me^4/\hbar^2)$; this is the binding energy of the electron in the ground state of the hydrogen atom.

3.6 The Superposition Principle in Quantum Mechanics

One of the most fundamental principles of quantum mechanics is the principle of linear superposition of states or, for short, the *superposition principle*. It states that a quantum-mechanical system which can take on the discrete states ψ_n $(n \in \mathbb{N})$ is also able to occupy the state

$$\psi = \sum_n a_n \psi_n \ . \tag{3.57}$$

The probability density is then given by

$$w = \psi\psi^* = \sum_{n,m} a_n a_m^* \psi_n \psi_m^* \ .$$

These physical circumstances correspond to the mathematical fact that every possible wave function ψ can be expanded in terms of an orthogonal complete set of functions ψ_n. We have made use of this fact in (3.42).

If a quantum mechanical system can be in a sequence of states φ_f, characterized by an arbitrary physical quantity f, the state

$$\psi = \int c_f \varphi_f \, \mathrm{d}f \tag{3.58}$$

is also realized. Thus the wave equation for ψ must be a linear differential equation (Chap. 6). The superposition principle can only be satisfied as follows: if the ψ_n are solutions of the linear fundamental equation, a linear combination of type (3.57) will also be a solution because of the linearity of the equation.

EXAMPLE

3.4 Superposition of Plane Waves, Momentum Probability

The representation of a wave field $\psi(\boldsymbol{r}, t)$ by superposition of de Broglie waves,

$$\psi_{\boldsymbol{p}}(\boldsymbol{r}, t) = \frac{1}{(2\pi\hbar)^{3/2}} \exp\left[\frac{\mathrm{i}}{\hbar}(\boldsymbol{p}\cdot\boldsymbol{r} - Et)\right] \tag{1}$$

is an example of such a superposition. The normalization factor in (1) results from

$$\int_{-\infty}^{\infty} \psi_{\boldsymbol{p}}^* \psi_{\boldsymbol{p}'} \, \mathrm{d}^3 r = \lim_{g\to\infty} N^2 \int_{-g}^{g} \exp\left[-\frac{\mathrm{i}}{\hbar}(\boldsymbol{p}-\boldsymbol{p}')\cdot\boldsymbol{r}\right] \mathrm{d}^3 r$$

Example 3.4

$$= \lim_{g\to\infty} N^2 2\frac{\sin[g(p_x-p'_x)/\hbar]}{(p_x-p'_x)/\hbar}$$
$$\times 2\frac{\sin[g(p_y-p'_y)/\hbar]}{(p_y-p'_y)/\hbar}2\frac{\sin[g(p_z-p'_z)/\hbar]}{(p_z-p'_z)/\hbar}$$
$$= N^2(2\pi)^3\delta\left(\frac{\boldsymbol{p}-\boldsymbol{p}'}{\hbar^3}\right)$$
$$= N^2(2\pi/\hbar)^3\delta(\boldsymbol{p}-\boldsymbol{p}') ,$$

with

$$\delta(x) = \frac{1}{\pi}\lim_{g\to\infty}\frac{\sin(gx)}{x} \quad \text{and} \quad \delta(ax) = a^{-1}\delta(x) .$$

The δ functions play an important role in various mathematical treatments encountered in quantum mechanics; they will be discussed in greater detail in Chap. 5 (Exercises 5.1, 5.2 and 5.4).

Since we are considering the dynamics of a free particle (no discrete momenta) we do not normalize to unity, but to the delta function, i.e.

$$\int_{-\infty}^{\infty} \psi_{\boldsymbol{p}'}\psi^*_{\boldsymbol{p}}\,\mathrm{d}^3r = \delta(\boldsymbol{p}-\boldsymbol{p}') . \tag{2}$$

With this normalization we get

$$N = (2\pi\hbar)^{-3/2} .$$

The wave function for an arbitrary state $\psi(\boldsymbol{r},t)$ can be expanded into de Broglie waves (1) according to

$$\psi(\boldsymbol{r},t) = \int_{-\infty}^{\infty} c(\boldsymbol{p},t)\psi_{\boldsymbol{p}}(\boldsymbol{r},t)\,\mathrm{d}^3p , \tag{3}$$

where $c(\boldsymbol{p},t)$ are the expansion coefficients of the wave field $\psi(\boldsymbol{r},t)$ in terms of the plane waves [the expansion coefficients correspond to the amplitudes with which the particular states, represented by de Broglie waves, are contained in the state $\psi(\boldsymbol{r},t)$; cf. (3.58)].

Now we are able to show that (3) is simply a factorization of $\psi(\boldsymbol{r},t)$ in a threefold Fourier integral. The Fourier formula reads:

$$\psi(\boldsymbol{r},t) = \frac{1}{(2\pi)^3}\int_{-\infty}^{\infty} \varphi(\boldsymbol{p},t)\,\mathrm{e}^{+\mathrm{i}\boldsymbol{k}\cdot\boldsymbol{r}}\,\mathrm{d}^3k . \tag{4}$$

$\varphi(\boldsymbol{p},t)$ is the Fourier transform of the wave function $\psi(t,r)$. We insert $\boldsymbol{k}=\boldsymbol{p}/\hbar$ in (4):

$$\psi(\boldsymbol{r},t) = \frac{1}{(2\pi)^3}\int_{-\infty}^{\infty} \varphi(\boldsymbol{p},t)\exp\left(\mathrm{i}\frac{\boldsymbol{p}\cdot\boldsymbol{r}}{\hbar}\right)\frac{\mathrm{d}^3p}{\hbar^3} .$$

Example 3.4

In a similar way we get for the Fourier transform

$$\varphi(\boldsymbol{p},t)=\int\limits_{-\infty}^{\infty}\psi(\boldsymbol{r},t)\exp\left(-\mathrm{i}\frac{\boldsymbol{p}\cdot\boldsymbol{r}}{\hbar}\right)\mathrm{d}^3r\ .$$

Comparing (3) and (4), we find

$$\varphi(\boldsymbol{p},t)=\sqrt{(2\pi\hbar)^3}c(\boldsymbol{p},t)\exp\left(-\mathrm{i}\frac{E_pt}{\hbar}\right)\ .$$

Now using (2), it can easily be proved that

$$\begin{aligned}\int\limits_{-\infty}^{+\infty}|c(\boldsymbol{p},t)|^2\mathrm{d}^3p&=(2\pi\hbar)^{-3}\int\limits_{-\infty}^{+\infty}|\varphi(\boldsymbol{p},t)|^2\mathrm{d}^3p\\
&=(2\pi\hbar)^{-3}\iiint\limits_{-\infty}^{+\infty}\mathrm{d}^3p\,\mathrm{d}^3r\,\mathrm{d}^3r'\\
&\quad\times\exp\left[-\mathrm{i}\frac{\boldsymbol{p}}{\hbar}\cdot(\boldsymbol{r}-\boldsymbol{r}')\right]\psi^*(\boldsymbol{r}',t)\psi(\boldsymbol{r},t)\\
&=\iint\limits_{-\infty}^{+\infty}\mathrm{d}^3r\,\mathrm{d}^3r'\,\delta(\boldsymbol{r}-\boldsymbol{r}')\psi^*(\boldsymbol{r}',t)\psi(\boldsymbol{r},t)\\
&=\int\limits_{-\infty}^{+\infty}|\psi(\boldsymbol{r},t)|^2\mathrm{d}^3r=1\end{aligned}$$

is valid. The probability of finding a momentum in the interval p_x, $p_x+\mathrm{d}p_x$; p_y, $p_y+\mathrm{d}p_y$; p_z, $p_z+\mathrm{d}p_z$ is given by the expansion coefficients $c(\boldsymbol{p},t)$. We obtain the following expression for the probability:

$$\mathrm{d}W(\boldsymbol{p},t)=|c(\boldsymbol{p},t)|^2\mathrm{d}^3p\ ,$$

and for the *probability density in momentum space*:

$$w(\boldsymbol{p},t)=|c(\boldsymbol{p},t)|^2\ .$$

3.7 The Heisenberg Uncertainty Principle

Among other things, the wave character of matter [i.e. that in quantum mechanics particles are guided by the field $\psi(x,t)$] manifests itself in the fact that there is a direct connection between position and momentum determination in microscopic physics, namely, we are not able to measure the exact position and momentum of a particle simultaneously. The amount of the uncertainty is given by the ***Heisenberg** uncertainty principle.*

Let us first demonstrate the existence of the uncertainty principle. To do this connection we consider the one-dimensional wave packet (3.19a, b), illustrated in Fig. 3.5,

$$\psi(x,t) = 2c(k_0)\frac{\sin[\Delta k(v_g t - x)]}{v_g t - x}\mathrm{e}^{\mathrm{i}(\omega_0 t - k_0 x)} \qquad (3.19c)$$

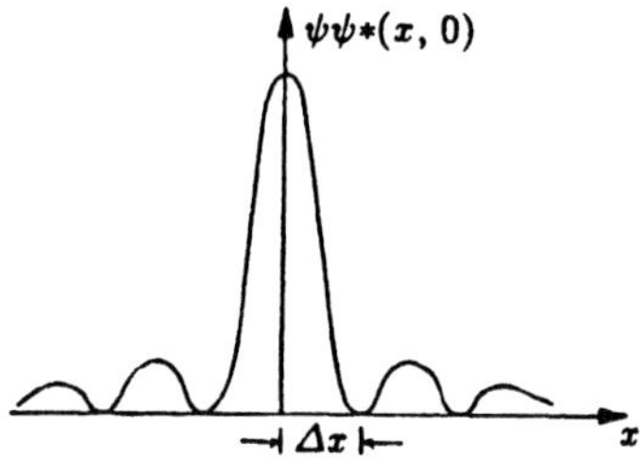

Fig. 3.5. The probability density of the wave packet (3.19c) at time $t = 0$

at time $t = 0$. The extension of the wave group can be characterized by the quantity Δx, i.e. the distance of the first minimum from the maximum. The condition for minima is

$$|\psi|^2 = 4c^2\frac{\sin^2 \Delta kx}{x^2} = 0 \ .$$

Thus we get for the first minimum

$$\Delta k \Delta x = \pi \ .$$

Inserting the momentum according to de Broglie, we get as an estimate for the Heisenberg uncertainty principle between position and momentum,

$$\Delta p \Delta x \approx \pi \hbar \ . \qquad (3.59)$$

This equation means that the simultaneous determination of position and momentum in microscopic physics is not arbitrarily exactly possible; both quantities are always related by the above relation.

Heisenberg's uncertainty principle is a consequence of the wave character of the particles (more exactly: of the guiding field of the particles). Using the superposition principle, the probability field is a wave packet superposed of waves with a definite momentum (*plane waves*). The particle guided by this wave packet can be found with a high probability within Δx. It is said to be *localized* in Δx. For such a localization Δx, a great number of plane waves with momenta near $\hbar k_0$, i.e. a momentum packet of width $\hbar \Delta k$, is required. In classical physics, uncertainty relations of a similar form appear in processes involving waves. The transmission of a spatially limited electromagnetic signal by a sender occurs in the form of a wave packet containing waves of all frequencies (momenta). To get a wave with a single frequency, the sender must transmit for as long as possible (indefinitely), because the process of switching on and off contributes other frequencies. Therefore the wave spreads throughout space and no determination of its position is possible.

After this rather illustrative consideration, we will now derive the Heisenberg uncertainty principle in an exact way. Our starting point is an arbitrary particle state which is described by the wave function $\psi(x)$. Furthermore, we assume that ψ is normalized to unity, and we restrict the calculation at first to one dimension only.

In deriving the uncertainty principle, we first have to determine a measure for the uncertainty, i.e. we have to define a measure for the deviation of p_x, or x, from their respective mean values

$$\overline{p_x} = \int \psi^*(x)\left(-\mathrm{i}\hbar\frac{\partial}{\partial x}\right)\psi(x)\,\mathrm{d}x \quad \text{and} \quad \bar{x} = \int \psi^*(x)x\psi(x)\,\mathrm{d}x \ .$$

Here, we use the mean-square deviations (dispersions) $\overline{\Delta p_x^2}$ and $\overline{\Delta x^2}$, which are defined by

$$\overline{\Delta p_x^2} = \overline{(p_x-\bar{p}_x)^2} = \overline{p_x^2} - \bar{p}_x^2 \ , \quad \overline{\Delta x^2} = \overline{(x-\bar{x})^2} = \overline{x^2} - \bar{x}^2 \ . \tag{3.60}$$

For the following calculation we choose a suitable coordinate system: we assume the origin to be fixed in the point $\bar{x}$ and let it move with the centre of the distribution $\bar{x}$ so that at any time $\bar{x}=0$ is valid. Then we have

$$\bar{x} = 0 \quad \text{and} \quad \overline{p_x} = 0 \ .$$

From the relation for the dispersions (mean-square deviations) (3.60) we get

$$\overline{\Delta x^2} = \overline{x^2} \quad \text{and} \quad \overline{\Delta p_x^2} = \overline{p_x^2} \ . \tag{3.61}$$

The mean values are easily calculated, i.e.

$$\begin{aligned} \overline{x^2} &= \int \psi^* x^2 \psi\,\mathrm{d}x, \\ \overline{p_x^2} &= \int \psi^*\left(-\hbar^2\frac{\partial^2}{\partial x^2}\right)\psi\,\mathrm{d}x = -\hbar^2\int \psi^*\frac{\partial^2\psi}{\partial x^2}\,\mathrm{d}x \ . \end{aligned} \tag{3.62}$$

To establish a connection between the quantities x^2 and p_x^2, we consider the integral

$$I(\alpha) = \int\limits_{-\infty}^{\infty} \left|\alpha x\psi(x) + \frac{\mathrm{d}\psi(x)}{\mathrm{d}x}\right|^2 \mathrm{d}x \ , \quad \alpha \in \mathbb{R} \ . \tag{3.63}$$

The integrand is an absolute square. Therefore $I(\alpha)$ is always greater than, or equal to, zero. We multiply out and get

$$\begin{aligned} I(\alpha) &= \alpha^2 \int\limits_{-\infty}^{\infty} x^2\,|\psi|^2\,\mathrm{d}x + \alpha\int\limits_{-\infty}^{\infty} x\left(\frac{\mathrm{d}\psi^*}{\mathrm{d}x}\psi + \psi^*\frac{\mathrm{d}\psi}{\mathrm{d}x}\right)\mathrm{d}x \\ &\quad + \int\limits_{-\infty}^{\infty} \frac{\mathrm{d}\psi^*}{\mathrm{d}x}\frac{\mathrm{d}\psi}{\mathrm{d}x}\,\mathrm{d}x \ . \end{aligned} \tag{3.64}$$

It is helpful to introduce the following abbreviations:

$$A = \int_{-\infty}^{\infty} x^2 \, |\psi|^2 \, \mathrm{d}x = \overline{\Delta x^2} \ ;$$

$$B = -\int_{-\infty}^{\infty} x \left(\frac{\mathrm{d}\psi^*}{\mathrm{d}x} \psi + \psi^* \frac{\mathrm{d}\psi}{\mathrm{d}x} \right) \mathrm{d}x = -\int_{-\infty}^{\infty} x \frac{\mathrm{d}}{\mathrm{d}x} (\psi^* \psi) \, \mathrm{d}x$$

$$= -x\psi^*\psi \Big|_{-\infty}^{\infty} + \int_{-\infty}^{\infty} \psi^* \psi \, \mathrm{d}x = 1 \ , \tag{3.65a}$$

because ψ vanishes at the boundaries of the integration;

$$C = \int_{-\infty}^{\infty} \frac{\mathrm{d}\psi^*}{\mathrm{d}x} \frac{\mathrm{d}\psi}{\mathrm{d}x} \mathrm{d}x = \psi^* \frac{\mathrm{d}\psi}{\mathrm{d}x} \Big|_{-\infty}^{\infty} - \int_{-\infty}^{\infty} \psi^* \frac{\mathrm{d}^2\psi}{\mathrm{d}x^2} \mathrm{d}x$$

$$= \frac{1}{\hbar^2} \int_{-\infty}^{\infty} \psi^* \left(-\hbar^2 \frac{\mathrm{d}^2}{\mathrm{d}x^2} \right) \psi \, \mathrm{d}x = \frac{1}{\hbar^2} \overline{\Delta p_x^2} \ . \tag{3.65b}$$

With the abbreviations (3.65), the integral (3.64) can be written

$$I(\alpha) = A\alpha^2 - B\alpha + C \geq 0 \ .$$

As this polynomial of second order in α is positive definite according to (3.63), the discriminant must necessarily be negative or vanish. $I(\alpha)$ must be positive for all α. Therefore the roots of the quadratic equation $I(\alpha) = 0$ must be complex. Thus the relation

$$B^2 - 4CA \leq 0$$

is necessarily fulfilled. Inserting in this inequality the values for A, B and C denoted in (3.65), we obtain the uncertainty relation for momentum and position in the form

$$\overline{(\Delta p_x)^2} \ \overline{(\Delta x)^2} \geq \frac{\hbar^2}{4} \ . \tag{3.66}$$

The experimentally proved wave nature of particles (meaning the existence of a guiding field) alone obviously implies that the momentum and coordinates of a particle cannot be simultaneously determined; these observables can never be measured simultaneously with arbitrary precision. We will see in the following that this principle is also valid for other pairs of physical quantities, provided that the product of their dimensions has the dimension of action (see, however, Sect. 4.7).

Using a number of typical examples for simultaneous measurements of particle momentum and position, we shall now illustrate the uncertainty principle.

EXAMPLE

3.5 Position Measurement with a Slit

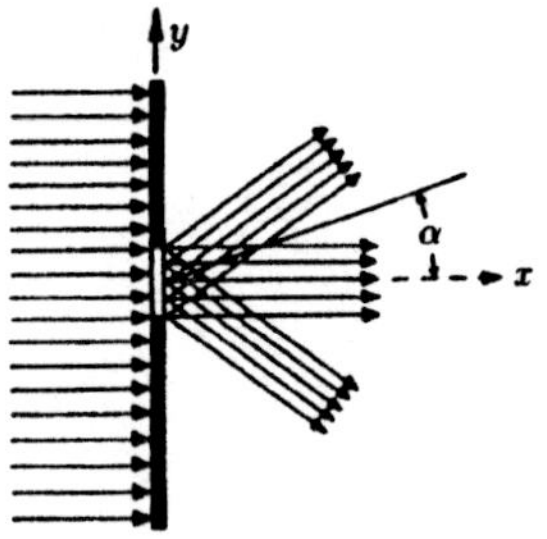

Localization of a particle by a slit

We observe a de Broglie wave moving in the x direction through a slit perpendicular to it with a width $d = \Delta y$ (see Figure). The corresponding interference pattern is visible on a screen standing parallel to, and behind, the slit. Since the momentum in the y direction is given by $p_y = 0$, we would expect that once the particle has passed the slit, a simultaneous determination of its momentum and position in the y direction is possible. However, the diffraction of the wave at the slit causes an additional component of momentum in the y direction. As the diffraction is symmetric, we normally have $\bar{p}_y = 0$. At diffraction angle α, corresponding to the first diffraction minimum, the path a light beam travels is longer by $\lambda/2$ than a nondiffracted beam (see Figure below). Then the greatest intensity is to be expected between $-\alpha$ and $+\alpha$, and we make use of this angle as a measure of the momentum uncertainty. The relation for α reads:

$$\lambda = d \sin\alpha \ . \tag{1}$$

The projection of the momentum on the y axis is given by

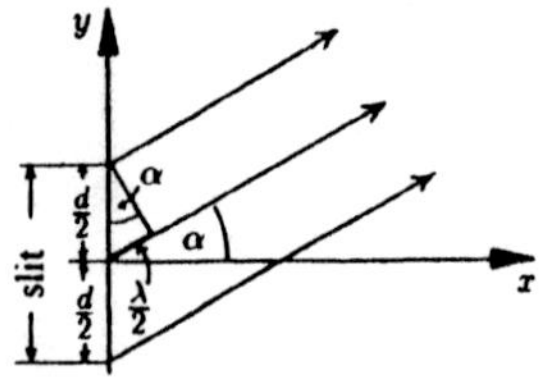

The geometry of the diffraction at a slit. In the first diffraction minimum, a ray from the middle of the slit differs from a ray from the edges by $\lambda/2$

$$p \sin\alpha = \Delta p_y = \frac{2\pi\hbar}{\lambda} \sin\alpha \ .$$

Inserting $\sin\alpha = \lambda \Delta p_y / 2\hbar\pi$ in (1) yields

$$\lambda = d \frac{\lambda \Delta p_y}{2\hbar\pi} \ .$$

The uncertainty principle $\Delta p_y \Delta y = 2\pi\hbar$ follows from this, i.e. the more precisely the particle position d is determined, the less exact the determination of its momentum will be. In other words: the smaller the slit, the more the particle will be diffracted.

EXERCISE

3.6 Position Measurement by Enclosing a Particle in a Box

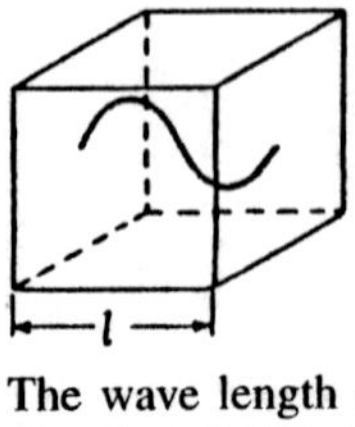

The wave length of the enclosed particle is $\lambda \sim l$

We will try to determine the position of a particle exactly by enclosing it in a box and letting the side $l = \Delta x$ of the box shrink ($l = \Delta x \to 0$). The uncertainty of the particle's momentum is given by $\Delta p \sim \hbar / l$ because standing waves fitting into the box have a wavelength of the order of l (see Figure).

It then follows for the kinetic energy that

$$E_{\text{kin}} = \frac{\Delta p^2}{2m} \sim \frac{\hbar^2}{2ml^2} \ .$$

Exercise 3.6

Hence, as the box becomes smaller, the kinetic energy and momentum will grow according to the uncertainty principle. The result of this "Gedankenexperiment" has been experimentally confirmed. Electrons in atoms have energies of 10–100 eV, their atomic diameter being 10^{-8}–10^{-9} cm, while nucleons have energies of the order of 1 MeV, the size of a nucleus being $\sim 10^{-12}$ cm, which confirms the uncertainty principle. Let us check the latter explicitly. To do so, we need some numerical values: nuclear diameter $\sim 10^{-12}$ cm, nucleon mass $m_n c^2 \sim 938$ MeV, $\hbar c \sim 197 \times 10^{-13}$ cm MeV. From the uncertainty principle it follows that

$$\Delta p \approx \frac{\hbar}{\Delta x} \quad \text{and} \quad \Delta E = \frac{(\Delta p)^2}{2m} \approx \frac{\hbar^2}{2m} \cdot \frac{1}{(\Delta x)^2} \ .$$

Inserting the values given above, we get for the order of magnitude of the kinetic energy of the nucleon:

$$\Delta E \approx 0.2 \,\text{MeV} \ .$$

EXAMPLE

3.7 Position Measurement with a Microscope

We consider a beam of light perpendicular to the x axis and illuminating the object to be observed. From the theory of the microscope it is known that the x coordinate of a particle can be measured with a precision of $\Delta x \approx \lambda / \sin \varepsilon$, where ε is the angle illustrated in the Figure. The resolution limit, Δx, is calculated with the aid of the following argument: for the lattice constant Δx to become visible, at least the first diffraction maximum must be observable through the lens, i.e. $\Delta x \sin \varepsilon = \lambda$; from this it follows that, for given angle ε and given wavelength λ, only quantities $\Delta x \approx \lambda / \sin \varepsilon$ can be resolved.

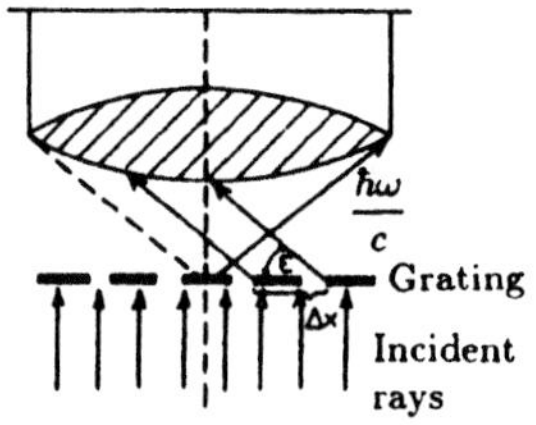

The resolution of a microscope

The particle's image is produced by a photon which is scattered by the particle and moves through the lens into the microscope. According to the Compton effect, the momenta change in the scattering process. The particle suffers a recoil momentum of the order of $\hbar\omega/c$. The momentum is not known exactly because of the arbitrary direction of the photons in the cone with the angle 2ε. Therefore the momentum transfer to the particle has to lie in the range

$$\Delta p_x = p \sin \varepsilon \approx \frac{\hbar\omega}{c} \sin \varepsilon \ .$$

The product of position and momentum uncertainties yields

$$\Delta x \Delta p_x \approx \lambda \frac{\hbar\omega}{c} = h \ ,$$

which is again the uncertainty relation.

EXAMPLE

3.8 Momentum Measurement with a Diffraction Grating

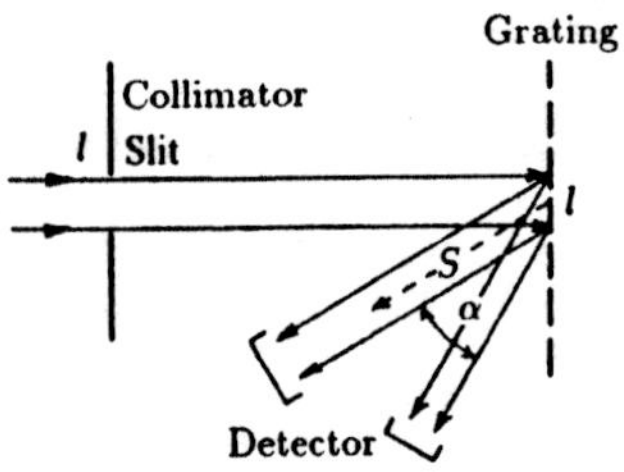

Principle of momentum measurement with a grating

We want to make use of the arrangement sketched above of collimator slit and diffraction grating to determine the momentum of a matter wave. A particle beam of width l is collimated by a slit and impinges on the grating. The grating constant of the grating is d. Thus the number of grating lines which are important for the diffraction are $N = l/d$. A grating is able to separate two waves of different wavelength if the following condition is valid (resolving power):

$$\frac{\Delta\lambda}{\lambda} = \frac{1}{N} \ .$$

This will be discussed in the next example.

For an exact determination of the position, we should fix the detector directly at that position of the grating from which the particle is scattered. But this does not help much because there, all waves of different wavelengths still cover the same spatial region. The detector must be fixed at least at a distance from the grating where the beams of different wavelength separate. Let $\alpha(\alpha \ll 1)$ denote the angle both beams enclose. Then this minimal distance is given by $\Delta s = l/\alpha$. Thus, with

$$p = \frac{h}{\lambda} \quad \text{and} \quad \Delta p = p \quad \text{resolving power}$$
$$= \frac{h}{\lambda}\frac{\Delta\lambda}{\lambda} = \frac{h}{\lambda}\frac{1}{N} \ ,$$

it follows that

$$\Delta p \Delta s = \frac{h}{\lambda}\frac{\Delta\lambda}{\lambda}\frac{l}{\alpha} = \frac{h}{\lambda N}\frac{l}{\alpha} = h\frac{d}{\lambda}\frac{1}{\alpha} \ .$$

In order to get a diffraction, d and λ must at least be of the same order of magnitude. α is small; thus we have

$$\Delta p \Delta s > h \ .$$

A Supplementary Remark. An exact momentum measurement of a particle could be made by scattering a monochromatic wave from this particle. Momentum and energy conservation is valid, so the particle momenta are determined, before and after scattering, by the measured frequencies, using the momentum-frequency relation. But since a monochromatic (plane) wave stretches throughout space, we do not get any information about the position. To determine the position of the particle, we should scatter a wave packet which is spatially restricted. On the other hand, it contains all frequency (momentum) components, thus leading to the uncertainty relation again.

EXAMPLE

3.9 Physical Supplement: The Resolving Power of a Grating

Let us consider a grating consisting of an infinite number of slits at a distance d (see Figure). Investigating all beams coming from corresponding positions of the slits of the grating (for example from the left edge of each slit) and moving in a direction defined by the angle β, we observe that the total intensity in general vanishes. To such a beam will generally correspond a similar beam which comes from a slit at a greater distance and has a phase difference of exactly 180° with respect to the former one. The beams will only overlap in a constructive way if the difference in wavelength $d(\sin\alpha - \sin\beta)$ between two neighbouring beams is exactly a multiple of the wavelength. Thus, for an infinitely extended lattice, the maxima of intensity only occur under the angle β with

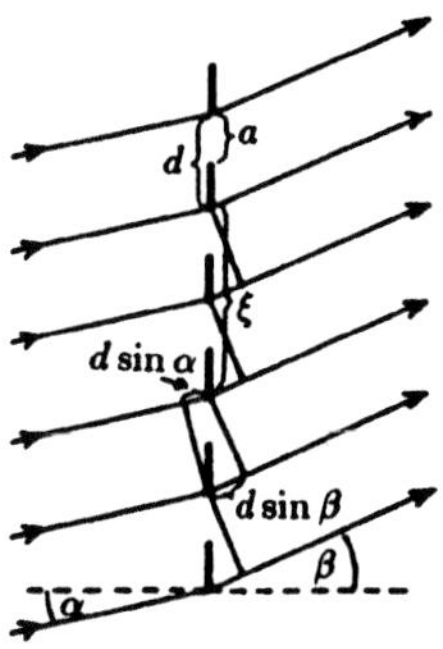

Incoming and diffracted beams at a grating

$$d(\sin\alpha - \sin\beta) = m\lambda \;, \quad m \in \mathbb{N}_0 \tag{1}$$

(m is called the order of the maximum). Here we have neglected the interference structure from the superposition of the beams coming through a single slit; however, they contribute to the diffraction pattern, too. In the following, we want to calculate the complete structure of intensity behind the grating, which now may have a finite number of slits instead of an infinite number.

We consider a grating of *slit width a, line distance d* and *number of slits (line number) N*. Let us now calculate the amplitude in the direction β in the case of an incident angle α (see Figure). If two beams overlap with a phase difference η, the resulting amplitude will be proportional to the complex number $\mathrm{e}^{\mathrm{i}\eta}$. Our aim is to integrate over all phase differences, those of the rays coming through the same slit and those of rays coming through different slits. The phase difference of two rays is given by

$$\eta = k\xi(\sin\alpha - \sin\beta) = 2\pi\frac{\xi(\sin\alpha - \sin\beta)}{\lambda} \;, \tag{2}$$

where ξ is the spatial separation of those positions at which the rays pass the grating. Thus the amplitude u is given by

$$u \sim \int_0^a + \int_d^{a+d} + \dots + \int_{(N-1)d}^{(N-1)d+a} \exp\left[\mathrm{i}2\pi\frac{\xi(\sin\alpha - \sin\beta)}{\lambda}\right] \mathrm{d}\xi \;. \tag{3}$$

We substitute

$$\gamma \equiv \pi\frac{a(\sin\alpha - \sin\beta)}{\lambda} \quad \text{and} \quad \delta \equiv \pi\frac{d(\sin\alpha - \sin\beta)}{\lambda} \;. \tag{4}$$

Example 3.9

Performing the integrals yields

$$\int\limits_{nd}^{nd+a} \exp\left\{\mathrm{i}\left[\frac{2\pi}{\lambda}(\sin\alpha - \sin\beta)\right]\xi\right\} \mathrm{d}\xi = -\mathrm{i}\frac{\lambda}{2\pi(\sin\alpha - \sin\beta)}$$

$$\times \left(\exp\left\{\mathrm{i}\frac{2\pi}{\lambda}(\sin\alpha - \sin\beta)\xi\right\}\right)\Big|_{nd}^{nd+a}$$

$$= \frac{-\mathrm{i}\lambda}{2\pi(\sin\alpha - \sin\beta)}\left\{\exp\left[\mathrm{i}\frac{2\pi}{\lambda}(\sin\alpha - \sin\beta)nd\right]\right\}$$

$$\times \left\{\exp\left[\mathrm{i}\frac{2\pi}{\lambda}(\sin\alpha - \sin\beta)a\right] - 1\right\}$$

$$= \frac{-\mathrm{i}a}{2\gamma}(\mathrm{e}^{\mathrm{i}2n\delta})(\mathrm{e}^{\mathrm{i}2\gamma} - 1) = \frac{-\mathrm{i}a}{2\gamma}(\mathrm{e}^{\mathrm{i}(2n\delta+2\gamma)} - \mathrm{e}^{\mathrm{i}2n\delta}) \; .$$

Therefore the wave field behind the grating reads:

$$u \sim \frac{1}{\gamma}\big[-1 + \mathrm{e}^{2\mathrm{i}\gamma} - \mathrm{e}^{2\mathrm{i}\delta} + \mathrm{e}^{2\mathrm{i}(\delta+\gamma)} - \mathrm{e}^{4\mathrm{i}\delta}$$

$$+ \mathrm{e}^{2\mathrm{i}(\delta+\gamma)} - + \ldots\big] = \frac{1}{\gamma}(\mathrm{e}^{2\mathrm{i}\gamma} - 1)\sum_{n=0}^{N-1} \mathrm{e}^{2\mathrm{i}n\delta}$$

$$= \frac{1}{\gamma}(\mathrm{e}^{2\mathrm{i}\gamma} - 1)\frac{\mathrm{e}^{2\mathrm{i}N\delta} - 1}{\mathrm{e}^{2\mathrm{i}\delta} - 1} \; . \tag{5}$$

Thus the intensity is

$$I \sim uu^* = 4\frac{\sin^2\gamma}{\gamma^2}\frac{\sin^2 N\delta}{\sin^2\delta} \; . \tag{6}$$

The second factor yields the principal maxima at $\delta = m\pi$, i.e. at

$$d(\sin\alpha - \sin\beta) = m\lambda \;\; (m \in \mathbb{N}_0) \tag{7}$$

[see (1)]. The first factor $(\sin^2\gamma)/\gamma^2$ provides the interference pattern of a single slit, which is superposed on the interference pattern (see next Figure, dashed lines).

The condition $\partial I/\partial\delta = 0$ provides further, less intense, secondary maxima, which are separated from the principal maxima and from each other by the dark spots at $\delta = m'(\pi/N)(m' \in \mathbb{N})$ (see Figure). The greater N is, the sharper the principal maximum will be and the closer the minima of intensity will lie to the principal maximum. Thus, *d determines the position of the principal maxima, N their sharpness, a the intensity of the principal maxima of first, second, third, … order.*

If, for example, $d = 2a$, we do not have any principal maxima of even order. If we consider gratings with a complicated slit structure, whose permeability is not a simple box function but for example sinusoidal (sinusoidal grating), the permeability function plays the role of the slit width.

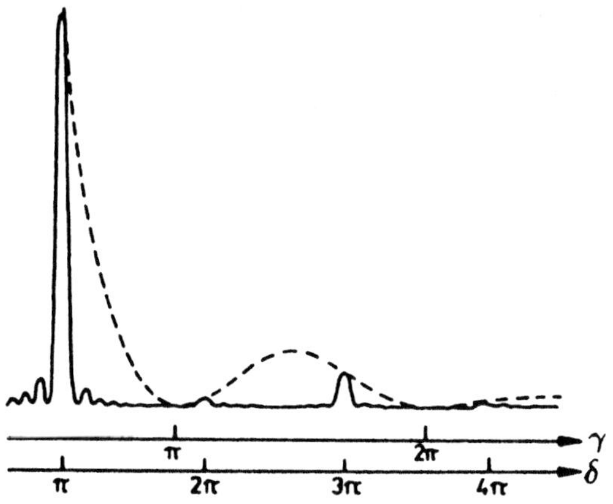

Intensity distribution of the diffraction by a grating with $N = 60$, $d/a = 7/4$

The *resolving power* of a grating is defined by its ability to separate two principal maxima. Two principal maxima (for example, belonging to different wavelengths) lying at different positions, can still be separated if one maximum just coincides with the dark region $\delta = \pi/N$ of the other one. Thus for both maxima the following relation must be valid:

$$|\Delta\delta| = \frac{\pi}{N} \; . \tag{8}$$

$\Delta\delta$ can be transformed into a wavelength difference $\Delta\lambda$: with (4) we have

$$|\Delta(\sin\alpha - \sin\beta)| = \frac{\lambda}{Nd} \; , \tag{9}$$

and with (7)

$$m\,|\Delta\lambda| = \frac{\lambda}{N} \; .$$

The resolving power is then given by

$$\frac{\lambda}{|\Delta\lambda|} = mN \; .$$

Put into words: resolution equals order of maximum times line number; the greater the number of slits, the better the resolution of the grating. To separate, for example, the two neighbouring yellow sodium vapour lines ($\Delta\lambda = 6$ Å at $\lambda = 5893$ Å), we need a resolving power of 1000; thus we can separate lines of first order with a grating of 1000 lines. On the other hand, if we are satisfied with weaker maxima of second order, 500 lines will be enough.

EXERCISE

3.10 Properties of a Gaussian Wave Packet

Let a wave packet be described at time $t = 0$ by

$$\psi(x,0) = A\exp\left(-\frac{x^2}{2a^2} + \mathrm{i}k_0 x\right) \tag{1}$$

(a *Gaussian wave packet*).

Problem. (a) Express $\psi(x,0)$ as a superposition of plane waves.

(b) What is the approximate relation between the width of the wave packet in configuration (x) space and its width in k space?

(c) Using the dispersion relation for de Broglie waves, calculate the function $\psi(x,t)$ for any time t.

(d) Discuss $|\psi(x,t)|^2$.

(e) How must the constant A be chosen according to the probability interpretation, so that $\psi(x,t)$ describes the motion of a particle?

Solution. (a) We obtain the frequency spectrum of a wave packet $\psi(x)$ by forming the Fourier transform $\alpha(k)$ of the wave function:

$$\begin{aligned}\alpha(k) &= \frac{1}{\sqrt{2\pi}}\int\limits_{-\infty}^{\infty}\psi(x,0)\exp(-\mathrm{i}kx)\,\mathrm{d}x\\ &= \frac{A}{\sqrt{2\pi}}\int\limits_{-\infty}^{\infty}\exp\left(-\frac{x^2}{2a^2} + \mathrm{i}k_0x - \mathrm{i}kx\right)\mathrm{d}x\ .\end{aligned}$$

This integral is solved by completing the square to give a complete error integral,

$$\int\limits_{-\infty}^{\infty}\exp(-\xi^2)\,\mathrm{d}\xi = \sqrt{\pi}\ .$$

Completing the square yields

$$\alpha(k) = \frac{A}{\sqrt{2\pi}}\int\limits_{-\infty}^{\infty}\exp\left[-\left(\frac{x}{\sqrt{2}a} + \frac{\mathrm{i}a(k-k_0)}{\sqrt{2}}\right)^2\right]\exp\left(-\frac{a^2(k-k_0)^2}{2}\right)\mathrm{d}x\ .$$

Now we replace the exponent by $-\xi^2$ and obtain

$$\alpha(k) = \frac{A}{\sqrt{2\pi}}\int\limits_{-\infty}^{\infty}\exp(-\xi^2)\sqrt{2}a\exp\left(-\frac{a^2(k-k_0)^2}{2}\right)\mathrm{d}\xi$$

Exercise 3.10

$$= \frac{A}{\sqrt{2\pi}}\sqrt{2}a\exp\left(-\frac{a^2(k-k_0)^2}{2}\right)\sqrt{\pi}$$

$$= Aa\exp\left(-\frac{a^2(k-k_0)^2}{2}\right) . \tag{2a}$$

The coefficients $\alpha(k)$ denote the portion of the partial wave with wave number k in the Gaussian wave packet. As a superposition of plane waves, the Gaussian wave packet has the form

$$\psi(x,k) = \frac{1}{\sqrt{2\pi}}\int_{-\infty}^{\infty}\alpha(k)\,\mathrm{e}^{\mathrm{i}kx}\,\mathrm{d}k . \tag{2b}$$

(b) In (1), the width of the Gaussian function is approximately $\Delta x \approx a$. The width of the distribution function of the plane waves in (2a) is given by the Gaussian function $\exp[-(k-k_0)^2a^2/2]$:

$$\Delta k \approx 1/a .$$

Thus the uncertainty principle involving both quantities is $\Delta x \Delta k \sim 1$.

(c) The general form of a wave function is

$$\psi(x,t) = \frac{1}{\sqrt{2\pi}}\int_{-\infty}^{\infty}\alpha(k)\exp[\mathrm{i}(kx-\omega t)]\,\mathrm{d}x .$$

The dispersion relation for de Broglie waves reads

$$\omega(k) = \frac{\hbar k^2}{2m}.$$

Inserting $\alpha(k)$ from part (a) of this exercise, we get

$$\psi(x,t) = \frac{Aa}{\sqrt{2\pi}}\int_{-\infty}^{\infty}\exp\left(-\frac{a^2(k-k_0)^2}{2}+\mathrm{i}kx-\mathrm{i}\frac{\hbar k^2}{2m}t\right)\mathrm{d}k .$$

Now we again complete the square, use the error integral and obtain the time-dependent wave function,

$$\psi(x,t) = \frac{A}{\sqrt{1+\mathrm{i}(\hbar t/ma^2)}}\exp\left(\frac{x^2-2\mathrm{i}a^2k_0x+\mathrm{i}(a^2\hbar k_0^2/m)t}{2a^2[1+\mathrm{i}(\hbar t/ma^2)]}\right) .$$

(d) The following holds:

$$|\psi(x,t)|^2 = \frac{|A|^2}{\sqrt{1+(\hbar t/ma^2)^2}}\exp\left(-\frac{[x-(hk_0/m)t]^2}{a^2[1+(\hbar t/ma^2)^2]}\right) .$$

Exercise 3.10

The maximum of this Gaussian function is at the position $x = \hbar k_0 t/m$. The maximum moves with the group velocity $v = \hbar k_0/m$. But the wave packet "flattens": at $t = 0$ the width of $|\psi|^2$ is just a, and at a later time (formally speaking: at an earlier time as well) its width is given by $a' = a\sqrt{1+(\hbar t/ma^2)^2}$.

(e) Independently of time, the normalization condition for a particle has to be

$$1 = \int\limits_{-\infty}^{\infty} |\psi(x, t=0)|^2 \, dx = |A|^2 a \int\limits_{-\infty}^{\infty} \exp(-\xi^2)\, d\xi = |A|^2 a\sqrt{\pi} \ .$$

Thus it follows for $|A|$

$$|A| = \frac{1}{(a\sqrt{\pi})^{1/2}} \ .$$

As this condition is valid only for the absolute value of A, the phase of the wave remains undetermined.

EXERCISE

3.11 Normalization of Wave Functions

For the $1s$ and $2s$ electrons, the unnormalized wave functions for the hydrogen atom (to be calculated in Chap. 9) are

$$\psi_{1s}(r, \vartheta, \varphi) = \psi_{1s}(r) = e^{-\varrho} \ ,$$
$$\psi_{2s}(r, \vartheta, \varphi) = \psi_{2s}(r) = \left(1 - \frac{\varrho}{2}\right) e^{-\varrho/2} \ ,$$

with $\varrho = r/a$ and the Bohr radius $a = \hbar^2/me^2$.

Problem. (a) Prove their orthogonality and normalize.

(b) Sketch $|\psi|^2$ and $4\pi r^2|\psi|^2$ for both cases. What is the meaning of $|\psi|^2$ and $4\pi r^2|\psi|^2$?

Solution. (a) The normalization condition for $\tilde{\psi}_{1s}$ and $\tilde{\psi}_{2s}$, which only differ by a factor of ψ_{1s} and ψ_{2s}, are given by

$$\int \tilde{\psi}^*_{1s}(r)\tilde{\psi}_{1s}(r)\, d^3r = \int |\tilde{\psi}_{1s}|^2\, d^3r = \int |\tilde{\psi}_{2s}|^2\, d^3r = 1 \ .$$

Moreover, the orthogonality

$$\int \psi^*_{1s}\psi_{2s}\, d^3r = 0$$

has to be proved. To this end, we calculate the three integrals which appear by using spherical coordinates and the relation

$$\int\limits_0^{\infty} x^{\nu-1} e^{-\mu x}\, dx = \frac{1}{\mu^\nu}(\nu-1)! \quad (\nu \in \mathbb{N}_0) \ ,$$

obtaining *Exercise 3.11*

$$\int |\psi_{1s}|^2 \mathrm{d}^3r = 4\pi \int_0^\infty r^2 \mathrm{d}r \, \mathrm{e}^{-2r/a} = 4\pi \left(\frac{a}{2}\right)^3 2! = \pi a^3 \ ,$$

$$\int |\psi_{2s}|^2 \mathrm{d}^3r = 4\pi \int_0^\infty r^3 \mathrm{d}r \left(1 - \frac{r}{2a}\right)^2 \mathrm{e}^{-r/a}$$

$$= 4\pi \int_0^\infty \mathrm{d}r \ \mathrm{e}^{-r/a} \left(r^2 - \frac{r^3}{a} + \frac{r^4}{4a^2}\right)$$

$$= 4\pi \left(2!a^3 - \frac{1}{a}a^4 3! + \frac{1}{4a^2}a^5 4!\right) = 8\pi a^3 \ ,$$

$$\int \psi_{1s}^* \psi_{2s} \mathrm{d}^3r = 4\pi \int_0^\infty r^2 \mathrm{d}r \left(1 - \frac{r}{2a}\right) \mathrm{e}^{-3r/2a}$$

$$= 4\pi \left[\left(\frac{2a}{3}\right)^3 2! - \frac{1}{2a}\left(\frac{2a}{3}\right)^4 3!\right] = 0 \ .$$

Thus the normalized wave functions are

$$\tilde{\psi}_{1s} = \frac{1}{\sqrt{\pi a^3}}\psi_{1s} \quad \text{and} \quad \tilde{\psi}_{2s} = \frac{1}{\sqrt{8\pi a^3}}\psi_{2s} \ .$$

(b) The probability of finding an electron with the (normalized) wave function ψ in the volume element $\mathrm{d}V$ at the position $\boldsymbol{r}$ is simply $|\psi(\boldsymbol{r})|^2 \mathrm{d}V$. The normalization condition $\int |\psi|^2 \mathrm{d}V = 1$ expresses the fact that the probability of finding the electron anywhere is just 1. The probability of finding the electron in a spherical shell with radius r and thickness $\mathrm{d}r$ is

$$\int_{\substack{\text{spherical}\\ \text{shell}}} |\psi|^2 \mathrm{d}V = 4\pi r^2 \, |\psi(r)|^2 \, \mathrm{d}r \ ,$$

if the wave function is independent of the angles ϑ and φ (as in our case). This illustrates the meaning of the second expression.

For the 1s and 2s electrons, the functions look similar to those in the figure on the next page, where we have consciously omitted the scale on the abscissa.

Probability density (*above*) and probability density in a spherical shell (*below*) of the two hydrogen wave functions

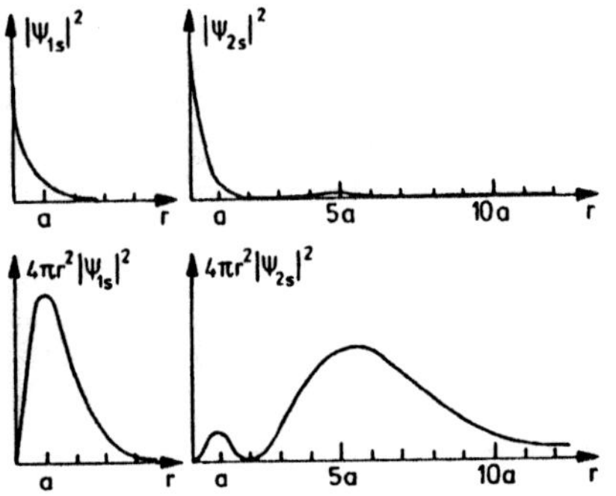

EXERCISE

3.12 Melons in Quantum Land

In Quantum Land, a strange land, where $\hbar = 10^4$ erg s, melons with a very hard peel are grown; they have a diameter of approximately 20 cm and contain seeds with a mass of around 0.1 g.

Problem. Why do we have to be careful when cutting open melons grown in Quantum Land? Are such melons visible? How big is the recoil of a melon at the reflection of a "visual" photon of 628 nm wavelength?

Solution. From the uncertainty relation $\Delta p \Delta x \approx \hbar$ follows for the momentum uncertainty of the melon seeds $\Delta p \approx 10^4$ erg s/10 cm $= 10^8$ g cm s^{-1} and therefore their velocity uncertainty is $\Delta v = \Delta p/m = 100$ m s^{-1}. The seeds leave the melon with this (mean) velocity when it is cut open.

A photon of wavelength $\lambda = 628$ nm has the momentum $p = \hbar(2\pi/\lambda) = 10^9$ g cm s^{-1} and the energy $E = pc = 3 \times 10^{19}$ erg. The mass of the melon is $m \sim (4\pi/3)R^3 \times 1$ g cm$^{-3} \sim 4$ kg, its rest energy is $mc^2 \sim 3.6 \times 10^{24}$ erg; hence we can calculate nonrelativistically. Let the collision be elastic. The momentum of the melon after the collision is approximately $p_\mathrm{M} = 2p = 2 \times 10^9$ g cm s^{-1}. This corresponds to a velocity $v_\mathrm{M} = 5$ km s^{-1} which is less than the escape velocity from the Earth. By the time this photon is seen, the melon is already situated elsewhere ($\Delta p_\mathrm{M} \Delta x_\mathrm{M} \approx \hbar$!). The absorption of such a photon would probably be rather unpleasant for a human being.

3.8 Biographical Notes

DE BROGLIE, Prince Louis Victor, French physicist, 1892–1987, professor of theoretical physics at the Institut Henri Poincaré. He founded with his Ph.D. thesis "Recherches sur la Théorie des Quants" (1924) the

theory of matter waves (de Broglie waves) and was awarded the Nobel Prize in Physics for it in 1929. Later he worked mainly on the development of the quantum theory of elementary particles (neutrino theory of light, wave theory of elementary particles) and proposed a new method for the treatment of wave equations with higher spin, the so-called *fusion method*.

DAVISSON, Clinton Joseph, American physicist, *Bloomington (IL), 22.10.1881, † Charlottesville (VA), 1.2.1958. From 1917 to 1946 D. was a scientist at Bell Telephone Laboratories; then, until 1954, he was a professor at the University of Virginia in Charlottesville. In 1927 D. and L.H. Germer measured electron diffraction by crystals, a decisive proof of the wave nature of matter. In 1937 he was awarded the Nobel Prize in Physics.

LAUE, Max von, German physicist, *Pfaffendorf (near Koblenz, Germany), 9.10.1879, † Berlin, 24.4.1960. v.L. was a student of M. Planck, a professor in Zürich, Frankfurt, Berlin and, from 1946, Director of the Institut für Physikalische Chemie und Elektrochemie in Berlin-Dahlem. v.L. was the first director of the Institut für Theoretische Physik in Frankfurt (from 1914 until 1919), his successor being M. Born. His proposal to irradiate crystals with X-rays was performed by Walther Friedrich and Paul Knipping in late April 1912. L.'s immediate explanation of the X-ray interferences detected in this experiment won him the Nobel Prize in Physics in 1914. With this, the wave nature of X-rays as well as the spatial grating structure of crystals was established. As early as 1911, v.L. wrote a book about the theory of relativity, which was widely read, and in which he later included general relativity. He also worked on the applications of relativity, e.g. on thermodynamics. Further treatises covered superconductivity, glow-electron emission and the mechanism of amplifier valves. After 1933 v.L. tried, often successfully, to oppose the influence of national socialism on science in Germany.

BORN, Max, German physicist, *Breslau, Germany (now Wrocław, Poland) 12.12.1882, † Göttingen 5.1.1970. B. was a professor in Berlin (1915), Frankfurt (1919) and Göttingen (1921); he emigrated to Cambridge in 1933 and then became Tait Professor of Natural Philosophy in Edinburgh in 1936. From 1954 on, B. lived in retirement in Bad Pyrmont (Germany). B. first devoted himself to relativity and the physics of crystals. From about 1922 on, he worked on the foundation of a new theory of atoms and succeeded in 1925, together with his students W. Heisenberg and P. Jordan, in the creation of matrix mechanics. In Göttingen, B. founded an important school of theoretical physics. In 1926 he interpreted Schrödinger's wave functions as probability amplitudes, thus introducing the statistical point of view into modern physics. For this he was belatedly awarded the Nobel Prize in 1954.

HILBERT, David, *Königsberg, Germany (now Kaliningrad, Russia) 23.1.1862, † Göttingen 14.2.1943. H., son of a lawyer, studied in Königsberg and Heidelberg and became a professor in Königsberg in 1886. From 1895 on, he contributed to making Göttingen a world centre of mathematical research. The most important living mathematician, H. proved himself as a world-wide authority in his famous talk given in Paris in

1900, where he proposed 23 mathematical problems which interest mathematicians even today. H. contributed to many fields that have deeply influenced modern mathematical research, e.g. on the theory of invariants, group theory and the theory of algebraic manifolds. His investigations on number theory culminated in 1897 in his report, "Die Theorie der algebraischen Zahlkörper" and in his proof of Warring's Problem. In the field of geometry he introduced strictly axiomatic concepts in "Die Grundlagen der Geometrie" (1899). His works on the theory of integral equations and on the calculus of variations strongly influenced modern analysis. H. also worked successfully on problems of theoretical physics, especially on kinetic gas theory and relativity. As a consequence of the development of set theory and the problems arising in the foundations of mathematics, H. created his proof theory and thereby became one of the leaders of the axiomatic branch of the foundation of mathematics.

HEISENBERG, Werner Karl, German physicist, *Würzburg 5.12.1901, † München 1.2.1976. From 1927–41 he was Professor of Theoretical Physics in Leipzig and Berlin; in 1941, professor at, and director of, the Max-Planck-Institute für Physik in Berlin, Göttingen and, from 1955 on, in München. In his search for the correct description of atomic phenomena, H. formulated his *positivistic principle* in July of 1925: it asserts that only quantities which are in principle observable are allowed to be taken into account. Thus the more intuitive ideas of the older Bohr-Sommerfeld quantum theory have to be rejected. At the same time H. provided the foundation for the new Göttinger matrix mechanics in his rules for multiplication of quadratic schemata, which he developed together with M. Born and P. Jordan in Sept. of 1925. In close collaboration with N. Bohr he was able to show the deeper physical- or philosophical-background of the new formalism. The Heisenberg uncertainty principle of 1927 became the basis of the *Copenhagen interpretation* of quantum theory. In 1932 H. was awarded the Nobel Prize in Physics "for the Creation of Quantum Mechanics". After the discovery of the neutron by J. Chadwick in 1932, H. realized that this new particle together with the proton must be considered as constituents of atomic nuclei. On this basis he developed a theory of the structure of atomic nuclei and introduced, in particular, the concept of isospin. From 1953 on, H. worked on a unifying theory of matter (often called a *world formula*). The aim of this theory is to describe all existing particles and their conversion processes using the conservation laws, which express the symmetry properties of the laws of nature. A nonlinear spinor equation is supposed to describe all elementary particles.

JACOBI, Carl Gustav Jakob, *10.12.1804 in Potsdam as the son of a banker, † 18.2.1851 in Berlin. J. became an instructor in Berlin after his studies in 1824 and 1827/42 was professor in Königsberg (Prussia). After an extended travel to Italy, which was to cure his impaired health, J. lived in Berlin as a university man. He became famous because of his work "Fundamenta nova theoriae functiorum ellipticarum" (1829). In 1832, J. found out that hyperelliptic functions can be inverted by functions of several variables. J. also made fundamental contributions to algebra, to elimination theory, and to the theory of partial differential equations, e.g. in his "Vorlesungen über Dynamik", which were published in 1866.

4. Mathematical Foundations of Quantum Mechanics I

4.1 Properties of Operators

We have already used average values of position and momentum of a particle and seen that we can get the average value of an observable F [represented by an operator function $\hat{F}(\hat{x}, \hat{p})$] in a state ψ by

$$\langle \hat{F} \rangle \equiv \bar{F} = \int \psi^* \hat{F} \psi \, \mathrm{d}V \ , \tag{4.1}$$

where $\hat{F}$ is the operator which is somehow related to F. In a first approach we are now going to deal with operators from a more general point of view. After this we shall determine a class of operators which is very important in quantum mechanics.

Let U and W be two sets of functions. We define a continuous mapping $\hat{L}$: $U \to W$ with $\hat{L}(u) = w(u \in U;\, w \in W)$, and call $\hat{L}$ an operator. The operator $\hat{L}$ relates a function $u \in U$ to a new function $w \in W$. Symbolically we write this relation as a product of the operator $\hat{L}$ with the function u:

$$\hat{L}(u) = \hat{L}u = w \ .$$

An operator with the property

$$\hat{L}(\alpha_1 u_1 + \alpha_2 u_2) = \alpha_1 \hat{L} u_1 + \alpha_2 \hat{L} u_2 \ , \tag{4.2}$$

where u_1, u_2 are arbitrary functions and α_1, α_2 are arbitrary constants, is called a *linear operator*.

We can see that the position operator $\hat{x} = x$ and the momentum operator $\hat{p}_x = \mathrm{i}\hbar\partial/\partial x$ are linear operators. A typical nonlinear operator is, for instance, the square-root operator, as $\sqrt{\alpha_1 u_1 + \alpha_2 u_2} \neq \alpha_1 \sqrt{u_1} + \alpha_2 \sqrt{u_2}$ obviously holds. Furthermore, a linear operator is *self-adjoint* or *Hermitian* if

$$\int \psi_1^* \hat{L} \psi_2 \, \mathrm{d}V = \int (\hat{L}\psi_1)^* \psi_2 \, \mathrm{d}V \ , \tag{4.3}$$

where ψ_1, ψ_2 are arbitrary square-integrable functions, whose derivatives vanish at the boundaries of the region of integration.

In quantum mechanics we require that all operators be self-adjoint and linear; in this case, the superposition principle holds. Of course, linear operators do not violate the superposition principle. In order to be able to describe meaningful and measurable quantities with our operators, we must demand that their average values be real. This property is guaranteed by Hermitian (self-adjoint) operators. We can show this in the following expression:

$$\bar{L} = \int \psi^* \hat{L} \psi \, \mathrm{d}V = \int (\hat{L}\psi)^* \psi \, \mathrm{d}V = \left[\int \psi^* (\hat{L}\psi) \, \mathrm{d}V \right]^* = \bar{L}^* \ , \tag{4.4}$$

and therefore the mean value is real.

4.2 Combining Two Operators

We define the sum $\hat{A} + \hat{B} = \hat{C}$ of two operators,

$$\hat{C}\psi = (\hat{A} + \hat{B})\psi = \hat{A}\psi + \hat{B}\psi \tag{4.5}$$

and their product $\hat{A}\hat{B} = \hat{C}$,

$$\hat{C}\psi = (\hat{A}\hat{B})\psi = \hat{A}(\hat{B}\psi) \ . \tag{4.6}$$

Equation (4.6) means that $\hat{B}$ acts on ψ first, and then $\hat{A}$ acts on the new function $(\hat{B}\psi)$. If $\hat{A}$ and $\hat{B}$ are Hermitian, we notice immediately that $\hat{A} + \hat{B}$ is Hermitian, too. The product operator $\hat{C}$ requires more care.

It is important to realize that the product of two operators in general does not commute, i.e. $\hat{A}\hat{B} - \hat{B}\hat{A} \neq 0$. Hence the order of the operators is important: in general, $\hat{A}(\hat{B}\psi) \neq \hat{B}(\hat{A}\psi)$. For instance, $\hat{p}_x x\psi \neq x\hat{p}_x\psi$ since $-\mathrm{i}\hbar\partial/\partial x(x\psi) \neq x(-\mathrm{i}\hbar(\partial\psi/\partial x))$.

Two operators commute if, and only if,

$$\hat{A}\hat{B} - \hat{B}\hat{A} = 0 \ . \tag{4.7}$$

We call this expression a *commutator* and write it as

$$\hat{A}\hat{B} - \hat{B}\hat{A} = [\hat{A}, \hat{B}]_- \ . \tag{4.8}$$

In analogy, we define an *anticommutator* as

$$\hat{A}\hat{B} + \hat{B}\hat{A} = [\hat{A}, \hat{B}]_+ \ . \tag{4.9}$$

Now we are going to obtain an answer to the question: under which conditions is the product $\hat{A}\hat{B}$ of two Hermitian operators Hermitian, too? We rewrite the product $\hat{A}\hat{B}$ as

$$\hat{A}\hat{B} = \tfrac{1}{2}[\hat{A}, \hat{B}]_+ + \tfrac{1}{2}[\hat{A}, \hat{B}]_- \ , \tag{4.10}$$

and we will now show that the $\frac{1}{2}[\hat{A},\hat{B}]_+$ part is always Hermitian, whereas the $\frac{1}{2}[\hat{A},\hat{B}]_-$ part never is. Let us start with the following relation:

$$\begin{aligned}\frac{1}{2}\int\psi_1^*[\hat{A},\hat{B}]_\pm\psi_2\,\mathrm{d}V &= \frac{1}{2}\int\psi_1^*(\hat{A}\hat{B}\pm\hat{B}\hat{A})\psi_2\,\mathrm{d}V\\ &= \frac{1}{2}\int(\hat{A}\psi_1)^*\hat{B}\psi_2\,\mathrm{d}V\pm\frac{1}{2}\int(\hat{B}\psi_1)^*\hat{A}\psi_2\,\mathrm{d}V\\ &= \frac{1}{2}\int(\hat{B}\hat{A}\psi_1)^*\psi_2\,\mathrm{d}V\pm\frac{1}{2}\int(\hat{A}\hat{B}\psi_1)^*\psi_2\,\mathrm{d}V\\ &= \frac{1}{2}\int(\hat{B}\hat{A}\pm\hat{A}\hat{B})^*\psi_1^*\psi_2\,\mathrm{d}V\\ &= \frac{1}{2}\int[\hat{B},\hat{A}]_\pm^*\psi_1^*\psi_2\,\mathrm{d}V\ .\end{aligned}\tag{4.11}$$

As $\hat{A}\hat{B}+\hat{B}\hat{A}=\hat{B}\hat{A}+\hat{A}\hat{B}$, the $\frac{1}{2}[\hat{A},\hat{B}]_+$ part is always Hermitian and as $\hat{A}\hat{B}-\hat{B}\hat{A}=-(\hat{B}\hat{A}-\hat{A}\hat{B})$, the $\frac{1}{2}[\hat{A},\hat{B}]_-$ part is only Hermitian if it vanishes. Hence, the product $\hat{A}\hat{B}$ of commuting Hermitian operators is Hermitian. As any operator commutes with itself, $\hat{A}^n$ is Hermitian if $\hat{A}$ is, and so is $\hat{A}^n\hat{B}^m$, if $\hat{A}$ and $\hat{B}$ are Hermitian and commuting.

4.3 Bra and Ket Notation

The integral $\int_{-\infty}^{+\infty}\psi_1^*\psi_2\,\mathrm{d}V$ can be considered a scalar product of the square-integrable functions ψ_1 and ψ_2. Usually the following shorthand notation is used:

$$\langle\psi_1|\psi_2\rangle=\int\limits_{-\infty}^{\infty}\psi_1^*\psi_2\,\mathrm{d}V\ .\tag{4.12}$$

This is interpreted to be a product of two elements $\langle\psi_1|$ and $|\psi_2\rangle$. The element $\langle\psi_1|$ is called a "bra" and $|\psi_2\rangle$ is called a "ket",[1] together forming the "bra-ket" (bracket). Both are vectors (state vectors) in a linear vector space. By using this notation, many relations in quantum mechanics can be expressed more succinctly than by using integral representation.

The state vectors are vectors of a complex linear vector space with an orthonormal base. Every expression in integral representation is related to an expression in ***Dirac*** notation. For instance, the orthonormality relation reads

$$\int\psi_m^*\psi_n\,\mathrm{d}V=\langle\psi_m|\psi_n\rangle=\delta_{mn}\ .\tag{4.13}$$

[1] This denotation originates from the famous physicist P. A. M. Dirac, whose contribution to relativistic quantum mechanics we will learn about in another volume of this series: W. Greiner: *Relativistic Quantum Mechanics – Wave Equations*, 3rd ed. (Springer, Berlin, Heidelberg 2000)

Obviously, $|\psi\rangle^* = \langle\psi|$ holds. We can write the expectation value of an operator $\hat{L}$ as

$$\langle\psi|\hat{L}|\psi\rangle = \int \psi^* \hat{L} \psi \, dV \, , \tag{4.14}$$

and the Hermiticity of $\hat{L}$ is denoted by

$$\langle\psi|\hat{L}|\psi\rangle = \langle\hat{L}\psi|\psi\rangle \, . \tag{4.15}$$

4.4 Eigenvalues and Eigenfunctions

We can obtain more information about a Hermitian operator $\hat{L}$ and that which is physically related to it if, besides the known mean value $\bar{L}$, we can get an expression for the mean-square deviation $\overline{(\Delta L)^2}$. First, it is necessary to find a quantum-mechanical operator describing $\overline{(\Delta L)^2}$. This is straightforward; in fact, we obtain the *deviation from the mean* value by

$$\Delta\hat{L} = \hat{L} - \bar{L} \, , \tag{4.16}$$

and hence the square of the deviation as

$$(\Delta\hat{L})^2 = (\hat{L} - \bar{L})^2 \, . \tag{4.17}$$

The *mean-square deviation* can be expressed by

$$\overline{(\Delta L)^2} = \int \psi^* (\Delta\hat{L})^2 \psi \, dV \, , \tag{4.18}$$

and it must be nonnegative. Indeed, from

$$\overline{(\Delta\hat{L})^2} = \int_{-\infty}^{\infty} \psi^* (\Delta\hat{L})^2 \psi \, dV \tag{4.19}$$

and the Hermiticity of $\Delta\hat{L}$, it follows that

$$\overline{(\Delta\hat{L})^2} = \int_{-\infty}^{\infty} (\Delta\hat{L}\psi)^* (\Delta\hat{L}\psi) \, dV = \int_{-\infty}^{\infty} |\Delta\hat{L}\psi|^2 \, dV \geq 0 \, . \tag{4.20}$$

As the integrand is a nonnegative function, the integral is positive definite and hence $\overline{(\Delta L)^2}$ is positive definite, too.

Now we search for those states ψ_L for which the quantity L has a constant value, i.e. for which the deviation ΔL of L vanishes. For states of this kind, $\overline{\Delta L}^2 = 0$ holds, and we obtain

$$\int |\Delta\hat{L}\psi_L|^2 \, dV = 0 \, . \tag{4.21}$$

The integrand is a real function which cannot be negative (as it is the absolute value of a complex function). Hence

$$\Delta\hat{L}\psi_L = 0 \ . \tag{4.22}$$

We can write this relation using the definition of ΔL as

$$(\hat{L} - \bar{L})\psi_L = 0 \ , \tag{4.23}$$

and since we may put $\bar{L} = L$ in the state ψ_L,

$$\hat{L}\psi_L = L\psi_L \tag{4.24}$$

holds.

An equation of this kind is called an *eigenvalue equation*. We call ψ_L an *eigenfunction* and L an *eigenvalue* of the operator $\hat{L}$. In general, an operator $\hat{L}$ has several eigenfunctions ψ_{L_ν} with eigenvalues L_ν. The eigenvalues L_ν can form a *discrete spectrum* $L_1, L_2, L_3, \ldots$ or a *continuous spectrum*. In the latter case, the eigenvalues L will take on any value within an interval $L_n \leq L \leq L_{n+1}$. We soon will encounter operators with discrete, continuous and mixed spectra (see Fig. 4.1).

Now we are going to examine some general properties of eigenfunctions. For this, let us investigate the eigenfunctions of Hermitian operators only and restrict ourselves to the case of the discrete spectrum. We can show that eigenfunctions belonging to two different eigenvalues are orthogonal. Let ψ_n and ψ_m be eigenfunctions to the eigenvalues L_n and L_m, respectively, i.e.

$$\hat{L}\psi_m = L_m\psi_m \quad \text{and} \quad \hat{L}\psi_n = L_n\psi_n \ . \tag{4.25}$$

We take the complex conjugate of the first equation and find, as the eigenvalues are real,

$$\hat{L}^*\psi_m^* = L_m^*\psi_m^* = L_m\psi_m^* \ . \tag{4.26}$$

Now multiplying the second equation in (4.25) by ψ_m^* and the complex conjugate of the first equation by ψ_n yields

$$\psi_m^*\hat{L}\psi_n = L_n\psi_n\psi_m^* \ , \quad \psi_n\hat{L}^*\psi_m^* = L_m\psi_m^*\psi_n \ . \tag{4.27}$$

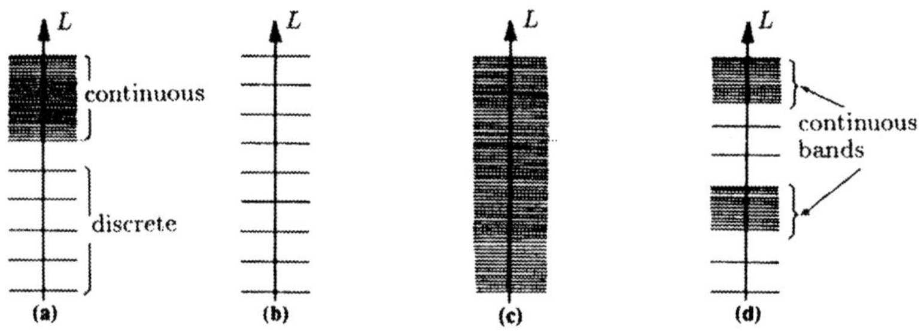

Fig. 4.1. (**a**) General spectrum; (**b**) totally discrete spectrum; (**c**) totally continuous spectrum; (**d**) spectrum with continuous bands, as occur, for example, in the energy spectrum of a crystal lattice

The difference between these two equations is

$$\psi_m^* \hat{L}\psi_n - \psi_n \hat{L}^* \psi_m^* = \psi_n \psi_m^* (L_n - L_m) \ . \tag{4.28}$$

If we integrate over the entire volume, we obtain

$$\int_{-\infty}^{\infty} \psi_m^* \hat{L}\psi_n \, \mathrm{d}V - \int_{-\infty}^{\infty} \psi_n \hat{L}^* \psi_m^* \, \mathrm{d}V = (L_n - L_m) \int_{-\infty}^{\infty} \psi_n \psi_m^* \, \mathrm{d}V \ . \tag{4.29}$$

As $\hat{L}$ is a Hermitian operator, both integrals on the left-hand side are equal, and therefore

$$0 = (L_n - L_m) \int_{-\infty}^{\infty} \psi_n \psi_m^* \, \mathrm{d}V \ . \tag{4.30}$$

We required that $L_n \neq L_m$; hence

$$0 = \int_{-\infty}^{\infty} \psi_n \psi_m^* \, \mathrm{d}V \ , \tag{4.31}$$

which is the desired result and proves that ψ_n and ψ_m are orthogonal.

As the eigenfunctions of a discrete spectrum are square-integrable, they can be normalized to unity:

$$\int \psi_n \psi_n^* \, \mathrm{d}V = 1 \ . \tag{4.32}$$

Then we can combine relations (4.31) and (4.32) as

$$\int_{-\infty}^{\infty} \psi_n \psi_m^* \, \mathrm{d}V = \delta_{nm} \ . \tag{4.33}$$

Hence the system of eigenfunctions is an orthonormal function system.

In general, there are several eigenfunctions for one eigenvalue L_n; we call them *degenerate states*. To be more precise, if a different eigenfunctions $\psi_{n1}, \dots, \psi_{na}$ belong to the eigenvalue L_n, we speak of an *a-fold degeneracy*. Physically, this degeneracy describes the possibility that a certain value of the observable L can be realized in different states.

We have proved that eigenfunctions of a discrete spectrum with different eigenvalues are mutually orthogonal. If we have degeneracy, the functions ψ_{nk} are related to the same eigenvalue L_n: $\hat{L}\psi_{nk} = L_n \psi_{nk}$, with $k = 1, \dots, a$; hence, they are in general not orthogonal. But there is always the possibility of finding orthogonal functions in this case, too, as we shall now show.

Let us assume that the eigenfunctions ψ_{nk} $(k = 1, \dots, a)$, related to the eigenvalue L_n, are linearly independent, i.e. if $\sum_{k=1}^{a} a_k \psi_{nk} = 0$, then $a_k = 0$ holds for all k. If we could not infer $a_k = 0$ for all k, we would be able to express at least one function by a linear combination of the others, and the number of eigenfunctions would be smaller than a. If the set of ψ_{nk} is orthogonal, we can use it to describe a certain state. If it is not orthogonal, we transform this set into a new set, i.e.

$$\varphi_{n\alpha} = \sum_{k=1}^{a} a_{\alpha k} \psi_{nk} \ , \quad \alpha = 1 \ldots a \ . \tag{4.34}$$

This transformation is linear; thus the functions $\varphi_{n\alpha}$ are also eigenfunctions of the operator $\hat{L}$ of the eigenvalue L_n. We now require orthogonality of the new functions $\varphi_{n\alpha}$:

$$\int\limits_{-\infty}^{\infty} \varphi^*_{n\alpha} \varphi_{n\beta} \, \mathrm{d}V = \delta_{\alpha\beta} \ .$$

The conditions that have to be fulfilled by the coefficients $a_{\alpha k}$ in order to describe a transformation to an orthogonal function system are

$$\sum_{k=1}^{a} \sum_{k'=1}^{a} a^*_{\alpha k} a_{\beta k'} s_{kk'} = \delta_{\alpha\beta} \ , \quad \text{with} \tag{4.35}$$

$$s_{kk'} = \int\limits_{-\infty}^{\infty} \psi^*_{nk} \psi_{nk'} \, \mathrm{d}V \ .$$

The coefficients $a_{\alpha k}$ are determined by analogy with geometry. We consider the functions ψ_{nk} as vectors in an a-dimensional function space and $s_{kk'}$ as scalar products of these vectors. Then we can regard transformation (4.34) as a basis transformation from an oblique-angled to an orthogonal coordinate system.

Hence applying this procedure to the case of a degenerate spectrum, we can obtain an orthonormal set of eigenfunctions. A practical method is E. ***Schmidt***'s *orthogonalization method*, familiar from geometry (vector calculus). In the first step we take one vector (state), for instance ψ_{n1}, and define the normalized wave function $\varphi_{n1} = \psi_{n1}/\sqrt{\langle\psi_{n1}|\psi_{n1}\rangle}$.

In the next step we construct a vector $\varphi_{n2} = \alpha\varphi_{n1} + \beta\psi_{n2}$ and require $\langle\varphi_{n1}|\varphi_{n2}\rangle = \alpha \langle\varphi_{n1}|\varphi_{n1}\rangle + \beta \langle\varphi_{n1}|\psi_{n2}\rangle = 0$. It follows that $\alpha/\beta = -\langle\varphi_{n1}|\psi_{n2}\rangle$. Apart from this condition, normalization is required, i.e. $\langle\varphi_{n2}|\varphi_{n2}\rangle = 1$. From these two conditions follow α and β. The third step is the construction of $\varphi_{n3} = \alpha'\varphi_{n1} + \beta'\varphi_{n2} + \gamma\psi_{n3}$. Again, orthogonality of this vector (state) to φ_{n1} and φ_{n2} and normalization are required. Hence there are three conditions to determine α, β, γ etc.

The next steps are now straightforward. We note that this transformation is defined only up to an orthogonal transformation. If the functions ψ_{nk} are already orthogonal, then $s_{kk'} = \delta_{kk'}$ and

$$\sum_{k=1}^{a} a^*_{\alpha k} a_{\beta k} = \delta_{\alpha\beta} \tag{4.36}$$

holds. This is the condition for an orthogonal transformation.

In the case of *continuous spectra* we cannot numerate eigenvalues and eigenfunctions. Instead we parametrize the eigenfunctions and take the eigenvalues as parameters. Then the equation

$$\hat{L}\psi_n(x) = L_n\psi_n(x) \tag{4.37}$$

becomes

$$\hat{L}\psi(x, L) = L\psi(x, L) \ , \tag{4.38}$$

if x denotes all coordinates appearing in the wave function ψ (for instance $x = x, y, z$). From the wave functions which are not orthogonal we can define ***Weyl's eigendifferentials***:

$$\Delta\psi(x, L) = \int_{L}^{L+\Delta L} \psi(x, L)\,\mathrm{d}L \ . \tag{4.39}$$

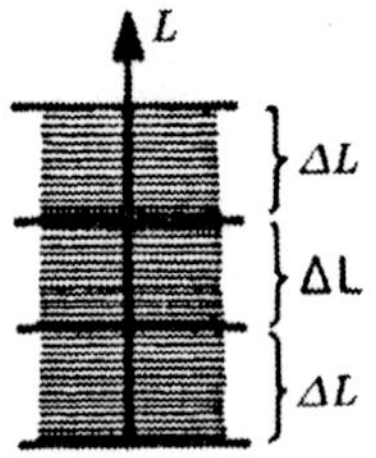

Fig. 4.2. Chopping of the continuous spectrum by integration of the function $\psi(x, L)$ over intervals ΔL leads to H. Weyl's eigendifferentials

They divide up the continuous spectrum of the eigenvalues L into discrete regions of size ΔL (see Fig. 4.2). The eigendifferentials are orthogonal and can be normalized. (See the mathematical addendum in the next chapter.)

EXAMPLE

4.1 Hermiticity of the Momentum Operator

We show that the momentum operator $\hat{p}_x = -\mathrm{i}\hbar\partial/\partial x$ is Hermitian:

$$\begin{aligned}\overline{p_x} &= \int_{-\infty}^{\infty} \psi_1^*\hat{p}_x\psi_2\,\mathrm{d}V = \int_{-\infty}^{\infty} \psi_1^*\left(-\mathrm{i}\hbar\frac{\partial}{\partial x}\right)\psi_2\,\mathrm{d}V \\ &= -\mathrm{i}\hbar\int_{-\infty}^{\infty} \psi_1^*\left(\frac{\partial}{\partial x}\psi_2\right)\mathrm{d}V \\ &= -\mathrm{i}\hbar[\psi_2\psi_1^*]_{-\infty}^{\infty} + \mathrm{i}\hbar\int_{-\infty}^{\infty} \psi_2\frac{\partial}{\partial x}\psi_1^*\,\mathrm{d}V \ .\end{aligned} \tag{1}$$

As ψ_1 and ψ_2 are square-integrable functions,

$$[\psi_1, \psi_2^*]_{-\infty}^{+\infty} = 0 \tag{2}$$

holds, and we obtain

$$\overline{p_x} = \mathrm{i}\hbar\int_{-\infty}^{\infty} \psi_2\frac{\partial}{\partial x}\psi_1^*\,\mathrm{d}V = \int_{-\infty}^{\infty} (\hat{p}_x\psi_1)^*\psi_2\,\mathrm{d}V \ . \tag{3}$$

This proves that $\hat{p}_x$ obeys the Hermiticity relation (4.4).

EXAMPLE

4.2 The Commutator of Position and Momentum Operators

We compute the commutator $[\hat{p}_x, \hat{x}]$. Since

$$\hat{p}_x\hat{x}\psi = -\mathrm{i}\hbar\frac{\partial}{\partial x}(x\psi) = -\mathrm{i}\hbar\left(\psi\frac{\partial x}{\partial x}+x\frac{\partial\psi}{\partial x}\right) = -\mathrm{i}\hbar\left(\psi+x\frac{\partial\psi}{\partial x}\right) ,$$

and

$$\hat{x}\hat{p}_x\psi = x\left(-\mathrm{i}\hbar\frac{\partial\psi}{\partial x}\right) = -\mathrm{i}\hbar x\frac{\partial}{\partial x}\psi ,$$

we easily obtain

$$\hat{p}_x\hat{x} - \hat{x}\hat{p}_x = [\hat{p}_x, \hat{x}] = -\mathrm{i}\hbar .$$

EXERCISE

4.3 Computation Rules for Commutators

Problem. Let $\hat{L}, \hat{L}_1, \hat{L}_2, \hat{L}_3, \hat{M} : H \to H$ be linear operators in a complex linear space and a be a scalar. Let $\hat{E}$ denote the identity operator. Show (with the help of the definition of a commutator) the following identities:

$$[\hat{L}, \hat{M}]_- = -[\hat{M}, \hat{L}]_- \quad (1)$$

$$[\hat{L}, \hat{L}]_- = 0 \quad (2)$$

$$[\hat{L}, a\hat{M}]_- = a[\hat{L}, \hat{M}]_- \quad (3)$$

$$[\hat{L}, a\hat{E}]_- = 0 \quad (4)$$

$$[\hat{L}_1 + \hat{L}_2, \hat{M}]_- = [\hat{L}_1, \hat{M}]_- + [\hat{L}_2, \hat{M}] \quad (5)$$

$$[\hat{L}_1\hat{L}_2, \hat{M}]_- = [\hat{L}_1, \hat{M}]_-\hat{L}_2 + \hat{L}_1[\hat{L}_2, \hat{M}]_- \quad (6)$$

$$[\hat{M}, \hat{L}_1\hat{L}_2]_- = [\hat{M}, \hat{L}_1]_-\hat{L}_2 + \hat{L}_1[\hat{M}, \hat{L}_2]_- \quad (7)$$

$$[\hat{L}_1, [\hat{L}_2, \hat{L}_3]_-]_- + [\hat{L}_2, [\hat{L}_3, \hat{L}_1]_-]_- + [\hat{L}_3, [\hat{L}_1, \hat{L}_2]_-]_- = 0 . \quad (8)$$

Solution. The first five relations are trivial (they follow directly from the definitions $[\hat{L}, \hat{M}] = \hat{L}\hat{M} - \hat{M}\hat{L}$). The proof of the other relations is also simple, but it is important to pay attention to the order of the factors:

$$\begin{aligned}[\hat{L}_1\hat{L}_2, \hat{M}] &= \hat{L}_1\hat{L}_2\hat{M} - \hat{M}\hat{L}_1\hat{L}_2\\ &= \hat{L}_1\hat{M}\hat{L}_2 - \hat{M}\hat{L}_1\hat{L}_2 + \hat{L}_1\hat{L}_2\hat{M} - \hat{L}_1\hat{M}\hat{L}_2\\ &= [\hat{L}_1, \hat{M}]\hat{L}_2 + \hat{L}_1[\hat{L}_2, \hat{M}] , \end{aligned} \quad (6)$$

$$\begin{aligned}[\hat{M}, \hat{L}_1\hat{L}_2] &= -[\hat{L}_1\hat{L}_2, \hat{M}] \overset{(6)}{=} -[\hat{L}_1, \hat{M}]\hat{L}_2 - \hat{L}_1[\hat{L}_2, \hat{M}]\\ &= \hat{L}_1[\hat{M}, \hat{L}_2] + [\hat{M}, \hat{L}_1]\hat{L}_2 , \end{aligned} \quad (7)$$

Exercise 4.3

$$\begin{aligned}[\hat{L}_1,[\hat{L}_2,\hat{L}_3]] &= [\hat{L}_1,\hat{L}_2\hat{L}_3]-[\hat{L}_1,\hat{L}_3\hat{L}_2] \\ &\overset{(7)}{=} [\hat{L}_1,\hat{L}_2]\hat{L}_3+\hat{L}_2[\hat{L}_1,\hat{L}_3]-[\hat{L}_1,\hat{L}_3]\hat{L}_2-\hat{L}_3[\hat{L}_1,\hat{L}_2] \\ &= -[\hat{L}_3,[\hat{L}_1,\hat{L}_2]]-[\hat{L}_2,[\hat{L}_3,\hat{L}_1]] \ . \end{aligned} \tag{8}$$

The last equation is also called the *Jacobi identity*.

EXAMPLE

4.4 Momentum Eigenfunctions

The equation for the eigenvalues of the momentum operator is

$$\hat{p}_x\psi_{p_x}(x)=p_x\psi_{p_x}(x) \quad \text{or} \quad -\mathrm{i}\hbar\frac{\mathrm{d}\psi_{p_x}(x)}{\mathrm{d}x}=p_x\psi_{p_x}(x) \quad \text{or}$$

$$\frac{\mathrm{d}\psi_{p_x}(x)}{\mathrm{d}x}=\mathrm{i}\frac{p_x}{\hbar}\psi_{p_x} \ .$$

We infer for every p_x within $-\infty\leq p_x\leq\infty$:

$$\psi_{p_x}(x)=C\exp\left(\mathrm{i}\frac{p_x}{\hbar}x\right)=C\,\mathrm{e}^{\mathrm{i}kx} \ .$$

C is a (at first arbitrary) constant; we will calculate its value in Example 5.1. The spectrum of momentum is continuous: there is an eigenfunction for every momentum p_x; we recognize this eigenfunction as a part of the well-known de Broglie wave [see Eqs. (3.3) and (3.37)].

4.5 Measurability of Different Observables at Equal Times

We know from Heisenberg's uncertainty principle that it is impossible to measure the coordinates and momentum of a particle simultaneously and exactly [see (3.59) ff.]. The value of an observable is unambiguously defined if the wave function is an eigenfunction of its related operator, i.e.

$$\hat{L}\psi_n=L_n\psi_n \ . \tag{4.40}$$

Then, in state ψ_n, the observable L is well defined, i.e. it has precisely the value L_n and its mean square deviation $(\Delta L)^2$ is zero. In general, ψ is not an eigenfunction of another operator $\hat{M}$. Hence, we cannot infer information about the observable M from the wave function ψ_n. Only if ψ_n is an eigenfunction of $\hat{M}$, too, can we measure both M and L sharply, i.e.

$$\hat{L}\psi_n = L_n\psi_n \quad \text{and} \quad \hat{M}\psi_n = M_n\psi_n \tag{4.41}$$

for all ψ_n. As both equations hold, we obtain $[\hat{L}, \hat{M}]_-\psi_n = 0$, because $\hat{M}\hat{L}\psi_n = L_n\hat{M}\psi_n = L_n M_n\psi_n$ and $\hat{L}\hat{M}\psi_n = M_n\hat{L}\psi_n = M_n L_n\psi_n$. By subtraction we get $(\hat{M}\hat{L} - \hat{L}\hat{M})\psi_n = 0$. The set of eigenfunctions ψ_n of the Hermitian operator $\hat{L}$ is complete. Therefore an arbitrary function $\psi(x)$ can be expanded into $\psi_n(x)$, i.e.

$$\psi(x) = \sum_n c_n\psi_n(x) \ .$$

Obviously, it follows that

$$(\hat{M}\hat{L} - \hat{L}\hat{M})\psi(x) = 0 \ .$$

Because $\psi(x)$ is arbitrary, we have the operator equation $\hat{M}\hat{L} - \hat{L}\hat{M} = 0$.

We have thus found that two observables are measurable simultaneously if their commutator, acting on a common eigenfunction, vanishes. In the other direction we obtain the following result: if $[\hat{L}, \hat{M}]_- = 0$, then for every ψ, $\hat{L}\hat{M}\psi = \hat{M}\hat{L}\psi$. If ψ is an eigenfunction of $\hat{L}$, we obtain $\hat{L}(\hat{M}\psi) = L(\hat{M}\psi)$ and $\psi' = \hat{M}\psi$ is an eigenfunction of $\hat{L}$, too. If L is not degenerate, we can infer $\hat{M}\psi = M\psi$; i.e. $\psi' = \hat{M}\psi$ is a multiple of ψ (here, $M\psi$).

In the case of degeneracy, $\psi' = \hat{M}\psi$ can be a linear combination of f degenerate eigenfunctions $\psi_k (k = 1, 2, \dots, f)$ of the eigenvalue L. Then we have

$$\psi'_k = \sum_{k'=1}^{f} M_{kk'}\psi_{k'} \ , \quad k = 1, 2, \dots, f \ . \tag{4.42}$$

Thus we cannot repeat the conclusion used above. But as the choice of the original wave function is arbitrary (we recall that $\hat{L}\hat{M}\psi = \hat{M}\hat{L}\psi$ must hold for *all* possible ψ), we can use a linear combination,

$$\varphi = \sum_{k'=1}^{f} a_{k'}\psi_{k'} \ , \tag{4.43}$$

as the initial wave function, instead of ψ_k. Of course

$$\hat{L}\varphi = L\varphi \tag{4.44}$$

holds, too. Now we choose the coefficients a_k in order to obtain

$$\hat{M}\varphi = M\varphi \ . \tag{4.45}$$

We can do this because inserting φ into this equation gives

$$\sum_{k'=1}^{f} a_{k'}\hat{M}\psi_{k'} = M\sum_{k'=1}^{f} a_{k'}\psi_{k'} \ . \tag{4.46}$$

After multiplication by the bra vector $\langle k|$ (corresponding to the operation $\int \psi_k^* \dots \mathrm{d}x$) and using the orthogonality condition $\langle \psi_k|\psi_{k'}\rangle = \delta_{kk'}$ we obtain

$$\sum_{k'=1}^{f} \langle k|\hat{M}|k'\rangle a_{k'} = M a_k \ . \tag{4.47}$$

Let us abbreviate the matrix elements $\langle k|M|k'\rangle$ by $M_{kk'} \equiv \langle k|\hat{M}|k'\rangle$. Since we obtained a linear homogeneous system of equations for the $a_{k'}$, its coefficient determinant must vanish, i.e.

$$\begin{vmatrix} M_{11}-M \ , & M_{12} \ , & \cdots M_{1f} \\ M_{21} \ , & M_{22}-M \ , & \cdots M_{2f} \\ \vdots & \vdots & \cdots \vdots \\ \vdots & \vdots & \cdots \vdots \\ \vdots & \vdots & \cdots \vdots \\ \vdots & \vdots & \cdots \vdots \\ M_{f1} \ , & M_{f2} \ , & \cdots M_{ff}-M \end{vmatrix} = 0 \ . \tag{4.48}$$

The solution of this equation gives the eigenvalues M. We thus see that in the case of degeneracy of the eigenfunction ψ_k of L, we can also construct the wave functions $\varphi = \sum_k a_k \psi_k$, *which are simultaneously eigenfunctions of* $\hat{L}$ *and* $\hat{M}$.

4.6 Position and Momentum Operators

If we start with a wave function $\psi = \psi(\boldsymbol{r})$, the position operator is the space vector itself:

$$\hat{\boldsymbol{r}} = \boldsymbol{r} \ . \tag{4.49}$$

Its components are

$$\hat{x} = x \ , \quad \hat{y} = y \ , \quad \hat{z} = z \ . \tag{4.49a}$$

The operator of momentum is expressed as

$$\hat{\boldsymbol{p}} = -\mathrm{i}\hbar\nabla \ , \tag{4.50}$$

and its components are

$$\hat{p}_x = -\mathrm{i}\hbar\frac{\partial}{\partial x} \ , \quad \hat{p}_y = -\mathrm{i}\hbar\frac{\partial}{\partial y} \ , \quad \hat{p}_z = -\mathrm{i}\hbar\frac{\partial}{\partial z} \ . \tag{4.50a}$$

The commutators are

$$\begin{aligned} &[\hat{x}, \hat{p}_x]_- = [\hat{y}, \hat{p}_y]_- = [\hat{z}, \hat{p}_z]_- = \mathrm{i}\hbar \ , \\ &[\hat{x}, \hat{p}_y]_- = [\hat{x}, \hat{p}_z]_- = [\hat{y}, \hat{p}_x]_- = [\hat{y}, \hat{p}_z]_- = [\hat{z}, \hat{p}_x]_- = [\hat{z}, \hat{p}_y]_- = 0 \ . \end{aligned} \tag{4.51}$$

Hence there is an uncertainty relation between the coordinates and their canonical conjugate momenta (x and p_x, y and $p_y \dots$). They cannot be exactly measured simultaneously. (See the following section, where this will be discussed in detail.) On the other hand, e.g. the $\hat{x}$ operator and the $\hat{p}_y$ operator do commute. Hence these two observables can be measured simultaneously as accurately as desired. Their common eigenstates are

$$\sqrt{\delta(x-x_0)}\exp\left(\frac{\mathrm{i}}{\hbar}p_y y\right)\;, \quad \text{etc.} \tag{4.52}$$

For the definition of the $\delta(x)$ function we refer to Chap. 5.

4.7 Heisenberg's Uncertainty Relations for Arbitrary Observables

We are now in a position to consider the uncertainty relations in a more general way. Let two physical quantities be described by *Hermitian operators* $\hat{A}$ and $\hat{B}$ [e.g. $\hat{A}=\hat{x}$ is the position operator and $\hat{B}=\hat{p}_x=-\mathrm{i}\hbar(\partial/\partial x)$ is the momentum operator]. The commutator of the two operators is written as

$$[\hat{A},\hat{B}]_- = \hat{A}\hat{B}-\hat{B}\hat{A}=\mathrm{i}\hat{C}\;, \tag{4.53}$$

where $\hat{C}$ is called the *remainder of commutation* (*commutation rest*). $\hat{C}$ can be zero; then $\hat{A}$ and $\hat{B}$ commute. In general $\hat{C}$ is a *Hermitian operator*, because we know from above that

$$\begin{aligned}\int \psi_1^*[\hat{A},\hat{B}]_-\psi_2\,\mathrm{d}x &= \int \psi_1^*(\hat{A}\hat{B}-\hat{B}\hat{A})\psi_2\,\mathrm{d}x\\ &= \int [(\hat{B}^*\hat{A}^*-\hat{A}^*\hat{B}^*)\psi_1^*]\psi_2\,\mathrm{d}x\\ &= -\int [(\hat{A}\hat{B}-\hat{B}\hat{A})\psi_1]^*\psi_2\,\mathrm{d}x\;.\end{aligned} \tag{4.54}$$

Hence

$$\int \psi_1^*\mathrm{i}\hat{C}\psi_2\,\mathrm{d}x = -\int (\mathrm{i}\hat{C}\psi_1)^*\psi_2\,\mathrm{d}x \quad \text{or}$$

$$\int \psi_1^*\hat{C}\psi_2\,\mathrm{d}x = \int (\hat{C}\psi_1)^*\psi_2\,\mathrm{d}x\;. \tag{4.55}$$

The physical quantities corresponding to the operators $\hat{A}$ and $\hat{B}$ in an arbitrary state ψ have the mean values

$$\bar{A} = \int \psi^*\hat{A}\psi\,\mathrm{d}x \quad \text{and} \quad \bar{B} = \int \psi^*\hat{B}\psi\,\mathrm{d}x\;. \tag{4.56}$$

As before in (3.60) and (4.16), we introduce operators for the deviation from the mean value,

$$\Delta\hat{A} = \hat{A} - \bar{A} \quad \text{and} \quad \Delta\hat{B} = \hat{B} - \bar{B} \ , \tag{4.57}$$

and recognize that $\Delta\hat{A}$ and $\Delta\hat{B}$ obey the same commutation relations as $\hat{A}$ and $\hat{B}$, namely:

$$[\Delta\hat{A}, \Delta\hat{B}]_- = \mathrm{i}\hat{C} \ .$$

In analogy to our considerations about the uncertainty relation of $\hat{p}_x$ and $\hat{x}$ [see (3.63)], we examine the integral

$$I(\alpha) = \int \left|(\alpha\Delta\hat{A} - \mathrm{i}\Delta\hat{B})\psi\right|^2 \mathrm{d}x \geq 0 \ , \tag{4.58}$$

which depends on a real parameter α. As $\Delta\hat{A}$ and $\Delta\hat{B}$ are Hermitian, we can write

$$\begin{aligned} I(\alpha) &= \int (\alpha\Delta\hat{A} - \mathrm{i}\Delta\hat{B})^*\psi^*(\alpha\Delta\hat{A} - \mathrm{i}\Delta\hat{B})\psi \,\mathrm{d}x \\ &= \int \psi^*(\alpha\Delta\hat{A} + \mathrm{i}\Delta\hat{B})(\alpha\Delta\hat{A} - \mathrm{i}\Delta\hat{B})\psi \,\mathrm{d}x \\ &= \int \psi^*[\alpha^2(\Delta\hat{A})^2 + \mathrm{i}\alpha(\Delta\hat{B}\Delta\hat{A} - \Delta\hat{A}\Delta\hat{B}) + (\Delta\hat{B})^2]\psi \,\mathrm{d}x \\ &= \int \psi^*[\alpha^2(\Delta\hat{A})^2 + \alpha\hat{C} + (\Delta\hat{B})^2]\psi \,\mathrm{d}x \geq 0 \ . \end{aligned} \tag{4.59}$$

Now we denote by $\langle|(\Delta A)^2|\rangle \equiv \overline{(\Delta A)^2}$, $\langle|\bar{C}|\rangle \equiv \bar{C}$, $\langle|\Delta\hat{B}^2|\rangle \equiv \overline{(\Delta B)^2}$ the mean values of the squares of deviation, or of the commutation rest $\hat{C}$. We can therefore write the last equation as

$$\overline{(\Delta A)^2}\left[\alpha + \frac{\bar{C}}{2\overline{(\Delta A)^2}}\right]^2 + \overline{(\Delta B)^2} - \frac{(\bar{C})^2}{4\overline{(\Delta A)^2}} \geq 0 \ . \tag{4.60}$$

As this holds for every real α, we have

$$\overline{(\Delta B)^2} - \frac{(\bar{C})^2}{4\overline{(\Delta A)^2}} \geq 0 \quad \text{or} \quad \overline{(\Delta A)^2}\ \overline{(\Delta B)^2} \geq \frac{(\bar{C})^2}{4} \ . \tag{4.61}$$

This is *Heisenberg's uncertainty principle* in its most general form. Obviously *it holds for all physical quantities with noncommuting operators*. For commuting operators ($\hat{C} = 0$) we have no uncertainty relation for the corresponding physical quantities. They can be exactly measured simultaneously. From (4.51) above, we know that $[\hat{p}_x, \hat{x}] = -\mathrm{i}\hbar$. Hence the uncertainty relation for these quantities is $\overline{(\Delta p_x)^2}\ \overline{(\Delta x)^2} \geq \hbar^2/4$, which coincides with the result we got earlier [see (3.66)].

In Chap. 6 we will prove that the *energy operator* is $\hat{E} = +\mathrm{i}\hbar(\partial/\partial t)$, and that the commutation relation

$$[\hat{E}, t]_- = \mathrm{i}\hbar \tag{4.62}$$

holds. Hence there also exists an uncertainty relation between energy and time, i.e.

$$\overline{(\Delta E)^2}\;\overline{(\Delta t)^2} \geq \frac{\hbar^2}{4} \; . \tag{4.63}$$

We will obtain similar results for the angular-momentum operators in the next section.

4.8 Angular-Momentum Operators

We want to derive an operator for angular momentum. For this we insert the operators $\hat{\boldsymbol{r}}$ and $\hat{\boldsymbol{p}}$ into the classical definition of the angular momentum $\boldsymbol{L} = \boldsymbol{r} \times \boldsymbol{p}$, and obtain the operator equation

$$\hat{\boldsymbol{L}} = \hat{\boldsymbol{r}} \times \hat{\boldsymbol{p}} = -\mathrm{i}\hbar(\boldsymbol{r} \times \nabla) \; .$$

Expressing the cross-product in Cartesian coordinates yields with (4.50):

$$\begin{aligned}
\hat{L}_x &= \hat{y}\hat{p}_z - \hat{z}\hat{p}_y = -\mathrm{i}\hbar\left(y\frac{\partial}{\partial z} - z\frac{\partial}{\partial y}\right) \; , \\
\hat{L}_y &= \hat{z}\hat{p}_x - \hat{x}\hat{p}_z = -\mathrm{i}\hbar\left(z\frac{\partial}{\partial x} - x\frac{\partial}{\partial z}\right) \; , \\
\hat{L}_z &= \hat{x}\hat{p}_y - \hat{y}\hat{p}_x = -\mathrm{i}\hbar\left(x\frac{\partial}{\partial y} - y\frac{\partial}{\partial x}\right) \; .
\end{aligned} \tag{4.64}$$

As the factors of the various products are all commuting Hermitian operators, $\hat{L}$ is also Hermitian [see (4.10) and (4.11)]. By straightforward calculation, we obtain the commutation relations of the angular-momentum components,

$$\begin{aligned}
&\hat{L}_x\hat{L}_y - \hat{L}_y\hat{L}_x = \mathrm{i}\hbar\hat{L}_z \; , \quad \hat{L}_y\hat{L}_z - \hat{L}_z\hat{L}_y = \mathrm{i}\hbar\hat{L}_x \; , \\
&\hat{L}_z\hat{L}_x - \hat{L}_x\hat{L}_z = \mathrm{i}\hbar\hat{L}_y \; ,
\end{aligned} \tag{4.65}$$

which are frequently written in shorthand notation as

$$\hat{\boldsymbol{L}} \times \hat{\boldsymbol{L}} = \mathrm{i}\hbar\hat{\boldsymbol{L}} \quad \text{or also} \quad [\hat{L}_i, \hat{L}_j]_- = \mathrm{i}\hbar\varepsilon_{ijk}\hat{L}_k \; . \tag{4.66}$$

Here, ε_{ijk} is the totally antisymmetric tensor in three dimensions, i.e.

$$\varepsilon_{ijk} = \begin{cases} +1 & \text{if } i, j, k \text{ is an even permutation of } 1, 2, 3 \\ -1 & \text{if } i, j, k \text{ is an odd permutation of } 1, 2, 3 \\ 0 & \text{if two or more indices are equal} \; . \end{cases}$$

By way of example, we test the first relation (4.65) and get

$$\begin{aligned}\hat{L}_x\hat{L}_y-\hat{L}_y\hat{L}_x &= (y\hat{p}_z-z\hat{p}_y)(z\hat{p}_x-x\hat{p}_z)-(z\hat{p}_x-x\hat{p}_z)(y\hat{p}_z-z\hat{p}_y)\\ &= y(\hat{p}_z z)\hat{p}_x+\underline{yz\hat{p}_z\hat{p}_x}-\underline{yx\hat{p}_z\hat{p}_z}-\underline{z^2\hat{p}_y\hat{p}_x}+zx\hat{p}_y\hat{p}_z\\ &\quad -\underline{zy\hat{p}_x\hat{p}_z}+\underline{z^2\hat{p}_x\hat{p}_y}+\underline{xy\hat{p}_z\hat{p}_z}-x(\hat{p}_z z)\hat{p}_y-xz\hat{p}_z\hat{p}_y\\ &= -\mathrm{i}\hbar y\hat{p}_x+\mathrm{i}\hbar x\hat{p}_y=\mathrm{i}\hbar(x\hat{p}_y-y\hat{p}_x)=\mathrm{i}\hbar\hat{L}_z\ .\end{aligned}\tag{4.67}$$

The terms underscored in similar fashion cancel out. Thus the components of angular momentum are not measurable at the same time, because the relations (4.65) are of the structure of (4.53) with a commutation rest. The square of the angular-momentum operator is

$$\hat{L}^2=\hat{L}_x^2+\hat{L}_y^2+L_z^2\ .$$

It commutes with all components of the angular-momentum operator, i.e.

$$[\hat{L}^2,\hat{L}_x]_-=[\hat{L}^2,\hat{L}_y]_-=[\hat{L}^2,\hat{L}_z]_-=0\ .\tag{4.68}$$

By way of example, we compute the first commutator and get

$$\begin{aligned}[\hat{L}^2,\hat{L}_x]_- &= [\hat{L}_x^2+\hat{L}_y^2+\hat{L}_z^2,\hat{L}_x]_-=[\hat{L}_y^2,\hat{L}_x]_-+[\hat{L}_z^2,\hat{L}_x]_-\\ &= (\hat{L}_y^2\hat{L}_x-\hat{L}_x\hat{L}_y^2)+(\hat{L}_z^2\hat{L}_x-\hat{L}_x\hat{L}_z^2)\ .\end{aligned}\tag{4.69}$$

Using (4.65), the first term becomes

$$\begin{aligned}\hat{L}_y^2\hat{L}_x-\hat{L}_x\hat{L}_y^2 &= \hat{L}_y(\hat{L}_y\hat{L}_x)-\hat{L}_x\hat{L}_y^2=\hat{L}_y(-\mathrm{i}\hbar\hat{L}_z+\hat{L}_x\hat{L}_y)-\hat{L}_x\hat{L}_y^2\\ &= -\mathrm{i}\hbar\hat{L}_y\hat{L}_z+(-\mathrm{i}\hbar\hat{L}_z+\hat{L}_x\hat{L}_y)\hat{L}_y-\hat{L}_x\hat{L}_y^2\\ &= -\mathrm{i}\hbar(\hat{L}_y\hat{L}_z+\hat{L}_z\hat{L}_y)\ ,\end{aligned}\tag{4.70}$$

and, similarly, we get for the second term

$$\begin{aligned}\hat{L}_z^2\hat{L}_x-\hat{L}_x\hat{L}_z^2 &= \hat{L}_z(\hat{L}_z\hat{L}_x)-\hat{L}_x\hat{L}_z^2=\hat{L}_z(\mathrm{i}\hbar\hat{L}_y+\hat{L}_x\hat{L}_z)-\hat{L}_x\hat{L}_z^2\\ &= \mathrm{i}\hbar\hat{L}_z\hat{L}_y+(\hat{L}_z\hat{L}_x)\hat{L}_z-\hat{L}_x\hat{L}_z^2\\ &= \mathrm{i}\hbar\hat{L}_z\hat{L}_y+(\mathrm{i}\hbar\hat{L}_y+\hat{L}_x\hat{L}_z)\hat{L}_z-\hat{L}_x\hat{L}_z^2\\ &= \mathrm{i}\hbar(\hat{L}_z\hat{L}_y+\hat{L}_y\hat{L}_z)\ .\end{aligned}\tag{4.71}$$

The sum of both terms, (4.70) and (4.71), is zero. Hence, $[\hat{L}^2,\hat{L}_x]=0$. Similarly one proves the second and third relation (4.68).

It is convenient to write the angular momentum in spherical coordinates. By the transformation

$$x=r\sin\vartheta\cos\varphi\ ,\qquad y=r\sin\vartheta\sin\varphi\ ,\qquad z=r\cos\vartheta\ ,\tag{4.72}$$

we obtain for the Cartesian coordinates of the angular-momentum operator:

$$\hat{L}_x = \mathrm{i}\hbar\left(\sin\varphi\frac{\partial}{\partial\vartheta} + \cot\vartheta\cos\varphi\frac{\partial}{\partial\varphi}\right) ,$$
$$\hat{L}_y = \mathrm{i}\hbar\left(-\cos\varphi\frac{\partial}{\partial\vartheta} + \cot\vartheta\sin\varphi\frac{\partial}{\partial\varphi}\right) ,$$
$$\hat{L}_z = -\mathrm{i}\hbar\frac{\partial}{\partial\varphi} . \tag{4.73}$$

The equations

$$r^2 = x^2 + y^2 + z^2 , \quad \cos\vartheta = \frac{z}{r} , \quad \tan\varphi = \frac{y}{x}$$

hold, and hence,

$$\begin{aligned}\hat{L}_z = &-\mathrm{i}\hbar\left(x\frac{\partial}{\partial y} - y\frac{\partial}{\partial x}\right) = -\mathrm{i}\hbar\left\{r\sin\vartheta\cos\varphi\left[\frac{\partial r}{\partial y}\frac{\partial}{\partial r} + \frac{\partial\vartheta}{\partial y}\frac{\partial}{\partial\vartheta} + \frac{\partial\varphi}{\partial y}\frac{\partial}{\partial\varphi}\right]\right.\\ &\left.-r\sin\vartheta\sin\varphi\left[\frac{\partial r}{\partial x}\frac{\partial}{\partial r} + \frac{\partial\vartheta}{\partial x}\frac{\partial}{\partial\vartheta} + \frac{\partial\varphi}{\partial x}\frac{\partial}{\partial\varphi}\right]\right\}\\ = &-\mathrm{i}\hbar r\sin\vartheta\left\{\cos\varphi\left[\sin\vartheta\sin\varphi\frac{\partial}{\partial r} + \frac{\cos\vartheta\sin\varphi}{r}\frac{\partial}{\partial\vartheta} + \frac{\cos\varphi}{r\sin\vartheta}\frac{\partial}{\partial\varphi}\right]\right.\\ &\left.-\sin\varphi\left[\sin\vartheta\cos\varphi\frac{\partial}{\partial r} + \frac{\cos\vartheta\cos\varphi}{r}\frac{\partial}{\partial\vartheta} - \frac{\sin\varphi}{r\sin\vartheta}\frac{\partial}{\partial\varphi}\right]\right\}\\ = &-\mathrm{i}\hbar\frac{\partial}{\partial\varphi}\end{aligned} \tag{4.74}$$

and

$$\begin{aligned}\hat{L}^2 &= \hat{L}_x^2 + \hat{L}_y^2 + \hat{L}_z^2\\ &= -\hbar^2\left\{(\sin^2\varphi + \cos^2\varphi)\frac{\partial^2}{\partial\vartheta^2} + \cot\vartheta\frac{\partial}{\partial\vartheta} + \cot^2\vartheta\frac{\partial^2}{\partial\varphi^2} + \frac{\partial^2}{\partial\varphi^2}\right\}\\ &= -\hbar^2\left\{\frac{1}{\sin\vartheta}\frac{\partial}{\partial\vartheta}\left(\sin\vartheta\frac{\partial}{\partial\vartheta}\right) + \frac{1}{\sin^2\vartheta}\frac{\partial^2}{\partial\varphi^2}\right\} = -\hbar^2\Delta_{\vartheta,\varphi} ,\end{aligned} \tag{4.75}$$

where we denote by $\Delta_{\vartheta,\varphi}$ that part of the Laplacian acting on the variables ϑ and φ only. In this context we also write down the eigenfunctions of $\hat{L}^2$:

$$\hat{L}^2 Y_{lm}(\vartheta,\varphi) = L^2 Y_{lm}(\vartheta,\varphi) . \tag{4.76a}$$

These are the *spherical harmonics*, as will be proved in two different ways in Examples 4.8 (p. 88) and 4.9 (p. 96). We are familiar with spherical harmonics from electrodynamics; they are related to the Legendre polynomials by

$$Y_{lm}(\vartheta,\varphi) = \sqrt{\frac{(l-m)!(2l+1)}{4\pi(l+m)!}}\,P_l^m(\cos\vartheta)\,\mathrm{e}^{\mathrm{i}m\varphi} . \tag{4.77}$$

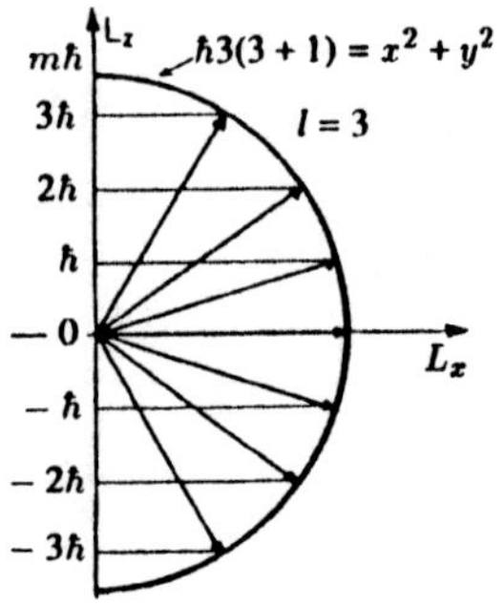

Fig. 4.3. The quantum numbers $m\hbar$ characterize the quantization of the z component of angular momentum. One sometimes speaks in this context of "quantization of direction"

The Legendre polynomials are

$$P_l^m(x) = \frac{(-1)^m}{2^l l!}(1-x^2)^{m/2}\frac{\mathrm{d}^{l+m}}{\mathrm{d}x^{l+m}}(x^2-1)^l \ , \tag{4.78}$$

with $l \geq m \geq -l$.

In the eigenvalue equation (4.76a), the quantity L^2 can be expressed in terms of l by

$$L^2 = \hbar^2 l(l+1) \ , \quad l = 0,\ 1,\ 2,\ 3,\ \dots\ , \tag{4.79}$$

so that (4.76a) becomes

$$\hat{L}^2 Y_{lm}(\vartheta, \varphi) = \hbar^2 l(l+1) Y_{lm}(\vartheta, \varphi) \ . \tag{4.76b}$$

By our choice of coordinate system, the z component of the angular momentum was given preference, as the Y_{lm} are also eigenfunctions of $\hat{L}_z$:

$$\hat{L}_z Y_{lm} = \hbar m Y_{lm} \ , \quad m = -l, -l+1, \dots, 0, \dots, l \ . \tag{4.80}$$

This can immediately be confirmed by (4.73), (4.77), and (4.78). Obviously *the spectrum of* $\hat{L}^2$ and $\hat{L}_z$ is always *discrete*. Because $\hat{L}^2$ and $\hat{L}_z$ commute (4.68), they can be measured simultaneously. The simultaneous eigenfunctions are the $Y_{lm}(\vartheta, \varphi)$. Every eigenvalue $\hbar^2 l(l+1)$ of $\hat{L}^2$ is $(2l+1)$-fold degenerate because to every l there are $2l+1$ eigenfunctions Y_{lm} $(l \geq m \geq -l)$.

Indeed we can infer from (4.77) and (4.80) the z projection of an angular momentum $\boldsymbol{L}$ with absolute value $\hbar\sqrt{l(l+1)}$. It takes $2l+1$ different values $m\hbar$. This is illustrated in Fig. 4.3. The angle between the angular momentum and the direction of quantization (e.g. defined by a weak magnetic field) can have only certain values:

$$\cos\vartheta = \frac{m}{\sqrt{l(l+1)}} \ . \tag{4.81}$$

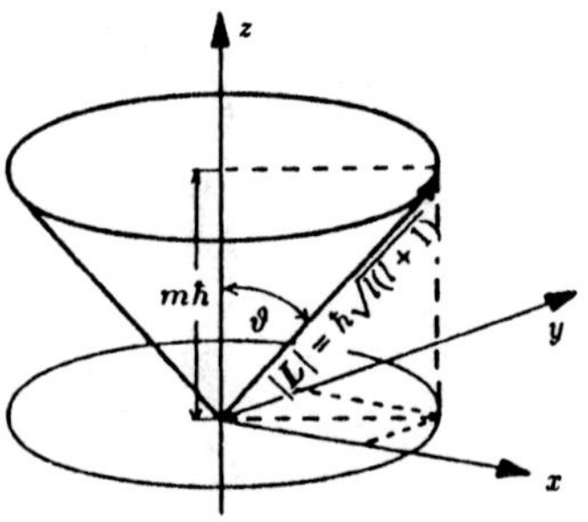

Fig. 4.4. A "sharp" angular momentum L, its z component and the components L_x and L_y, which are not sharp. The vector L precesses in an eigenstate of L^2 and L_z around the z axis on a cone

This is sometimes called *quantization of direction* and means nothing more than the quantization of the z component of angular momentum, i.e. $\hat{L}_z$. The thus obtained results can be interpreted in a pictorial way (see Fig. 4.4); the angular-momentum vector $\boldsymbol{L}$ precesses on a cone around the direction of quantization (z axis). As a result, the x and y components of angular momentum are not constant in time. This illustrates the uncertainty relations between L_z and L_x and between L_z and L_y [see (4.65)].

4.9 Kinetic Energy

We obtain the operator of kinetic energy in Cartesian coordinates by analogy from $T = p^2/2m$ as

$$\hat{T} = \frac{\hat{p}^2}{2m} = -\frac{\hbar^2}{2m}\left(\frac{\partial^2}{\partial x^2} + \frac{\partial^2}{\partial y^2} + \frac{\partial^2}{\partial z^2}\right) = -\frac{\hbar^2}{2m}\Delta \ . \tag{4.82}$$

It reads in polar coordinates:

$$\begin{aligned}\hat{T} &= -\frac{\hbar^2}{2m}\left[\frac{1}{r^2}\frac{\partial}{\partial r}\left(r^2\frac{\partial}{\partial r}\right) + \frac{1}{r^2}\Delta_{\vartheta,\varphi}\right] = -\frac{\hbar^2}{2m}\frac{1}{r^2}\frac{\partial}{\partial r}\left(r^2\frac{\partial}{\partial r}\right) - \frac{\hbar^2}{2mr^2}\Delta_{\vartheta,\varphi} \\ &= \hat{T}_{\mathrm{r}} + \frac{\hat{L}^2}{2mr^2} \ . \end{aligned} \tag{4.82a}$$

Here $\hat{T}_{\mathrm{r}}$ can be interpreted as the operator of kinetic energy for a motion along the radial direction and $\hat{L}^2/2mr^2$ as the operator of kinetic energy for the rotational motion. From the above relation it follows immediately that $[\hat{T}, \hat{L}^2]_- = 0$. Therefore the kinetic energy and the square of angular momentum can be measured simultaneously.

4.10 Total Energy

Corresponding to the Hamiltonian of classical mechanics, we define the *Hamiltonian operator* as the operator of total energy:

$$\hat{H} = \hat{T} + \hat{V} \equiv \frac{\hat{p}^2}{2m} + V(\boldsymbol{r}) \ . \tag{4.83}$$

If we take for granted that the potential energy is only a function of distance, i.e. $\hat{V} = V(r)$ (central potential), we have $[\hat{H}, \hat{L}^2] = 0$; the square of angular momentum and the total energy can be measured simultaneously. Equally, $[\hat{H}, \hat{L}_z] = 0$.

Since $\hat{T} = \hat{p}^2/2m$ and $\hat{V} = V(r)$ do not commute, no statement is possible concerning the exact values of potential and kinetic energy, even if we know the total energy. Solely for the mean values of these quantities do we have the so-called *virial theorem* $\langle T\rangle = +\langle \boldsymbol{r}\cdot\nabla V\rangle$, the proof of which we defer for the moment (see Example 8.2).

EXERCISE

4.5 Proof of an Operator Inequality

Problem. Let $\hat{A}$ and $\hat{B}$ be Hermitian operators and $\hat{C} = -\mathrm{i}[\hat{A}, \hat{B}]_- = -\mathrm{i}(\hat{A}\hat{B} - \hat{B}\hat{A})$, $\hat{D} = \{\hat{A}, \hat{B}\} = \hat{A}\hat{B} + \hat{B}\hat{A}$. Prove the following relation for the expectation values:

$$\bar{\hat{A}^2}\bar{\hat{B}^2} \geq \frac{1}{4}[(\bar{\hat{C}})^2 + (\bar{\hat{D}})^2] .$$

Solution. Let $\varphi(x, t)$ be an arbitrary state; $\lambda \in \mathbb{C}$, $\lambda = \alpha + \mathrm{i}\beta$, a complex number. We define

$$\begin{aligned}
0 \leq I(\lambda) &= \int |(\hat{A} + \mathrm{i}\lambda\hat{B})\varphi|^2 \mathrm{d}x \\
&= \int \varphi^*(\hat{A} - \mathrm{i}\lambda^*\hat{B})(\hat{A} + \mathrm{i}\lambda\hat{B})\varphi \mathrm{d}x \\
&= \int \varphi^*\hat{A}^2\varphi \mathrm{d}x + |\lambda|^2 \int \varphi^*\hat{B}^2\varphi \mathrm{d}x + \int \varphi^*(\hat{A}\hat{B}\mathrm{i}\lambda - \hat{B}\hat{A}\mathrm{i}\lambda^*)\varphi \mathrm{d}x \\
&= \bar{\hat{A}^2} + \bar{\hat{B}^2}|\lambda|^2 - \alpha\bar{\hat{C}} - \beta\bar{\hat{D}} .
\end{aligned}$$

With

$$\begin{aligned}
&\bar{\hat{B}^2}[\alpha - \bar{\hat{C}}/2\bar{\hat{B}^2}]^2 = \alpha^2\bar{\hat{B}^2} - \alpha\bar{\hat{C}} + \bar{\hat{C}}^2/4\bar{\hat{B}^2} \quad \text{and} \\
&\bar{\hat{B}^2}[\beta - \bar{\hat{D}}/2\bar{\hat{B}^2}]^2 = \beta^2\bar{\hat{B}^2} - \beta\bar{\hat{D}} + \bar{\hat{D}}^2/4\bar{\hat{B}^2} ,
\end{aligned}$$

we now have

$$\bar{\hat{A}^2} + \bar{\hat{B}^2}[\alpha - \bar{\hat{C}}/2\bar{\hat{B}^2}]^2 + \bar{\hat{B}^2}[\beta - \bar{\hat{D}}^2/2\bar{\hat{B}^2}]^2 - \bar{\hat{C}}^2/4\bar{\hat{B}^2} - \bar{\hat{D}}^2/4\bar{\hat{B}^2} \geq 0 .$$

But α, β can be chosen arbitrarily, i.e.

$$\bar{\hat{A}^2}\bar{\hat{B}^2} \geq \frac{1}{4}(\bar{\hat{C}}^2 + \bar{\hat{D}}^2)$$

must hold, which was to be demonstrated.

EXERCISE

4.6 The Difference Between Uncertainty Relations

Problem. Discuss the "uncertainty relation"

$$\Delta E \Delta t \sim \hbar .$$

What is the fundamental difference compared with *Exercise 4.6*

$$\Delta x \Delta p \sim \hbar \ ?$$

Solution. A (free) wave packet with a width Δx in configuration space has a distribution around a certain momentum p_0 with a width Δp in momentum space, where $\Delta x \Delta p \sim \hbar$ holds. Its group velocity is $v = \partial E/\partial p|_{p=p_0}$. The time at which the particle passes a point x_0, is uncertain by $\Delta t \approx \Delta x/v$. On the other hand, the particle has an uncertainty in energy,

$$\Delta E = \left. \frac{\partial E}{\partial p} \right|_{p=p_0} \Delta p = v \Delta p \ ,$$

because $E = E(p)$. Therefore we obtain

$$\hbar \approx \Delta x \Delta p \approx \Delta t v \Delta E / v = \Delta t \Delta E \ .$$

This is the origin of the "energy–time uncertainty relation". Therefore, if we want to measure the energy of a state with satisfactory accuracy, a sufficiently long time is needed. If this is not possible (for example, because of the finite "lifetime" of a state), the energy of the state remains uncertain. From this point of view, the uncertainty relations $\Delta p \Delta x \sim \hbar$ and $\Delta E \Delta t \sim \hbar$ are equivalent. But the physical interpretation is entirely different. A measuring apparatus can measure a given observable of a physical system at different times (e.g. position, momentum or energy). Then the time is given by the hand of a macroscopic clock, which is connected with the apparatus. Hence this time is *not an observable* of the quantum-mechanical system itself, but a *parameter*, which is described by a real number t.

EXERCISE

4.7 Expansion of an Operator

Let f be a function $f(z) : \mathbb{C} \to \mathbb{C}$, which can be expanded into a Taylor series $f(z) = \sum_{n=0}^{\infty} a_n z^n$. Then the operator $\hat{f}(\hat{A})$ can be defined by

$$\hat{f}(\hat{A}) = \sum_{n=0}^{\infty} a_n \hat{A}^n$$

for an "appropriate" operator $\hat{A}$.

Problem. (a) Why is this definition incomplete? What is actually the meaning of

$$\lim_{n \to \infty} \hat{S}_n = \hat{S} \ ?$$

Exercise 4.7

(b) Prove that

$$\hat{T}(\boldsymbol{a}) = \exp(\mathrm{i}\hat{\boldsymbol{p}}\cdot\boldsymbol{a}/\hbar) \quad \text{with} \quad \hat{\boldsymbol{p}} = -\mathrm{i}\hbar\nabla$$

is the translation operator; i.e. we have for suitable functions $\psi(\boldsymbol{x})$

$$\hat{T}(\boldsymbol{a})\varphi(\boldsymbol{x}) = \varphi(\boldsymbol{x}+\boldsymbol{a}) \ .$$

Solution. (a) The definition is incomplete, since we have not explained under what circumstances a sequence $\hat{S}_n (= \sum_{\nu=0}^{n} a_\nu \hat{A}^\nu)$ of operators, which may be defined on the full Hilbert space H, converges to an operator $\hat{S}$. Unfortunately there are several nonequivalent convergence notions concerning operators. On the one hand we can say that $\hat{S}_n \to \hat{S}$ means that for arbitrary vectors, $\varphi_\nu(\boldsymbol{x}) \in H$ the functions (vectors) relation $\hat{S}_n\varphi_\nu(\boldsymbol{x}) \to \hat{S}\varphi_\nu(\boldsymbol{x})$ holds in this Hilbert space. On the other hand, we can assign a norm

$$O = \sup_{\varphi_\nu(\boldsymbol{x})\in H} \frac{\|\hat{O}\varphi_\nu(\boldsymbol{x})\|}{\|\varphi_\nu(\boldsymbol{x})\|}$$

to an operator $\hat{O}$ and define that $\hat{S}_n \to \hat{S}$, if $\|\hat{S}_n - \hat{S}\| \to O$ holds in $\mathbb{R}$. Here we cannot go further into these problems (or others, like: What is a Hilbert space? What actually is the momentum operator?). To *us*, an exhaustive study of functional analysis seems to be indispensable for a mathematical comprehension of quantum mechanics.

(b) As indicated in (a), we will not concern ourselves with mathematical "subtleties".

Let $\psi(\boldsymbol{x})$ be expandable in its Taylor series. Then we have:

$$\begin{aligned}
\hat{T}(\boldsymbol{a})\psi(\boldsymbol{x}) &= \exp\left(\mathrm{i}\frac{\boldsymbol{p}\cdot\boldsymbol{a}}{\hbar}\right)\psi(\boldsymbol{x}) \\
&= \sum_{n=0}^{\infty} \frac{1}{n!}\left[\mathrm{i}\frac{(-\mathrm{i}\hbar\nabla)\cdot\boldsymbol{a}}{\hbar}\right]^n \psi(\boldsymbol{x}) \\
&= \sum_{n=0}^{\infty} \frac{(\boldsymbol{a}\cdot\nabla)^n}{n!}\psi(\boldsymbol{x}) = \psi(\boldsymbol{x}+\boldsymbol{a}) \ ,
\end{aligned}$$

as the penultimate expression is simply the shorthand notation for the Taylor expansion of the function $\psi(\boldsymbol{x}+\boldsymbol{a})$ at the point $\boldsymbol{x}$.

EXAMPLE

4.8 Legendre Polynomials

The so-called "special functions" of mathematical physics are solutions of specified linear differential equations of second order, which frequently recur. We

Example 4.8

will discuss a small selection of special functions in this and the following examples, namely the ***Legendre*** functions, the associated Legendre functions and the spherical harmonics. There are several possible ways to represent these functions:

(1) as special solutions of specified differential equations (Laplace equation, force-free Schrödinger equation);
(2) by means of recurrence formulae; or
(3) by means of a generating function (i.e. an $|\boldsymbol{r}-\boldsymbol{r}'|^{-1}$ expansion),

to mention only a few. A valuable aid to studying the Legendre functions (polynomials) is the generating function; therefore we place it at the beginning of our considerations.

The Legendre Polynomials and Their Generating Function

We have frequently met the term $|\boldsymbol{r}-\boldsymbol{r}'|^{-1}$ in solving potential problems:

$$|\boldsymbol{r}-\boldsymbol{r}'|^{-1} = \frac{1}{\sqrt{|\boldsymbol{r}|^2+|\boldsymbol{r}'|^2-2|\boldsymbol{r}||\boldsymbol{r}'|\cos\vartheta}} \tag{1}$$

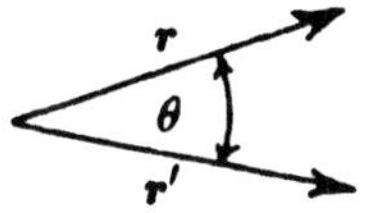

Now we want to expand the root into a power series of the ratio r to r'. To that end, we designate $r_<$ as the smaller, and $r_>$ as the greater, of both values of r and r'. Then certainly $r_</r_> < 1$ holds and we obtain

$$\begin{aligned}\frac{1}{\sqrt{r^2+r'^2-2rr'\cos\vartheta}} &= \frac{1}{\sqrt{r_>^2\left\{1-2\frac{r_<}{r_>}\cos\vartheta+\left(\frac{r_<}{r_>}\right)^2\right\}}} \\ &= \frac{1}{r_>}\left\{1+\frac{r_<}{r_>}\cos\vartheta+\frac{1}{2}(3\cos^2\vartheta-1)\left(\frac{r_<}{r_>}\right)^2\pm\ldots\right\} .\end{aligned} \tag{2}$$

The ϑ-dependent coefficients appearing here define the Legendre polynomials:

$$\begin{aligned}\frac{1}{r_>\sqrt{1-2\frac{r_<}{r_>}\cos\vartheta+\left(\frac{r_<}{r_>}\right)^2}} &= \frac{1}{r_>}\sum_{l=0}^{\infty}\left(\frac{r_<}{r_>}\right)^l P_l(\cos\vartheta) \\ &= \sum_{l=0}^{\infty}\frac{r_<^l}{r_>^{l+1}}P_l(\cos\vartheta) .\end{aligned} \tag{3}$$

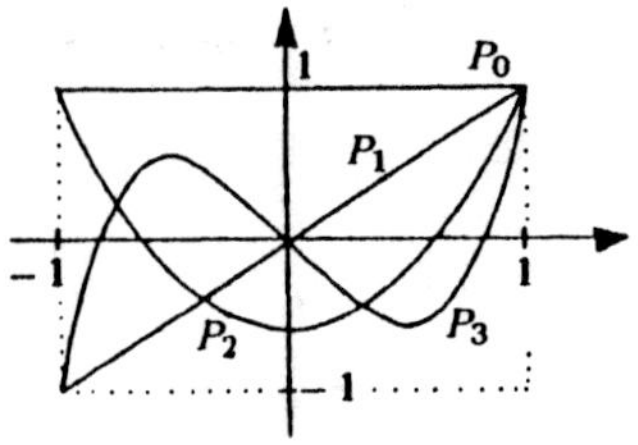

Legendre polynomials of lowest order

If we write $\cos\vartheta = x$, we find for the $P_l(x)$ (see figure):

$$\begin{aligned} P_0(x) &= 1 \ , \\ P_1(x) &= x \ , \\ P_2(x) &= \frac{1}{2}(3x^2 - 1) \ , \\ P_3(x) &= \frac{1}{2}(5x^3 - 3x) \ , \\ P_4(x) &= \frac{1}{8}(35x^4 - 30x^2 + 3) \ , \end{aligned} \tag{4}$$

or generally according to *Rodriguez's formula*, which we will prove in the next section,

$$P_l(x) = \frac{1}{2^l l!}\frac{\mathrm{d}^l}{\mathrm{d}x^l}(x^2 - 1)^l \ . \tag{5}$$

Mathematical Properties of Legendre Polynomials

We recognize that the Legendre polynomials can be introduced as expansion coefficients of a power series:[2]

$$(1 - 2xt + t^2)^{-1/2} = \sum_{l=0}^{\infty} P_l(x)t^l \ , \tag{6}$$

with $x = \cos\vartheta$ and $|t| < 1$.

We call the function $(1 - 2xt + t^2)^{-1/2}$ the *generating function* of the Legendre polynomials. By means of the generating function, we now calculate a *recurrence formula* for the Legendre polynomials. To that end we define

$$F(t, x) = (1 - 2xt + t^2)^{-1/2} = \sum_{l=0}^{\infty} P_l(x)t^l \ . \tag{7}$$

The first derivative with respect to t yields

$$\begin{aligned} \frac{\partial F}{\partial t} &= \frac{x - t}{1 - 2xt + t^2}F \ , \\ (1 - 2xt + t^2)\sum_{l=0}^{\infty} lt^{l-1}P_l(x) &= (x - t)\sum_{l=0}^{\infty} t^l P_l(x) \ . \end{aligned} \tag{8}$$

Comparing now the same power of t on both sides of the equation, we easily obtain

$$(l + 1)P_{l+1} - (2l + 1)xP_l + lP_{l-1} = 0 \ . \tag{9}$$

[2] The series converges for $|t| < 1$ and $\vartheta \in [0, \pi]$ and can be differentiated term by term arbitrarily often with respect to r and ϑ, whereby the series so obtained converges uniformly with respect to (t, ϑ) in $[-t_0, t_0] \times [0, \pi]$ for arbitrary $|t_0| < 1$.

Example 4.8

A second *recurrence formula* may be obtained from $F(t,x)$ by differentiating with respect to x,

$$(1-2xt+t^2)\frac{\partial F}{\partial x}=tF\ , \tag{10}$$

and by an analogous procedure,

$$P_l'(x)-2xP_{l-1}'(x)+P_{l-2}'(x)=P_{l-1}(x)\ , \tag{11}$$

with $'=\partial/\partial x$. From these recurrence formulae we easily find the relations

$$\begin{aligned} &P_{l+1}'-xP_l'=(l+1)P_l\ ,\\ &xP_l'-P_{l-1}'=lP_l\ ,\\ &P_{l+1}'-P_{l-1}'=(2l+1)P_l\ ,\\ &(x^2-1)P_l'=lxP_l-lP_{l-1}\ . \end{aligned} \tag{12}$$

If we now inspect the generating function for $x=1$, we find that

$$F(t,1)=\frac{1}{1-t}=1+t+t^2+t^3+\ldots=\sum_{l=0}^{\infty}t^lP_l(1) \tag{13}$$

and therefore

$$P_l(1)=1\ . \tag{14}$$

Analogously, we find for $x=0$ that

$$F(t,0)=\frac{1}{\sqrt{1+t^2}}=1-\frac{1}{2}t^2\pm\ldots\equiv\sum_{l=0}^{\infty}t^lP_l(0)\ , \tag{15}$$

and therefore

$$P_l(0)=\begin{cases}0 & \text{for}\quad l\ \text{odd}\\ \dfrac{(l-1)!!(-1)^{l/2}}{2^{l/2}(l/2)!} & \text{for}\quad l\ \text{even}\end{cases}\ . \tag{16}$$

The double factorial, characterized by two exclamation marks, is the product over the odd numbers, e.g. $7!!=1\times3\times5\times7$.

Next we derive the so-called *Rodriguez recurrence formula*,

$$P_l(x)=\frac{1}{2^l l!}\frac{\mathrm{d}^l}{\mathrm{d}x^l}(x^2-1)^l\ . \tag{17}$$

To do so, we use (6) and find the representation

$$P_l(x)=\frac{1}{l!}\frac{\mathrm{d}^l}{\mathrm{d}t^l}(1-2xt+t^2)^{-1/2}\bigg|_{t=0}\ . \tag{18}$$

Example 4.8

Now we expand the generating function $(1-2xt+t^2)^{-1/2}$ of P_l into powers of t:

$$(1-2xt+t^2)^{-1/2}=\sum_n \binom{-1/2}{n}(-2xt)^n(1+t^2)^{-(1/2)-n}$$
$$=\sum_{n,m}\binom{-1/2}{n}\binom{-(1/2)-n}{m}(-2x)^n t^{n+2m} \; . \qquad (19)$$

Therefrom we obtain

$$\frac{\mathrm{d}^l}{\mathrm{d}t^l}(1-2xt+t^2)^{-1/2}=\sum_{n,m}\binom{-1/2}{n}\binom{-(1/2)-n}{m}$$
$$\times\frac{(n+2m)!}{(n-l+2m)!}(-2x)^n t^{n-l+2m} \; .$$

Here the sum contains only those terms for which $n+2m\geq l$ holds. For $t=0$, only a contribution from the terms of the sum with $m=(l-n)/2$ remains. Thus we obtain

$$P_l(x)=\sum_n \binom{-1/2}{n}\binom{-n-(1/2)}{(l/2)-(n/2)}(-2x)^n$$
$$=\sum_n(-1)^{(l-n)/2}\frac{(l+n)!}{2^l\left(\frac{l+n}{2}\right)!\left(\frac{l-n}{2}\right)!n!}x^n \; . \qquad (20)$$

If $n=2m-l$ is inserted, we get

$$P_l(x)=\sum_m\frac{(-1)^{l-m}}{2^l m!(l-m)!}\frac{2m!}{(2m-l)!}x^{2m-l}$$
$$=\frac{1}{2^l l!}\frac{\mathrm{d}^l}{\mathrm{d}x^l}\sum_m\binom{l}{m}(-1)^{l-m}x^{2m} \; . \qquad (21)$$

By means of the binomial theorem, we now can easily see that

$$P_l(x)=\frac{1}{2^l l!}\frac{\mathrm{d}^l}{\mathrm{d}x^l}(x^2-1)^l$$

holds, which was to be demonstrated.

From Rodriguez's formula (17) immediately follows a *symmetry of the Legendre polynomials*:

$$P_l(-x)=(-1)^l P_l(x) \; . \qquad (22)$$

A further important property of Legendre polynomials is their *orthogonality*. Thus we consider

$$I_{mn}=\int_{-1}^{+1}P_m(x)P_n(x)\,\mathrm{d}x \quad (m<n)$$
$$=\frac{1}{2^{m+n}}\frac{1}{m!n!}\int_{-1}^{+1}\left[\frac{\mathrm{d}^m}{\mathrm{d}x^m}(x^2-1)^m\right]\left[\frac{\mathrm{d}^n}{\mathrm{d}x^n}(x^2-1)^n\right]\mathrm{d}x \; . \qquad (23)$$

Partial integration yields

Example 4.8

$$I_{mn} = \frac{(-1)^n}{2^{m+n}m!n!}\int_{-1}^{+1}\left[\frac{\mathrm{d}^{m+n}}{\mathrm{d}x^{m+n}}(x^2-1)^m\right](x^2-1)^n\,\mathrm{d}x\ . \tag{24}$$

From this we obtain $I_{mn}=0$ for $m<n$, since

$$\frac{\mathrm{d}^{m+n}}{\mathrm{d}x^{m+n}}(x^2-1)^m = 0\ . \tag{25}$$

For the case $m=n$ we get

$$\begin{aligned} I_{nn} &= \frac{(-1)^n}{2^{2n}(n!)^2}\int_{-1}^{+1}(x^2-1)^n\frac{\mathrm{d}^{2n}}{\mathrm{d}x^{2n}}(x^2-1)^n\,\mathrm{d}x \\ &= \frac{(-1)^n}{2^{2n}(n!)^2}\int_{-1}^{+1}\Big[(x^2-1)^n 2n(2n-1)(2n-2)\dots[2n-(2n-1)]\Big]\,\mathrm{d}x \\ &= \frac{(-1)^n(2n)!}{2^{2n}(n!)^2}\int_{-1}^{+1}(x^2-1)^n\,\mathrm{d}x\ . \end{aligned} \tag{26}$$

By means of the variable transformation $x=2u-1$, we find

$$I_{nn} = \frac{(-1)^n 2(2n)!}{(n!)^2}\int_0^1 u^n(u-1)^n\,\mathrm{d}u = \frac{2}{2n+1}\ . \tag{27}$$

If we combine the two results, we obtain

$$\int_{-1}^{+1}P_m(x)P_n(x)\,\mathrm{d}x = \frac{2}{2n+1}\delta_{mn}\ . \tag{28}$$

These are the *orthogonality relations* for Legendre polynomials.

As a further point of interest, directly demonstrable by taking the orthogonality relation as a basis, every function $f(x)$, which is continuous and bounded in the interval $-1\le x\le 1$, can be expanded in a series of Legendre polynomials:

$$f(x) = \sum_{n=0}^{\infty} c_n P_n(x)\ . \tag{29}$$

Example 4.8

The function $f(x)$ can be expanded into $P_n(x)$ if the coefficients c_n of the expansion can be found uniquely. To that end we calculate

$$\int_{-1}^{+1} P_m(x)f(x)\,\mathrm{d}x = \sum_{n=0}^{\infty} c_n \int_{-1}^{+1} P_m(x)P_n(x)\,\mathrm{d}x$$
$$= \sum_{n=0}^{\infty} c_n \frac{2}{2n+1}\delta_{mn} = c_m \frac{2}{2m+1} , \tag{30a}$$

and thus

$$c_m = \frac{2m+1}{2}\int_{-1}^{+1} P_m(x)f(x)\,\mathrm{d}x$$
$$= \frac{(-1)^m}{2^{m+1}}\frac{2m+1}{m!}\int_{-1}^{+1}(x^2-1)^m \frac{\mathrm{d}^m}{\mathrm{d}x^m} f(x)\,\mathrm{d}x . \tag{30b}$$

In particular it can be immediately shown that

$$\delta(x-x') = \sum_{n=0}^{\infty}\frac{2n+1}{2}P_n(x')P_n(x) \tag{31}$$

holds. Hereby, the completeness of Legendre polynomials was the requirement for this procedure. In fact, according to Weierstrass's approximation theorem[3], every function $f(x)$, which is continuous on a compact interval, can be uniformly approximated by polynomials, i.e. the set of functions $\{1, x, x^2, x^3, \dots\}$ is complete on this interval. The Legendre polynomials are obtained by application of E. Schmidt's orthogonalization procedure, which of course in no way affects the completeness.

The Legendre polynomials are solutions of the so-called *Legendre differential equation*:

$$(1-x^2)\frac{\mathrm{d}^2}{\mathrm{d}x^2}P_n(x) - 2x\frac{\mathrm{d}}{\mathrm{d}x}P_n(x) + n(n+1)P_n(x) = 0 . \tag{32}$$

Now, with the aid of recurrence formulae (12) derived earlier, the above proposition can easily be proved:

$$(x^2-1)\frac{\mathrm{d}}{\mathrm{d}x}P_n(x) = nxP_n(x) - nP_{n-1}(x) ,$$
$$2x\frac{\mathrm{d}}{\mathrm{d}x}P_n(x) + (x^2-1)\frac{\mathrm{d}^2}{\mathrm{d}x^2}P_n(x) = nP_n(x) + nx\frac{\mathrm{d}}{\mathrm{d}x}P_n(x) - n\frac{\mathrm{d}}{\mathrm{d}x}P_{n-1}(x) . \tag{33}$$

[3] See E. Isaacson, H.B. Keller: *Analysis of Numerical Methods* (Wiley, New York 1966) Chap. 5.

Thus we obtain

Example 4.8

$$(1-x^2)\frac{\mathrm{d}^2}{\mathrm{d}x^2}P_n(x) = (2-n)x\frac{\mathrm{d}}{\mathrm{d}x}P_n(x) - nP_n(x) + n\frac{\mathrm{d}}{\mathrm{d}x}P_{n-1}(x) \ . \tag{34}$$

Insertion into the differential equation yields

$$\begin{aligned} 0 &= 2x\frac{\mathrm{d}}{\mathrm{d}x}P_n(x) - nx\frac{\mathrm{d}}{\mathrm{d}x}P_n(x) - nP_n(x) + n\frac{\mathrm{d}}{\mathrm{d}x}P_{n-1} \\ &\quad - 2x\frac{\mathrm{d}}{\mathrm{d}x}P_n + n^2P_n + nP_n \end{aligned} \tag{35}$$

and finally

$$x\frac{\mathrm{d}}{\mathrm{d}x}P_n(x) - \frac{\mathrm{d}}{\mathrm{d}x}P_{n-1}(x) = nP_n(x) \ , \tag{36}$$

i.e. one of the first recurrence formulae in (12), which proves the above proposition.

We do *not* obtain the Legendre differential equation from the Laplace equation in spherical coordinates after a separation of variables, but the *associated Legendre differential equation* (for the angle ϑ, $x = \cos\vartheta$):

$$(1-x^2)\frac{\mathrm{d}^2P(x)}{\mathrm{d}x^2} - 2x\frac{\mathrm{d}P(x)}{\mathrm{d}x} + \left[n(n+1) - \frac{m^2}{1-x^2}\right]P(x) = 0 \ . \tag{37}$$

For the case $m = 0$, this differential equation transforms into the Legendre differential equation. The general solution of (37) is the *associated Legendre polynomial* $P_n^m(x)$:

$$P_n^m(x) = (1-x^2)^{m/2}\frac{\mathrm{d}^m}{\mathrm{d}x^m}P_n(x) \tag{38}$$

$$= \frac{1}{2^n n!}(1-x^2)^{m/2}\frac{\mathrm{d}^{n+m}}{\mathrm{d}x^{n+m}}(x^2-1)^n \ . \tag{39}$$

The following orthogonality relation (n is replaced by l here) can easily be derived:

$$\int_{-1}^{+1} P_l^m(x)P_{l'}^{m'}(x)\,\mathrm{d}x = \frac{(l+m)!}{(l-m)!}\frac{2}{2l+1}\delta_{ll'}\delta_{mm'} \ . \tag{40}$$

Usually every function of an orthogonal system is furnished with a factor so as to yield the value one for the integral over the square of each function. We then say that these functions are *normalized*. A system of normalized orthogonal functions is termed *orthonormal*. We easily derive from (40) the normalization factor

$$\left\{\frac{(2l+1)(l-m)!}{2(l+m)!}\right\}^{1/2} \ . \tag{41}$$

Example 4.8

It will prove useful to define the associated Legendre polynomials for negative values of m, too. Since the differential equation (37) transforms into itself when m is replaced by $-m$,

$$P_l^{-m}(x) = \frac{1}{2^l l!}(1-x^2)^{-m/2}\frac{\mathrm{d}^{l-m}}{\mathrm{d}x^{l-m}}(x^2-1)^l$$

is also a solution of the general Legendre differential equation. This solution is a polynomial in x of order l and is continuous for $x = \pm 1$, too. Therefore the solutions P_l^m and P_l^{-m} can differ only by a factor for fixed l and m ($0 \leq m \leq 1$):

$$P_l^{-m}(x) = A P_l^m \ . \tag{42}$$

We now determine the constant A by setting $x = 1$ and dividing by $(1-x^2)^{m/2}$. With

$$\begin{aligned}(1-x^2)^{-m}\frac{\mathrm{d}^{l-m}}{\mathrm{d}x^{l-m}}(x^2-1)^l|_{x=1} &= (-1)^m 2^{l-m}\frac{l!}{m!} \\ &= A 2^{l-m}\frac{l!}{m!}\frac{(l+m)!}{(l-m)!} \\ &= A\frac{\mathrm{d}^{l+m}}{\mathrm{d}x^{l+m}}(x^2-1)^l|_{x=1}\end{aligned}$$

we find

$$A = (-1)^m\frac{(l-m)!}{(l+m)!} \ ,$$

and therefore

$$P_l^{-m}(x) = (-1)^m\frac{(l-m)!}{(l+m)!}P_l^m(x) \ . \tag{43}$$

In the next example we will introduce spherical harmonics. To that end, we make a short digression to the Laplace equation. This digression will help us to understand better the physical significance of the differential equation discussed before. Furthermore, we will see that in order to construct the spherical harmonics, in addition to the Legendre polynomials $P_n(x)$, the associated Legendre polynomials $P_n^m(x)$ are needed, too.

EXAMPLE

4.9 Mathematical Supplement: Spherical Harmonics

The Laplace Equation in Spherical Coordinates. A scalar potential U, outside a charge distribution, satisfies the Laplace equation[4]

$$\Delta U(x,y,z) = \left(\frac{\partial^2}{\partial x^2}+\frac{\partial^2}{\partial y^2}+\frac{\partial^2}{\partial z^2}\right)U(x,y,z) = 0 \ . \tag{1}$$

If spherical coordinates, (r, ϑ, φ) with

Example 4.9

$$x = r\sin\vartheta\cos\varphi\ , \quad y = r\sin\vartheta\sin\varphi,$$
$$z = r\cos\vartheta, \tag{2}$$

are introduced, we find:

$$\Delta U = \left\{\frac{\partial^2}{\partial r^2} + \frac{2}{r}\frac{\partial}{\partial r} + \frac{1}{r^2}\left(\frac{\partial^2}{\partial\vartheta^2} + \cot\vartheta\frac{\partial}{\partial\vartheta} + \frac{1}{\sin^2\vartheta}\frac{\partial^2}{\partial\varphi^2}\right)\right\}U = 0\ . \tag{3}$$

This can be further simplified to

$$\left\{\frac{1}{r}\frac{\partial^2}{\partial r^2}r + \frac{1}{r^2\sin\vartheta}\frac{\partial}{\partial\vartheta}\left(\sin\vartheta\frac{\partial}{\partial\vartheta}\right) + \frac{1}{r^2\sin^2\vartheta}\frac{\partial^2}{\partial\varphi^2}\right\}U = 0\ . \tag{4}$$

This differential operator divides into a radial part and an angular part $\hat{\boldsymbol{L}}^2$:

$$\left\{\frac{1}{r}\frac{\partial^2}{\partial r^2}r - \frac{\hat{\boldsymbol{L}}^2}{r^2}\right\}U(r,\vartheta,\varphi) = 0\ , \quad \text{with} \tag{5}$$

$$\hat{\boldsymbol{L}}^2 = -\left\{\frac{1}{\sin\vartheta}\frac{\partial}{\partial\vartheta}\sin\vartheta\frac{\partial}{\partial\vartheta} + \frac{1}{\sin^2\vartheta}\frac{\partial^2}{\partial\varphi^2}\right\}\ . \tag{6}$$

We now state that $\hat{\boldsymbol{L}}^2$ is proportional (up to the factor $\hbar$) to the square of the well-known angular-momentum operator [see (4.73)–(4.75)]

$$\hat{\boldsymbol{L}} = -\mathrm{i}(\boldsymbol{r}\times\nabla)\ . \tag{7}$$

We have been directed automatically towards this operator in the course of this chapter [(4.64) pp.]. Here we interpret it solely as a mathematical tool, which allows us to formulate some operations more concisely.

The operator $\hat{\boldsymbol{L}}$ acts only on the angles ϑ and φ. It has the components

$$\hat{L}_x = -\mathrm{i}\left(y\frac{\partial}{\partial z} - z\frac{\partial}{\partial y}\right) = -\mathrm{i}\left(\sin\varphi\frac{\partial}{\partial\vartheta} + \cos\varphi\cot\vartheta\frac{\partial}{\partial\varphi}\right)\ ,$$
$$\hat{L}_y = -\mathrm{i}\left(z\frac{\partial}{\partial x} - x\frac{\partial}{\partial z}\right) = -\mathrm{i}\left(\cos\varphi\frac{\partial}{\partial\vartheta} - \sin\varphi\cot\vartheta\frac{\partial}{\partial\varphi}\right)\ ,$$
$$\hat{L}_z = -\mathrm{i}\left(x\frac{\partial}{\partial y} - y\frac{\partial}{\partial x}\right) = -\mathrm{i}\frac{\partial}{\partial\varphi}\ . \tag{8}$$

[4] In the following we will introduce spherical coordinates as we did in (4.73)–(4.75) for the angular-momentum operator and draw attention to the fact that we already know the (ϑ, φ)-dependent part of the differential equation from (4.73)–(4.75) on. It will be considered again in the discussion of the "hydrogen problem" [cf. (9.11)]. See also chapters on potential theory in J.D. Jackson: *Classical Electrodynamics*, 2nd ed. (Wiley, New York 1975) and W. Greiner: *Classical Electrodynamics* (Springer, New York 1998).

Example 4.9

We can easily verify that

$$\hat{L}^2 = \hat{L}_x^2 + \hat{L}_y^2 + \hat{L}_z^2 \,, \tag{9}$$

holds [see (4.75)]. Earlier in this chapter we became acquainted with the angular-momentum operators $\hat{L}_i$. The operators $\hat{\boldsymbol{L}}_i$ differ by a factor $\hbar$ from those operators $\hat{L}_i$, i.e. $\hat{L}_i = \hbar\hat{\boldsymbol{L}}_i$.[5]

In order to solve the Laplace equation we use a *separation of variables procedure*:

$$U(r,\vartheta,\varphi) = R(r)\Theta(\vartheta,\varphi) \,. \tag{10}$$

Thereby we obtain

$$\frac{r(\partial^2/\partial r^2)(rR(r))}{R(r)} = \frac{\hat{\boldsymbol{L}}^2\Theta}{\Theta} = l(l+1) \,, \tag{11}$$

where we have chosen the factor $l(l+1)$ as separation constant without restriction of generality. We now have

$$r^2\frac{\partial^2}{\partial r^2}R(r) + 2r\frac{\partial}{\partial r}R(r) = l(l+1)R(r) \quad \text{and} \tag{12}$$

$$\hat{\boldsymbol{L}}^2\Theta(\vartheta,\varphi) = l(l+1)\Theta(\vartheta,\varphi) \,. \tag{13}$$

If now we choose

$$\Theta = P(\vartheta)E(\varphi) \,, \tag{14}$$

for the angular part, we find with the new separation constant $-m^2$:

$$\frac{1}{\sin\vartheta}\frac{(\mathrm{d}/\mathrm{d}\vartheta)[\sin\vartheta(\mathrm{d}/\mathrm{d}\vartheta)]P(\vartheta)}{P(\vartheta)} - \frac{m^2}{\sin^2\vartheta} = -l(l+1) \,,$$

$$\frac{(\partial^2/\partial\varphi^2)E(\varphi)}{E(\varphi)} = -m^2 \,. \tag{15}$$

A solution of the φ-dependent part is

$$E(\varphi) = c\,\mathrm{e}^{\mathrm{i}m\varphi} \,; \quad m \in \mathbb{G} \,, \tag{16}$$

where we required that $E(\varphi)$ be periodic with 2π, corresponding to the chosen symmetry. For the radial equation one obtains

$$R(r) = c_1 r^l + c_2 r^{-l-1} \,. \tag{17}$$

[5] The algebra of angular-momentum operators is investigated in detail in W. Greiner, B. Müller: *Quantum Mechanics – Symmetries*, 2nd ed. (Springer, Berlin, Heidelberg 1994).

Example 4.9

We find for the ϑ-dependent part with the abbreviations

$$\cos\vartheta \equiv x \quad \text{and} \quad \sin\vartheta = \sqrt{1-x^2}\ ,$$
$$\frac{\partial}{\partial\vartheta} \equiv {}' = -\sqrt{1-x^2}\frac{\partial}{\partial x}\ ,$$
$$\frac{\partial^2}{\partial\vartheta^2} \equiv {}'' = (1-x^2)\frac{\partial^2}{\partial x^2} - x\frac{\partial}{\partial x}\ , \tag{18}$$

the transformed differential equation (*associated Legendre differential equation*),

$$(1-x^2)P'' - 2xP' + \left\{l(l+1) - \frac{m^2}{1-x^2}\right\}P = 0\ , \tag{19}$$

with the associated Legendre polynomials $P_n^m(\cos\vartheta)$ as solutions (cf. Example 4.8). Thus we obtain

$$\Theta(\vartheta,\varphi) = C_l^m P_l^m(\cos\vartheta)\,\mathrm{e}^{\mathrm{i}m\varphi}$$
$$m = 0, \pm 1, \pm 2, \dots\ , \quad l \geq |m|\ , \quad l \in \mathbb{N}\ . \tag{20}$$

Concerning the fact that l is an integer, we refer the reader to Example 4.8. The functions thus characterized by two integers l and m are the *spherical harmonics*

$$\Theta(\vartheta,\varphi) = Y_{lm}(\vartheta,\varphi) = C_l^m P_l^m(\cos\vartheta)\,\mathrm{e}^{\mathrm{i}m\varphi}\ . \tag{21}$$

In general the constant C_l^m is fixed in such a way that the spherical harmonics are normalized. To that end we recall that

$$\int\limits_{-1}^{+1} \mathrm{d}x'\,P_l^m(x')P_{l'}^m(x') = \frac{2(l+m)!}{(2l+1)(l-m)!}\delta_{ll'} \tag{22}$$

holds [cf. (38–40) in the preceding example] and calculate easily

$$\int\limits_0^{2\pi} \mathrm{d}\varphi\,\mathrm{e}^{\mathrm{i}(m-m')\varphi} = 2\pi\delta_{mm'}\ . \tag{23}$$

Then we have

$$\int\limits_\Omega \mathrm{d}\Omega\,Y_{l'm'}^*(\vartheta,\varphi)Y_{lm}(\vartheta,\varphi) = |C_{lm}|^2\frac{4\pi}{2l+1}\frac{(l+m)!}{(l-m)!}\delta_{ll'}\delta_{m'm}\ . \tag{24}$$

Customarily, in the literature, the normalization constant C_{lm} is fixed so that

$$\int\limits_\Omega \mathrm{d}\Omega\,Y_{l'm'}^*(\vartheta,\varphi)Y_{lm}(\vartheta,\varphi) = \delta_{ll'}\delta_{mm'} \tag{25}$$

Example 4.9

holds and thus

$$C_{lm} = \sqrt{\frac{2l+1}{4\pi}\frac{(l-m)!}{(l+m)!}} \,, \tag{26}$$

$$Y_{lm}(\vartheta, \varphi) = \sqrt{\frac{2l+1}{4\pi}\frac{(l-m)!}{(l+m)!}} P_l^m(\cos\vartheta)\,\mathrm{e}^{\mathrm{i}m\varphi} \,. \tag{27}$$

We point out to the reader that for negative m,

$$P_l^{-m}(\cos\vartheta) = (-1)^m \frac{(l-m)!}{(l+m)!} P_l^m(\cos\vartheta) \tag{28}$$

holds. Thereupon we find a *symmetry of the spherical harmonics*:

$$Y_{l-m}(\vartheta, \varphi) = (-1)^m Y_{lm}^*(\vartheta, \varphi) \,. \tag{29}$$

One of the most important properties of spherical harmonics is that every bounded function $f(\vartheta, \varphi)$ defined on the surface of a sphere can be expanded in a series of the $Y_{lm}(\vartheta, \varphi)$:

$$f(\vartheta, \varphi) = \sum_{l=0}^{\infty} \sum_{m=-l}^{+l} d_l^m Y_{lm}(\vartheta, \varphi) \,. \tag{30}$$

Using the orthonormality of the $Y_{lm}(\vartheta, \varphi)$, we determine the expansion coefficients d_l^m:

$$d_l^m = \int_{\Omega} \mathrm{d}\Omega\, f(\vartheta, \varphi) Y_{lm}^*(\vartheta, \varphi) \,. \tag{31}$$

EXAMPLE

4.10 The Addition Theorem of Spherical Harmonics

In the following we shall prove that

$$P_l(\cos\gamma) = \frac{4\pi}{2l+1} \sum_{m=-l}^{l} Y_{lm}^*(\vartheta', \varphi') Y_{lm}(\vartheta, \varphi) \tag{1}$$

holds, where

$$\cos\gamma = \cos\vartheta \cos\vartheta' + \sin\vartheta \sin\vartheta' \cos(\varphi - \varphi') \,.$$

This relation is termed the *addition theorem of spherical harmonics*. In order to prove it, we expand a Legendre polynomial of order l for the angle γ in spherical harmonics. Let $\boldsymbol{x}'$ be fixed in space. Then $P_l(\cos\gamma)$ is only a function of ϑ, φ with ϑ', φ' as parameters. Therefore $P_l(\cos\gamma)$ can be expanded:

Example 4.10

$$P_l(\cos\gamma) = \sum_{l'=0}^{\infty} \sum_{m=-l'}^{l'} A_{l'm}(\vartheta', \varphi') Y_{l'm}(\vartheta, \varphi) \ . \tag{2}$$

Comparison with (1) shows that only terms with $l' = l$ seem to appear. In order to understand why this is so, we assume that $\boldsymbol{x}$ coincides with the z axis. Then $P_l(\cos\gamma)$ satisfies the equation [cf. Example 4.9, Eqs. (5), (13), (20)]

$$\nabla'^2 P_l(\cos\gamma) + \frac{l(l+1)}{r^2} P_l(\cos\gamma) = 0 \tag{3}$$

(differential equation for spherical harmonics, γ is the usual polar angle), which can be checked easily if ∇'^2 is written in spherical coordinates. If we now rotate the vector $\boldsymbol{x}$ to its old position, then ∇'^2 passes to ∇^2, and r remains unchanged ($\nabla \cdot \nabla$ is a scalar product and therefore rotationally invariant). Therefore P_l still satisfies (3); consequently, P_l is itself a spherical harmonic of order l. Thus P_l can only be represented as a linear combination of Y_{lm} of the same order l, and our separation of variables equation reduces to:

$$P_l(\cos\gamma) = \sum_{m=-l}^{l} A_{lm}(\vartheta', \varphi') Y_{lm}(\vartheta, \varphi), \tag{4}$$

with

$$A_{lm}(\vartheta', \varphi') = \int_{\Omega} Y^*_{lm}(\vartheta, \varphi) P_l(\cos\gamma)\, \mathrm{d}\Omega \ . \tag{5}$$

Now we want to determine the coefficients A_{lm}. To that end we examine the $Y^*_{lm}(\vartheta, \varphi)$. By expanding the $Y^*_{lm}(\vartheta, \varphi)$ into a linear combination of spherical harmonics with angle γ and β:

$$Y^*_{lm}(\vartheta, \varphi) = \sum_{m'=-l}^{l} c_{lm'} Y_{lm'}(\gamma, \beta) \ , \tag{6}$$

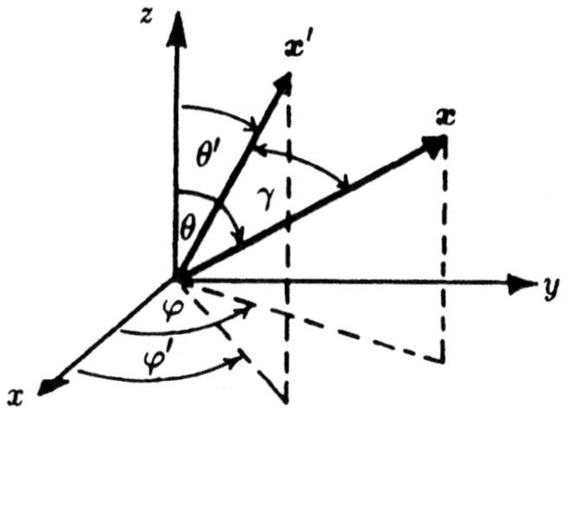

The definition of angles in the addition theorem

Example 4.10

with

$$c_{lm'} = \int\limits_{\Omega} Y^*_{lm}(\vartheta, \varphi) Y^*_{lm'}(\gamma, \beta)\, \mathrm{d}\Omega \ . \tag{7}$$

Here, γ and β are functions of $\vartheta, \vartheta', \varphi, \varphi'$. If now we choose $m' = 0$, we obtain

$$Y^*_{l0}(\gamma, \beta) = \left\{ \frac{2l+1}{4\pi} \right\}^{1/2} P_l(\cos\gamma) \ ; \tag{8}$$

then

$$c_{l0} = \left\{ \frac{2l+1}{4\pi} \right\}^{1/2} \int\limits_{\Omega} P_l(\cos\gamma) Y^*_{lm}(\vartheta, \varphi)\, \mathrm{d}\Omega \ . \tag{9}$$

Comparison with (5) yields

$$A_{lm}(\vartheta', \varphi') = \left\{ \frac{4\pi}{2l+1} \right\}^{1/2} c_{l0} \ . \tag{10}$$

We now seek an equation for the c_{l0} and therefore inspect (6) for $\gamma = 0$:

$$\begin{aligned} &Y^*_{lm}(\vartheta, \varphi) = \sum_{m'=-l}^{l} c_{lm'} Y_{lm'}(\gamma, \beta) \ , \quad \text{with} \\ &(\vartheta, \varphi) = [\vartheta(\gamma, \beta), \varphi(\gamma, \beta)] \ . \end{aligned} \tag{11}$$

We obtain for $\gamma = 0$:

$$Y^*_{lm}(\vartheta, \varphi) = \left\{ \frac{2l+1}{4\pi} \right\}^{1/2} c_{l0}\bigg|_{\gamma=0} \ , \quad \text{or} \tag{12}$$

$$c_{l0} = \left\{ \frac{4\pi}{2l+1} \right\}^{1/2} Y^*_{lm}(\vartheta[\gamma, \beta], \varphi[\gamma, \beta])\bigg|_{\gamma=0} \ . \tag{13}$$

If we insert this into (10),

$$A_{lm}(\vartheta', \varphi') = \frac{4\pi}{2l+1} Y^*_{lm}(\vartheta', \varphi') \tag{14}$$

follows, since ϑ and φ pass to ϑ' and φ' for $\gamma \to 0$. Thus proposition (1) is proved if we insert this result into our separation of variables equation (4). Often it turns out to be more advantageous to express (1) in terms of the P_l^m. If we bear in mind that

$$P_l^{-m} = (-1)^m \frac{(l-m)!}{(l+m)!} P_l^m \ , \quad \text{then} \tag{15}$$

Example 4.10

$$
\begin{aligned}
P_l(\cos\gamma) &= P_l(\cos\vartheta)P_l(\cos\vartheta') \\
&\quad + 2\sum_{m=1}^{l}\frac{(l-m)!}{(l+m)!}P_l^m(\cos\vartheta)P_l^m(\cos\vartheta')\cos[m(\varphi-\varphi')] \qquad (16)
\end{aligned}
$$

results. For $\gamma = 0$, we find a formula concerning the squares of Y_{lm}:

$$
\sum_{m=-l}^{l}[Y_{lm}(\vartheta,\varphi)]^2 = \frac{2l+1}{4\pi} \ . \qquad (17)
$$

Here we have used (14) from Example 4.8. The properties of the spherical harmonics derived here are extraordinarily important. We will encounter the spherical harmonics Y_{lm} time and again and come to appreciate them.

4.11 Biographical Notes

HERMITE, Charles, French mathematician, *Dieuze 24.12.1822, †Paris 14.1.1902. H. grew up in a comfortable, bourgeois family as the son of a textile merchant. At an early age he became so involved in research that he had difficulties passing his obligatory examinations. He studied for just one year at the Ecole Polytechnique and only with help of friends became qualified to teach in 1847. His scientific results, mostly on elliptic functions, modular functions, theory of numbers and invariant theory, were recognized only late. H. coordinated the ideas of Gaussian arithmetic, Abel's and Jacobi's elliptic functions and Cayley's and Sylvester's algebraic invariant theory and developed them further. Not until 1870 did he become a professor at the Sorbonne. In 1873 he proved the transcendency of e. He was in correspondence with many famous contemporaries and was determined to break down national barriers in the scientific struggle. He was a teacher and promoter of Stieltjes, Darboux, Borel, Poincaré and others.

SCHMIDT, Erhard, German mathematician, *Dorpat (Tartu), †Berlin 6.12.1959. S. studied in Berlin and Göttingen, graduated in 1905 and became a professor in Zürich in 1908, 1909 in Erlangen, 1911 in Breslau (Wrocław) and 1917 in Berlin. He worked primarily on integral equations and isoperimetric problems.

WEYL, Claus Hugo Hermann, German mathematician, *Elmshorn 9.11.1885, †Zürich 9.12.1955, became a professor at the ETH Zürich in 1913, 1930 in Göttingen; went to the Institute for Advanced Study in Princeton, USA in 1933. After working on the theory of differential and integral equations, W. connected topological considerations with the concept of Riemannian surfaces and made great progress in the theory of uniformalization. His fundamental publication, *Raum, Zeit, Materie* (1918, 1961, lectures), originated from meetings with A. Einstein. He developed an integral method for the representation of mathematical groups used in quantum mechanics, which contrasted with the infinitesimal method used by S. Lie and E. Cartan. W. supported intuitionism (a method for a constructive foundation of mathematics) and tried to closely combine mathematics, physics and philosophy in his own work. He was first in considering local gauge invariance as a general principle in theoretical physics.

LEGENDRE, Adrien Marie, French mathematician, *Paris 18.9.1752, †Paris 10.1.1833. L. greatly shared in the foundation and development of number theory and geodesy. He also made important contributions to elliptic integrals, to the foundations and methods of Euclidean geometry, to variation calculus and to theoretical astronomy; for example, he first applied the *method of least squares* and calculated extensive tables. L. concerned himself with many problems in which Gauss was interested, too, but never reached the latter's perfection. From 1775 on, L. worked as a professor at various Parisian universities and published outstanding, influential textbooks.

DIRAC, Paul Adrien Maurice, *Bristol 8.8.1902, †Bristol 1984. D. studied at Bristol, Cambridge and at several foreign universities. He was appointed professor of mathematics in 1932. D. is one of the founders of quantum mechanics. The mathematical equivalent created by him consists essentially of a noncommutative algebra for the calculation of the properties of the electron; he predicted the existence of the positron in 1928 and contributed fundamentally to quantum field theory. D. was awarded the Nobel prize in 1933.

5. Mathematical Supplement

5.1 Eigendifferentials and the Normalization of Eigenfunctions for Continuous Spectra

We begin our discussion with the eigenvalue equation

$$\hat{L}\psi(x, L) = L\psi(x, L) \ , \qquad (5.1)$$

which is supposed to have a continuous spectrum with the eigenvalues L and the eigenfunctions $\psi(x, L)$. Now we integrate (5.1) with respect to L over the small interval ΔL, and obtain

$$\hat{L}\Delta\psi(x, L) = \int\limits_{L}^{L+\Delta L} \tilde{L}\psi(x, \tilde{L})\, d\tilde{L} \ , \qquad (5.2)$$

where

$$\Delta\psi(x, L) = \int\limits_{L}^{L+\Delta L} \psi(x, \tilde{L})\, d\tilde{L} \qquad (5.3)$$

is called the *eigendifferential of the operator* $\hat{L}$, introduced (as mentioned earlier) by the great mathematician H. Weyl. The eigendifferential is a special wave group which has only a finite extension in space (in the x domain), similar to the previously studied wave groups; hence, it vanishes at infinity and therefore can be seen in analogy to bound states. Now we show that indeed the functions $\psi(x, L)$ are not orthogonal, but the eigendifferentials $\Delta\psi(x, L)$ are. Furthermore, because the $\Delta\psi(x, L)$ have finite spatial extension, they can be normalized. Then in the limit $\Delta L \to 0$, a meaningful normalization of the functions $\psi(x, L)$ themselves follows: the normalization on δ functions.

To begin with, we form the complex conjugate expressions of (5.1) and (5.2), i.e.

$$\hat{L}^*\psi^*(x, L') = L'\psi^*(x, L') \ , \qquad (5.4)$$

$$\hat{L}^*\Delta\psi^*(x, L') = \int\limits_{L'}^{L'+\Delta L'} \tilde{L}'\psi^*(x, \tilde{L}')\, d\tilde{L}' \ , \qquad (5.5)$$

where we have renamed the continuous eigenvalue in L'. Multiplication of (5.2) by $\Delta\psi^*(x, L')$, and of (5.5) by $\Delta\psi(x, L)$, and subsequent subtraction yields

$$\int \mathrm{d}x[\Delta\psi^*(x, L')\hat{L}\Delta\psi(x, L) - \Delta\psi(x, L)\hat{L}^*\Delta\psi^*(x, L')]$$

$$= \int \mathrm{d}x \int\limits_{L}^{L+\Delta L} \mathrm{d}\tilde{L} \int\limits_{L'}^{L'+\Delta L'} \mathrm{d}\tilde{L}'(\tilde{L} - \tilde{L}')\psi^*(x, \tilde{L}')\psi(x, \tilde{L}) \ .$$

Since $\hat{L}$ is Hermitian, the left-hand side vanishes. As the intervals ΔL and $\Delta L'$ should be small, we can place $(L - L')$ outside of the triple integral, by using the mean-value theorem from integral calculus, and obtain

$$(L - L')\int \mathrm{d}x\Delta\psi^*(x, L')\Delta\psi(x, L) = 0 \ . \tag{5.6}$$

In the case where the intervals ΔL and $\Delta L'$ do not overlap (see Fig. 5.1), $L \neq L'$ holds, and from (5.6) the *orthogonality of the eigendifferentials* follows:

$$\int \mathrm{d}x\Delta\psi^*(x, L)\Delta\psi(x, L') = 0 \ , \quad \text{for } L \neq L' \ . \tag{5.7}$$

The situation is different when the intervals ΔL and $\Delta L'$ overlap (see Fig. 5.1).

First we show that the integral

$$N = \int \mathrm{d}x\Delta\psi^*(x, L)\Delta\psi(x, L) \tag{5.8}$$

is small, of the order of ΔL. We can see this by writing for (5.8):

$$N = \int \mathrm{d}x\Delta\psi^*(x, L)\Delta\psi(x, L) = \int \mathrm{d}x\Delta\psi^*(x, L) \int\limits_{L}^{L+\Delta L} \psi(x, \tilde{L})\mathrm{d}\tilde{L}$$

$$= \int \mathrm{d}x\Delta\psi^*(x, L) \int\limits_{L_1}^{L_2} \psi(x, \tilde{L})\mathrm{d}\tilde{L} \ , \tag{5.9}$$

where L_1 and L_2 are chosen in such a way that the interval $(L, L + \Delta L)$ is located inside of the interval (L_1, L_2) (see Fig. 5.1). As a result of the orthogonality (5.7) of the eigendifferentials, the contribution of the intervals (L_1, L) and $(L + \Delta L, L_2)$ to the total integral vanishes in the last step of (5.9). If we let ΔL tend to zero, it will become apparent that N tends to zero like $\Delta\psi^*(x, L)$; hence it is proportional to ΔL. Therefore we can always achieve, by an appropriate normalization of $\Delta\psi(x, L)$, that

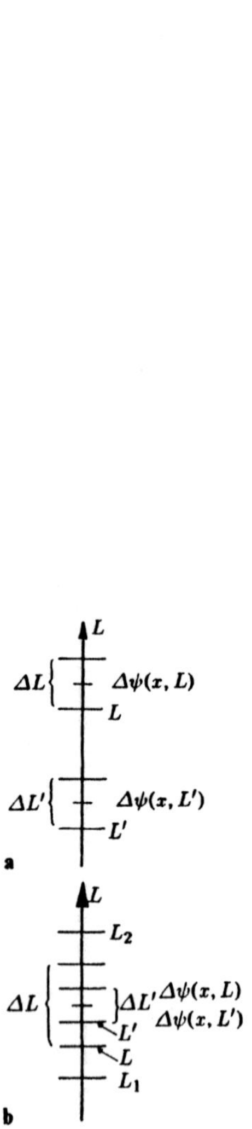

Fig. 5.1a,b. Nonoverlapping (**a**) and overlapping (**b**) intervals in the eigenvalue spectrum

$$\lim_{\Delta L \to 0} \frac{N}{\Delta L} = 1 \ , \tag{5.10}$$

i.e.

$$\int dx \Delta\psi^*(x,L)\Delta\psi(x,L) = \Delta L$$

for $\Delta L \to 0$.

We now combine results (5.7) and (5.10) in the following *orthogonality condition for eigendifferentials*:

$$\int dx \Delta\psi^*(x,L')\Delta\psi(x,L) = \begin{cases} \Delta L & \text{for overlapping intervals} \\ & (L, L+\Delta L) \text{ and } (L', L'+\Delta L') \\ 0 & \text{for nonoverlapping intervals} \\ & (L, L+\Delta L) \text{ and } (L', L'+\Delta L') \ . \end{cases}$$

This allows a further transformation for small ΔL, namely:

$$\begin{aligned} \int dx \Delta\psi^*(x,L')\Delta\psi(x,L) &= \int dx \Delta\psi^*(x,L') \int\limits_{L}^{L+\Delta L} \psi(x,\tilde{L})\,d\tilde{L} \\ &= \int dx \Delta\psi^*(x,L')\psi(x,L)\Delta L = \begin{cases} \Delta L & \text{for overlapping intervals} \\ 0 & \text{for nonoverlapping intervals} \ . \end{cases} \end{aligned} \tag{5.11}$$

After division by ΔL this becomes

$$\int dx \Delta\psi^*(x,L')\psi(x,L) = \begin{cases} 1 & \text{if the point } L' = L \text{ lies within} \\ & \text{the interval } (L',\ L'+\Delta L') \\ 0 & \text{if } L \text{ does not lie within the interval} \\ & (L',\ L'+\Delta L') \ . \end{cases}$$

This can also be written as

$$\int\limits_{L'}^{L'+\Delta L'} d\tilde{L}' \int dx\, \psi^*(x,\tilde{L}')\psi(x,L) = \begin{cases} 1 & \text{if } L = L' \text{ in the limit } \Delta L' \to 0 \\ 0 & \text{if } L \neq L' \text{ in the limit } \Delta L' \to 0 \ . \end{cases} \tag{5.12}$$

The expression

$$\int dx \psi^*(x,L')\psi(x,L) = \delta(L-L') \tag{5.13}$$

obviously must be Dirac's *delta function*, already familiar to us from electrodynamics, for which, according to (5.12),

$$\int\limits_{L'}^{L'+\Delta L'} d\tilde{L}'\delta(L-\tilde{L}') = \begin{cases} 1 & \text{if } L \text{ lies within the interval } \Delta L' \\ 0 & \text{if } L \text{ does not lie within the interval } \Delta L' \ , \end{cases} \tag{5.14}$$

holds. From this relation, we immediately get the familiar property

$$\int_a^b f(\tilde{L}')\delta(L-\tilde{L}')\,\mathrm{d}\tilde{L}' = \begin{cases} f(L) & \text{if } L \text{ is located in } (a,b) \\ 0 & \text{if } L \text{ is not located in } (a,b)\ , \end{cases} \tag{5.15}$$

for, according to (5.14), $\delta(L-\tilde{L}')$ has to be localized as a function of $\tilde{L}'$ around the value L so as to always yield for the integral $\int_{L'}^{L'+\Delta L'} \ldots \mathrm{d}\tilde{L}'$ over L the value one (see Fig. 5.2).

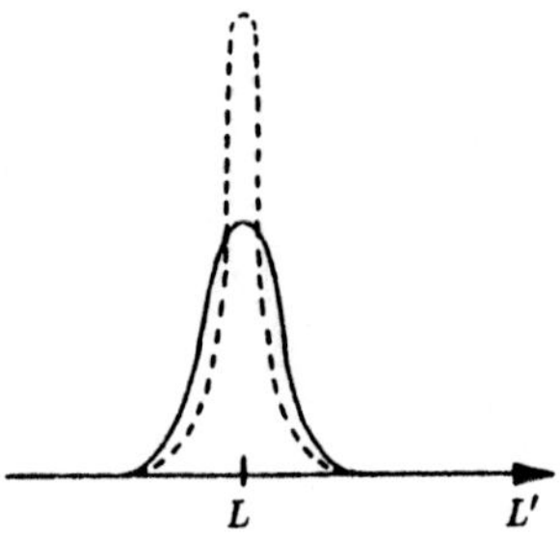

Fig. 5.2. Two functions which approximate the $\delta(L-\tilde{L}')$ function

Normalization (5.13) for the functions $\psi(x, L)$ of the continuous spectrum has grown out of normalization (5.11) for the eigendifferentials. Instead of speaking of the orthonormalization of eigendifferentials, we can say according to (5.13): *the functions* $\psi(x, L)$ *of the continuous spectrum are normalized on* δ *functions*. Hence this normalization on δ functions in the continuous spectrum corresponds exactly to the normalization on *Kronecker's* $\delta_{\mu\nu}$ *in the discrete spectrum* (cf. Sect. 16.4).

5.2 Expansion into Eigenfunctions

We make the mathematical *assumption* that all eigenfunctions of an operator $\hat{L}$, which we call $\psi_n(x)$ and which belong to the eigenvalues L_n, constitute a *complete set of functions*. By this we mean that each arbitrary function $\psi(x)$ can be expanded in terms of these eigenfunctions $\psi_n(x)$:

$$\psi(x) = \sum_n a_n \psi_n(x)\ . \tag{5.16}$$

We can easily determine the a_n because of the orthogonality of the ψ_n: $\langle\psi_n|\psi_m\rangle = \delta_{nm}$. Multiplying (5.16) on both sides by $\psi_m^*(x)$ and integrating over x yields

$$\int \psi(x)\psi_m^*(x)\,\mathrm{d}x = \sum_n a_n \int \psi_n(x)\psi_m^*(x)\,\mathrm{d}x = \sum_n a_n\delta_{nm} = a_m\ . \tag{5.17}$$

We should notice the analogy between expansion (5.16) and the expansion of a vector $\boldsymbol{A} = \sum_i a_i \boldsymbol{e}_i$ in an orthonormal vector basis $\boldsymbol{e}_i$. Therefore the expansion coefficients a_n in (5.16) can also be interpreted as components of the vector (state) ψ in the basis ψ_n. If we insert (5.17) into (5.16), we obtain

$$\psi(x) = \sum_n \left(\int \psi(x')\psi_n^*(x')\,\mathrm{d}x'\right)\psi_n(x) = \int\left(\sum_n \psi_n^*(x')\psi_n(x)\right)\psi(x')\,\mathrm{d}x'\ . \tag{5.18}$$

In order for this identity to hold for each arbitrary function $\psi(x)$, obviously

$$\sum_n \psi_n^*(x')\psi_n(x) = \delta(x-x') \tag{5.19}$$

must be valid. This is the so-called *closure relation*, which also follows directly by expansion of the δ function in the $\psi_n(x)$:

$$\delta(x-x') = \sum_n a_n \psi_n(x) \ , \quad a_n = \int \psi_n^*(x)\delta(x-x')\,\mathrm{d}x = \psi_n^*(x') \ ,$$

and therefore

$$\delta(x-x') = \sum_n \psi_n^*(x')\psi_n(x) \ . \tag{5.20}$$

If a function as singular as the $\delta(x-x')$ function can be expanded in terms of the set $\psi_n(x)$, it must be complete (or closed); hence the name "closure relation".

EXAMPLE

5.1 Normalization of the Eigenfunctions of the Momentum Operator $\hat{p}_x$

We know the eigenfunctions of the momentum operator

$$\psi_{p_x}(x) = C_{p_x} \exp\left(\mathrm{i}\frac{p_x x}{\hbar}\right) \ ,$$

where C_{p_x} is a normalization constant, which here has to be determined and can, in principle, depend on p_x. p_x is the eigenvalue of momentum with the continuous spectrum $-\infty \le p_x \le \infty$. In order to determine C_{p_x}, we form, according to (5.13), the integral

$$\begin{aligned}
&\int\limits_{-\infty}^{\infty} \psi_{p'_x}^*(x)\psi_{p_x}(x)\,\mathrm{d}x \\
&\quad = C_{p'_x}^* C_{p_x} \int\limits_{-\infty}^{\infty} \exp\left(-\mathrm{i}\frac{(p'_x - p_x)x}{\hbar}\right)\mathrm{d}x \\
&\quad = C_{p'_x}^* C_{p_x} \hbar \lim_{n\to\infty} \int\limits_{-n}^{n} \exp\left(-\mathrm{i}\frac{(p'_x - p_x)x}{\hbar}\right)\frac{\mathrm{d}x}{\hbar} \\
&\quad = C_{p'_x}^* C_{p_x} \hbar \lim_{n\to\infty} \frac{2\sin(p'_x - p_x)n}{p'_x - p_x} \ .
\end{aligned}$$

In Example 5.2 we will show that

$$\lim_{n\to\infty} \frac{\sin(nx)}{\pi x} = \delta(x)$$

is valid. Therefore the above expression can now be written as follows:

$$\int\limits_{-\infty}^{\infty} \psi_{p'_x}^*(x)\psi_{p_x}(x)\,\mathrm{d}x = C_{p'_x}^* C_{p_x} 2\pi\hbar\delta(p'_x - p_x) \ .$$

Example 5.1

In order to achieve normalization (5.13) on δ functions,

$$|C_{p_x}|^2 2\pi\hbar = 1 \ , \quad \text{i.e.} \quad C_{p_x} = \frac{1}{\sqrt{2\pi\hbar}}$$

must hold. A possible phase factor $e^{i\varphi(p_x)}$, which does not affect anything, was set equal to one in the last step. Consequently, the orthonormalized momentum eigenfunctions are

$$\psi_{p_x} = \frac{1}{\sqrt{2\pi\hbar}} \exp\left(i\frac{p_x x}{\hbar}\right) \ .$$

The three-dimensional generalization of this obviously reads

$$\begin{aligned}
\psi_{\boldsymbol{p}}(\boldsymbol{r}) &= \psi_{p_x}(x)\psi_{p_y}(y)\psi_{p_z}(z) \\
&= \frac{1}{\sqrt{(2\pi\hbar)^3}} \exp\left(i\frac{p_x x + p_y y + p_z z}{\hbar}\right) \\
&= \frac{1}{\sqrt{(2\pi\hbar)^3}} \exp\left(i\frac{\boldsymbol{p}\cdot\boldsymbol{r}}{\hbar}\right) \ .
\end{aligned}$$

These are, as is immediately clear, normalized as

$$\begin{aligned}
\int_{-\infty}^{\infty} \psi^*_{\boldsymbol{p}'}(\boldsymbol{r})\psi_{\boldsymbol{p}}(\boldsymbol{r})\,d^3r &= \int_{-\infty}^{\infty} \psi^*_{p'_x}(x)\psi_{p_x}\,dx \\
&\times \int_{-\infty}^{\infty} \psi^*_{p'_y}(y)\psi_{p_y}(y)\,dy \int_{-\infty}^{\infty} \psi^*_{p'_z}(z)\psi_{p_z}(z)\,dz \\
&= \delta(p'_x - p_x)\delta(p'_y - p_y)\delta(p'_z - p_z) \equiv \delta(\boldsymbol{p}' - \boldsymbol{p}) \ .
\end{aligned}$$

The last step comprises the definition of the three-dimensional δ function.

EXAMPLE

5.2 A Representation of the δ Function

In accordance with the definition

$$\delta(x) = 0 \quad \text{for} \quad x \neq 0 \quad \text{and}$$

$$\int_{-\infty}^{\infty} \delta(x)\,dx = \int_{-\varepsilon}^{\varepsilon} \delta(x)\,dx = 1$$

as well as

$$\int_{-\infty}^{\infty} f(x)\delta(x)\,dx = f(0) \ ,$$

the δ function must be an extremely singular function.[1] As already frequently mentioned, we can visualize that $\delta(x)$ equals zero everywhere except at $x = 0$, where it takes on such large values, that the area between $\delta(x)$ and the x axis yields exactly the value one (see besides figure).

The δ function can be represented in a more formal way by a set of analytic functions $\varphi_n(x)$, so that

$$\delta(x) = \lim_{n\to\infty} \varphi_n(x) \ .$$

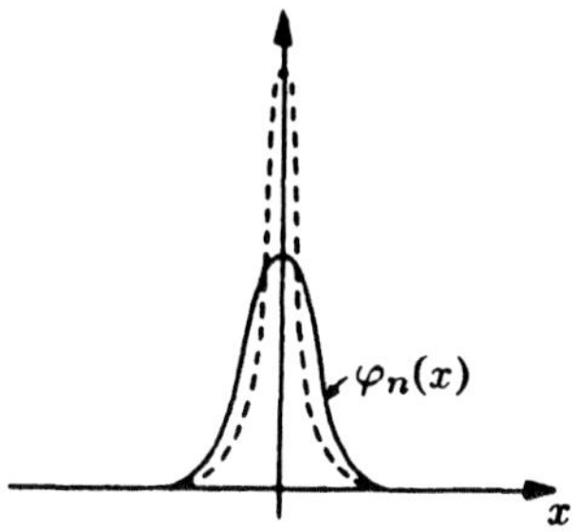

Two approximation functions $\varphi_n(x)$, with $\int_{-\infty}^{\infty} \varphi_n(x)\,dx = 1$, which tend to the $\delta(x)$-function for $n \to \infty$

These functions $\varphi_n(x)$ must have the property (see figure) of constantly increasing at $x = 0$ for large values of n and decreasing continually for $x \neq 0$, so that

$$\int_{-\infty}^{\infty} \varphi_n(x)\,dx = 1$$

remains valid for every n. There are many sets of functions which satisfy these conditions. In this case we speak of *various representations of the δ function*. An especially advantageous representation is given by the functions

$$\varphi_n(x) = \frac{\sin nx}{\pi x} \ , \tag{1}$$

where n is a positive number ($n \in \mathbb{N}_0$). Obviously

$$\varphi_n(0) = \frac{n}{\pi} \tag{2}$$

holds.

In addition, the $\varphi_n(x)$ oscillate with the period $2\pi/n$ and have decreasing amplitudes for $|x| \to \infty$. Moreover, for all φ_n

$$\int_{-\infty}^{\infty} \varphi_n(x)\,dx = \int_{-\infty}^{\infty} \frac{\sin nx}{\pi x}\,dx = 1 \tag{3}$$

holds. This is obvious for the limit $n \to \infty$. Then $\varphi_n(0) = n/\pi$ tends to ∞, and the oscillation period $n\Delta x = 2\pi$, i.e. $\Delta x = 2\pi/n$, tends to zero, so that the contribution of the rapidly oscillating function to the integral vanishes in the domain $x \neq 0$. Then only the contribution from the surroundings of the point $x = 0$ remains (see figure). This yields

$$\varphi(0)\,\Delta x = \frac{n}{\pi}\frac{\pi}{n} = 1 \ .$$

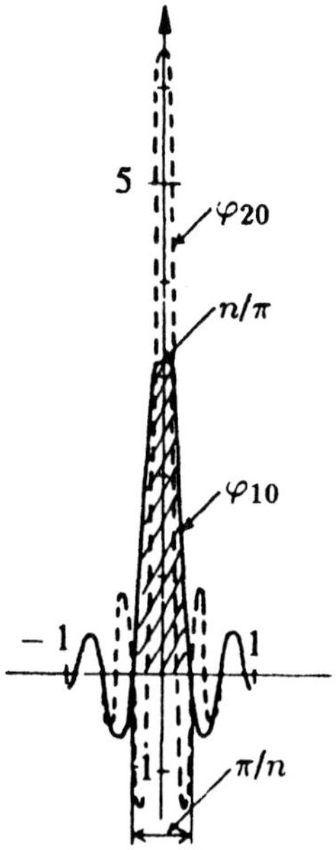

The functions $\varphi_{n=10}(x)$ and $\varphi_{n=20}(x)$

[1] A stringent mathematical foundation of the δ function was given by Laurent Schwartz. We draw special attention to the elegant paper by C. Schmieden, D. Laugwitz: Mathematische Zeitschrift **69**, 1–39 (1958).

Example 5.2

Hence functions (1) have all properties of the δ function in the limit $n \to \infty$, and we can write

$$\delta(x) = \lim_{n\to\infty} \frac{\sin nx}{\pi x} \ .$$

We list (without proof) some other representations of the δ function:

$$\text{(a)}\ \ \delta(x-x_0) = \frac{1}{\pi} \lim_{\tau\to\infty} \frac{1-\cos\tau(x-x_0)}{\tau(x-x_0)^2} \ .$$

$$\text{(b)}\ \ \delta(x-x_0) = \frac{1}{\pi} \lim_{\varepsilon\to 0} \frac{\varepsilon}{(x-x_0)^2+\varepsilon^2} \ .$$

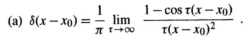

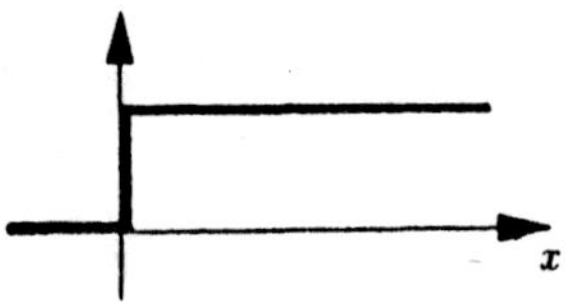

Heaviside's step function

(c) Let $\theta(x)$ be *Heaviside's step function* (see figure)

$$\theta(x) = \begin{cases} 1 & \text{for } x > 0 \\ 0 & \text{for } x \leq 0 \ . \end{cases}$$

Then the relation

$$\delta(x-x_0) = \lim_{\varepsilon\to 0} \frac{\theta(x-x_0+\varepsilon)-\theta(x-x_0)}{\varepsilon} = \frac{\mathrm{d}\theta(x-x_0)}{\mathrm{d}x}$$

holds. This means: the distribution δ is the derivative of the Heaviside distribution. The integral relation

$$\theta(x-x_0) = \int_{-\infty}^{x-x_0} \delta(x'-x_0)\,\mathrm{d}x' = \begin{cases} 0 & \text{for } x < x_0 \\ 1 & \text{for } x > x_0 \end{cases}$$

is immediately clear.

$$\text{(d)}\ \ \delta(x-x_0) = \lim_{b\to 0} \frac{1}{\sqrt{\pi}b} \exp\left[-\frac{(x-x_0)^2}{b^2}\right] \ .$$

$$\text{(e)}\ \ \delta(x-x_0) = \lim_{b\to 0} \frac{1}{\sqrt{\pi}b} \frac{1}{1+[(x-x_0)^2]/b^2} \ .$$

EXAMPLE

5.3 Cauchy's Principal Value

A further, very useful, relation follows from the identity

$$\frac{1}{x\pm\mathrm{i}\varepsilon} = \frac{x\mp\mathrm{i}\varepsilon}{x^2+\varepsilon^2} = \frac{x}{x^2+\varepsilon^2} \mp \frac{\mathrm{i}\varepsilon}{x^2+\varepsilon^2} \ . \qquad (1)$$

Example 5.3

If we inspect integrals of the form

$$\lim_{\varepsilon\to 0}\int_{-\infty}^{\infty}\frac{f(x)}{x\pm \mathrm{i}\varepsilon}\,\mathrm{d}x=\lim_{\varepsilon\to 0}\int_{-\infty}^{\infty}f(x)\frac{x}{x^2+\varepsilon^2}\,\mathrm{d}x$$

$$\mp\mathrm{i}\lim_{\varepsilon\to 0}\int_{-\infty}^{\infty}f(x)\frac{\varepsilon}{x^2+\varepsilon^2}\,\mathrm{d}x\ , \tag{2}$$

the last term on the right-hand side yields, according to relation (b) stated in Example 5.2,

$$\mp\mathrm{i}\lim_{\varepsilon\to 0}\int_{-\infty}^{\infty}f(x)\frac{\varepsilon}{x^2+\varepsilon^2}\,\mathrm{d}x=\mp\mathrm{i}\pi\int_{-\infty}^{\infty}f(x)\delta(x)\,\mathrm{d}x=\mp\mathrm{i}\pi f(0)\ . \tag{3}$$

The first term in (2) can also be rewritten as

$$\lim_{\varepsilon\to 0}\int_{-\infty}^{\infty}f(x)\frac{x}{x^2+\varepsilon^2}\,\mathrm{d}x$$

$$=\lim_{\varepsilon\to 0}\int_{-\infty}^{-\varepsilon}f(x)\frac{\mathrm{d}x}{x}+\lim_{\varepsilon\to 0}\int_{\varepsilon}^{\infty}f(x)\frac{\mathrm{d}x}{x}+\lim_{\varepsilon\to 0}\int_{-\varepsilon}^{\varepsilon}\frac{f(x)x}{x^2+\varepsilon^2}\,\mathrm{d}x$$

$$=P\int_{-\infty}^{\infty}\frac{f(x)}{x}\,\mathrm{d}x+f(0)\lim_{\varepsilon\to 0}\int_{-\varepsilon}^{\varepsilon}\frac{x\,\mathrm{d}x}{x^2+\varepsilon^2}\ . \tag{4}$$

Here, P designates *Cauchy's principal value*,

$$P\int_{-\infty}^{\infty}\frac{f(x)}{x}\,\mathrm{d}x=\lim_{\varepsilon\to 0}\left[\int_{-\infty}^{-\varepsilon}\frac{f(x)}{x}\,\mathrm{d}x+\int_{\varepsilon}^{\infty}\frac{f(x)}{x}\,\mathrm{d}x\right]\ . \tag{5}$$

The second term in (4) vanishes because the integrand is an odd function even in the limit $\varepsilon\to 0$. Therefore we now can write (2) as

$$\lim_{\varepsilon\to 0}\int_{-\infty}^{\infty}\frac{f(x)}{x\pm\mathrm{i}\varepsilon}=P\int_{-\infty}^{\infty}\frac{f(x)}{x}\,\mathrm{d}x\mp\mathrm{i}\pi f(0)\ . \tag{6}$$

This can be summarized symbolically in the simple, often-used formula

$$\lim_{\varepsilon\to 0}\frac{1}{x\pm\mathrm{i}\varepsilon}=P\frac{1}{x}\mp\mathrm{i}\pi\delta(x)\ . \tag{7}$$

EXERCISE

5.4 The δ Function as the Limit of Bell-Shaped Curves

Problem. Show that the δ function can be represented as the limit of the "bell-shaped curves",

$$y(x,\varepsilon) = \pi^{-1}\varepsilon[x^2+\varepsilon^2]^{-1}\,(\varepsilon > 0) \ .$$

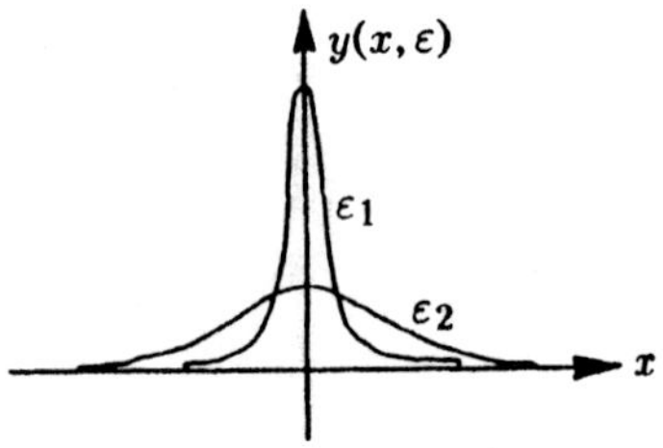

Two bell-shaped curves

Solution. The bell-shaped curves become narrower and higher with decreasing ε (see figure with $\varepsilon_1 < \varepsilon_2$). We have

$$\lim_{\varepsilon\to 0} y(x,\varepsilon) = \begin{cases} 0 & \text{for } x \neq 0 \\ \infty & \text{for } x = 0 \ , \end{cases}$$

but the areas below the curves always have the value

$$\int_{-\infty}^{+\infty} y(x,\varepsilon)\,\mathrm{d}x = \frac{1}{\pi}\arctan\frac{x}{\varepsilon}\bigg|_{-\infty}^{+\infty} = 1 \ ,$$

independently of ε. Now we examine the integral

$$F(\varepsilon) = \int_{-\infty}^{+\infty} f(x)y(x,\varepsilon)\,\mathrm{d}x$$

for a continuous, bounded function $f(x)$ as function of the parameter ε. We substitute $x = \varepsilon\xi$ and obtain

$$F(\varepsilon) = \int_{-\infty}^{+\infty} f(\varepsilon\xi)g(\xi)\,\mathrm{d}\xi \ , \quad \text{with}$$

$$g(\xi) = \frac{1}{\pi}\frac{1}{\xi^2+1} \quad \text{and} \quad \int_{-\infty}^{+\infty} g(\xi)\,\mathrm{d}\xi = 1 \ .$$

This integral now converges uniformly according to our assumptions, for an $M \in \mathbb{R}$ independent of ε exists with $|f(\varepsilon\xi)g(\xi)| \leq Mg(\xi)$. Now we have a theorem at hand which then guarantees that $F(\varepsilon)$ is continuous. Hence we have $\lim_{\varepsilon\to 0} F(\varepsilon) = F(0)$ and thus

$$\lim_{\varepsilon\to 0}\int_{-\infty}^{+\infty} f(x)y(x,\varepsilon)\,\mathrm{d}x = \lim_{\varepsilon\to 0} F(\varepsilon) = F(0)$$

$$= \int_{-\infty}^{+\infty} f(0)g(\xi)\,\mathrm{d}\xi = \int_{-\infty}^{+\infty} f(x)\delta(x)\,\mathrm{d}x$$

Exercise 5.4

for arbitrary continuous bounded functions f. Consequently we can write

$$\delta(x) = \text{``}\lim_{\varepsilon \to 0} y(x, \varepsilon)\text{''} \; .$$

The quotation marks should remind us that the limit $\varepsilon \to 0$ may not be performed before the integration over a test function.

6. The Schrödinger Equation

In classical mechanics it is possible to calculate, for example, the vibrational modes of a string, membrane or resonator by solving a wave equation, subject to certain boundary conditions. At the very beginning of the development of quantum mechanics, one was faced with the problem of finding a differential equation describing discrete states of an atom. It was not possible to deduce exactly such an equation from old and well-known physical principles; instead, one had to search for parallels in classical mechanics and try to deduce the desired equation on the basis of plausible arguments. Such an equation, not derived but guessed at intuitively, would then be a postulate of the new theory, and its validity would have to be checked by experiment. This equation for the calculation of quantum-mechanical states is called the ***Schrödinger** equation*; let us now "derive" it.

In relativistic classical mechanics, time coordinates and spatial coordinates as well as energy and momentum are treated as the four components of a four-vector, i.e.

$$x_\nu = (\boldsymbol{r}, \mathrm{i}ct) \ , \quad p_\nu = \left(\boldsymbol{p}, \mathrm{i}\frac{E}{c}\right) \ , \quad \nu = 1, 2, 3, 4 \ . \tag{6.1}$$

By enlarging the operator representation of the three-dimensional momentum to a four-dimensional, relativistic covariant vector-operator we get

$$\left(\hat{\boldsymbol{p}}, \frac{\mathrm{i}}{c}\hat{E}\right) = -\mathrm{i}\hbar \left\{\frac{\partial}{\partial x_\nu}\right\} = -\mathrm{i}\hbar \left(\frac{\partial}{\partial x}, \frac{\partial}{\partial y}, \frac{\partial}{\partial z}, \frac{\partial}{\partial(\mathrm{i}ct)}\right) \ . \tag{6.2}$$

Both sides of this equation are four-vectors. By comparison, the energy is replaced by the following operator:

$$\hat{E} = \mathrm{i}\hbar\frac{\partial}{\partial t} \ . \tag{6.3}$$

We remember here that we already had an operator for the energy in (4.83), namely the Hamiltonian $\hat{H}$ of a particle. Obviously we have two operators for the energy. Both $\hat{E}$ and the Hamiltonian $\hat{H}$ describe the total energy and can therefore be set equal. This generates the Schrödinger equation.

$$\hat{E}\psi(\boldsymbol{r}, t) = \hat{H}\psi(\boldsymbol{r}, t) \quad \text{or} \quad \mathrm{i}\hbar\frac{\partial}{\partial t}\psi(\boldsymbol{r}, t) = \hat{H}\psi(\boldsymbol{r}, t) \quad \text{with}$$

$$\hat{H} = -\frac{\hbar^2}{2m}\Delta + V(r) \ . \tag{6.4}$$

Using the wave function of a free particle (de Broglie wave),

$$\psi(\boldsymbol{r},t) = A\exp\left[-\frac{\mathrm{i}}{\hbar}(Et-\boldsymbol{p}\cdot\boldsymbol{r})\right]$$
$$= A\exp\left(\frac{\mathrm{i}}{\hbar}\sum_\nu p_\nu x_\nu\right) = A\exp\left(\frac{\mathrm{i}}{\hbar}\hat{p}\hat{x}\right) , \tag{6.5}$$

we find that the operator $\hat{E}$ has the total energy E as an eigenvalue.

The Schrödinger equation (6.4) is not a relativistic equation. Indeed, starting from

$$E^2 = \boldsymbol{p}^2c^2 + m_0^2c^4 \tag{6.6}$$

for the energy of a free relativistic particle, the free *Klein–Gordon equation* follows, i.e.

$$-\hbar^2\frac{\partial^2}{\partial t^2}\psi(\boldsymbol{r},t) = (-\hbar^2c^2\Delta + m_0^2c^4)\psi(\boldsymbol{r},t) . \tag{6.7}$$

The Schrödinger equation and the Klein–Gordon equation are linear differential equations; this means that with ψ_1 and ψ_2, the function defined by $\psi = a\psi_1 + b\psi_2$ is a solution, too. This is the mathematical formulation for the principle of superposition, which was discussed in Chap. 3. Schrödinger's equation is of first order in time and second order in space; the Klein–Gordon equation is of second order in both space and time. We suppose that the wave function at time t_0 contains all the information about how the state propagates if there are no external perturbations. Only Schrödinger's equation as a first-order differential equation in time satisfies this requirement. The Klein–Gordon equation, being important in relativistic quantum mechanics, needs to be reinterpreted. The Schrödinger equation (6.4) contains the imaginary unit i as a factor, which implies that oscillating solutions are possible.

It is separable into time and space, if the Hamiltonian $\hat{H} = \hat{H}(\boldsymbol{r},\boldsymbol{p})$ is not explicitly time dependent:

$$\psi(\boldsymbol{r},t) = \psi(\boldsymbol{r})f(t)$$

and therefore

$$\mathrm{i}\hbar\psi(\boldsymbol{r})\frac{\partial}{\partial t}f(t) = [\hat{H}\psi(\boldsymbol{r})]\,f(t) . \tag{6.8}$$

[Since $\psi(\boldsymbol{r},t)$ and $\psi(\boldsymbol{r})$ are two different functions, this should not lead to any misunderstanding.] After separating the variables, one finds the equation

$$\mathrm{i}\hbar\,\frac{\dot{f}(t)}{f(t)} = \frac{\hat{H}\psi(\boldsymbol{r})}{\psi(\boldsymbol{r})} = \mathrm{const} = E . \tag{6.9}$$

This means for the time-dependent function

$$f(t) = \mathrm{const}\,\exp\left(-\mathrm{i}\frac{Et}{\hbar}\right) . \tag{6.10}$$

The function with the spatial argument $\psi(\boldsymbol{r})$ solves the *stationary Schrödinger equation*

$$\hat{H}\psi(\boldsymbol{r}) = E\psi(\boldsymbol{r}) \ . \tag{6.11}$$

The wave function $\psi(\boldsymbol{r}, t)$ is periodic in time, with the phase factor $\exp[-\mathrm{i}(Et/\hbar)]$, and that is why the densities $\psi^*\psi$ and also, as we shall see, the currents are time independent. Equation (6.4) is an eigenvalue equation of the Hamiltonian, with E being the real energy eigenvalue. The general solutions of (6.4) are oscillating functions in time,

$$\psi_n(\boldsymbol{r}, t) = \psi_n(\boldsymbol{r}) \exp\left(-\mathrm{i}\frac{E_n t}{\hbar}\right) \ , \tag{6.12}$$

with the normalization

$$\int \psi_n^*(\boldsymbol{r}, t)\psi_n(\boldsymbol{r}, t)\,\mathrm{d}V = \int \psi_n^*(\boldsymbol{r})\psi_n(\boldsymbol{r})\,\mathrm{d}V = 1 \ . \tag{6.13}$$

Any stationary state corresponds to well-defined energy and to an infinite stability in time. It has the character of a standing wave because the density of probability given by $\psi^*\psi$ is time independent. This is not true for a linear superposition of stationary states.

EXERCISE

6.1 A Particle in an Infinitely High Potential Well

Problem. A particle of mass m is captured in a box limited by

$$0 \le x \le a \ ; \quad 0 \le y \le b \ ; \quad 0 \le z \le c \ .$$

The corresponding potential is given by

$$V = \begin{cases} 0 & \text{if} \quad 0 < x < a \ ; \quad 0 < y < b \ ; \quad 0 < z < c \\ \infty & \text{elsewhere} \ . \end{cases}$$

(See figure.)

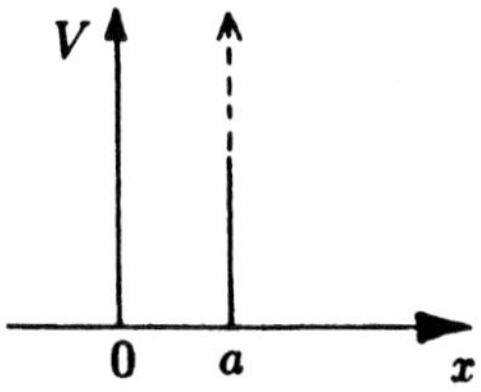

The potential along the x axis

Solution. The Hamiltonian is given by

$$\hat{H} = \hat{T} + \hat{V} = -\frac{\hbar^2}{2m}\Delta(x, y, z) + V(x, y, z) \ . \tag{1}$$

Inside the box: Here we have the potential $V = 0$ and the following stationary Schrödinger equation:

$$-\frac{\hbar^2}{2m}\left(\frac{\partial^2}{\partial x^2} + \frac{\partial^2}{\partial y^2} + \frac{\partial^2}{\partial z^2}\right)\psi(x, y, z) = E\psi(x, y, z) \ . \tag{2}$$

Exercise 6.1

We are going to solve this problem using the well-known separation of variables procedure:

$$\psi(x,y,z) = \psi_1(x)\psi_2(y)\psi_3(z) \ . \tag{3}$$

This leads to three separate equations connected by the constants of separation $-k_1^2$ (the constants squared are chosen to be negative, this does not conflict with the general case, since the constants themselves are allowed to be imaginary):

$$\begin{aligned} &\frac{1}{\psi_i(x_i)}\frac{\mathrm{d}^2\psi_i(x_i)}{\mathrm{d}x_i^2} = -k_i^2 \ , \quad i=1,2,3 \ , \\ &x_1 = x \ , \quad x_2 = y \ , \quad x_3 = z \ . \end{aligned} \tag{4}$$

The solutions of these different equations are simply

$$\psi_i(x_i) = \text{const}\ \sin(k_i x_i + \delta_i) \ .$$

The total solution inside the potential well is therefore given by

$$\psi(x,y,z) = A\sin(k_1 x+\delta_1)\sin(k_2 y+\delta_2)\sin(k_3 z+\delta_3) \ , \tag{5}$$

where A is a normalization factor and δ_i are phases which must be determined.

Outside the potential well: The wave function must vanish here, because V is infinitely large; otherwise there would be an infinitely large potential energy, since $\langle\psi|V(\boldsymbol{r})|\psi\rangle$ diverges. Since the wave functions have to be smooth, we get two sets of boundary conditions:

$$\psi(x=0,y,z) = \psi(x,y=0,z) = \psi(x,y,z=0) = 0 \ ,$$

and

$$\psi(x=a,y,z) = \psi(x,y=b,z) = \psi(x,y,z=c) = 0 \ . \tag{6}$$

The first set requires $\delta_1=\delta_2=\delta_3=0$. The other set gives a quantization condition:

$$ak_1 = n_1\pi \ , \quad bk_2 = n_2\pi \ , \quad ck_3 = n_3\pi$$

and therefore

$$k_1 = n_1\frac{\pi}{a} \ , \quad k_2 = n_2\frac{\pi}{b} \ , \quad k_3 = n_3\frac{\pi}{c} \ . \tag{7}$$

Here, $n_1, n_2, n_3 = \pm1;\ \pm2;\ \pm3;\ \ldots$ are independent quantum numbers. The possibility of choosing $n_i=0$ must be excluded, because the corresponding wave function (see the end of this exercise) would vanish everywhere. The total energy is

$$E = \frac{\hbar^2}{2m}(k_1^2+k_2^2+k_3^2) \ ; \tag{8}$$

Exercise 6.1

it can have only discrete values, namely

$$E_{n_1n_2n_3} = \frac{\hbar^2}{2m}\left[\left(n_1\frac{\pi}{a}\right)^2 + \left(n_2\frac{\pi}{b}\right)^2 + \left(n_3\frac{\pi}{c}\right)^2\right] . \tag{9}$$

This discrete energy spectrum is converted into a *quasi-continuum* when the mass m or the extension of the box becomes very large. The lowest energy value

$$E_{111} = \frac{\hbar^2}{2m}\left[\left(\frac{\pi}{a}\right)^2 + \left(\frac{\pi}{b}\right)^2 + \left(\frac{\pi}{c}\right)^2\right] \tag{10}$$

is not zero, as one would expect classically. This is the first example of *non-vanishing zero-point energy* (see the extended discussion in connection with the harmonic oscillator, Chap. 7).

The solution inside the box

$$\psi_{n_1n_2n_3}(\boldsymbol{r}) = A\sin(k_1x)\sin(k_2y)\sin(k_3z) \tag{11}$$

must yield a unit total probability, which means that

$$\begin{aligned} 1 &= \int \psi^*_{n_1n_2n_3}\psi_{n_1n_2n_3}\,\mathrm{d}V \\ &= |A|^2\int_0^a \sin^2(k_1x)\,\mathrm{d}x\int_0^b \sin^2(k_2y)\,\mathrm{d}y\int_0^c \sin^2(k_3z)\,\mathrm{d}z \\ &= \frac{abc}{8}|A|^2 , \end{aligned} \tag{12}$$

and therefore the normalization factor equals

$$|A| = \sqrt{\frac{2}{a}\frac{2}{b}\frac{2}{c}} . \tag{13}$$

The energy spectrum is shown in the next figure.

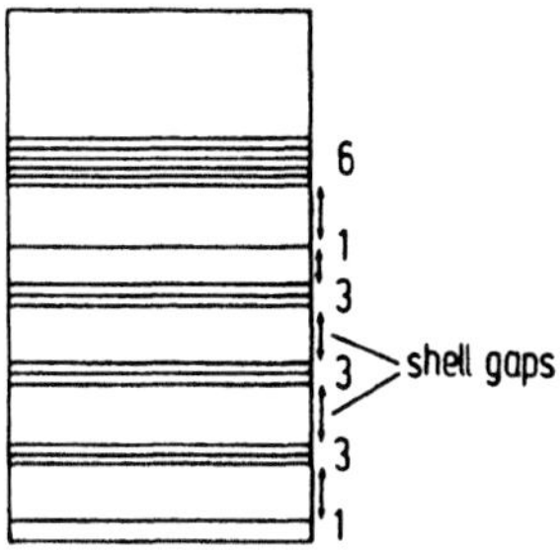

Energy levels for a particle within a rectangular box

We took $a < b < c$; therefore the level E_{211} is energetically higher than the levels E_{121} and E_{112}, and the relation $E_{211} > E_{121} > E_{112}$ holds. In the case in which a, b, and c do not differ too much, all these levels are close together. Then we speak of a *triplet* (in general a *multiplet*), of states. For $a = b = c$, the particle moves in a cube, and all states belonging to a triplet are degenerated in energy. We have then

$$E_{211} = E_{121} = E_{112} . \tag{14}$$

The wave functions of the three states are

$$\begin{aligned} E_{211}: \quad \psi_{211} &= \sqrt{\frac{8}{a^3}}\sin\frac{2\pi}{a}x\,\sin\frac{\pi}{a}y\,\sin\frac{\pi}{a}z , \\ E_{121}: \quad \psi_{121} &= \sqrt{\frac{8}{a^3}}\sin\frac{\pi}{a}x\,\sin\frac{2\pi}{a}y\,\sin\frac{\pi}{a}z , \\ E_{112}: \quad \psi_{112} &= \sqrt{\frac{8}{a^3}}\sin\frac{\pi}{a}x\,\sin\frac{\pi}{a}y\,\sin\frac{2\pi}{a}z . \end{aligned} \tag{15}$$

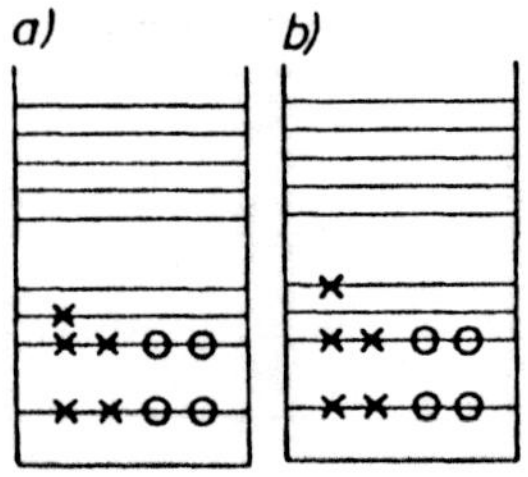

Nuclei in **(a)** the ground state, **(b)** the excited state; (×) stands for a proton, (o) stands for a neutron

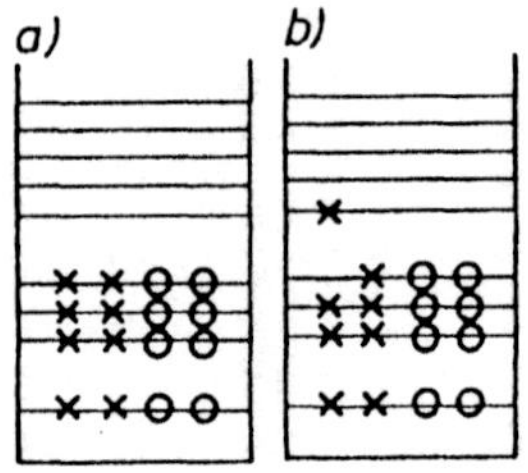

Magic nuclei in **(a)** the ground state, **(b)** the excited state. Comparison with the diagram above shows a much higher excitation energy

If we break this degeneracy slightly, the volume is approximately that of a cube and the three levels are close together in energy, as we just pointed out. For states of higher energy we observe an equivalent phenomenon. For instance, there are two triplets (because of a slight break in the cube's symmetry) close together, namely ψ_{221}, ψ_{122}, ψ_{212} and ψ_{311}, ψ_{131}, ψ_{113}, followed by one single state (a singlet), namely ψ_{222}. Such multiplet structures of states are identified with "*shells*". *Shell models* explaining shell structures are important in atomic and nuclear physics. For example, in nuclear physics all nucleons in a nucleus are supposed to be in a potential well. Of course this potential well is spherically symmetric, but for small nuclei a boxlike potential is an acceptable approximation. Because of the spin of the proton and the neutron (compare later Chap. 12) and the Pauli principle (compare later Chap. 14), only two protons and two neutrons can be put into each level. We start by filling the lowest energy levels individually, because the system prefers the state of lowest energy. Here, the "last" particle determines the most "visible" properties. If this last state is inside a multiplet, then a small excitation energy suffices to lift that particle into a higher state of energy; the nucleus is then easily excitable.

If we consider a nucleus that contains just the number of protons and neutrons to fill a shell, then much more energy will be required to excite a nucleon into the first excited state. Such nuclei are particularly stable, because they can only be destroyed (i.e. strongly excited) if large energy gaps are overcome. In such a case we speak of *magic nuclei* (comparable with filled electron shells in the atoms of *inert gases*) or double *magic nuclei*.[1]

EXERCISE

6.2 A Particle in a One-Dimensional Finite Potential Well

Problem. Solve the one-dimensional Schrödinger equation for a finite potential well described by the following potential (see figure)

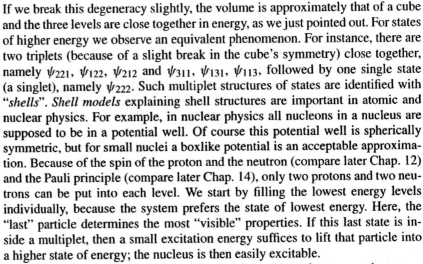

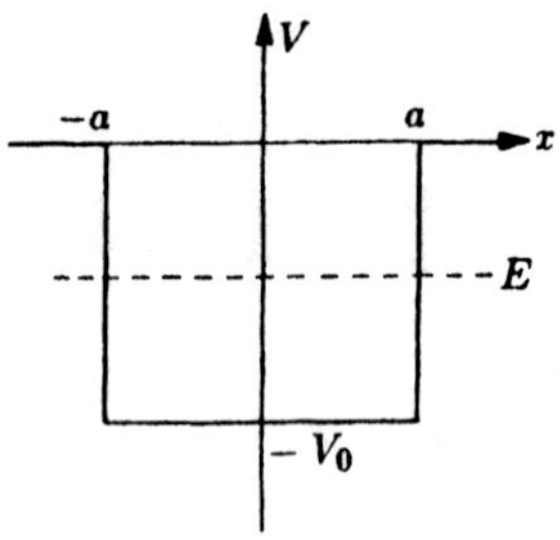

$$V(x) = \begin{cases} -V_0 & \text{if} \quad |x| \le a \\ 0 & \text{if} \quad |x| > a \ . \end{cases}$$

Consider bound states ($E < 0$) only.

Solution. (a) The wave functions for $|x| < a$ and $|x| > a$. The corresponding Schrödinger equation is given by

$$-\frac{\hbar^2}{2m}\psi''(x) + V(x)\psi(x) = E\psi(x) \quad \text{where}$$

$$\psi'' \equiv \frac{d^2\psi}{dx^2} \ . \tag{1}$$

[1] For an extensive discussion of the nuclear shell model, see J. M. Eisenberg and W. Greiner: *Nuclear Theory*, Vol. 1, *Nuclear Models (Collective and Single Particle Phenomena)*, 3rd ed. (North-Holland, Amsterdam 1987).

We define, for the sake of brevity, *Exercise 6.2*

$$\kappa^2 = -\frac{2mE}{\hbar^2} \;, \quad k^2 = \frac{2m(E+V_0)}{\hbar^2} \tag{2}$$

and get:

(1) if $x < -a$: $\quad \psi_1'' - \kappa^2\psi_1 = 0 \;,$

$$\psi_1 = A_1 \exp(\kappa x) + B_1 \exp(-\kappa x) \;; \tag{3a}$$

(2) if $-a \le x \le a$: $\quad \psi_2'' + k^2\psi_2 = 0 \;,$

$$\psi_2 = A_2 \cos(kx) + B_2 \sin(kx) \;; \tag{3b}$$

(3) if $x > a$: $\quad \psi_3'' - \kappa^2\psi_3 = 0 \;,$

$$\psi_3 = A_3 \exp(\kappa x) + B_3 \exp(-\kappa x) \;. \tag{3c}$$

(b) Formulation of boundary conditions. The normalization of the bound state requires the vanishing of the solution at infinity. This means that $B_1 = A_3 = 0$. Furthermore, $\psi(x)$ should be continuously differentiable. All particular solutions are fitted in such a way that ψ as well as its first derivative ψ' are smooth at that value of x corresponding to the border between the inside and outside. The second derivative ψ'' contains the jump required by the particular box-type potential of this Schrödinger equation. All this together leads to

$$\begin{aligned} &\psi_1(-a) = \psi_2(-a) \;, \quad \psi_2(a) = \psi_3(a) \;, \\ &\psi_1'(-a) = \psi_2'(-a) \;, \quad \psi_2'(a) = \psi_3'(a) \;. \end{aligned} \tag{4}$$

(c) The eigenvalue equations. From (4) we obtain four linear and homogeneous equations for the coefficients A_1; A_2; B_2; B_3; namely

$$\begin{aligned} &A_1 \exp(-\kappa a) = A_2 \cos(ka) - B_2 \sin(ka) \;, \\ &\kappa A_1 \exp(-\kappa a) = A_2 k \sin(ka) + B_2 k \cos(ka) \;, \\ &B_3 \exp(-\kappa a) = A_2 \cos(ka) + B_2 \sin(ka) \;, \\ &-\kappa B_3 \exp(-\kappa a) = -A_2 k \sin(ka) + B_2 k \cos(ka) \;. \end{aligned} \tag{5}$$

By addition and subtraction of these equations, we get a more lucid form of the system of equations, which is easy to solve:

$$\begin{aligned} &(A_1 + B_3)\exp(-\kappa a) = 2A_2 \cos(ka) \\ &\kappa(A_1 + B_3)\exp(-\kappa a) = 2A_2 k \sin(ka) \\ &(A_1 - B_3)\exp(-\kappa a) = -2B_2 \sin(ka) \\ &\kappa(A_1 - B_3)\exp(-\kappa a) = 2B_2 k \cos(ka) \;. \end{aligned} \tag{6}$$

Assuming that $A_1 + B_3 \neq 0$ and $A_2 \neq 0$, the first two equations yield

$$\kappa = k \tan(ka) \;. \tag{7}$$

Inserting this in one of the last two equations gives

$$A_1 = B_3 \;; \quad B_2 = 0 \;. \tag{8}$$

Exercise 6.2

Hence, as a result, we have a symmetric solution with $\psi(x) = \psi(-x)$. We then speak of *positive parity*.

Almost identical calculations lead for $A_1 - B_3 \neq 0$ and for $B_2 \neq 0$ to

$$\kappa = -k\cot(ka) \quad \text{and} \quad A_1 = -B_3 \ ; \quad A_2 = 0 \ . \tag{9}$$

The thus-obtained wave function is an antisymmetric one, corresponding to *negative* parity.

(d) Qualitative solution of the eigenvalue problem. The equations connecting κ and k, which we have already obtained, are conditions for the energy eigenvalue. Using the short forms

$$\xi = ka \ , \quad \eta = \kappa a \ , \tag{10}$$

we get from the definition (2)

$$\xi^2 + \eta^2 = \frac{2mV_0a^2}{\hbar^2} = r^2 \ . \tag{11}$$

On the other hand, using (7) and (9) we get the equations

$$\eta = \xi\tan(\xi) \ , \quad \eta = -\xi\cot(\xi) \ .$$

Therefore the desired energy values can be obtained by constructing the intersection of those two curves with the circle defined by (11), within the (ξ, η) plane (see next figure).

At least one solution exists for arbitrary values of the parameter V_0, in the case of positive parity, because the tan function intersects the origin. For negative parity, the radius of the circle needs to be larger than a minimum value so that the two curves can intersect. The potential must have a certain depth in connection with a given size a and a given mass m, to permit a solution with negative

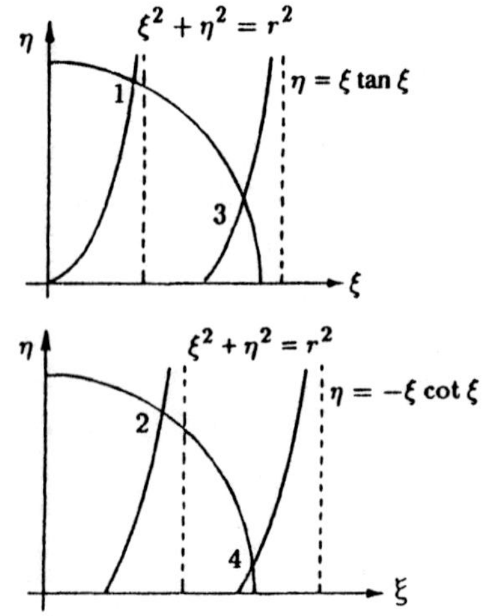

The intersections of these curves determine the energy eigenvalues

parity. The number of energy levels increases with V_0, a and mass m. For the case $mVa^2 \to \infty$, the intersections are found at

$$\begin{aligned} \tan(ka) &= \infty \quad \text{corresponding to} \quad ka = \frac{2n-1}{2}\pi \ , \\ -\cot(ka) &= \infty \quad \text{corresponding to} \quad ka = n\pi \ , \\ & n = 1, 2, 3, \ldots \end{aligned} \tag{12}$$

or, combined:

$$k(2a) = n\pi \ . \tag{13}$$

For the energy spectrum this means that

$$E_n = \frac{\hbar^2}{2m}\left(\frac{n\pi}{2a}\right)^2 - V_0 \ . \tag{14}$$

On enlarging the potential well and/or the particle's mass m, the difference between two neighbouring energy eigenvalues will decrease. The lowermost state ($n = 1$) is not located at $-V_0$, but a little higher. This difference is called the *zero-point energy.* We will come back to it later when discussing the harmonic oscillator (see Chap. 7).

(e) The shape of the wave function is shown for the discussed solutions in the two figures.

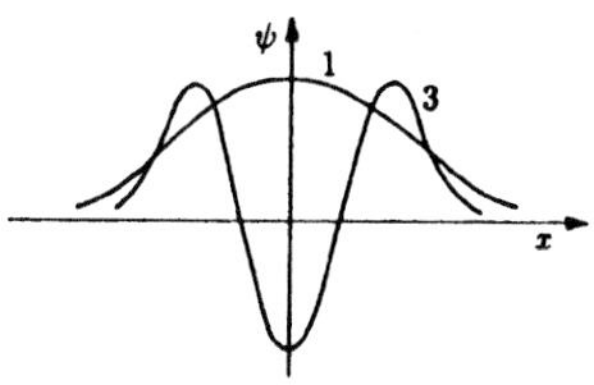

Wave functions with positive parity; they are symmetric relative to the origin

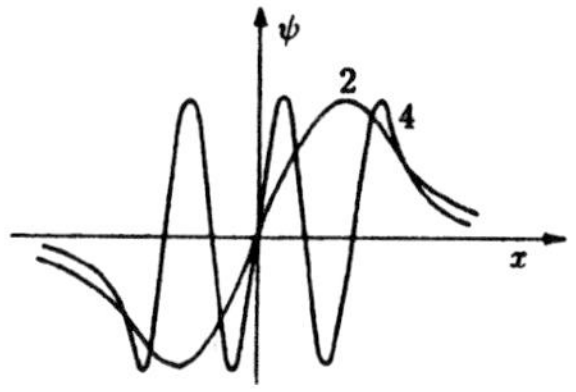

Wave functions with negative parity; they are antisymmetric relative to the origin

EXERCISE

6.3 The Delta Potential

Suppose we have a potential of the form

$$V(x) = -V_0\delta(x) \ ; \quad V_0 > 0 \ ; \quad x \in \mathbb{R} \ .$$

The corresponding wave function $\psi(x)$ is supposed to be smooth.

Problem. (a) Search for the bound states ($E < 0$) which are localized at this potential.

(b) Calculate the scattering of an incoming plane wave at this potential and find the *coefficient of reflection*

$$R = \left.\frac{|\psi_{\text{ref}}|^2}{|\psi_{\text{in}}|^2}\right|_{x=0} \ ,$$

where ψ_{ref}, ψ_{in} are the reflected and incoming waves, respectively.

Hint: To evaluate the behaviour of $\psi(x)$ at $x = 0$, integrate the Schrödinger equation over the interval $(-\varepsilon, +\varepsilon)$ and consider the limit $\varepsilon \to 0$.

Exercise 6.3

Solution. (a) The Schrödinger equation is given by

$$\left[-\frac{\hbar^2}{2m}\frac{\mathrm{d}^2}{\mathrm{d}x^2}-V_0\delta(x)\right]\psi(x)=E\psi(x)\ . \tag{1}$$

Away from the origin we have a differential equation of the form

$$\frac{\mathrm{d}^2}{\mathrm{d}x^2}\psi(x)=-\frac{2mE}{\hbar^2}\psi(x)\ . \tag{2}$$

The wave functions are therefore of the form

$$\psi(x)=A\,\mathrm{e}^{-\beta x}+B\,\mathrm{e}^{\beta x}\quad \text{if}\quad x>0\quad \text{or}\quad x<0\ , \tag{3}$$

with $\beta=\sqrt{-2mE/\hbar^2}\in\mathbb{R}$. As $|\psi|^2$ must be integrable, there cannot be an exponentially increasing part. Furthermore the wave function should be continuous at the origin. Hence,

$$\begin{aligned}\psi(x)&=A\,\mathrm{e}^{\beta x}\ ;\quad (x<0)\ ,\\ \psi(x)&=A\,\mathrm{e}^{-\beta x}\ ;\quad (x>0)\ .\end{aligned} \tag{4}$$

Integrating the Schrödinger equation from $-\varepsilon$ to $+\varepsilon$, we get

$$-\frac{\hbar^2}{2m}[\psi'(\varepsilon)-\psi'(-\varepsilon)]-V_0\psi(0)=E\int\limits_{-\varepsilon}^{+\varepsilon}\psi(x)\,\mathrm{d}x\approx 2\varepsilon E\psi(0)\ . \tag{5}$$

Inserting now result (4) and taking the limit $\varepsilon\to 0$, we have

$$-\frac{\hbar^2}{2m}(-\beta A-\beta A)-V_0A=0 \tag{6}$$

or $E=-m(V_0^2/2\hbar^2)$. Clearly (though surprisingly) there is only one energy eigenvalue. The normalization constant is found to be $A=\sqrt{mV_0/\hbar^2}$.

(b) The wave function of a plane wave is described by

$$\psi(x)=A\,\mathrm{e}^{\mathrm{i}kx}\ ,\quad k^2=\frac{2mE}{\hbar^2}\ . \tag{7}$$

It moves from the left-hand to the right-hand side and is reflected at the potential. If B and C are the amplitudes of the reflected and transmitted waves, respectively, we get

$$\begin{aligned}\psi(x)&=A\,\mathrm{e}^{\mathrm{i}kx}+B\,\mathrm{e}^{-\mathrm{i}kx}\ ;\quad (x<0)\ ,\\ \psi(x)&=C\,\mathrm{e}^{\mathrm{i}kx}\ ;\quad (x>0)\ .\end{aligned} \tag{8}$$

Conditions of continuity and the relation $\psi'(\varepsilon)-\psi'(-\varepsilon)=-f\psi(0)$ with $f=2mV_0/\hbar^2$ give

$$\left.\begin{aligned}A+B&=C\\ \mathrm{i}k(C-A+B)&=-fC\end{aligned}\right\}\Rightarrow\left\{\begin{aligned}B&=-\frac{f}{f+2\mathrm{i}k}A\ ,\\ C&=\frac{2\mathrm{i}k}{f+2\mathrm{i}k}A\ .\end{aligned}\right. \tag{9}$$

Exercise 6.3

The desired coefficient of reflection is therefore

$$R=\left.\frac{|\psi_{\mathrm{ref}}|^2}{|\psi_{\mathrm{in}}|^2}\right|_{x=0}=\frac{|B|^2}{|A|^2}=\frac{m^2V_0^2}{m^2V_0^2+\hbar^4k^2}\ . \tag{10}$$

If the potential is extremely strong ($V_0\to\infty$) $R\to 1$, i.e. the whole wave is reflected.

The *coefficient of transmission* is, on the other hand,

$$T=\left.\frac{|\psi_{\mathrm{trans}}|^2}{|\psi_{\mathrm{in}}|^2}\right|_{x=0}=\frac{|C|^2}{|A|^2}=\frac{\hbar^4k^2}{m^2V_0^2+\hbar^4k^2}\ . \tag{11}$$

If the potential is very strong, ($V_0\to\infty$) $T\to 0$, i.e. the transmitted wave vanishes.

Obviously, $R+T=1$ as is to be expected.

EXERCISE

6.4 Distribution Functions in Quantum Statistics

Let a quantum-mechanical system have the energy eigenvalues ε_i which are degenerate g_i times and are each occupied with n_i particles in such a way that

$$\sum_i n_i = N(\approx 10^{23}) \tag{1}$$

is the total number of indistinguishable particles and

$$\sum_i n_i\varepsilon_i = E \tag{2}$$

is the total energy (for example, an ideal gas enclosed in a box).

Problem. (a) A state can be occupied by one fermion only, but with an unlimited number of bosons. This is the consequence of the *Pauli principle*. (We will return to it later in Chap. 14.) Prove that it is possible to distribute n_i particles over g_i states with

$$\text{(I)}\quad W_i^{\mathrm{FD}}=\binom{g_i}{n_i}=\frac{g_i!}{n_i!(g_i-n_i)!}\qquad\text{(Fermi–Dirac statistics)}$$

possibilities in the case of fermions, with

$$\text{(II)}\quad W_i^{\mathrm{BE}}=\binom{g_i+n_i-1}{n_i}\qquad\text{(Bose–Einstein statistics)}$$

Exercise 6.4

possibilities in the case of bosons, and with

$$\text{(III)} \quad W_i^{\mathrm{B}} = g_i^{n_i} \quad \text{(Boltzmann statistics)}$$

possibilities for classical, i.e. distinguishable, particles.

(b) We define $\{n_i\} = \{n_1, n_2, n_3, \dots\}$ as a distribution of particles with a "weight" $W\{n_i\}$, which is simply the number of possibilities of distributing exactly n_i particles in energy levels ε_i. Naturally the one most likely to be realized, i.e. the most probable distribution, is the one with the greatest weight.

Derive the variational principle from these remarks and by considering the constant number of particles and the constant total energy:

$$\delta[\ln(W\{\langle n_i\rangle\}) - \alpha N - \beta E] = 0 \; . \tag{3}$$

Here, the parameters α, β are Lagrange multipliers and $\{\langle n_i\rangle\}$ is the distribution searched for. Prove that in the case of $g_i \gg n_i \gg 1$ the average number of occupation, namely $\langle n_i\rangle$, for the level with index i is given by:

$$\langle n_i\rangle = g_i[\exp(\varepsilon_i - \mu)/kT + \delta]^{-1} \; , \tag{4}$$

with

$$\delta = \begin{cases} +1 & \text{for fermions} \\ -1 & \text{for bosons} \\ 0 & \text{for classical, distinguishable particles} \; . \end{cases} \tag{5}$$

Hint: Use Stirling's formula for calculating $n!$ for large values of n

$$n! = \sqrt{2\pi} n^{(n+1/2)} \mathrm{e}^{-n} \approx \left(\frac{n}{\mathrm{e}}\right)^n \; , \tag{6}$$

and then insert $x_i \in \mathbb{R}$ for $n_i \in \mathbb{N}$, i.e. change from a discrete n_i to a continuous variable x_i.

(c) Draw a diagram for the Fermi–Dirac distribution $\langle n_E\rangle^{\mathrm{FD}}$ as a function of the energy E at $T \approx 0$. How is the parameter μ to be interpreted?

Solution. (a) In order to understand the different statistics, we first consider the problem of n_i indistinguishable balls that are to be distributed into g_i boxes.

(I) *Fermi–Dirac statistics*. There are n_i distinguishable balls to be distributed into g_i boxes:

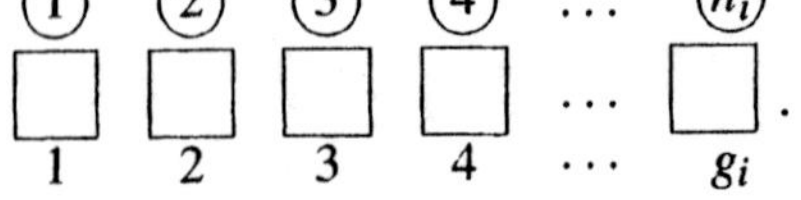

(7)

Each box can contain only one ball (the Fermi–Dirac case!). The first one could be placed in one of the g_i boxes. For the second one there are then $g_i - 1$ possibilities left, because one box is already occupied by the first ball. For the last ball,

Exercise 6.4

exactly $(g_i - n_i + 1)$ possibilities are found to exist if $g_i > n_i$. The total number of possibilities is given by the product

$$\begin{aligned} &g_i(g_i-1)(g_i-2)\dots(g_i-n_i+1)\\ &\quad=\frac{g_i(g_i-1)\dots(g_i-n_i+1)(g_i-n_i)\dots 1}{(g_i-n_i)!}\\ &\quad=\frac{g_i!}{(g_i-n_i)!} \end{aligned} \tag{8}$$

So far, we have assumed that the particles are distinguishable; if they are not distinguishable, however, there are several identical combinations. For example, the combination

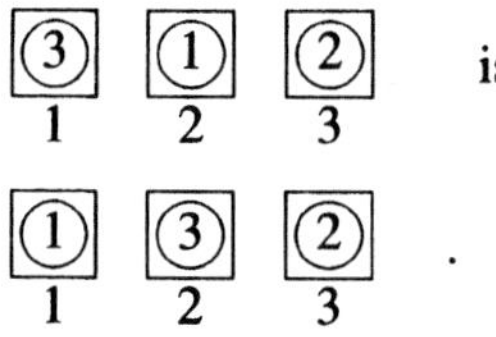

is identical with (9)

This means that we have overestimated the number of possibilities of distribution up to now by the number of permutations among the n_i particles. The permutations of n_i elements raise a factor $n_i!$, so that finally

$$W_i^{\mathrm{FD}} = \frac{g_i!}{n_i!(g_i-n_i)!} = \binom{g_i}{n_i}\;. \tag{10}$$

(II) *Bose–Einstein statistics.* Each box is now able to contain arbitrarily many, indistinguishable balls (the Bose–Einstein case!). Here the following method will work. There are

$$n+1 = \binom{n+1}{n}$$

possibilities of placing n ***Bose*** particles in two degenerate states. In three degenerate states there are

$$\binom{n+2}{2} = \binom{n+1}{1} + \binom{n}{1} + \binom{n-1}{1} + \dots + \binom{2}{1} + \binom{1}{1}$$

possibilities. Generally it holds that

$$W_i^{\mathrm{BE}} = \binom{g_i+n_i-1}{n_i} = \binom{g_i+n_i-1}{g_i-1} = \frac{(g_i+n_i-1)!}{(g_i-1)!n_i!}\;.$$

We want to reflect on this once more in a different way. For each particle and box, a slip of paper is marked with $K_1, \dots, K_{g_i}$ for the boxes, and with $B_1, \dots, B_{n_i}$ for the balls. Now slip K_1 is set aside and all other slips (both kinds, those labelled K_μ and those labelled B_μ) are placed in an urn; there should be $g_i + n_i - 1$

Exercise 6.4

left. Now the slips are taken out again, one by one and in an arbitrary sequence, and placed at the right of K_1; for example:

$$K_1 B_8 K_7 B_3 B_1 K_3 K_2 B_4 \dots \quad . \tag{11}$$

This can be interpreted as follows: balls located between two boxes are supposed to be in the left one. In our example, it would be ball number 8 that has to be in box number 1, whereas balls 3 and 1 belong to box number 7, no ball to box 3, ball 4 into box 2, and so on.

From part (I) we already know that there are $(g_i + n_i - 1)!$ arrangements possible [arrangements as shown in example (11)]. On the other hand the positions of the balls and their boxes in that row are unimportant. Indeed there are no new arrangements if the n_i balls are exchanged among themselves. The same is valid if the $(g_i - 1)K$ slips are exchanged. Thus there are

$$W_i^{\mathrm{BE}} = \frac{(g_i + n_i - 1)!}{n_i!(g_i - 1)!} = \binom{g_i + n_i - 1}{n_i} \tag{12}$$

different arrangements.

(III) *Boltzmann statistics:* Suppose we have two balls and g_i boxes. There are exactly g_i possibilities of distributing ball 1. If more than one ball is allowed in a single box, there are also g_i possibilities of distributing ball 2. Altogether there are g_i^2 arrangements. Analogously, for n_i balls there are

$$W_i^{\mathrm{B}} = g_i^{n_i} \tag{13}$$

arrangements.

(b) As we have indistinguishable particles in both the Fermi–Dirac statistics and Bose–Einstein statistics, the number of distribution arrangements for n_i particles into energy levels ε_i, if we have the particle distribution $\{n_1, \dots, n_m\}$, is given by the product

$$W^{\mathrm{FD}}\{n_1, n_2, n_3, \dots, n_m\} = \prod_{i=1}^{m} W_i^{\mathrm{FD}} \tag{14}$$

$$W^{\mathrm{BE}}\{n_1, n_2, n_3, \dots, n_m\} = \prod_{i=1}^{m} W_i^{\mathrm{BE}} \quad . \tag{15}$$

Let us turn now to ***Boltzmann*** statistics with distinguishable particles, where things are found to be more complicated. We are obliged to set $N = \sum_{i=1}^{m} n_i$ particles into the levels in such a way that there are n_i particles in each one. The number of possible ways of doing this is calculated in the following way. We start with $N!$ possibilities, disregarding the group structure containing m groups. We then correct this number by factors $n_j!$ coming from the arbitrary occupation in each group. This yields

$$N!/n_1!n_2!\dots n_m! \tag{16}$$

possibilities. Hence,

Exercise 6.4

$$W^{\mathrm{B}}\{n_1, n_2, n_3, \ldots, n_m\} = \frac{N!}{n_1! n_2! n_3! \cdots n_m!} \prod_{i=1}^{m} g_i^{n_i} , \tag{17}$$

which we can also write as

$$W^{\mathrm{B}}\{n_i\} = N! \prod_{i=1}^{m} \frac{g^{n_i}}{n_i!} . \tag{18}$$

The next step is to calculate the maximum of these various distributions $W\{n_i\}$, in order to find the particular distribution with the greatest weight. The maximum of $W^{\mathrm{FD}}\{n_i\}$ agrees with the maximum of $\ln(W^{\mathrm{FD}}\{n_i\})$, which is easier to handle mathematically. Therefore using Stirling's formula (6), we obtain

$$\begin{aligned} \ln W^{\mathrm{FD}}\{n_i\} &= \sum_i \ln \frac{g_i!}{n_i!(g_i - n_i)!} \\ &\approx \sum_i \ln \frac{(2\pi)^{-1/2} g_i^{g_i+1/2} \mathrm{e}^{-g_i}}{n_i^{n_i+1/2} \mathrm{e}^{-n_i} (g_i - n_i)^{g_i - n_i + 1/2} \mathrm{e}^{-g_i + n_i}} \\ &\approx \sum_i \left[\ln \frac{1}{\sqrt{2\pi}} + g_i \ln g_i - n_i \ln n_i - (g_i - n_i) \ln(g_i - n_i) \right] . \end{aligned} \tag{19}$$

In the last two steps we have made use of $g_i \gg 1$, $n_i \gg 1$. To find the maxima of a distribution we admit continuous values for n_i, so that $n_i \to x_i \in \mathbb{R}$, and introduce two Lagrange multipliers[2] β and α to incorporate the conditions $E = \sum_i n_i \varepsilon_i$ and $N = \sum_i n_i$, respectively. The variational principle yields

$$\begin{aligned} &\delta \left[\ln W^{\mathrm{FD}}\{x_j\} - \beta \sum_j \varepsilon_j x_j - \alpha \sum_j x_j \right] \\ &= \sum_i \delta x_i \left[\frac{\partial}{\partial x_i} \ln W^{\mathrm{FD}}\{x_j\} - \beta \varepsilon_i - \alpha \right] \\ &= \sum_i \delta x_i \left(\ln \frac{g_i - x_i}{x_i} - \beta \varepsilon_i - \alpha \right) . \end{aligned} \tag{20}$$

The last step is found as a result of (19), thus:

$$\begin{aligned} &\frac{\partial}{\partial x_i} \sum_j [-x_j \ln x_j - (g_j - x_j) \ln(g_j - x_j)] \\ &= \delta_{ij} [-\ln x_j - 1 + \ln(g_j - x_j) + 1] \\ &= \ln(g_i - x_i) - \ln x_i = \ln \frac{g_i - x_i}{x_i} . \end{aligned} \tag{21}$$

[2] See H. Goldstein: *Classical Mechanics*, 2nd ed. (Addison-Wesley, Reading, MA 1980) or W. Greiner: *Theoretische Physik*, Vol. 2, *Mechanik*, 4th ed. (Verlag Harri Deutsch, Thun, Frankfurt a.M. 1981).

Exercise 6.4

A necessary condition for the existence of the maximum is that the terms in parentheses in (20) vanish; therefore we deduce that

$$\frac{g_i - x_i}{x_i} = \mathrm{e}^{\beta\varepsilon_i+\alpha} \Leftrightarrow x_i = (g_i - x_i)(\mathrm{e}^{\beta\varepsilon_i+\alpha})^{-1} \tag{22}$$

and then

$$\langle x_i\rangle_{\mathrm{FD}} \equiv \langle n_i\rangle_{\mathrm{FD}} = \frac{g_i}{\mathrm{e}^{\beta\varepsilon_i+\alpha}+1} \ . \tag{23}$$

Finally, it is convenient to rewrite the parameters β and α in terms of the customary (and more physical) quantities T and μ, so that

$$\langle x_i\rangle_{\mathrm{FD}} \equiv \langle n_i\rangle_{\mathrm{FD}} = g_i\{\exp[(\varepsilon_i - \mu)/kT]+1\}^{-1}$$

with

$$\beta = \frac{1}{kT} \ , \quad \alpha = -\frac{\mu}{kT} = -\mu\beta \ . \tag{24}$$

Coming back to the case of Bose–Einstein statistics, we proceed analogously. From

$$\begin{aligned} \ln W^{\mathrm{BE}}\{n_i\} &= \sum_i \ln \frac{(g_i+n_i-1)!}{n_i!(g_i-1)!} \\ &\approx \sum_i \ln \Big\{(2\pi)^{-1/2}(g_i+n_i-1)^{g_i+n_i-1/2}\mathrm{e}^{-g_i-x_i+1} \\ &\quad \times \big[n_i^{n_i+1/2}\mathrm{e}^{-n_i}(g_i-1)^{g_i-1/2}\mathrm{e}^{-g_i+1}\big]^{-1}\Big\} \\ &\approx \sum_i \Big\{ \ln(2\pi)^{-1/2} + (g_i+n_i)\ln(g_i+n_i) \\ &\quad - n_i \ln n_i - g_i \ln g_i \Big\} \ , \end{aligned} \tag{25}$$

it follows that

$$\frac{\partial}{\partial x_i} \ln W^{\mathrm{BE}}\{x_i\} = \ln(g_i+x_i)+1-\ln x_i - 1 = \ln\frac{g_i+x_i}{x_i} \ , \tag{26}$$

and therefore,

$$\begin{aligned} &\delta\left[\ln W^{\mathrm{BE}}\{x_j\} - \beta\sum_j \varepsilon_j x_j - \alpha\sum_j x_j\right] \\ &\quad = \sum_i \delta x_i \left(\ln\frac{g_i+x_i}{x_i} - \beta\varepsilon_i - \alpha\right) \ . \end{aligned} \tag{27}$$

Thus we deduce that

$$\langle x_i\rangle_{\mathrm{BE}} \equiv \langle n_i\rangle_{\mathrm{BE}} = g_i\{\exp[(\varepsilon_i - \mu)/kT]-1\}^{-1} \ . \tag{28}$$

In the case of Boltzmann statistics, we again proceed analogously and find

Exercise 6.4

$$
\begin{aligned}
\ln W^{\mathrm{B}}\{n_i\} &= \ln(N!) + \sum_i \ln\left(\frac{g_i^{n_i}}{n_i!}\right) \\
&= \ln N! + \sum_i (n_i \ln g_i - \ln n_i!) \\
&\approx \ln N! + \sum_i (n_i \ln g_i - \ln[\sqrt{2\pi} n_i^{n_i+1/2} \mathrm{e}^{-n_i}]) \\
&\approx \ln N! + \sum_i (n_i \ln g_i - \ln \sqrt{2\pi} - n_i \ln n_i + n_i) \\
&= \ln N! + \sum_i (n_i - \ln \sqrt{2\pi} + n_i \ln g_i - n_i \ln n_i) \ , \qquad (29)
\end{aligned}
$$

so that

$$
\frac{\partial}{\partial x_i} \ln W^{\mathrm{B}}\{x_i\} = 1 + \ln g_i - \ln x_i - 1 = \ln \frac{g_i}{x_i} \ . \qquad (30)
$$

Here, the variational principle yields

$$
\begin{aligned}
&\delta\left[\ln W^{\mathrm{B}}\{x_j\} - \beta \sum_j \varepsilon_j x_j - \alpha \sum_j x_j\right] \\
&\quad = \sum_i \delta x_i \left[\ln\left(\frac{g_i}{x_i}\right) - \beta\varepsilon_i - \alpha\right] \ , \qquad (31)
\end{aligned}
$$

and thus, finally,

$$
(x_i)_{\mathrm{B}} \equiv \langle n_i \rangle_{\mathrm{B}} = g_i \exp[-(\varepsilon_i - \mu)/kT] \ . \qquad (32)
$$

The Lagrange multipliers, which were introduced in (24), are fixed by the conditions

$$
E = \sum_i \varepsilon_i x_i \quad \text{and} \quad N = \sum_i x_i \ . \qquad (33)
$$

Therefore

$$
E = E(\mu, T) \ , \quad N = N(\mu, T)
$$

are functions of μ and T; the chemical potential $\mu = \mu(N, T)$ is a function of the particle number N and of the temperature T. The interpretation of the parameter T is suggested by comparison of e.g. the first equation in (33) in the case of Bose–Einstein statistics with Planck's radiation law [see Example 2.2, Eq. (13)]. Hence, it is plausible that T will also be the temperature in the case of Fermi–Dirac statistics and Boltzmann statistics. For a stationary number of particles and fixed temperature T, μ is fixed, too. The quantity $K \ln(W\{\langle n_i \rangle\}) = S$ is called

Exercise 6.4

the entropy of the system, which coincides with the concept of entropy in thermodynamics. The entropy S gives information about the system's disorder. The bigger S, the "more wildly" the particles are spread over the levels. From (23), (28) and (32) we finally get the form for the distribution functions corresponding to the various statistics. In compact form, $\langle n_i \rangle$ can be given as a function of temperature as

$$\langle n_i \rangle = g_E \{\exp[(\varepsilon_i - \mu)/kT] + \delta\}^{-1} , \tag{34}$$

$$\text{using} \quad \delta = \begin{cases} +1 & \text{in the case of fermions} \\ -1 & \text{in the case of bosons} \\ 0 & \text{in the case of classical particles .} \end{cases}$$

(c) In this special case we have from (23)

$$\langle n_E \rangle^{\mathrm{FD}} = g_E \{\exp[(E - \mu)/kT] + 1\}^{-1} \tag{35}$$

and therefore

$$\lim_{T\to 0} \langle n_E \rangle^{\mathrm{FD}} = \begin{cases} 0 \; , & E > \mu \\ \frac{1}{2} \; , & E = \mu \equiv \Theta(E-\mu) \\ 1 \; , & E < \mu \end{cases}$$

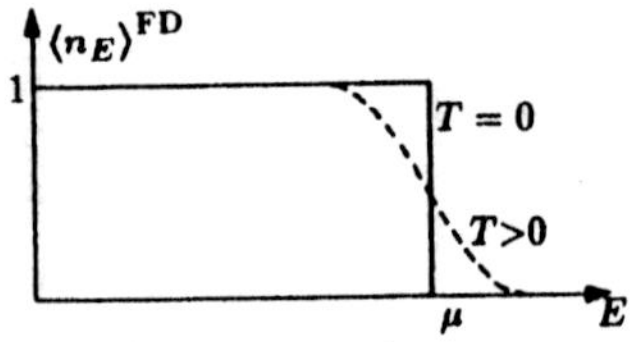

The Fermi–Dirac distribution as a function of the energy E

(solid line in the figure).

For $T \gtrsim 0$, corresponding to an energy kT, the Fermi level is "smeared" over the region $\Delta E \sim kT$ (see figure).

EXERCISE

6.5 The Fermi Gas

From the foregoing we know that the allowed energy levels of noninteracting particles of mass m in a three-dimensional, infinitely deep potential well with edges of length l are given by (see Exercise 6.1)

$$E = \frac{\pi^2 \hbar^2}{2ml^2}(n_x^2 + n_y^2 + n_z^2) \; ; \quad n_x, n_y, n_z \in \mathbb{N} \; . \tag{1}$$

For electrons in a metal or for gas molecules in a container, the value of l is supposed to be so large that the energy spectrum may be regarded as continuous.

Problem. (a) Show that the number of states ΔN of energies in the interval between E and $E + \Delta E$ is given by

$$\Delta N = \frac{1}{4\pi^2}\left(\frac{\sqrt{2m}}{\hbar}\right)^3 V E^{1/2} \Delta E \; . \tag{2}$$

Exercise 6.5

(b) Calculate the ***Fermi*** energy $\varepsilon_\mathrm{f} = \mu\ (T = 0)$ for an ideal Fermi gas of N particles. Make use of the fact that each energy level is twice degenerate because of the spin degree of freedom.

Solution. (a) Instead of counting the states (n_x, n_y, n_z) of equal energy (which would not be easy), we use the *quasi-continuous approximation* and determine the number of points inside a spherical shell of radius

$$n = \sqrt{n_x^2 + n_y^2 + n_z^2} \tag{3}$$

and thickness Δn in the first octant in (n_x, n_y, n_z)-space, since $n_x, n_y, n_z \in \mathbb{N}$. This is simply the "volume"

$$\Delta N = \tfrac{1}{8}(4\pi n^2)\Delta n = \tfrac{\pi}{2} n^2 \Delta n \ . \tag{4}$$

According to the assumption $n^2 = E(2ml^2/\pi^2\hbar^2)$, we have

$$n = \frac{l\sqrt{2m}}{\pi\hbar} E^{1/2} \ , \tag{5}$$

$$\Delta n = \frac{\mathrm{d}n}{\mathrm{d}E}\Delta E = \frac{l\sqrt{2m}}{2\pi\hbar} E^{-1/2}\Delta E \quad \text{and} \tag{6}$$

$$\Delta N = [(2m)^{3/2} V/(4\pi^2\hbar^3)] E^{1/2}\Delta E = C E^{1/2} \Delta E \ ,$$
$$C = (2m)^{3/2} V/(4\pi^2\hbar^3) \ , \quad V = l^3 \ . \tag{7}$$

(b) The energy of the discrete case is given by

$$E_T = \sum_i \langle n_i \rangle_T \varepsilon_i \ . \tag{8}$$

If the energies are continuous, this expression changes into

$$E_T = \sum_{i=0}^{\infty} \frac{\langle n_i \rangle}{\Delta\varepsilon_i} \varepsilon_i \Delta\varepsilon_i = \sum_{i=0}^{\infty} g(E_i) f(E_i) E_i \Delta E_i$$
$$\xrightarrow[\Delta E_i \to 0]{} \int_0^\infty g(E) f(E) E \,\mathrm{d}E \ . \tag{9}$$

Here, $g(E)$ is the number of states in the energy interval ΔE; $f(E)$ describes the fraction of *occupied* states. In fact, we are able to interpret the energy interval $(E, E + \Delta E)$ in the continuous case (for sufficiently small ΔE) as *one* level with degeneracy $g_{i=E} = g(E)\Delta E = s\Delta N$ ($s = 1$ for bosons, $s = 2$ for fermions of spin $\frac{1}{2}$). The expression $f(E) = \langle n_E \rangle / g(E)\Delta E$ is called the *distribution function* and $g(E)$ the *density of states*.

Because of spin degeneracy ($s = 2$), we have from (7) for fermions:

$$g(E) = 2CE^{1/2} \ ,$$
$$f^{\mathrm{FD}}(E) = [\exp(E - \mu)/kT + 1]^{-1} \ . \tag{10}$$

Exercise 6.5

For $T=0$, $f^{\mathrm{FD}}_{T=0}(E)=\theta(E-\varepsilon_{\mathrm{f}})$ [see Exercise 6.4; $\theta(E-\varepsilon_{\mathrm{f}})$ is the Heaviside step function] and the number of particles N is assumed to be constant, i.e.

$$N=\sum_{i=0}^{\infty}\langle n_i\rangle \to \int_0^{\infty} f^{\mathrm{FD}}_{T=0}(E)g(E)\,\mathrm{d}E$$

$$=\int_0^{\infty}\theta(E-\varepsilon_{\mathrm{f}})2CE^{1/2}\,\mathrm{d}E=2C\int_0^{\varepsilon_{\mathrm{f}}}E^{1/2}\,\mathrm{d}E=\frac{4}{3}C\varepsilon_{\mathrm{f}}^{3/2}$$

$$\Rightarrow \varepsilon_{\mathrm{f}}=\left(\frac{3}{4}\frac{N}{C}\right)^{2/3}=(3\pi^2)^{3/2}\frac{\hbar^2}{2m}\left(\frac{N}{V}\right)^{2/3}\ . \tag{11}$$

Application: an electron gas in a metal can be considered a Fermi gas. The mean energy at temperature $T=0$ is then

$$\bar{E}=\frac{\int_0^{\varepsilon_{\mathrm{f}}}EE^{1/2}\,\mathrm{d}E}{\int_0^{\varepsilon_{\mathrm{f}}}E^{1/2}\,\mathrm{d}E}=\frac{3}{5}\varepsilon_{\mathrm{f}}\ . \tag{12}$$

From this quantity we can calculate the pressure p of the electron gas (*zero-point pressure*). It is possible to define the pressure as the work $\mathrm{d}A$ that has to be done to reduce the volume by $\mathrm{d}V$: $\mathrm{d}A=p|\mathrm{d}V|$. This work is equivalent to the increase of contained energy, which is $N\bar{E}$. Then, without changing the total number of electrons during compression, we simply obtain from (12) and (11):

$$p=N\left|\frac{\mathrm{d}\bar{E}}{\mathrm{d}V}\right|=\frac{2}{5}\frac{N}{V}\varepsilon_{\mathrm{f}}\ . \tag{13}$$

Taking silver as an example, we get a zero-point pressure of approximately 2×10^5 times the atmospheric pressure. This enormous pressure just means that the compression of the electron gas requires a large amount of work. The result is a typical quantum-mechanical effect: if all electrons were in the lowest state, then only the much smaller amount of work involved in raising the zero-level energy of each electron would be required, approaching zero with increasing volume. The *Pauli principle* causes occupation of higher states, the mutual distance of which increases with a decrease in the volume as a consequence of the uncertainty principle. Therefore the Pauli principle is essential to the explanation of the limited compressibility of solids.

EXERCISE

6.6 An Ideal Classical Gas

Problem. Show that for the ideal classical gas

$$E = \tfrac{3}{2} NkT$$

is valid, where E is the total energy and N the total number of particles. What is the interpretation of the temperature T in this case?

Hint: Use $\int_0^\infty x^{t-1}\,\mathrm{e}^{-x}\,\mathrm{d}x = \Gamma(t)$ together with $\Gamma(t+1) = t\Gamma(t)$ and $\Gamma(\frac{3}{2}) = \sqrt{\pi}$.

Solution. For the ideal Boltzmann gas we have [see Exercise 6.4, Eq. (34)]:

$$f^{\mathrm{B}}(E) = \mathrm{e}^{(\mu - E)/kT} \quad \text{and} \quad g(E) = CE^{1/2} \ . \tag{1}$$

Then

$$\begin{aligned} E(T,\mu) &= \int\limits_0^\infty f^{\mathrm{B}}(E) g(E) E\,\mathrm{d}E = C\,\mathrm{e}^{\mu/kT} \int\limits_0^\infty E^{3/2}\,\mathrm{e}^{-E/kT}\,\mathrm{d}E \\ &= C\,\mathrm{e}^{\mu/kT} (kT)^{5/2} \int\limits_0^\infty x^{3/2}\,\mathrm{e}^{-x}\,\mathrm{d}x \\ &= \Gamma(5/2) C (kT)^{5/2}\,\mathrm{e}^{\mu/kT} \quad \text{and} \end{aligned} \tag{2}$$

$$\begin{aligned} N &= \int\limits_0^\infty f^{\mathrm{B}}(E) g(E)\,\mathrm{d}E = C\,\mathrm{e}^{\mu/kT} (kT)^{3/2} \int\limits_0^\infty x^{1/2}\,\mathrm{e}^{-x}\,\mathrm{d}x \\ &= \Gamma(3/2) C (kT)^{3/2}\,\mathrm{e}^{\mu/kT} \ . \end{aligned} \tag{3}$$

It follows from (2) and (3) that

$$E/N = \tfrac{3}{2} kT \ , \quad \text{i.e.} \quad E(T) = \tfrac{3}{2} NkT \ . \tag{4}$$

The mean energy of a particle is $E/N = \frac{3}{2}kT$. This result elucidates the meaning of the parameter T, at least in the case of the classical ideal gas: T is directly proportional to the energy per particle.

EXERCISE

6.7 A Particle in a Two-Centred Potential

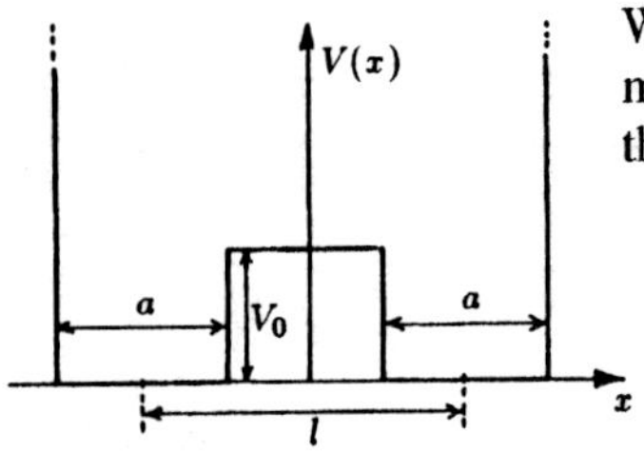

The schematic two-centred potential

We consider the following simple model of a diatomic (one-dimensional) molecule, consisting of two potential wells lying side by side (see figure); here, the potential is given by ($a < l$).

$$V(x) = \begin{cases} \infty & |x| > (l+a)/2 \\ 0 \quad \text{for} & (l-a)/2 < |x| < (l+a)/2 \\ V_0 & |x| < (l-a)/2 \ . \end{cases}$$

Particles of mass m described by the Schrödinger equation move in this potential. We wish to find the eigenvalues ($0 < E < V_0$) of the Hamiltonian operator and the corresponding wave functions. We shall then compare this result with the result for two atoms at large distance ($l \gg a$) and discuss why molecular binding is established.

We write down the Schrödinger equation,

$$E\psi(x) = \left[-\hbar^2 \frac{\Delta}{2m} + V(x) \right] \psi(x) \ ,$$

for the regions II, III, IV.

$$\begin{aligned} \text{II, IV}: &\quad -\frac{\hbar^2 \Delta}{2m}\psi(x) = E\psi(x) \ ; \\ \text{III}: &\quad \left(-\frac{\hbar^2 \Delta}{2m} + V_0 \right) \psi(x) = E\psi(x) \ . \end{aligned} \tag{1}$$

In regions I and V, the wave function has to vanish, $\psi_{\text{I}} = \psi_{\text{V}} = 0$. These equations give

$$\begin{aligned} &(\text{I}) \quad \psi = 0 && x < -\tfrac{1}{2}(l+a) \ , \\ &(\text{II}) \quad \psi = A' \sin kx + A'' \cos kx && -\tfrac{1}{2}(l+a) < x < -\tfrac{1}{2}(l-a) \ , \\ &(\text{III}) \quad \psi = B e^{\beta x} + C e^{-\beta x} && -\tfrac{1}{2}(l-a) < x < +\tfrac{1}{2}(l-a) \ , \\ &(\text{IV}) \quad \psi = D' \sin kx + D'' \cos kx && +\tfrac{1}{2}(l-a) < x < +\tfrac{1}{2}(l+a) \ , \\ &(\text{V}) \quad \psi = 0 && +\tfrac{1}{2}(l-a) < x \ , \end{aligned} \tag{2}$$

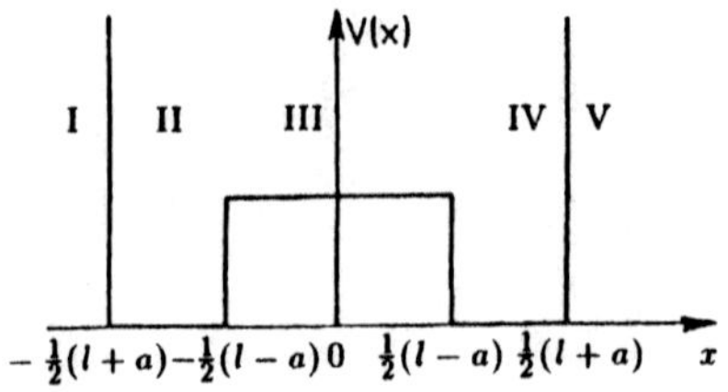

The different regions of the potential

with

$$k=\sqrt{\frac{2mE}{\hbar^2}} \quad \text{and} \quad \beta=\sqrt{\frac{2m(V_0-E)}{\hbar^2}} \ . \tag{3}$$

The wave function has to be continuous; in particular, we must have $\psi_{\text{I}}(-\frac{1}{2}(l-a))=\psi_{\text{II}}(-\frac{1}{2}(l-a))$, etc. Furthermore, the derivative has to be continuous for $|x|=\frac{1}{2}(l-a)$. Therefore we have to "join" the different solutions in an appropriate way, which will only be possible for certain energies $E(\sim k,\beta)$. The boundary condition at the points $x=\pm\frac{1}{2}(l+a)$ immediately implies

$$\begin{aligned} &(\text{II}) \quad \psi(x)=A\sin k\left[x+\tfrac{1}{2}(l+a)\right] \ , \\ &(\text{IV}) \quad \psi(x)=D\sin k\left[x-\tfrac{1}{2}(l-a)\right] \ . \end{aligned} \tag{4}$$

Now we still have four boundary conditions:

$$\begin{aligned} B\,\mathrm{e}^{-\beta/2(l-a)}+C\,\mathrm{e}^{+\beta/2(l-a)} &= A\sin ka \ , \\ B\beta\,\mathrm{e}^{-\beta/2(l-a)}-C\beta\,\mathrm{e}^{+\beta/2(l-a)} &= Ak\cos ka \ , \\ B\,\mathrm{e}^{+\beta/2(l-a)}+C\,\mathrm{e}^{-\beta/2(l-a)} &= -D\sin ka \ , \\ B\beta\,\mathrm{e}^{+\beta/2(l-a)}-C\beta\,\mathrm{e}^{-\beta/2(l-a)} &= Dk\cos ka \ , \end{aligned}$$

or

$$\begin{aligned} 2(B+C)\cosh\frac{\beta}{2}(l-a) &= (A-D)\sin ka \ , \\ 2(B-C)\beta\cosh\frac{\beta}{2}(l-a) &= (A+D)k\cos ka \ , \\ -2(B-C)\sinh\frac{\beta}{2}(l-a) &= (A+D)\sin ka \ , \\ -2(B+C)\beta\sinh\frac{\beta}{2}(l-a) &= (A-D)k\cos ka \ . \end{aligned} \tag{5}$$

These are four (partially decoupled) equations for the variables $B+C$, $B-C$, $A+D$ and $A-D$ that are not allowed to vanish all together. Now $\frac{1}{2}\beta(l-a)$ is always $\neq 0$; therefore we can divide by the hyperbolic functions,

$$\begin{aligned} B+C &= \frac{\sin ka}{2\cosh\frac{\beta}{2}(l-a)}(A-D) = -\frac{k\cos ka}{2\beta\sinh\frac{\beta}{2}(l-a)}(A-D) \ , \\ B-C &= \frac{-\sin ka}{2\sinh\frac{\beta}{2}(l-a)}(A+D) = \frac{k\cos ka}{2\beta\cosh\frac{\beta}{2}(l-a)}(A+D) \ . \end{aligned} \tag{6}$$

Hence,

$$\underbrace{\left\{A-D=B+C=0\right.}_{A_1}$$

Exercise 6.7

or

$$\underbrace{\left[\frac{1}{k}\tan ka = -\frac{1}{\beta}\coth\frac{\beta}{2}(l-a) \quad \text{and} \quad B+C = \frac{\sin ka}{2\cosh\frac{\beta}{2}(l-a)}(A-D)\right]\Bigg\}}_{A_2} \tag{7}$$

and

$$\underbrace{\Big\{A+D = B-C = 0}_{A_3}$$

or

$$\underbrace{\left[\frac{1}{k}\tan ka = -\frac{1}{\beta}\tanh\frac{\beta}{2}(l-a) \quad \text{and} \quad B-C = \frac{-\sin ka}{2\sinh\frac{\beta}{2}(l-a)}(A-D)\right]\Bigg\}}_{A_4}. \tag{8}$$

The conditions A_1 to A_4 are either true or false. Therefore we abbreviate the above statements using logic symbols:

$$(A_1 \vee A_2) \wedge (A_3 \vee A_4) = \text{true} \ ;$$

$\vee$ means "or" and $\wedge$ means "and".

Expanding gives

$$(A_1 \wedge A_3) \vee (A_1 \wedge A_4) \vee (A_2 \wedge A_3) \vee (A_2 \wedge A_4) = \text{true} \ .$$

First of all, we show that the conditions $(A_1 \wedge A_3)$ and $(A_2 \wedge A_4)$ cannot be true:

$$(A_1 \wedge A_3) = \text{true} \quad \Leftrightarrow \quad A = B = C = D = 0 \ . \tag{9}$$

Since all the variables are not allowed to vanish, we have

$$\begin{aligned} &(A_1 \wedge A_3) = \text{false} \\ &(A_2 \wedge A_4) = \text{true} \quad \Rightarrow \quad \tanh\frac{\beta}{2}(l-a) = \coth\frac{\beta}{2}(l-a) \ . \end{aligned} \tag{10}$$

The last equation is true only if $\beta = 0$; this case has also been excluded above. Hence, $(A_2 \wedge A_4) = \text{false}$, too. Our logical equation is therefore reduced to

$$(A_1 \wedge A_4) \vee (A_2 \wedge A_3) = \text{true} \ .$$

Inserting the definitions gives the following system of equations:

Exercise 6.7

$$\left\{\begin{aligned} A_1 - D_1 &= B_1 + C_1 = 0\ , \\ \tfrac{1}{k_1}\tan k_1 a &= -\tfrac{1}{\beta_1}\tanh\tfrac{\beta_1}{2}(l-a)\ , \\ B_1 &= \frac{-\sin k_1 a}{2\sinh\frac{\beta_1}{2}(l-a)} A_1\ , \end{aligned}\right\} \quad \text{or}$$

$$\left\{\begin{aligned} A_2 + D_2 &= B_2 - C_2 = 0\ , \\ \tfrac{1}{k_2}\tan k_2 a &= -\tfrac{1}{\beta_2}\coth\tfrac{\beta_2}{2}(l-a)\ , \\ B_2 &= \frac{-\sin k_2 a}{2\cosh\frac{\beta_2}{2}(l-a)} A_2 \end{aligned}\right\} \ . \tag{11}$$

The two middle equations yield the eigenvalues $E_1^{(i)}$ and $E_2^{(i)}$, respectively, as the solution of a complicated equation

$$\left\{k_i = (2mE_i/\hbar^2)^{1/2}\ , \quad \beta_i = [2m(V_0 - E_i)/\hbar^2]^{1/2}\right\}\ . \tag{12}$$

Now we drop the upper index, which numbers the different solutions. The wave functions in the regions II, III, IV are consequently

$$\begin{aligned} &\text{(II)} && \psi_1 = A_1 \sin k_1[x + \tfrac{1}{2}(l+a)]\ , \\ &&& \psi_2 = A_2 \sin k_2[x + \tfrac{1}{2}(l+a)]\ , \\ &\text{(III)} && \psi_1 = 2B_1 \sinh\beta_1 x\ , \\ &&& \psi_2 = 2B_2 \cosh\beta_2 x\ , \\ &\text{(IV)} && \psi_1 = A_1 \sin k_1[x - \tfrac{1}{2}(l+a)]\ , \\ &&& \psi_2 = -A_2 \sin k_2[x - \tfrac{1}{2}(l+a)]\ . \end{aligned} \tag{13}$$

Strictly speaking, we should now normalize the wave functions; this, however, we will not do. Since they belong to different eigenvalues, they are certainly orthogonal. The normalization determines the A_i and therefore also the B_i up to a phase.

We note that ψ_1 and ψ_2 are also eigenfunctions of the parity operator: ψ_1 possesses negative parity, ψ_2 positive parity.

In the case of large distances ($l \to \infty$), i.e. two distinct atoms, both equations determining the energy become identical,

$$\tan ka = -\frac{k}{\beta} \quad \text{or} \quad \tan\frac{a}{\hbar}\sqrt{2mE} = -\sqrt{\frac{E}{V_0 - E}}\ . \tag{14}$$

This equation is not analytically solvable, either. [If we analyse the extreme case $V_0 \to \infty$, we get two atoms which are separated by a very high potential wall and are therefore also independent. Here we have $\tan(a/\hbar)\sqrt{2mE} = 0$ (see Exercise 6.1).]

Exercise 6.7

Now we want to derive an approximate relation for the energy differences $E_1 - E = \Delta_1$ and $E_2 - E = \Delta_2$. E is the energy for two completely separated atoms ($l \to \infty$, $V_0 < \infty$); E_i ($i = 1, 2$) is the energy for two atoms at a finite but large distance l. We assume $\beta_i(l-a) \gg 1$, i.e. we examine the lower energy states. To abbreviate, we define $b = \frac{1}{2}(l-a)$; therefore $\beta_i b \gg 1$. k and β are assumed to have the same meaning as above ($l \to \infty$).

With the aid of Taylor's formula, it follows that

$$\begin{aligned}
k_i &= \frac{\sqrt{2m}}{\hbar}\sqrt{E+\Delta_i} = \frac{\sqrt{2mE}}{\hbar}\sqrt{1+\frac{\Delta_i}{E}} = k\left(1+\frac{\Delta_i}{2E}\right) , \\
\beta_i &= \frac{\sqrt{2m}}{\hbar}\sqrt{V_0 - E - \Delta_i} = \beta\sqrt{1-\frac{\Delta_i}{V_0-E}} \\
&= \beta\left(1-\frac{\Delta_i}{2(V_0-E)}\right) ,
\end{aligned} \tag{15}$$

if we drop expressions containing

$$\Delta_i^2 , \quad \Delta_i \mathrm{e}^{-2\beta b} \quad \text{and} \quad -\mathrm{e}^{-4\beta b} .$$

With this we get, using $\tan ka = -k/\beta$,

$$\begin{aligned}
\tan k_i a &= \tan\left(ka + ka\frac{\Delta_i}{2E}\right) = \tan ka + \frac{1}{\cos^2 ka} ka\frac{\Delta_i}{2E} \\
&= \tan ka + (1+\tan^2 ka) ka\frac{\Delta_i}{2E} \\
&= -\frac{k}{\beta} + \left(1+\frac{k^2}{\beta^2}\right) ka\frac{\Delta_i}{2E} \\
&= -\frac{k}{\beta} + \frac{ka}{2}\frac{\Delta_i}{E}\left(1+\frac{E}{V_0-E}\right) \\
&= -\frac{k}{\beta} + \frac{kaV_0}{2E(V_0-E)}\Delta_i ,
\end{aligned} \tag{16}$$

$$\begin{aligned}
\tanh \beta_1 b &= \tanh\left[\beta b - \beta b\frac{\Delta_1}{2(V_0-E)}\right] \\
&= \tanh \beta b - \frac{1}{\cosh^2 \beta b}\beta b\frac{\Delta_1}{2(V_0-E)} = \tanh \beta b \\
&= 1 + (\tanh \beta b - 1) = 1 + \frac{\sinh \beta b - \cosh \beta b}{\cosh \beta b} \\
&= 1 - 2\mathrm{e}^{-2\beta b} ,
\end{aligned} \tag{17}$$

$$\coth \beta_2 b = \coth \beta b = 1 + 2\mathrm{e}^{-2\beta b} .$$

Consequently

$$\frac{1}{k_i}\tan k_i a = \frac{1}{k}\left(1-\frac{\Delta_i}{2E}\right)\left[-\frac{k}{\beta} + \frac{kaV_0}{2E(V_0-E)}\Delta_i\right]$$

Exercise 6.7

$$= -\frac{1}{\beta} + \left[\frac{1}{2\beta E} + \frac{aV_0}{2E(V_0-E)}\right]\Delta_i$$
$$= -\frac{1}{\beta} + \frac{V_0(1+\alpha\beta)-E}{2\beta E(V_0-E)}\Delta_i \ ,$$

$$-\frac{1}{\beta_1}\tanh\beta_1 b = -\frac{1}{\beta}\left[1+\frac{\Delta_1}{2(V_0-E)}\right](1-2\mathrm{e}^{-2\beta b})$$
$$= -\frac{1}{\beta}+\frac{2}{\beta}\mathrm{e}^{-2\beta b} - \frac{\Delta_1}{2\beta(V_0-E)} \ ,$$
$$-\frac{1}{\beta_2}\coth\beta_2 b = -\frac{1}{\beta}-\frac{2}{\beta}\mathrm{e}^{-2\beta b} - \frac{\Delta_2}{2\beta(V_0-E)} \ . \tag{18}$$

Hence, the eigenvalue equation for E_1 reads

$$-\frac{1}{\beta}+\frac{V_0(1+\alpha\beta)-E}{2\beta E(V_0-E)}\Delta_1 = -\frac{1}{\beta}+\frac{2}{\beta}\mathrm{e}^{-2\beta b} - \frac{\Delta_1}{2\beta(V_0-E)} \tag{19}$$

with the solution

$$\Delta_1 = 4\frac{E(V_0-E)}{V_0(1+\alpha\beta)}\mathrm{e}^{-\beta(l-a)} \ . \tag{20}$$

Correspondingly, $\Delta_2 = -\Delta_1$. The energy E_1 exceeds the energy of a single atom; the energy E_2 is, however, smaller. Hence, we get the following figure:

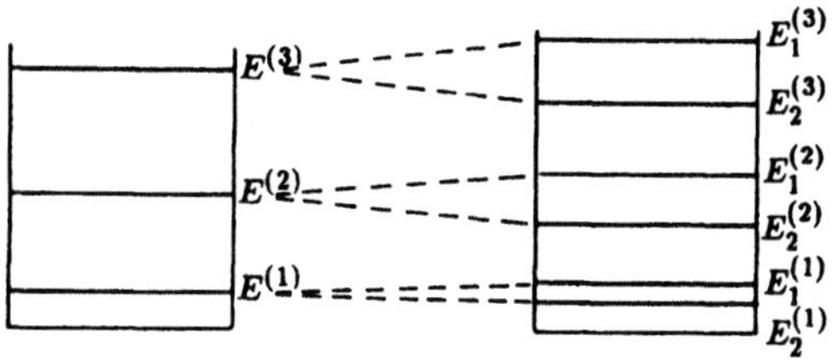

$l \to \infty$: two separated potential wells with energy levels $E^{(1)}$, $E^{(2)}$, … ; l large, but finite: the formerly degenerate energy levels split up; one level is lowered, the other one is elevated in energy

If we bring together two potential wells (atoms) with an electron in the state of lowest energy, we will decrease the energy, i.e. we get a bound state for the molecule.

Our model is, of course, somewhat unrealistic. For one thing, we should question how well atoms can be described by one-dimensional rectangular potential wells. Furthermore, we have to take into account that the two atomic nuclei are repulsive, so that binding is hampered. The two atoms will adjust themselves to a distance at which the force of attraction caused by the change of energy of the electron is as large as the repulsion of the two nuclei. In this state, the total energy of the system (2 atoms $+1\,\mathrm{e}^-$) is minimal.

6.1 The Conservation of Particle Number in Quantum Mechanics

In electrodynamics, the well-known continuity equation is valid:

$$\frac{\partial \varrho_\mathrm{e}}{\partial t} + \operatorname{div} \boldsymbol{j}_\mathrm{e} = 0 \ , \tag{6.14}$$

where ϱ_e is the charge density and $\boldsymbol{j}_\mathrm{e}$, the current density. This equation is the law of conservation for the electric charge: if the charge density in a volume element changes, then a current flows through the surface of the volume element (*Gauss' law*).

Now we try to find a similar relation for the number of particles in a region. Instead of the charge density, we consider the probability density $w = \psi^*\psi$. If we demand that no particles be created or annihilated, we also have a continuity equation:

$$\frac{\partial w}{\partial t} + \operatorname{div} \boldsymbol{j} = 0 \ . \tag{6.15}$$

Our aim is to deduce the particle current density $\boldsymbol{j}$. To this end, we begin with the time-dependent Schrödinger equation

$$\frac{\partial \psi}{\partial t} = \frac{1}{\mathrm{i}\hbar}\hat{H}\psi \ . \tag{6.16}$$

The complex-conjugated equation is

$$\frac{\partial \psi^*}{\partial t} = -\frac{1}{\mathrm{i}\hbar}\hat{H}^*\psi^* \ . \tag{6.17}$$

Multiplying the first equation by ψ^* from the left and the second by ψ and adding both, we get

$$\frac{\partial}{\partial t}(\psi^*\psi) + \frac{\mathrm{i}}{\hbar}(\psi^*\hat{H}\psi - \psi\hat{H}^*\psi^*) = 0 \ . \tag{6.18}$$

If we assume that the potential is independent of velocity and real, we are able to insert $\hat{H} = \hat{p}^2/2m + V(r)$ and obtain

$$\frac{\partial}{\partial t}(\psi^*\psi) + \frac{\mathrm{i}\hbar}{2m}(\psi\nabla^2\psi^* - \psi^*\nabla^2\psi) = 0 \ . \tag{6.19}$$

From the second expression in brackets we can extract a nabla or del operator:

$$\begin{aligned}\psi\nabla^2\psi^* - \psi^*\nabla^2\psi &= \psi\nabla^2\psi^* + \nabla\psi\nabla\psi^* - \nabla\psi\nabla\psi^* - \psi^*\nabla^2\psi \\ &= \nabla(\psi\nabla\psi^* - \psi^*\nabla\psi) \ .\end{aligned}$$

With this we have

$$\frac{\partial}{\partial t}(\psi^*\psi) + \frac{\mathrm{i}\hbar}{2m}\operatorname{div}(\psi\nabla\psi^* - \psi * \nabla\psi) = 0 \ . \tag{6.20}$$

This equation is of the form of the desired continuity equation, if we define the particle current density in the following way:

$$j = \frac{i\hbar}{2m}(\psi \nabla \psi^* - \psi^* \nabla \psi) \ . \tag{6.21}$$

Application of Gauss' law

$$\int_V (\mathrm{div}\, \boldsymbol{j})\, dV = \oint_F \boldsymbol{j} \cdot \boldsymbol{n}\, dF \tag{6.22}$$

leads to the integrated equation:

$$\frac{\partial}{\partial t} \int_V \psi^* \psi \, dV + \oint_F j_\mathrm{n} \, dF = 0 \ . \tag{6.23}$$

The particle flux through the surface of a region is equivalent to the variation of the particle density inside the region.

We had required the time-independent normalization of the wave function

$$\int_V \psi^* \psi \, dV = 1 \ ,$$

i.e. that the particle current through an infinitely distant surface vanish. Hence, by inspection of (6.23), only those states can be normalized to 1 whose current flux through an infinitely distant surface vanishes. If we want to calculate the mass current density or the electric current from the particle current density, we have to multiply the continuity equation by the mass (or charge), since the *mass* (or *charge*) *density* is given by

$$\varrho_\mathrm{m} = m\psi^* \psi \ , \quad \varrho_\mathrm{e} = e\psi^* \psi \ . \tag{6.24}$$

Therefore a law of conservation exists for the mass and charge of a system, too.

As an example for the calculation of a particle current, we take a plane wave $\psi = A \exp(\mathrm{i}\boldsymbol{k} \cdot \boldsymbol{x})$. From (6.21) we get

$$\boldsymbol{j} = A^2 \frac{\hbar \boldsymbol{k}}{m} = \psi^* \psi \frac{\boldsymbol{p}}{m} = w\boldsymbol{v} \ . \tag{6.25}$$

The close relation between particle current $\boldsymbol{j}$ and velocity $\boldsymbol{v}$ is immediately evident. Of course, the current through an arbitrary distant surface is not zero in this case, so the functions are not normalizable in this way; the plane waves have to be normalized to δ functions according to Chap. 5.

6.2 Stationary States

As we recall, in the case of a not explicitly time-dependent $\hat{H}$, we were able to separate the variables x and t of the time-dependent Schrödinger equation,

$$\mathrm{i}\hbar\frac{\partial}{\partial t}\psi(x,t)=\hat{H}\psi(x,t)\ . \tag{6.26}$$

With $\psi_n(x,t)=\psi_n(x)f_n(t)$ we got two differential equations,

$$\mathrm{i}\hbar\frac{\partial f_n}{\partial t}=E_n f_n(t)\ ,\quad \hat{H}\psi_n(x)=E_n\psi_n(x)\ . \tag{6.27}$$

From the first equation we obtained the time factor $f_n(t)=\exp[-\mathrm{i}(E_n/\hbar)t]$, which was normalized in such a way that $|f_n|^2=1$. Equation (6.27) is the stationary Schrödinger equation. With $E_n=\hbar\omega_n$, we have for the eigenfunctions of $\hat{H}$:

$$\psi_n(x,t)=\psi_n(x)\,\mathrm{e}^{-\mathrm{i}\omega_n t}\ . \tag{6.28}$$

The general solution $\Psi(x,t)$ of the time-dependent Schrödinger equation is a superposition of all $\psi_n(x,t)$:

$$\begin{aligned}\Psi(x,t)&=\sum_n C_n(0)\psi_n(x,t)\\&=\sum_n C_n(t)\psi_n(x)\quad\text{with}\quad C_n(t)=C_n(0)\,\mathrm{e}^{-\mathrm{i}\omega_n t}\ .\end{aligned} \tag{6.29}$$

The coefficients C_n are determined by the integral

$$C_n(0)=\int\Psi(x,0)\psi_n^*(x,0)\,\mathrm{d}x\ . \tag{6.30}$$

To prove this, let us first observe (6.29) at time $t=0$:

$$\Psi(x,0)=\sum_n C_n(0)\psi_n(x)\ . \tag{6.31}$$

Since the wave functions $\psi_n(x)$ are orthonormal, i.e. $\langle\psi_n|\psi_m\rangle=\delta_{mn}$, we can multiply both sides by $\psi_m^*(x)$ and integrate. This gives us

$$\begin{aligned}\int\Psi(x,0)\psi_m^*(x)\,\mathrm{d}x&=\sum_n C_n(0)\int\psi_n(x)\psi_n^*(x)\,\mathrm{d}x\\&=\sum_n C_n(0)\delta_{nm}=C_m(0)\ .\end{aligned}$$

This is precisely result (6.30); in particular, we refer to the analogy of the expansion (6.29) for the decomposition of an arbitrary vector $\boldsymbol{A}$ in terms of an orthonormalized basis $\boldsymbol{e}_i$:

$$\boldsymbol{A}=\sum_i a_i\boldsymbol{e}_i\ ,$$

where the components (coefficients of expansion)

$$a_i = \boldsymbol{A} \cdot \boldsymbol{e}_i$$

are scalar products of the vector $\boldsymbol{A}$ and the basis vectors $\boldsymbol{e}_i$. According to this, we can view (6.29) as a decomposition of a state $\Psi(x, t)$ in terms of the basis $\psi_n(x)$. The coefficients of expansion $C_n(t)$ are consequently the components of the state $\Psi(x, t)$ in terms of the *basis vectors* $\psi_n(x)$.

6.3 Properties of Stationary States

It holds that $\psi_n^*(x, t)\psi_n(x, t) = \psi_n^*(x)\psi_n(x)$, since the time factor is normalized. Therefore, the probability density is constant in time for stationary states:

$$w(x, t) = w(x) \ . \tag{6.32}$$

The current $j_n(x, t)$ is given by (6.21):

$$j_n(x, t) = \frac{\mathrm{i}\hbar}{2m}[\psi_n(x, t)\nabla\psi_n^*(x, t) - \psi_n^*(x, t)\nabla\psi_n(x, t)] \ . \tag{6.33}$$

Since the nabla operator does not affect the time factor, we also have

$$j_n(x, t) = j_n(x) \ , \tag{6.34}$$

i.e. the current of stationary states is constant in time as well. Now we can expand $\psi_n(x, t)$ in terms of eigenfunctions of any operator $\hat{A}$:

$$\psi_n(x, t) = \sum_A C_A(t)\psi_A(x) \ .$$

For stationary states, the probabilities $|C_A|^2$ for finding the value A of the observable described by the operator $\hat{A}$ are time independent if $\hat{A}$ is not explicitly time dependent, since

$$C_A(t) = \int \psi_A^*(x)\psi_n(x, t)\,\mathrm{d}x = \mathrm{e}^{-\mathrm{i}\omega_n t}\int \psi_A^*(x)\psi_n(x)\,\mathrm{d}x \ ,$$

where for $t = 0$

$$C_A(0) = \int \psi_A^*(x)\psi_n(x)\,\mathrm{d}x \ .$$

From these two equations the statement

$$|C_A(t)|^2 = |C_A(0)|^2 \tag{6.35}$$

follows.

EXERCISE

6.8 Current Density of a Spherical Wave

The spherical wave

$$\psi = \frac{e^{\pm i\boldsymbol{k}\cdot\boldsymbol{r}}}{r} \; ; \quad \boldsymbol{k} = k\frac{\boldsymbol{r}}{r}$$

is given.

Problem. (a) Calculate the probability current density $\boldsymbol{j}$ for the wave function.

(b) With $\boldsymbol{k} = k(\boldsymbol{r}/r)$ calculate the number of particles that flow per second through a sphere of radius r. What physical processes are described here by ψ?

Solution. (a) The probability current density of a wave function is defined as

$$\boldsymbol{j} = \frac{\hbar}{2\mathrm{i}m}(\psi^*\nabla\psi - \psi\nabla\psi^*) = \frac{\hbar}{m}\,\mathrm{Im}\{\psi^*\nabla\psi\} \; .$$

For the wave function

$$\psi = \exp(\pm\mathrm{i}\boldsymbol{k}\cdot\boldsymbol{r})/r \; ,$$

the gradient is

$$\begin{aligned}\nabla\psi &= \nabla\exp(\pm\mathrm{i}\boldsymbol{k}\cdot\boldsymbol{r})/r \\ &= \frac{1}{r}\nabla\exp(\pm\mathrm{i}\boldsymbol{k}\cdot\boldsymbol{r}) + \exp(\pm\mathrm{i}\boldsymbol{k}\cdot\boldsymbol{r})\nabla(1/r) \\ &= (\pm\mathrm{i}\boldsymbol{k} - \boldsymbol{r}/r^2)\exp(\pm\mathrm{i}\boldsymbol{k}\cdot\boldsymbol{r})/r \; .\end{aligned}$$

From this we have for the current density, with the help of $\boldsymbol{p} = \hbar\boldsymbol{k}$:

$$\boldsymbol{j} = \pm\frac{\hbar\boldsymbol{k}}{m}\frac{1}{r^2} = \pm\boldsymbol{v}\frac{1}{r^2} \; .$$

(b) The number of particles per second streaming through a unit surface is given by

$$N = \boldsymbol{j}\cdot\boldsymbol{n}\times 1\,\mathrm{s} = \boldsymbol{j}\cdot\frac{\boldsymbol{r}}{r}\times 1\,\mathrm{s} = \pm\frac{v}{r^2} \; .$$

Thus

$$N_\mathrm{S} = \pm 4\pi v$$

expresses the number of particles passing through the surface of the whole sphere.

Depending on the sign, the wave function describes either an emission or absorption process.

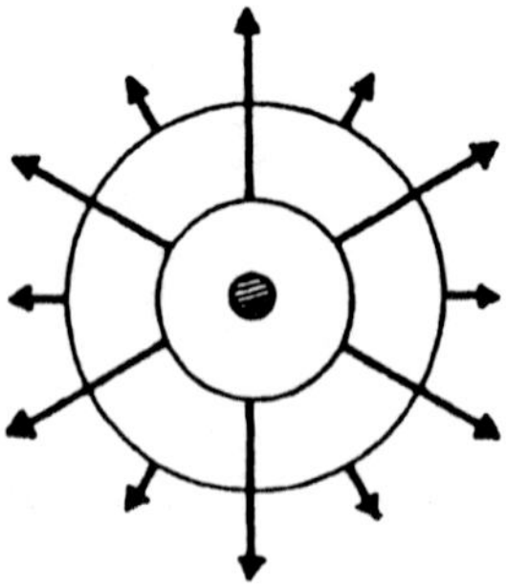

Emission of a spherical wave. The vectors illustrate the radially directed phase velocity of the wave

EXERCISE

6.9 A Particle in a Periodic Potential

Let $V(x)$ be a periodic potential with $V(x+a) = V(x)$.

Problem. (a) Show that the Hamiltonian

$$\hat{H} = -\frac{\hbar^2}{2m}\frac{\mathrm{d}^2}{\mathrm{d}x^2} + V(x) \tag{1}$$

commutes with the translation operator $\hat{T}(a)$, which has the property

$$\hat{T}(a)\psi(x) = \psi(x+a) \ . \tag{2}$$

(b) Deduce from the periodicity of the potential that the wave function has the form

$$\psi_k(x) = \mathrm{e}^{\mathrm{i}kx}\phi_k(x) \quad \text{(Bloch functions)} \ , \tag{3}$$

where $k \in \mathbb{R}$ and $\phi_k(x+a) = \phi_k(x)$. *Suggestion*: For two linear independent solutions $\psi_1(x)$, $\psi_2(x)$ of the energy eigenvalue E, which are simultaneously eigenfunctions of $\hat{T}(a)$, we have

$$W(x) \equiv \left[\psi_1(x)\frac{\mathrm{d}}{\mathrm{d}x}\psi_2(x) - \psi_2(x)\frac{\mathrm{d}}{\mathrm{d}x}\psi_1(x)\right] = \text{const} \ . \tag{4}$$

(c) Discuss the eigenvalue condition and determine the allowed range of energy in the potential

$$V(x) = -V_0 \sum_{n=-\infty}^{+\infty} \delta(x+na) \ ; \quad V_0 > 0 \ , \quad n \in \mathbb{Z} \tag{5}$$

[*Kronig–Penney* model of the energy levels (band structure) in solids.]

Solution. (a) The translation of a wave function $\psi(x)$ by a is given by

$$\psi(x+a) = \sum_n \frac{1}{n!} a^n \frac{\mathrm{d}^n}{\mathrm{d}x^n}\psi(x) = \sum_n \frac{1}{n!}\left(\frac{\mathrm{i}a}{\hbar}\right)^n \hat{p}^n \psi(x) \equiv \hat{T}(a)\psi(x) \ . \tag{6}$$

Here Taylor's formula has been used to expand $\psi(x+a)$. Obviously, the translation operator is $\hat{T}(a) = \exp(\mathrm{i}\hat{p}a/\hbar)$ with the momentum operator $\hat{p} = -\mathrm{i}\hbar(\mathrm{d}/\mathrm{d}x)$ and

$$-\frac{\hbar^2}{2m}\frac{\mathrm{d}^2}{\mathrm{d}x^2} = \frac{1}{2m}\hat{p}^2 \ . \tag{7}$$

From this we get directly

$$\left[\hat{T}(a) \, , \, \frac{1}{2m}\hat{p}^2\right] = 0 \ . \tag{8}$$

Exercise 6.9

Besides,

$$[\hat{T}(a), V(x)]\psi(x) = \hat{T}(a)V(x)\psi(x) - V(x)\hat{T}(a)\psi(x)$$
$$= V(x+a)\psi(x+a) - V(x)\psi(x+a) = 0 \;, \tag{9}$$

because $V(x+a) = V(x)$, since periodicity is assumed.

(b) The one-dimensional Schrödinger equation is an ordinary differential equation of second order. Therefore it possesses two linearly independent solutions for each eigenvalue E: $\psi_E^{(1)}$ and $\psi_E^{(2)}$. Since $[\hat{T}(a), \hat{H}(a)] = 0$ we can choose $\psi_E^{(1)}$, $\psi_E^{(2)}$ to be simultaneous eigenfunctions of $\hat{T}(a)$. So we have

$$\hat{T}(a)\psi_E^{(i)}(x) = \psi_E^{(i)}(x+a) = \lambda_E^{(i)}\psi_E^{(i)}(x) \quad i = 1, 2 \;. \tag{10}$$

The $\lambda_E^{(i)}$ are constant numbers that depend on the energy. Since $\hat{H}\psi_E^{(i)} = E\psi_E^{(i)}$, we get

$$\frac{\mathrm{d}}{\mathrm{d}x}[\psi_E^{(1)}(x)\psi_E^{(2)\prime}(x) - \psi_E^{(2)}(x)\psi_E^{(1)\prime}(x)] \equiv \frac{\mathrm{d}}{\mathrm{d}x}W(x)$$
$$= \psi_E^{(1)\prime}\psi_E^{(2)\prime} - \psi_E^{(2)\prime}\psi_E^{(1)\prime} + \psi_E^{(1)}\psi_E^{(2)\prime\prime} - \psi_E^{(2)}\psi_E^{(1)\prime\prime}$$
$$= \psi_E^{(1)}\frac{2m}{\hbar^2}(V(x)-E)\psi_E^{(2)} - \psi_E^{(2)}\frac{2m}{\hbar^2}(V-E)\psi_E^{(1)} = 0 \;. \tag{11}$$

Hence $W(x) = \text{const}$. Since $W(x) = W(x+a) = \lambda_1\lambda_2 W(x)$ is also valid, it follows that $\lambda_1\lambda_2 = 1$. Besides,

$$\psi_E^{(i)*}(x+a) = \hat{T}(a)\psi_E^{(i)*}(x) = \lambda_E^{(i)*}\psi_E^{(i)*}(x) \;; \tag{12}$$

i.e. $\lambda_E^{(i)*}$ are also eigenvalues of $\hat{T}(a)$. But they cannot differ from $\lambda_E^{(i)}$, because $\psi^{(1)}$ and $\psi^{(2)}$ are linearly independent; i.e. either

$$\lambda_E^{(1)*} = \lambda_E^{(1)} \;, \quad \lambda_E^{(2)*} = \lambda_E^{(2)} \Rightarrow \lambda_E^{(i)} \in \mathbb{R} \tag{13}$$

or

$$\lambda_E^{(1)*} = \lambda_E^{(2)} \;, \quad \lambda_E^{(2)*} = \lambda_E^{(1)} \Rightarrow |\lambda_E|^2 = 1 \;. \tag{14}$$

If, in the first case, we assume without restriction of generality that $\lambda_E^{(1)} > 1$, then $\psi_E^{(1)}(x)$ cannot be square integrable, because $\psi_E^{(1)}(x+na) = (\lambda_E^{(1)})^n\psi_E^{(1)}(x)$ due to (10) and $\psi_E^{(1)}(x)$ will increase ad infinitum for $x \to \infty$. Therefore the second case must be true. Let

$$\lambda_E^{(1)} = \mathrm{e}^{\mathrm{i}\alpha_E} \;, \quad \lambda_E^{(2)} = \mathrm{e}^{-\mathrm{i}\alpha_E} \;, \quad \alpha_E \in \mathbb{R} \tag{15}$$

($\alpha_E = 0$ includes the case $\lambda_E = 1$). If we define $k_E = \alpha_E/a$, we get

$$\psi_E^{(1)}(x+a) = \mathrm{e}^{\mathrm{i}k_E a}\psi_E^{(1)}(x) \quad \text{and} \quad \psi_E^{(2)}(x+a) = \mathrm{e}^{-\mathrm{i}k_E a}\psi_E^{(2)}(x) \;. \tag{16}$$

Exercise 6.9

In the decomposition

$$\psi_E^{(1)}(x) = \mathrm{e}^{\mathrm{i}k_E x}\phi_E^{(\mathrm{i})}(x) \tag{17}$$

we must have

$$\phi_E^{(1)}(x+a) = \phi_E^{(1)}(x) \ , \tag{18}$$

since

$$\mathrm{e}^{\mathrm{i}k_E a}\mathrm{e}^{\mathrm{i}k_E x}\phi_E^{(1)}(x) = \mathrm{e}^{\mathrm{i}k_E a}\psi_E^{(1)}(x) = \psi_E^{(1)}(x+a) = \mathrm{e}^{\mathrm{i}k_E a}\mathrm{e}^{\mathrm{i}k_E x}\phi_E^{(1)}(x+a)$$

(for $i = 2$, the proof is analogous). In general, we suppress the index E of the wave function and write

$$\psi_k(x) = \mathrm{e}^{\mathrm{i}kx}\phi_k(x) \ , \tag{19}$$

where ψ_k and ψ_{-k} are linearly independent and ϕ_k is periodic.

(c) In the range $0 < x < a$, it holds that $V(x) = 0$, and therefore

$$\psi_k(x) = A\,\mathrm{e}^{\mathrm{i}\kappa x} + B\,\mathrm{e}^{-\mathrm{i}\kappa x}$$

with

$$\kappa^2 = 2mE/\hbar^2 \ , \quad x \in (0, a) \ . \tag{20}$$

But now we have, because of (16),

$$\psi_k(x) = \mathrm{e}^{\mathrm{i}ka}\psi_k(x-a) \ ; \tag{21}$$

therefore with (20) we must have in the interval $x \in (a, 2a)$

$$\psi_k(x) = \mathrm{e}^{\mathrm{i}ka}[A\,\mathrm{e}^{\mathrm{i}\kappa(x-a)} + B\,\mathrm{e}^{-\mathrm{i}\kappa(x-a)}] \ . \tag{22}$$

The wave function ψ, but not the derivative of ψ, is continuous at a, as can be seen by integrating the Schrödinger equation from $a-\varepsilon$ to $a+\varepsilon$:

$$\begin{aligned} 0 &= \int\limits_{a-\varepsilon}^{a+\varepsilon} \mathrm{d}x \left[E\psi(x) + \frac{\hbar^2}{2m}\psi''(x) - V(x)\psi(x)\right] \\ &= \int\limits_{a-\varepsilon}^{a+\varepsilon} \mathrm{d}x \left[E\psi(x) + \frac{\hbar^2}{2m}\psi''(x) + V_0\delta(x-a)\psi(x)\right] \ . \end{aligned}$$

From this follows

$$\frac{\hbar^2}{2m}\psi'(x)\Big|_{a-\varepsilon}^{a+\varepsilon} + V_0\psi(a) = 0 \quad \text{or}$$

$$\frac{\hbar^2}{2m}[\psi'(a+\varepsilon) - \psi'(a-\varepsilon)] + V_0\psi(a) = 0 \ ,$$

Exercise 6.9

which is written in the limit $\varepsilon \to 0$ as

$$\frac{\hbar^2}{2m}[\psi'(a+0)-\psi'(a-0)]+V_0\psi(a)=0 \ . \tag{23}$$

The continuity of the wave function at $x=a$ yields

$$\psi(a-0)=\psi(a+0) \ . \tag{24}$$

With (22), (23) and (24) become

$$\begin{aligned} &A\,\mathrm{e}^{\mathrm{i}\kappa a}+B\,\mathrm{e}^{-\mathrm{i}\kappa a}=\mathrm{e}^{\mathrm{i}ka}(A+B) \ , \\ &\mathrm{e}^{\mathrm{i}ka}(\mathrm{i}\kappa A-\mathrm{i}\kappa B)-(\mathrm{i}\kappa A\,\mathrm{e}^{\mathrm{i}\kappa a}-\mathrm{i}\kappa B\,\mathrm{e}^{-\mathrm{i}\kappa a})+\frac{2m}{\hbar^2}V_0(A\,\mathrm{e}^{\mathrm{i}\kappa a}+B\,\mathrm{e}^{-\mathrm{i}\kappa a})=0 \ , \end{aligned} \tag{25}$$

or

$$\begin{bmatrix} \mathrm{e}^{\mathrm{i}\kappa a}-\mathrm{e}^{\mathrm{i}ka} & \mathrm{e}^{-\mathrm{i}\kappa a}-\mathrm{e}^{\mathrm{i}ka} \\ \mathrm{i}\kappa(\mathrm{e}^{\mathrm{i}ka}-\mathrm{e}^{\mathrm{i}\kappa a})+(2m/\hbar^2)V_0\,\mathrm{e}^{\mathrm{i}\kappa a} & -\mathrm{i}\kappa(\mathrm{e}^{\mathrm{i}ka}-\mathrm{e}^{-\mathrm{i}\kappa a})+(2mV_0/\hbar^2)\,\mathrm{e}^{-\mathrm{i}\kappa a} \end{bmatrix} \times \begin{pmatrix} A \\ B \end{pmatrix}=0 \ . \tag{26}$$

The vanishing of the determinant leads to the eigenvalue equation

$$\cos ka=\cos\kappa a-\frac{amV_0}{\hbar^2}\frac{\sin\kappa a}{\kappa a}\equiv f(\kappa a) \ . \tag{27}$$

This equation relates k of (16) and E. Instead of choosing E and calculating k, we can also choose k and calculate E graphically. Since $|\cos ka|\leq 1$, we have no solution of the eigenvalue equation for $|f(\kappa a)|>1$.

1st Case. $E<0$ (bound states),

$$\kappa=\mathrm{i}\beta \ , \quad \beta\in\mathbb{R}_+ \ , \quad \beta=\sqrt{|2mE/\hbar^2|} \ . \tag{28}$$

Now we have $\sin\mathrm{i}\beta=\mathrm{i}\sinh\beta$, $\cos\mathrm{i}\beta=\cosh\beta$ and

$$f(\mathrm{i}\beta a)=\cosh\beta a-\frac{amV_0}{\hbar^2}\frac{\sinh\beta a}{\beta a} \tag{29}$$

is a steeply monotonous increasing, even function that exceeds 1 at $\beta_0 a$ (see figure). Therefore

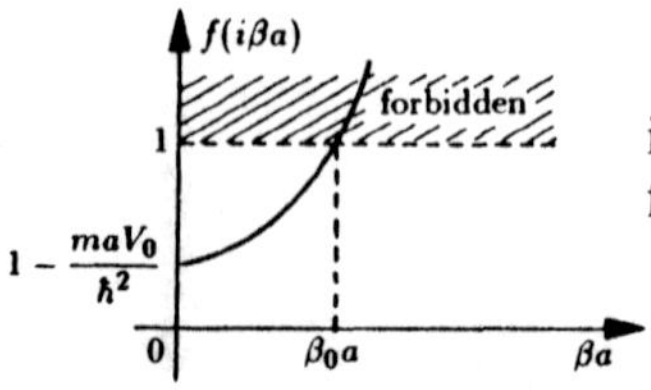

As soon as the function $f(\mathrm{i}\beta a)>1$, then (27) will have no solutions

$$|E|=\frac{\beta^2\hbar^2}{2m}<\frac{\beta_0^2\hbar^2}{2m}\equiv E_0 \ , \tag{30}$$

or, because E is assumed to be negative, $E>-E_0$. This is illustrated in the last figure below.

Exercise 6.9

2nd Case. $E > 0, \kappa \in \mathbb{R}_+$
According to (27) $f(\kappa a)$ is even and equal to 1 at $\kappa a = x$ with

$$\cos x - \frac{am V_0}{\hbar^2}\frac{\sin x}{x} = 1 \ , \ \text{i.e.} \tag{31}$$

$$-\frac{am V_0}{\hbar^2 x} 2 \sin\frac{x}{2}\cos\frac{x}{2} = 2\sin^2\frac{x}{2} \ . \tag{32}$$

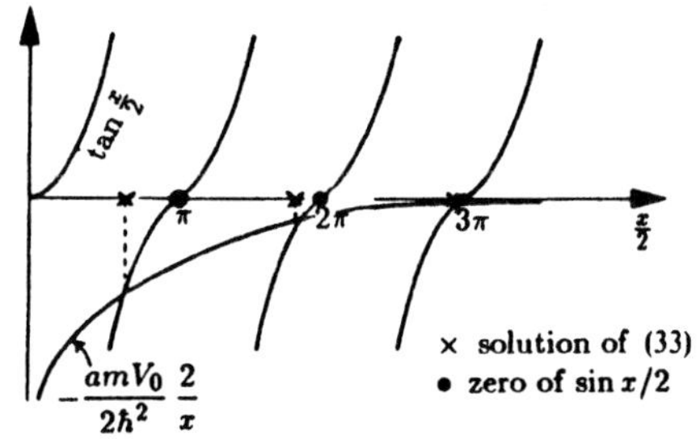

Graphical display of the solutions of (33): they are indicated by x_n and are shifted by $-\Delta(n\pi)$ from the zeros of the function $\sin(\frac{1}{2}x)$, which are located at $x/2 = n\pi, n = 0, 1, 2\ldots$

This is fulfilled for

$$\sin\frac{x}{2} = 0 \quad \text{or} \quad \tan\frac{x}{2} = -\frac{am V_0}{2\hbar^2}\frac{1}{x/2} \tag{33}$$

i.e. for $(x_1)_n = 2n\pi$ and $(x_2)_n = 2n\pi - \Delta(n\pi)$, where $n \in \mathbb{N}$ and $\lim_{n\to\infty} \Delta(n\pi) = 0$. Analogously we find the points at which $f(\kappa a)$ is equal to -1 at $(x_1')_n = (2n-1)\pi$ and $(x_2')_n = (2n-1)\pi - \Delta[(2n-1)\pi]$. Between $(x_2)_n$ and $(x_1)_n$, or $(x_2')_n$ and $(x_1')_n$, there are no allowed energy eigenvalues, as is obvious from the figure above.

The graphic representation of the energy dependence on the wave number k is characterized by "forbidden regions" that shrink for increasing k. If $f(n\pi) = (-1)^n$, then we obviously have $ka = n\pi$ for $\cos(ka) = (-1)^n$, i.e. we have "energy gaps" at those places (see top, right figure). Therefore the spectrum falls apart into "allowed" energy regions (named *energy bands*) and "forbidden" energy regions (*gaps*) (see last figure). Energy bands play an important role in the motion of electrons in periodic structures in solid state physics (conduction bands, valence bands).

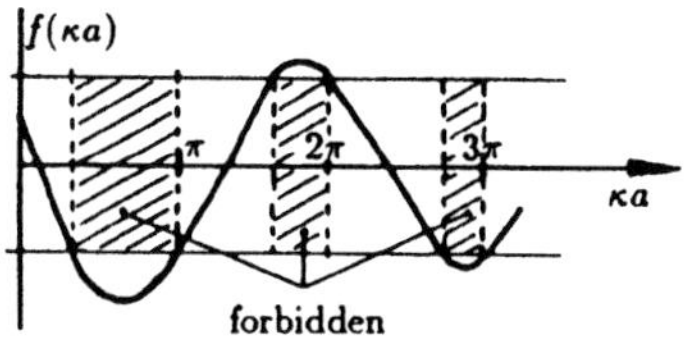

Allowed and forbidden regions

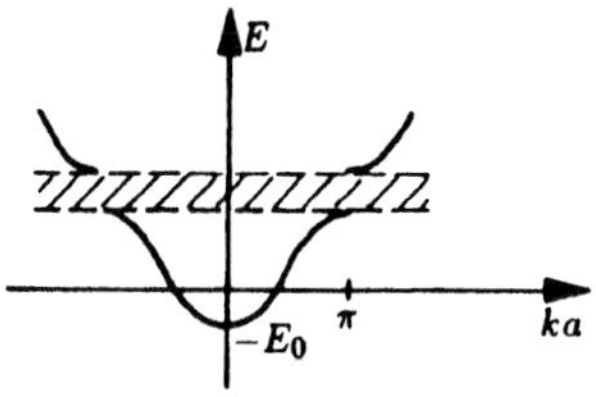

Energy bands with intervening energy gaps

6.4 Biographical Notes

SCHRÖDINGER, Erwin, Austrian physicist, *Vienna 12.8.1887, † Alpbach (Tirol) 4.1.1961, was a student of F. Hasenöhrl. As a professor in Zürich, S. worked on statistical thermodynamics, the theory of general relativity and the theory of colour vision. Excited by L. de Broglie's Ph. D thesis and A. Einstein's publications concerning Bose statistics, S. created *wave mechanics*. In December, 1925 he defined the *Klein–Gordon equation* and later, in January 1926, he invented the *Schrödinger equation*, which describes, in nonrelativistic approximation, the atomic eigenvalues. In March 1926, S. proved the mathematical equivalence of his theory with *matrix mechanics* (M. Born, W. Heisenberg and P. Jordan). S. always attacked the statistical interpretation of quantum theory (as did Einstein, von Laue and de Broglie), especially the "Copenhagen interpretation". In 1927 S. went to Berlin as Planck's successor and emigrated in 1933, as a convinced liberal, to Oxford. In the same year he was awarded, together with P.A.M. Dirac, the Nobel Prize in Physics. In 1936 S. went to the University of Graz, Austria, emigrating a second time when Austria was annexed. The Institute for Advanced Studies was founded in Dublin for him and others. In 1956, S. returned to Austria.

BOSE, Satyendra Nath, Indian physicist, *1.1.1894, † 4.2.1974 Calcutta. Together with Einstein, he set up a theory of quantum statistics *(Bose–Einstein statistics)* that differs from the classical Boltzmann statistics and from Fermi statistics, too. B. invented this statistics for photons; Einstein extended it to massive particles. B. was a professor in Dacca and Calcutta from 1926 to 1956.

BOLTZMANN, Ludwig, Austrian physicist, *Vienna 1844, † Duino near Trieste 1906. He studied physics at the university of Vienna where he was an assistant of Josef Stefan. B. became professor of mathematical physics at the university of Graz in 1869. He also taught at Vienna, Munich and Leipzig. Among his students were S. Arrhenius, W. Nernst, F. Hasenöhrl and L. Meitner. The young B. worked successfully on experimental physics (he proved the relationship between the refractive index and the dielectric constant for sulphur, which was required by Maxwell). Near the end of his life he occupied his mind with philosophical issues, but his main interest was always theoretical physics.

The central problem of his life's theoretical work was the reduction of thermodynamics to mechanics, requiring the solution of the contradiction between the reversibility of mechanical processes and the irreversibility of thermodynamical processes. He showed the relationship between the entropy S and the probability of a state W with the formula $S = k \ln W$ (k: Boltzmann's constant). This was the starting point of quantum theory both in the formulation of Max Planck in 1900 and in the expanded version of Albert Einstein (1905). Other important achievements of B. are the formulas for the energy distribution of atoms moving freely or in force fields (Maxwell–Boltzmann distribution) and the theoretical explanation of the law of the radiation power of a black body (Stefan–Boltzmann law, 1884).

B. was an exponent of the atom theory. The small response and even rejection that he received for it from many contemporary physicists disappointed him throughout his lifetime. He did not live to see the final victory of the atom theory introduced in 1905 by Einstein's theory of Brownian motion.

B. committed suicide at the age of 62.

FERMI, Enrico, Italian physicist, *Rome 29.9.1901, †Chicago 28.11.1954. F. was a professor in Florence and Rome before going to Columbia University in New York in 1939. There he stayed until 1946, when he went to Chicago. F. was mainly engaged in quantum mechanics. He discovered the conversion of nuclei by the bombardment of neutrons, and, beginning in 1934, was thus able to create many new synthetic radioactive substances that he thought were transuranic. F. formulated the statistics named after him (*Fermi statistics*) in his treatise "Sulla Quantizazione del gas perfetto monatomico" (Lincei Rendiconti 1935; Zeitschrift für Physik 1936). In 1938 he was awarded the Nobel Prize in Physics. During World War II, F. was substantially engaged in projects devoted to making use of atomic energy. Under his guidance the first nuclear chain reaction was performed at the Chicago nuclear reactor on 2.12.1942. In memory of F., the Enrico Fermi Prize was established in the United States.

7. The Harmonic Oscillator

As an application of the Schrödinger equation, we now calculate the states of a particle in an oscillator potential. From classical mechanics we know that such a potential is of greater importance, because many complicated potentials can be approximated in the vicinity of their equilibrium points by a harmonic oscillator. Expanding a potential $V(x)$ in one dimension in a Taylor series yields

$$\begin{aligned} V(x) &= V(a+(x-a)) \\ &= V(a)+V'(a)(x-a)+\frac{1}{2}V''(a)(x-a)^2+\dots \; . \end{aligned} \tag{7.1}$$

If a stable equilibrium exists for $x=a$, $V(x)$ has a minimum at $x=a$, i.e. $V'(a)=0$ and $V''(a)>0$. We can choose a as the origin of the coordinate system and set $V(a)=0$; then an oscillator potential is indeed a first approximation in the vicinity of $x=a$, i.e. in the vicinity of the equilibrium point.

In the following we shall consider the one-dimensional case. Then the classical Hamiltonian function of a particle with mass m oscillating with frequency ω takes the form

$$H=\frac{p_x^2}{2m}+\frac{m}{2}\omega^2x^2 \; , \tag{7.2}$$

and the corresponding quantum-mechanical Hamiltonian reads

$$\hat{H}=-\frac{\hbar^2}{2m}\frac{\mathrm{d}^2}{\mathrm{d}x^2}+\frac{m}{2}\omega^2x^2 \; . \tag{7.3}$$

Since the potential is constant in time, the time-independent (stationary) Schrödinger equation determines the stationary solutions ψ_n and the corresponding eigenvalues (energies) E_n. The stationary Schrödinger equation takes the form

$$-\frac{\hbar^2}{2m}\frac{\mathrm{d}^2}{\mathrm{d}x^2}\psi(x)+\frac{m}{2}\omega^2x^2\psi(x)=E\psi(x) \; . \tag{7.4}$$

Because of the importance of the harmonic oscillator and its solutions for quantum mechanics, we will now consider the method of solving this differential equation in detail. Using the abbreviations

$$k^2=\frac{2m}{\hbar^2}E \; , \quad \lambda=\frac{m\omega}{\hbar} \; , \tag{7.5}$$

we can rewrite the differential equation as

$$\frac{\mathrm{d}^2\psi}{\mathrm{d}x^2} + (k^2 - \lambda^2 x^2)\psi = 0 \; . \tag{7.6}$$

Equation (7.6) is known as ***Weber***'s *differential equation*. For further simplification, we introduce the transformation

$$y = \lambda x^2 \; , \tag{7.7}$$

and obtain

$$y\frac{\mathrm{d}^2\psi}{\mathrm{d}y^2} + \frac{1}{2}\frac{\mathrm{d}\psi}{\mathrm{d}y} + \left(\frac{\kappa}{2} - \frac{1}{4}y\right)\psi = 0 \; , \tag{7.8}$$

with

$$\kappa = \frac{k^2}{2\lambda} = \frac{\hbar k^2}{2m\omega} = \frac{E}{\hbar\omega} \; . \tag{7.9}$$

To rewrite (7.8) in standard form, we split off the asymptotic solution. The latter can be inferred by examining the dominant behaviour in terms linear in y for the asymptotic region $y \to \infty$. Hence, we try writing

$$\psi(y) = \mathrm{e}^{-y/2}\varphi(y) \; . \tag{7.10}$$

Using

$$\frac{\mathrm{d}\psi}{\mathrm{d}y} = \left[-\frac{1}{2}\varphi(y) + \frac{\mathrm{d}\varphi}{\mathrm{d}y}\right]\mathrm{e}^{-y/2} \quad \text{and} \quad \frac{\mathrm{d}^2\psi}{\mathrm{d}y^2} = \left[\frac{1}{4}\varphi(y) - \frac{\mathrm{d}\varphi}{\mathrm{d}y} + \frac{\mathrm{d}^2\varphi}{\mathrm{d}y^2}\right]\mathrm{e}^{-y/2} \, ,$$

the differential equation for $\varphi(y)$ follows from (7.8):

$$y\frac{\mathrm{d}^2\varphi}{\mathrm{d}y^2} + \left(\frac{1}{2} - y\right)\frac{\mathrm{d}\varphi}{\mathrm{d}y} + \left(\frac{\kappa}{2} - \frac{1}{4}\right)\varphi = 0 \; . \tag{7.11}$$

Before further examining (7.11), we shall digress to the field of hypergeometric functions. Our aim is to understand the basic mathematical features as well as possible without going into rigorous derivations; a heuristic treatment will suffice.

EXERCISE

7.1 Mathematical Supplement: Hypergeometric Functions

The Hypergeometric Differential Equation

The hypergeometric differential equation, expressed by ***C.F. Gauß*** in the form

$$z(1-z)\frac{\mathrm{d}^2\phi}{\mathrm{d}z^2}+[c-(a+b+1)z]\frac{\mathrm{d}\phi}{\mathrm{d}z}-ab\phi=0 \tag{1}$$

contains the three free parameters a, b, c and possesses a great variety of solutions. It has three nonessential singularities at $z = 0, 1, \infty$. To solve (1), we substitute the power series

$$\phi(z)=z^\sigma\sum_{\nu=0}^{\infty}c_\nu z^\nu$$

into the differential equation (1) and find the recurrence relation

$$\begin{aligned}&z(1-z)z^\sigma\sum_{\nu=0}^{\infty}c_\nu(\nu+\sigma)(\nu+\sigma-1)z^{\nu-2}\\&\quad+[c-(a+b+1)z]z^\sigma\sum_{\nu=0}^{\infty}c_\nu(\nu+\sigma)z^{\nu-1}-abz^\sigma\sum_{\nu=0}^{\infty}c_\nu z^\nu=0\ .\end{aligned} \tag{2}$$

Multiplying out the factors and re-ordering the terms yields

$$\begin{aligned}&c_0\sigma(c+\sigma-1)z^{\sigma-1}+\sum_{\nu=0}^{\infty}[c_{\nu+1}(\nu+\sigma+1)(\nu+c+\sigma)\\&\qquad\qquad-c_\nu(\nu+a+\sigma)(\nu+b+\sigma)]z^{\nu+\sigma}=0\ .\end{aligned} \tag{3}$$

For this expression to vanish identically, all the coefficients have to be equal to zero, i.e.

$$\sigma(c-1+\sigma)=0 \quad \text{(the “index equation”)} \quad \text{and} \tag{4}$$

$$c_{\nu+1}=\frac{(\nu+a+\sigma)(\nu+b+\sigma)}{(\nu+1+\sigma)(\nu+c+\sigma)}c_\nu\ . \tag{5}$$

One solution of (1) (if we set $c_0 = 1$) is therefore given by

$$\phi(z)=z^\sigma\sum_{\nu=0}^{\infty}\frac{(a+\sigma)_\nu(b+\sigma)_\nu}{(1+\sigma)_\nu(c+\sigma)_\nu}z^\nu\ , \tag{6}$$

using the abbreviations (Pochammer symbols)

$$\begin{aligned}&(a)_\nu=a(a+1)\dots(a+\nu-1)\ ,\\&(a)_0=1\ .\end{aligned} \tag{7}$$

Exercise 7.1

The radius of convergence can be inferred from the ratio test for convergence,

$$r = \lim_{\nu\to\infty} \left| \frac{c_\nu}{c_\nu + 1} \right| = 1 \ . \tag{8}$$

The index equation (4) yields two possible values for the exponent σ:

(1) $\sigma = 0$. The solution in this case is the *hypergeometric series*

$$\phi_1(z) = {}_2F_1(a, b; c; z) = \sum_{\nu=0}^{\infty} \frac{(a)_\nu (b)_\nu}{(c)_\nu} \frac{z^\nu}{\nu!} \ . \tag{9}$$

The indices appended to ${}_2F_1$ are related to a generalization of the hypergeometric series in the form

$${}_pF_q(\alpha_1, \ldots, \alpha_p; \beta_1, \ldots, \beta_q; z) = \sum_{\nu=0}^{\infty} \frac{(\alpha_1)_\nu (\alpha_2)_\nu \ldots (\alpha_p)_\nu}{(\beta_1)_\nu \ldots (\beta_q)_\nu} \frac{z^\nu}{\nu!} \ . \tag{10}$$

The solution (9) only makes sense if, in the series ${}_2F_1$, none of the denominators of the various terms of the series vanishes, i.e. the existence of ${}_2F_1$ implies the condition that $c \neq -n$, where $n = 0, 1, \ldots$. Then the series is holomorphic in the unit circle. When $a = -n$ or $b = -n$, the series terminates and defines a polynomial of nth degree. For example,

$${}_2F_1(-n, n+1; 1; x) = P_n(1 - 2x) \tag{11}$$

is a *Legendre polynomial* (see Examples 4.8–4.10). Further special cases are, among others, the Gegenbauer and Tschebycheff polynomials.[1]

(2) $\sigma = 1 - c$. According to (6) and (9), the second solution may be expressed by the hypergeometric function with changed parameters, namely

$$\phi_2(z) = z^{1-c} \, {}_2F_1(a + 1 - c, b + 1 - c; 2 - c; z) \ . \tag{12}$$

Note the factor z^{1-c} before the hypergeometric function ${}_2F_1$. The solution ϕ_2 only exists if $c \neq 2, 3, \ldots$.

The *general solution* of the hypergeometric differential equation is therefore

$$\phi(z) = A \, {}_2F_1(a, b; c; z) + B z^{1-c} \, {}_2F_1(a + 1 - c, b + 1 - c; 2 - c; z) \ , \tag{13}$$

under the condition that c is not a positive integer; otherwise there is only one single solution. The second independent fundamental solution then becomes more complicated.

[1] See, for example, George Arfken: *Mathematical Methods for Physicists*, 2nd ed. (Academic Press, New York 1970) or Milton Abramowitz and I.A. Stegun: *Handbook of Mathematical Functions* (Dover Publ., New York 1972).

Exercise 7.1

For the analytic extension of the solution beyond its region of convergence we use the appropriate formula

$$
\begin{aligned}
&{}_2F_1(a, b; c; z) \\
&\quad = \frac{\Gamma(c)\Gamma(b-a)}{\Gamma(b)\Gamma(c-a)}(-z)^{-a}\,{}_2F_1\left(a, 1-c+a; 1-b+a; \frac{1}{z}\right) \\
&\quad + \frac{\Gamma(c)\Gamma(a-b)}{\Gamma(a)\Gamma(c-b)}(-z)^{-b}\,{}_2F_1\left(b, 1-c+b; 1-a+b; \frac{1}{z}\right) . \qquad (14)
\end{aligned}
$$

From this, the asymptotic behaviour for $|z| \to \infty$ follows:

$$
{}_2F_1(a, b; c; z) = \frac{\Gamma(c)\Gamma(b-a)}{\Gamma(b)\Gamma(c-a)}(-z)^{-a} + \frac{\Gamma(c)\Gamma(a-b)}{\Gamma(a)\Gamma(c-b)}(-z)^{-b} . \qquad (15)
$$

The Confluent Hypergeometric Differential Equation

By analytical continuation of the unit circle to the entire complex plane we may infer another important differential equation from (1). Substituting the linear transformation $x = bz$ into (1) leads to

$$
x\left(1-\frac{x}{b}\right)\frac{\mathrm{d}^2\phi}{\mathrm{d}x^2} + \left[c-(a+1)\frac{x}{b}-x\right]\frac{\mathrm{d}\phi}{\mathrm{d}x} - a\phi = 0 . \qquad (16)
$$

In the limit $b \to \infty$, we get the ***Kummer** differential equation*:

$$
x\frac{\mathrm{d}^2\phi}{\mathrm{d}x^2} + (c-x)\frac{\mathrm{d}\phi}{\mathrm{d}x} - a\phi = 0 . \qquad (17)
$$

This equation has a nonessential singularity at $x = 0$ and an essential one at $x = \infty$, which arises through the amalgamation (confluence) of $z = 1$ and $z = \infty$.

The general solution of (17) is obtained by again expanding in a power series around $x = 0$. Therefore we have

$$
\phi(x) = A\,{}_1F_1(a; c, x) + Bx^{1-c}\,{}_1F_1(a-c+1; 2-c; x) , \qquad (18)
$$

with the *confluent hypergeometric function*

$$
{}_1F_1(a; c; x) = \sum_{\nu=0}^{\infty} \frac{(a)_\nu}{(c)_\nu}\frac{x^\nu}{\nu!} = 1 + \frac{a}{c}\frac{x}{1!} + \frac{a(a+1)}{c(c+1)}\frac{x^2}{2!} + \dots . \qquad (19)
$$

The solution (19) originates from (13) in the limit $b \to \infty$, with $x = bz$. This is quite obvious. The series (19) exists only on condition that $c \neq -n$. It converges for arbitrary values of x. The case $a = -n$ again yields a finite polynomial. Special cases are the Hermite and Laguerre polynomials.

The asymptotic behaviour for $|x| \to \infty$ is

$$
{}_1F_1(a; c; x) \to \frac{\Gamma(c)}{\Gamma(c-a)}\mathrm{e}^{-\mathrm{i}a\pi}x^{-a} + \frac{\Gamma(c)}{\Gamma(a)}\mathrm{e}^{x}x^{a-c} . \qquad (20)
$$

Exercise 7.1

For $a=-n$, polynomials of nth degree arise, in particular the ***Laguerre*** *polynomials*,

$$L_n^{(m)}(z)=\frac{(n+m)!}{n!m!}\,{}_1F_1(-n;m+1;z)\ , \tag{21}$$

and the *Hermite polynomials*,

$$\begin{aligned} H_{2n}(z)&=(-1)^n\frac{(2n)!}{n!}\,{}_1F_1\left(-n;\frac{1}{2};z^2\right) \quad \text{and}\\ H_{2n+1}(z)&=(-1)^n\frac{(2n+1)!}{n!}2z\,{}_1F_1\left(-n;\frac{3}{2};z^2\right)\ . \end{aligned} \tag{22}$$

We finally quote a useful integral formula for hypergeometric functions:[2]

$$\begin{aligned} &\int_0^\infty \mathrm{e}^{-st}t^{d-1}\,{}_AF_B[(a),(b);kt]\,{}_{A'}F_B[(a'),(b');k't]\,\mathrm{d}t\\ &=s^{-d}\Gamma(d)\sum_{m=0}^{\infty}\frac{(a)_m(d)_m k^2}{(b)_m m! s^m}\,{}_{A'+1}F_{B'}[(a'),d+m;b';k'/s] \end{aligned} \tag{23}$$

with the following notation:

$$\begin{aligned} &(a)_m=a(a+1)(a+2)\dots(a+m-1) \quad \text{and}\\ &{}_AF_B[(a),(b);z]\\ &={}_AF_B[a_1,a_2,\dots,a_A;b_1,b_2,\dots,b_B;z]\\ &=1+\frac{a_1a_2\dots a_A}{b_1b_2\dots b_B}\frac{z}{1!}+\frac{a_1(a_1+1)a_2(a_2+1)\dots a_A(a_A+1)}{b_1(b_1+1)b_2(b_2+1)\dots b_B(b_B+1)}\frac{z^2}{2!}+\dots\ . \end{aligned} \tag{24}$$

[2] See L.J. Slater: *Confluent Hypergeometric Functions* (Cambridge University Press, Cambridge 1960) p. 54.

7.1 The Solution of the Oscillator Equation

Comparing (7.11) with (7) of the foregoing example, we identify (7.11) as Kummer's differential equation. The general solution is given by (18) of the example:

$$\varphi(y) = A\,{}_1F_1(a;\,\tfrac{1}{2};\,y) + By^{1/2}\,{}_1F_1(a+\tfrac{1}{2};\,\tfrac{3}{2};\,y)\ , \tag{7.12}$$

where

$$a = -\left(\frac{\kappa}{2} - \frac{1}{4}\right)\ . \tag{7.13}$$

The solution of our physical problem is determined by the wave function in (7.10). Therefore the necessary square integrability of ψ implies that ψ has to vanish at infinity. Nevertheless, as we see from (20) of the example, both particular solutions, so long as they are not finite polynomials, behave for large values of y as follows:

$$\begin{aligned}&y \to \infty : \varphi(y) \to \text{const}\ \mathrm{e}^{y} y^{a-1/2}\ ; \quad \text{i.e.}\\ &\psi(y) = \mathrm{e}^{-y/2}\varphi(y) \to \text{const}\ \mathrm{e}^{y/2} y^{a-1/2}\ .\end{aligned} \tag{7.14}$$

This means that the normalization integral diverges. However, if for the hypergeometric series the condition for break-off (termination) is fulfilled, φ becomes a polynomial. Owing to the factor $\exp(-y/2)$ [see (7.14)], ψ will vanish at infinity. Therefore the requirement for normalization leads, in consequence of the condition for break-off (i.e. the hypergeometric functions terminate and become polynomials), to the *quantization of energy*. Let us now consider the two possible cases.

(1) $a = -n$ and $B = 0$ with $n = 0, 1, 2, \ldots$; i.e.

$$\frac{\kappa}{2} - \frac{1}{4} = n\ ,$$

with the eigenfunction

$$\psi_n(x) = N_n\,\mathrm{e}^{-(\lambda/2)x^2}\,{}_1F_1\left(-n;\,\tfrac{1}{2};\,\lambda x^2\right)\ , \tag{7.15}$$

and the energy

$$E_n = \hbar\omega(2n + \tfrac{1}{2})\ . \tag{7.16}$$

(2) $a + \frac{1}{2} = -n$; i.e.

$$\frac{\kappa}{2} - \frac{1}{4} = n + \frac{1}{2}\ ,$$

with the eigenfunction

$$\psi_n(x) = N_n\,\mathrm{e}^{-(\lambda/2)x^2}\,x\,{}_1F_1\left(-n;\,\tfrac{1}{2};\,\lambda x^2\right)\ , \tag{7.17}$$

and the energy

$$E_n = \hbar\omega[(2n+1)+\tfrac{1}{2}] \ . \tag{7.18}$$

Using (7.9) we find for the energy values:

$$\begin{aligned} E_n &= (2n+\tfrac{1}{2})\hbar\omega \quad \text{and} \\ E_n &= (2n+\tfrac{3}{2})\hbar\omega = [(2n+1)+\tfrac{1}{2}]\hbar\omega \ . \end{aligned}$$

Combining these two results, we obtain the discrete energy spectrum:

$$E_n = (n+\tfrac{1}{2})\hbar\omega \ , \quad n = 0, 1, 2 \dots \ . \tag{7.19}$$

As we see, the energy spectrum of the harmonic oscillator is equidistant with the spacing $\hbar\omega$ and has a finite value in the ground state ($n = 0$), the *zero-point energy* $\frac{1}{2}\hbar\omega$ (see Fig. 7.1).

The polynomials occurring in (7.15) and (7.17) are known as *Hermite polynomials*. With the usual normalization factor, they are defined as

$$\begin{aligned} H_{2n}(\xi) &= (-1)^n \frac{(2n)!}{n!} \, {}_1F_1\left(-n; \frac{1}{2}; \xi^2\right) \ , \\ H_{2n-1}(\xi) &= (-1)^n \frac{2(2n+1)!}{n!} \xi \, {}_1F_1\left(-n; \frac{3}{2}; \xi^2\right) \ . \end{aligned} \tag{7.20}$$

The eigenfunctions and energies (7.15)–(7.18) can then be written as

$$\begin{aligned} &\text{(a)} \quad \psi_n = N_n \, \mathrm{e}^{(-\lambda/2)x^2} H_{2n}(\sqrt{\lambda}x) \ , \quad E_n = (2n+\tfrac{1}{2})\hbar\omega ; \\ &\text{(b)} \quad \psi_n = N_n \, \mathrm{e}^{(-\lambda/2)x^2} H_{2n+1}(\sqrt{\lambda}x) \ , \\ &\qquad\ \ E_n = [(2n+1)+\tfrac{1}{2}]\hbar\omega \ , \quad n = 0, 1, 2 \dots \ ; \end{aligned}$$

and can finally be collected uniformly as

$$\begin{aligned} \psi_n &= N_n \, \mathrm{e}^{(-\lambda/2)x^2} H_n(\sqrt{\lambda}x) \ , \\ E_n &= (n+\tfrac{1}{2})\hbar\omega \ , \quad n = 0, 1, 2 \dots \ . \end{aligned} \tag{7.21}$$

For the Hermite polynomials we have the useful relation

$$H_n(\xi) = (-1)^n \, \mathrm{e}^{\xi^2} \frac{\mathrm{d}^n}{\mathrm{d}\xi^n} \mathrm{e}^{-\xi^2} \ , \tag{7.22}$$

which we will prove in the following example.

EXAMPLE

7.2 Mathematical Supplement: Hermite Polynomials

On the basis of the foregoing considerations, the functions $\exp[-(\lambda/2)x^2]$ $H_n(\sqrt{\lambda}x)$, i.e. the Hermite polynomials multiplied by $\exp[-(\lambda/2)x^2]$, obviously fulfil the differential equation (7.6) if

$$\begin{aligned} k^2 &= \frac{2m}{\hbar^2}E_n = \frac{2m}{\hbar^2}\hbar\omega(n+\tfrac{1}{2}) \\ &= \frac{2m\omega}{\hbar}(n+\tfrac{1}{2}) = 2\lambda(n+\tfrac{1}{2}) \ . \end{aligned}$$

Therefore substitution of $\exp[-(\lambda/2)x^2]H_n(\sqrt{\lambda}x)$ into (7.6) and straightforward calculation yield

$$\begin{aligned} &\frac{\mathrm{d}}{\mathrm{d}x}\exp[-(\lambda/2)x^2]H_n(\sqrt{\lambda}x) \\ &\quad = -\lambda x\exp[-(\lambda/2)x^2]H_n(\sqrt{\lambda}x)+\exp[-(\lambda/2)x^2]\frac{\mathrm{d}H_n(\sqrt{\lambda}x)}{\mathrm{d}x} \ ; \\ &\frac{\mathrm{d}^2}{\mathrm{d}x^2}\exp[-(\lambda/2)x^2]H_n(\sqrt{\lambda}x) \\ &\quad = (\lambda x)^2\exp[-(\lambda/2)x^2]H_n(\sqrt{\lambda}x)+\exp[-(\lambda/2)x^2]\frac{\mathrm{d}^2H_n(\sqrt{\lambda}x)}{\mathrm{d}x^2} \\ &\qquad -\lambda\exp[-(\lambda/2)x^2]H_n(\sqrt{\lambda}x)-2\lambda x\exp[-(\lambda/2)x^2]\frac{\mathrm{d}H_n(\sqrt{\lambda}x)}{\mathrm{d}x} \ . \end{aligned} \tag{1}$$

Thus

$$\begin{aligned} &(\lambda^2x^2-\lambda)H_n(\sqrt{\lambda}x)-2\lambda x\frac{\mathrm{d}H_n(\sqrt{\lambda}x)}{\mathrm{d}x}+\frac{\mathrm{d}^2H_n(\sqrt{\lambda}x)}{\mathrm{d}x^2} \\ &\quad +[2\lambda(n+\tfrac{1}{2})-\lambda^2x^2]H_n(\sqrt{\lambda}x)=0 \ , \end{aligned} \tag{2}$$

or

$$\frac{\mathrm{d}^2H_n(\sqrt{\lambda}x)}{\mathrm{d}x^2}-2\lambda x\frac{\mathrm{d}H_n(\sqrt{\lambda}x)}{\mathrm{d}x}+2\lambda nH_n(\sqrt{\lambda}x)=0 \ .$$

Introducing the variable $\xi=\sqrt{\lambda}x$, we obtain after division by λ

$$\frac{\mathrm{d}^2H_n(\xi)}{\mathrm{d}\xi^2}-2\xi\frac{\mathrm{d}H_n(\xi)}{\mathrm{d}\xi}+2nH_n(\xi)=0 \ , \quad n=0,1,2\dots \ . \tag{3}$$

This differential equation is the *defining differential equation* for the Hermite polynomials if n is a positive integer. From (3) a significantly more elegant and manageable formulation of the Hermite polynomials can be given, using the *generating function* $S(\xi,s)$, so that

$$S(\xi,s)=\mathrm{e}^{\xi^2-(s-\xi)^2}=\mathrm{e}^{-s^2+2s\xi}=\sum_{n=0}^{\infty}\frac{H_n(\xi)}{n!}s^n \ . \tag{4}$$

Example 7.2

Expanding the exponential function in terms of powers of s and ξ, we see that the coefficients of the powers s^n are polynomials in terms of ξ – the Hermite polynomials. This can be shown as follows: we have

$$\frac{\partial S}{\partial \xi} = 2s\,\mathrm{e}^{-s^2+2s\xi} = \sum_{n=0}^{\infty} \frac{2s^{n+1}}{n!} H_n(\xi) = \sum_{n=0}^{\infty} \frac{s^n}{n!} \frac{\partial H_n(\xi)}{\partial \xi} \;,$$

$$\begin{aligned} \frac{\partial S}{\partial s} &= (-2s+2\xi)\,\mathrm{e}^{-s^2+2s\xi} = \sum_{n=0}^{\infty} \frac{(-2s+2\xi)s^n}{n!} H_n(\xi) \\ &= \sum_{n=0}^{\infty} \frac{s^{n-1}}{(n-1)!} H_n(\xi) \;. \end{aligned} \tag{5}$$

Equating equal powers of s in the sums of these two equations, we obtain

$$\frac{\partial H_n(\xi)}{\partial \xi} = 2n H_{n-1}(\xi) \;, \quad H_{n+1}(\xi) = 2\xi H_n(\xi) - 2n H_{n-1}(\xi) \;. \tag{6}$$

Therefore it follows that

$$\frac{\partial H_n(\xi)}{\partial \xi} = 2\xi H_n(\xi) - H_{n+1}(\xi) \;, \tag{7}$$

and hence

$$\begin{aligned} \frac{\partial^2 H_n(\xi)}{\partial \xi^2} &= 2H_n(\xi) + 2\xi \frac{\partial H_n(\xi)}{\partial \xi} - \frac{\partial H_{n+1}(\xi)}{\partial \xi} \\ &= 2\xi \frac{\partial H_n(\xi)}{\partial \xi} + 2H_n(\xi) - (2n+2)H_n(\xi) \\ &= 2\xi \frac{\partial H_n(\xi)}{\partial \xi} - 2nH_n(\xi) \;. \end{aligned} \tag{8}$$

This is exactly differential equation (3), proving that the $H_n(\xi)$ appearing in the generating function (4) are indeed Hermite polynomials.

The recurrence formulas (6) may be used to calculate the H_n and their derivatives. Another explicit expression directly obtainable from the generating function is quite useful; let us now establish this important relation. From (4) it follows instantly that

$$\left.\frac{\partial^n S(\xi, s)}{\partial s^n}\right|_{s=0} = H_n(\xi) \;. \tag{9}$$

Now, for an arbitrary function $f(s-\xi)$, it also holds that

$$\frac{\partial f}{\partial s} = -\frac{\partial f}{\partial \xi} \; . \tag{10}$$

Thus

$$\frac{\partial^n S}{\partial s^n} = \mathrm{e}^{\xi^2} \frac{\partial^n \mathrm{e}^{-(s-\xi)^2}}{\partial s^n} = (-1)^n \mathrm{e}^{\xi^2} \frac{\partial^n}{\partial \xi^n} \mathrm{e}^{-(s-\xi)^2} \; . \tag{11}$$

Comparing (11) with (9) yields the very useful formula,

$$H_n(\xi) = (-1)^n \mathrm{e}^{\xi^2} \frac{\partial^n}{\partial \xi^n} \mathrm{e}^{-\xi^2} \; . \tag{12}$$

The $H_n(\xi)$ are polynomials of nth degree in ξ with the dominant term $2^n \xi^n$. The first five $H_n(\xi)$ calculated from (7.22) or (12) of the foregoing example are:

$$\begin{aligned} &H_0(\xi) = 1 \; , && H_1(\xi) = 2\xi \; , \\ &H_2(\xi) = 4\xi^2 - 2 \; , && H_3(\xi) = 8\xi^3 - 12\xi \; , \\ &H_4(\xi) = 16\xi^4 - 48\xi^2 + 12 \; . \end{aligned} \tag{7.23}$$

The eigenfunctions (7.21) were combined by introducing the abbreviation $\xi = \sqrt{\lambda}x$ and using the Hermite polynomials in a way that holds for both even and odd n, i.e.

$$\psi_n(x) = N_n \mathrm{e}^{(-1/2)\xi^2} H_n(\xi) \; , \quad \xi = \sqrt{\lambda}x \; . \tag{7.24}$$

The constant N_n, which depends on the index n, is determined by the normalization condition

$$\int_{-\infty}^{\infty} |\psi_n(x)|^2 \mathrm{d}x = 1 \; , \tag{7.25}$$

since we require the position probability to be 1 for the particle in the entire configuration space. Thus

$$\int_{-\infty}^{\infty} |\psi_n(x)|^2 \mathrm{d}x = \frac{1}{\sqrt{\lambda}} N_n^2 \int_{-\infty}^{\infty} \mathrm{e}^{-\xi^2} H_n(\xi)^2 \mathrm{d}\xi = 1 \; . \tag{7.26}$$

Using relation (12) of Example 7.2 to express one of the Hermite polynomials that appears in the integrand of the normalization integral, the evaluation of this integral becomes simply

$$\int_{-\infty}^{\infty} |\psi_n(x)|^2 \mathrm{d}x = (-1)^n \frac{N_n^2}{\sqrt{\lambda}} \int_{-\infty}^{\infty} H_n(\xi) \frac{\mathrm{d}^n}{\mathrm{d}\xi^n} \mathrm{e}^{-\xi^2} \mathrm{d}\xi \; . \tag{7.27}$$

By partial integration we obtain

$$\int_{-\infty}^{\infty} H_n(\xi)\frac{d^n}{d\xi^n}e^{-\xi^2}\,d\xi$$

$$=\left[\left(\frac{d^{n-1}}{d\xi^{n-1}}e^{-\xi^2}\right)H_n(\xi)\right]_{-\infty}^{\infty}-\int_{-\infty}^{\infty}\frac{dH_n}{d\xi}\frac{d^{n-1}}{d\xi^{n-1}}e^{-\xi^2}\,d\xi\ . \tag{7.28}$$

The first term is, because of (12) in Example 7.2, equal to $(-1)^{n-1}e^{-\xi^2}$ $H_{n-1}(\xi)H_n(\xi)$. It vanishes at infinity, due to the exponential function.

Having carried out partial integration n times, we are left with

$$\int_{-\infty}^{\infty} H_n(\xi)\frac{d^n}{d\xi^n}e^{-\xi^2}\,d\xi=(-1)^n\int_{-\infty}^{\infty}\frac{d^nH_n}{d\xi^n}e^{-\xi^2}\,d\xi\ . \tag{7.29}$$

Since $H_n(\xi)$ is a polynomial of nth order with the dominant term $2^n\xi^n$, for the nth derivative,

$$\frac{d^n}{d\xi^n}H_n(\xi)=2^nn! \tag{7.30}$$

holds.

From this we find that

$$\int_{-\infty}^{\infty} H_n(\xi)\frac{d^n}{d\xi^n}e^{-\xi^2}\,d\xi$$

$$=(-1)^n(2^n)n!\int_{-\infty}^{\infty}e^{-\xi^2}\,d\xi=(-1)^n(2^n)n!\sqrt{\pi}\ , \tag{7.31}$$

and for the normalization constant,

$$N_n=\sqrt{\sqrt{\frac{\lambda}{\pi}}\frac{1}{2^nn!}}\ .$$

The stationary states of the harmonic oscillator in quantum mechanics are therefore

$$\psi_n(x)=\sqrt{\frac{1}{2^nn!}\sqrt{\frac{\lambda}{\pi}}}\exp\left(-\frac{1}{2}\lambda x^2\right)H_n(\sqrt{\lambda}x)\ . \tag{7.32}$$

To discuss the solution, we take a look at the first three eigenfunctions of the linear harmonic oscillator (see Fig. 7.1):

$$n=0: \psi_0(x) = \sqrt[4]{\frac{\lambda}{\pi}} \exp\left(-\frac{1}{2}\lambda x^2\right) ,$$

$$n=1: \psi_1(x) = 2\sqrt{\frac{1}{2}\sqrt{\frac{\lambda}{\pi}}} \exp\left(-\frac{1}{2}\lambda x^2\right) \sqrt{\lambda}x ,$$

$$n=2: \psi_2(x) = \sqrt{\frac{1}{8}\sqrt{\frac{\lambda}{\pi}}} \exp\left(-\frac{1}{2}\lambda x^2\right) (4\lambda x^2 - 2) . \qquad (7.33)$$

From (7.22) and (7.24) it follows that, for space reflection, the eigenfunctions have the symmetry property

$$\psi_n(-x) = (-1)^n \psi_n(x) . \qquad (7.34)$$

This means

$$n \text{ even}: \quad \psi(-x) = \psi(x) \rightarrow \text{ parity } +1$$
$$n \text{ odd}: \quad \psi(-x) = -\psi(x) \rightarrow \text{ parity } -1 .$$

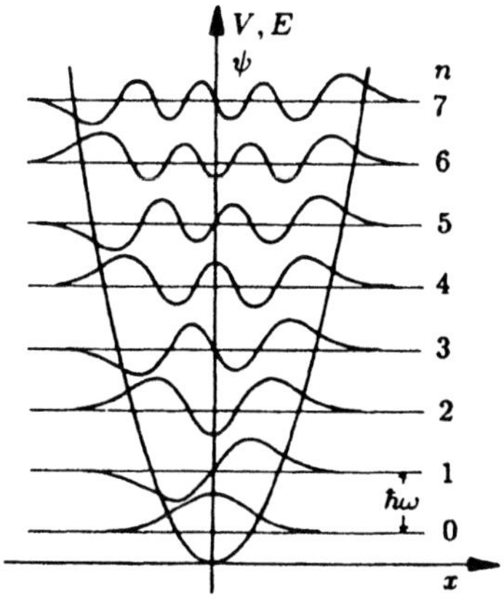

Fig. 7.1. Oscillator potential, energy levels and corresponding wavefunctions

For the lowest H_n, it can easily be shown that they possess precisely n different real zeros and $n+1$ extremal values (see Fig. 7.1). With respect to (12) in Example 7.2, we have

$$H_{n+1} = -\mathrm{e}^{\xi^2} \frac{\mathrm{d}}{\mathrm{d}\xi} (\mathrm{e}^{-\xi^2} H_n) . \qquad (7.35)$$

On the assumption that H_n possesses $n+1$ real extremal values, we can conclude the existence of $n+1$ extremal values for $\mathrm{e}^{-\xi^2} H_n$ (since $\mathrm{e}^{-\xi^2} \rightarrow 0$ for $\xi \rightarrow \infty$). The extremal values are identical with the zeros of the derivative $\mathrm{d}/\mathrm{d}\xi$; therefore H_{n+1} has precisely $n+1$ real zeros. This conclusion shows that the Hermite polynomials $H_n(\xi)$ – and, in consequence, the wave functions $\psi_n(\xi)$ – possess n different real zeros. This is a special case of a universally valid theorem which states that the principal quantum number of an eigenfunction is identical with the number of zeros.

In Fig. 7.1, some of the ψ_n are plotted together with an energy diagram. The energy eigenvalues are represented as horizontal lines with the quantum segments $E_n = (n+\frac{1}{2})\hbar\omega$. For each of the lines there is a corresponding eigenfunction $\psi_n(x)$ drawn on an arbitrary scale.

In addition, the figure contains the function of the potential energy

$$V(x) = \frac{1}{2}m\omega^2 x^2 . \qquad (7.36)$$

Thus we can make a comparison with the classical harmonic oscillator, which oscillates with a certain amplitude characterized by the vanishing kinetic energy at the turning point. Since $E = T + V$, the region of classically possible oscillations is bounded by the point of intersection of the parabola $V(x)$ and the straight line of total energy E. As a matter of fact, the figure shows that the extreme values of the function ψ are localized within the classical region, nevertheless, tails of the wave function extend to infinity.

The deviating behaviour becomes even more significant if we contemplate the position probability of the particle. Let T represent the period of revolution of the particle, then classically we have

$$w_{\text{cl}}(x)\,\mathrm{d}x = \frac{\mathrm{d}t}{T/2} = \frac{2\omega}{2\pi}\,\mathrm{d}t = \frac{\omega}{\pi}\frac{\mathrm{d}x}{\mathrm{d}x/\mathrm{d}t} \ . \tag{7.37}$$

The particle performs harmonic oscillations:

$$x = a\sin\omega t \ , \quad \frac{\mathrm{d}x}{\mathrm{d}t} = a\omega\cos\omega t = \omega a\sqrt{1-(x/a)^2} \ ; \tag{7.38}$$

hence,

$$w_{\text{cl}}(x)\,\mathrm{d}x = \frac{1}{\pi a}\frac{1}{\sqrt{1-(x/a)^2}}\,\mathrm{d}x \ . \tag{7.39}$$

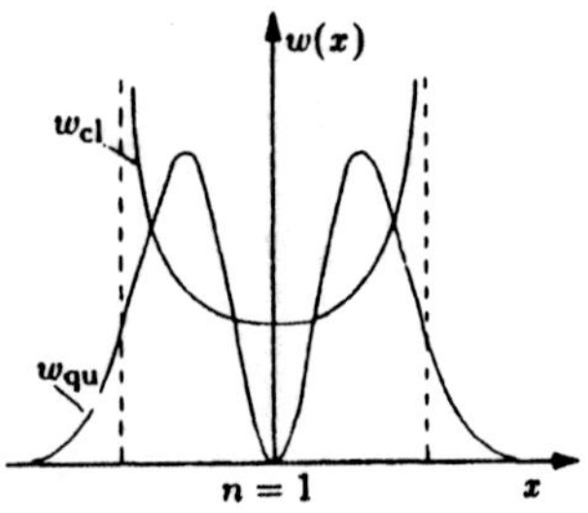

Fig. 7.2. Comparison of the probability density for finding a particle moving in an oscillator, classically and quantum-mechanically. The dashed lines denote the classical points of revolution

The amplitude a is obtained from the energy $E = \frac{1}{2}m\omega^2 a^2$, i.e. $a = \sqrt{2E/m\omega^2}$. Contrary to this, the quantum-mechanical probability for localizing a particle within an interval $x + \mathrm{d}x$ is given by (see Fig. 7.2):

$$w_{\text{qu}}(x)\,\mathrm{d}x = |\psi(x)|^2\,\mathrm{d}x \ , \tag{7.40}$$

which means, e.g. for $n = 1$ with respect to (7.33):

$$w_{\text{qu}}(x)\,\mathrm{d}x = |\psi_1(x)|^2\,\mathrm{d}x = 2\sqrt{\frac{\lambda}{\pi}}\,\mathrm{e}^{-\lambda x^2}\lambda x^2\,\mathrm{d}x \ . \tag{7.41}$$

It can easily be shown that $w_{\text{qu}}(x)$ has a minimum at $x = 0$ and a maximum at

$$x_{\text{max qu}} = \frac{\pm 1}{\sqrt{\lambda}} = \pm\sqrt{\frac{\hbar}{m\omega}} \ , \tag{7.42}$$

whereas classically, with $E = 3/2\hbar\omega$, it holds that

$$x_{\text{max cl}} = \pm a = \pm\sqrt{\frac{2E}{m\omega^2}} = \pm\sqrt{\frac{3\hbar}{m\omega}} \ . \tag{7.43}$$

The agreement between classical and quantum probabilities improves rapidly with increasing quantum number n. A plot for $n = 15$ is given in Fig. 7.3; for large quantum numbers (here $n = 15$), the mean value of the quantum distribution approximates the classical limit.

From the figures we perceive that beyond the region which is classically limited by the relation $E = T + V$, the probability density is not equal to zero. This is a consequence of the fact that T and V are noncommuting quantities, i.e. they do not have simultaneously exact values, since $V(x)$ is a function of space, whereas

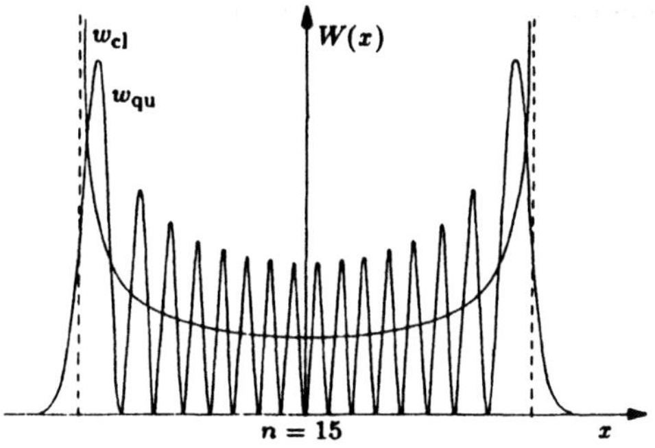

Fig. 7.3. Quantum-mechanical and classical probability densities for a particle in a harmonic oscillator potential with the energy $E = (15 + 1/2)\hbar\omega$, i.e. in the $n = 15$ state. The dashed vertical lines indicate the classical points of revolution

$T = p^2/2m$ is a function of momentum. Therefore, owing to the uncertainty relation $[\hat{p}, x]_- = -\mathrm{i}\hbar$, it is impossible to split the energy precisely in $E = T + V$. It would seem that localizing the particle beyond the classically permissible region implies a violation of energy conservation; however, this is not the case. If we try to localize the particle (i.e. concentrate its wave function) in the small tails of the function ψ, the uncertainty of momentum increases to a point where the new total energy exceeds the value of the potential energy $V(x)$. Thus, from the point of view of energy, the particle is allowed to adopt an x value beyond the classically permitted region. In any case, it is the wave character of the quantum-mechanical wave function which allows the penetration into potential wells and, finally, its tunnelling. This effect is analogous to the jumping of electromagnetic waves (light) over narrow slits.[3]

The behaviour described above is responsible for the *tunnel effect*, according to which a potential well of size V_0 can even be surmounted by particles with energy $E < V_0$. The tunnel effect appears, for example, in the case of field emission and α decay. It has recently received particular attention because of its practical application in the so-called tunnelling-electron microscope.[4] A further difference between the classical and quantum-mechanical oscillator is the state of minimum energy. Classically a particle can be in the state of equilibrium at $x = 0$, $p = 0$, $E = 0$. In quantum mechanics the smallest possible energy value is $E = \hbar\omega/2$, the *zero-point energy*.

This zero-point energy is a direct consequence of the uncertainty relation

$$\overline{\Delta x^2}\,\overline{\Delta p^2} \geq \frac{\hbar}{4} \,. \tag{7.44}$$

[3] See J.D. Jackson: *Classical Electrodynamics*, 2nd ed. (Wiley, New York 1980) and W. Greiner: *Classical Electrodynamics* (Springer, New York 1998)

[4] In 1986 G. Binnig and H. Rohrer received the Nobel Prize in physics for developing the tunnelling-electron microscope; see e.g. G. Binnig and H. Rohrer, Scientific American, Aug. 1985, p. 40.

Let us take a closer look at the expressions

$$\overline{\Delta p^2} = \overline{(p-\bar{p})^2} = \overline{p^2 - 2p\bar{p} + \bar{p}^2} = \overline{p^2} - 2\overline{p\bar{p}} + \overline{\bar{p}^2} = \overline{p^2} - \bar{p}^2 \ . \tag{7.45}$$

Analogously, $\overline{\Delta x^2} = \overline{x^2} - \bar{x}^2$. On the other hand, in a state with a fixed energy value, the mean values $\bar{p}$ and $\bar{x}$ are equal to zero since the integrand is an odd function:

$$\bar{x} = \int_{-\infty}^{\infty} \psi_n^*(x) x \psi_n(x)\,\mathrm{d}x = \int_{-\infty}^{\infty} |\psi_n(x)|^2 x\,\mathrm{d}x = 0 \ , \tag{7.46}$$

and

$$\begin{aligned} \bar{p}_x &= \int_{-\infty}^{\infty} \psi_n^*(x) \hat{p}_x \psi_n(x)\,\mathrm{d}x = -\mathrm{i}\hbar \int_{-\infty}^{\infty} \psi_n^*(x) \frac{\mathrm{d}}{\mathrm{d}x} \psi_n(x)\,\mathrm{d}x \\ &= -\frac{\mathrm{i}\hbar}{2} \int_{-\infty}^{\infty} \frac{\mathrm{d}}{\mathrm{d}x} |\psi_n(x)|^2\,\mathrm{d}x = -\frac{\mathrm{i}\hbar}{2} |\psi_n(x)|^2 g|_{-\infty}^{\infty} = 0 \ . \end{aligned} \tag{7.47}$$

Therefore

$$\overline{\Delta p^2} = \overline{p^2} \ , \quad \overline{\Delta x^2} = \overline{x^2} \ . \tag{7.48}$$

With this we may write the uncertainty relation in the form

$$\overline{p^2}\,\overline{x^2} \geq \frac{\hbar^2}{4} \ . \tag{7.49}$$

Now the average energy of the oscillator is

$$\overline{H} = \frac{\overline{p^2}}{2m} + \frac{m\omega^2}{2} \overline{x^2} \ . \tag{7.50}$$

Comparing both equations, we see that an increasing potential energy leads to a decrease of kinetic energy and vice versa.

Combining these equations, we obtain

$$\overline{H} \geq \frac{\overline{p^2}}{2m} + \frac{m\omega^2}{8} \frac{\hbar^2}{\overline{p^2}} \ . \tag{7.51}$$

The function $\overline{H} = \overline{H}(\overline{p^2})$ possesses a minimum at $\overline{p^2} = \frac{1}{2} m\omega\hbar$, which can easily be confirmed by evaluating the first and second derivative.

Since a state of fixed energy is characterized by $\overline{H} = E$, we obtain as minimum value of the possible energy eigenvalues

$$\min E \geq \frac{h\omega}{2} = E_0 \ .$$

The zero-point energy E_0 therefore is the smallest energy value that is compatible with the uncertainty relation.

It is possible to indicate the zero-point motion (which leads to the zero-point energy) by observing the dispersion of light in crystals. In solid bodies, atoms and molecules perform small oscillations; according to classical theory, they ought to diminish with decreasing temperature. These atomic oscillations are the cause of the light dispersion, which therefore should vanish with decreasing temperature. However, experiments demonstrate that the intensity of the scattered light converges towards a finite limit, showing that even at the absolute zero-point, atomic oscillations occur.

7.2 The Description of the Harmonic Oscillator by Creation and Annihilation Operators

The normalized eigenfunctions of the harmonic oscillator all take the form

$$\psi_n(\xi) = \frac{\sqrt[4]{\lambda}}{\sqrt{\sqrt{\pi}2^n n!}} \mathrm{e}^{-\xi^2/2} H_n(\xi) \;, \quad \xi = \sqrt{\lambda}x \;. \tag{7.52}$$

For the Hermite polynomials $H_n(\xi)$, the following recurrence relations hold (see Example 7.2):

$$\xi H_n = n H_{n-1} + \frac{1}{2} H_{n+1} \;, \quad \frac{\mathrm{d}}{\mathrm{d}\xi} H_n = 2n H_{n-1} \;. \tag{7.53}$$

From these formulae, connections between eigenfunctions of the harmonic oscillator, which belong to neighbouring quantum numbers, are obtainable:

$$\xi\psi_n = \sqrt{\frac{n}{2}}\psi_{n-1} + \sqrt{\frac{n+1}{2}}\psi_{n+1} \;, \tag{7.54}$$

$$\frac{\partial}{\partial\xi}\psi_n = 2\sqrt{\frac{n}{2}}\psi_{n-1} - \xi\psi_n \;. \tag{7.55}$$

Equation (7.55) is now rearranged, using (7.54), so that the right-hand sides of both equations resemble each other:

$$\frac{\partial}{\partial\xi}\psi_n = \sqrt{\frac{n}{2}}\psi_{n-1} - \sqrt{\frac{n+1}{2}}\psi_{n+1} \;. \tag{7.56}$$

By addition (or subtraction) of (7.54) and (7.56), we get the relations

$$\frac{1}{\sqrt{2}}\left(\xi + \frac{\partial}{\partial\xi}\right)\psi_n = \sqrt{n}\psi_{n-1} \;,$$

$$\frac{1}{\sqrt{2}}\left(\xi - \frac{\partial}{\partial\xi}\right)\psi_n = \sqrt{n+1}\psi_{n+1} \;. \tag{7.57}$$

With these equations we can now evaluate the neighbouring eigenfunctions ψ_{n-1} and ψ_{n+1} from the eigenfunctions ψ_n. For the sake of brevity, we define the operators

$$\frac{1}{\sqrt{2}}\left(\xi+\frac{\partial}{\partial\xi}\right)=\hat{a}\ ,\quad \frac{1}{\sqrt{2}}\left(\xi-\frac{\partial}{\partial\xi}\right)=\hat{a}^+\ , \tag{7.58}$$

and hence obtain, instead of (7.57),

$$\hat{a}\psi_n=\sqrt{n}\psi_{n-1}\ ,\quad \hat{a}^+\psi_n=\sqrt{n+1}\psi_{n+1}\ . \tag{7.59}$$

For the present we will call $\hat{a}$ the *lowering operator* and $\hat{a}^+$ the *raising operator* since the index n of the state ψ_n is lowered or raised, respectively. In the following we formulate a better interpretation of $\hat{a}$ and $\hat{a}^+$ and then also name the operators more appropriately.

7.3 Properties of the Operators $\hat{a}$ and $\hat{a}^+$

The operators $\hat{a}$ and $\hat{a}^+$ are adjoint to each other (i.e. not self-adjoint), since it holds that (with partial integration)

$$\int \psi^*\left(\xi\varphi+\frac{\partial\varphi}{\partial\xi}\right)\mathrm{d}\xi=\int\left(\xi\psi^*-\frac{\partial\psi^*}{\partial\xi}\right)\varphi\,\mathrm{d}\xi\ , \tag{7.60}$$

or, abbreviated,

$$\langle\psi|\hat{a}\varphi\rangle=\langle\hat{a}^+\psi|\varphi\rangle\ . \tag{7.61}$$

We have made use of the fact that the operators are real from their definition (7.58), i.e. $\hat{a}=\hat{a}^*$ and $\hat{a}^+=(\hat{a}^+)^*$.

The wave function ψ_n is an eigenfunction of the operator product $\hat{a}^+\hat{a}$ because

$$\hat{a}^+\hat{a}\psi_n=\sqrt{n}\hat{a}^+\psi_{n-1}=n\psi_n\ , \tag{7.62}$$

which can be checked with (7.59). The eigenvalue n is the index of the oscillator wave function ψ_n. We therefore define a *number operator* $\hat{N}$

$$\hat{N}=\hat{a}^+\hat{a}\ ,\quad \hat{N}\psi_n=n\psi_n\ . \tag{7.63}$$

The eigenvalues of $\hat{N}$ are n; the eigenfunctions are ψ_n. We get the commutator

$$[\hat{a},\hat{a}^+]_-=1 \tag{7.64}$$

easily by evaluating the two products according to (7.58).

By successively applying $\hat{a}^+$ on ψ, we are able to calculate all eigenfunctions, starting from the ground state. From (7.59) it follows that

$$\psi_n = \frac{1}{\sqrt{n}}\hat{a}^+\psi_{n-1} = \ldots = \frac{1}{\sqrt{n!}}(\hat{a}^+)^n\psi_0 \ . \tag{7.65}$$

Up to this point we have developed a formalism in terms of $\hat{a}$ and $\hat{a}^+$ that enables us to set up a differential equation for the ground state. With $n = 0$, we find from (7.58) and (7.59):

$$\hat{a}\psi_0 = 0 \quad \text{and} \quad \xi\psi_0 + \frac{\partial\psi_0}{\partial\xi} = 0 \ . \tag{7.66}$$

The substitution $\psi_0 \approx \mathrm{e}^{\alpha\xi^2}$ yields $\alpha = -\frac{1}{2}$. Thus the function of the ground state is, up to a normalization factor,

$$\psi_0 \approx \mathrm{e}^{-\xi^2/2} \ ,$$

which coincides with the solution of the Schrödinger equation for the harmonic oscillator, yielding for the normalized ground state (7.33)

$$\psi_0 = \sqrt[4]{\frac{\lambda}{\pi}}\,\mathrm{e}^{-\xi^2/2} \ . \tag{7.67}$$

7.4 Representation of the Oscillator Hamiltonian in Terms of $\hat{a}$ and $\hat{a}^+$

For the one-dimensional harmonic oscillator, the Hamiltonian is given by

$$\hat{H} = \hat{T} + \hat{V} = -\frac{\hbar^2}{2m}\frac{\partial^2}{\partial x^2} + \frac{1}{2}m\omega^2x^2 \ . \tag{7.68}$$

Corresponding to the introduction of the new variables $\xi = \sqrt{\lambda}x = \sqrt{(m\omega/\hbar)}x$ we can now define a new momentum operator

$$\hat{p}_\xi = -\mathrm{i}\frac{\partial}{\partial\xi} \Rightarrow \hat{p}_\xi^2 = -\frac{\partial^2}{\partial\xi^2} = -\frac{\hbar}{m\omega}\frac{\partial^2}{\partial x^2} \ , \tag{7.69}$$

so that the Hamiltonian becomes

$$\hat{H} = \frac{1}{2}\hbar\omega(\xi^2 + \hat{p}_\xi^2) = \frac{1}{2}\hbar\omega\left(\xi^2 - \frac{\partial^2}{\partial\xi^2}\right) \ . \tag{7.70}$$

From the relation

$$\xi^2 - \frac{\partial^2}{\partial\xi^2} = \hat{a}\hat{a}^+ + \hat{a}^+\hat{a} \ ,$$

which is easily verified with (7.58), and by use of the commutation property (7.64) and definition (7.63), we may infer a simple representation of the Hamiltonian, namely

$$\hat{H} = \hbar\omega(\hat{a}^+\hat{a} + \tfrac{1}{2}) = \hbar\omega(\hat{N} + \tfrac{1}{2}) \ . \tag{7.71}$$

From this we can evaluate the energy eigenvalues

$$\hat{H}\psi_n = \hbar\omega(\hat{N} + \tfrac{1}{2})\psi_n = \hbar\omega(n + \tfrac{1}{2})\psi_n = E_n\psi_n \ . \tag{7.72}$$

The energy eigenvalues are $E_n = \hbar\omega(n + \frac{1}{2})$, as calculated above.

7.5 Interpretation of $\hat{a}$ and $\hat{a}^+$

The ground state ψ_0 has the zero-point energy $E_0 = \hbar\omega/2$. Since the energy spectrum of the harmonic oscillator is equidistant, the state ψ_n possesses an energy value that is larger by the term $n\hbar\omega$. We will distribute this energy to n energy quanta $\hbar\omega$ (quanta of the oscillator field), called *phonons*. ψ_n is called an *n-phonon state*. In Dirac's notation it reads

$$\psi_n = |n\rangle \ . \tag{7.73}$$

The "kets" $|n\rangle$ contain the number of phonons. The zero-phonon state $|0\rangle$ is also called the *vacuum*. Using the notation above, the equations (7.59) become

$$\hat{a}\,|n\rangle = \sqrt{n}\,|n-1\rangle \quad , \quad \hat{a}^+\,|n\rangle = \sqrt{n+1}\,|n+1\rangle \ . \tag{7.74}$$

The following interpretation is appropriate: if acting on the wave function, the operator $\hat{a}$ *annihilates* one phonon at a time, whereas $\hat{a}^+$ *creates* one. Therefore, from now on we will call $\hat{a}$ and $\hat{a}^+$ the *annihilation operator* and *creation operator*, respectively. $\hat{N}$ is termed the *phonon number operator*, since its eigenvalues, given by the equation

$$\hat{N}\,|n\rangle = n\,|n\rangle \ , \tag{7.75}$$

are the numbers of phonons of the corresponding state.

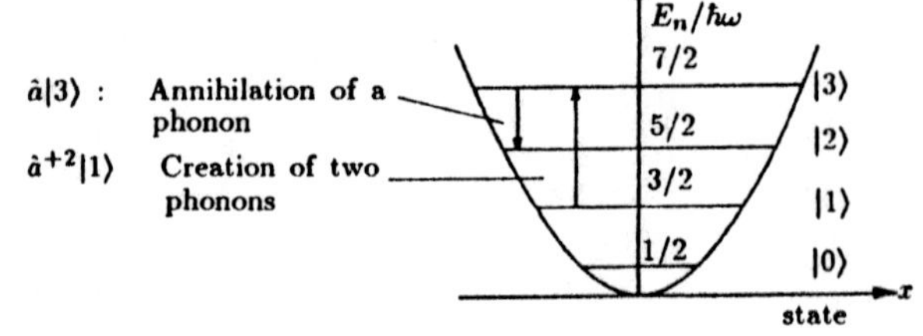

Fig. 7.4. Energy levels of the harmonic oscillator and the effect of the creation and annihilation operators

The introduction of the phonon representation is often referred to (somewhat imprecisely) as *second quantization*. The *quanta* of the wave field of the oscillator are exactly the *phonons*. This becomes clear if we consider the analogy with *photons*.

However, for a complete mathematical treatment of the second quantization of the electromagnetic field, it is necessary to use quantum field theoretical methods.[5]

EXAMPLE

7.3 The Three-Dimensional Harmonic Oscillator

Problem. Determine the eigenvalues and eigenfunctions of the Hamiltonian for the three-dimensional, spherically symmetric harmonic oscillator.

Solution. The Hamiltonian for the three-dimensional, spherically symmetric harmonic oscillator is of the form

$$\hat{H} = -\frac{\hbar^2}{2m}\nabla^2 + \frac{m\omega^2}{2}r^2 \ . \tag{1}$$

Because of the symmetry of the problem, we solve the stationary Schrödinger equation

$$\hat{H}\psi_{nl} = E_{nl}\psi_{nl} \tag{2}$$

in spherical coordinates. n, l are quantum numbers characterizing the eigenfunctions and will have to be further specified. The Laplacian in spherical coordinates is

$$\nabla^2 = \frac{\partial^2}{\partial r^2} + \frac{2}{r}\frac{\partial}{\partial r} - \frac{\hat{L}^2}{\hbar^2 r^2} \ , \tag{3}$$

where the angular-momentum operator $\hat{L}$ contains the derivatives with respect to the angles ϑ and φ [see (4.75)–(4.77) and (4.82)]. The eigenfunctions of $\hat{L}^2$ are the spherical harmonics [see (4.76)–(4.80) and Example 4.9]:

$$\hat{L}^2 Y_{lm}(\vartheta,\varphi) = \hbar^2 l(l+1) Y_{lm}(\vartheta,\varphi) \ . \tag{4}$$

To separate the angular and radial parts of the wave function ψ_{nlm}, we try writing

$$\psi_{nlm}(r,\vartheta,\varphi) = \frac{R_{nl}(r)}{r} Y_{lm}(\vartheta,\varphi) \ . \tag{5}$$

---

[5] See W. Greiner, J. Reinhardt: *Quantum Electrodynamics*, 2nd ed. (Springer, Berlin, Heidelberg 1994).

Example 7.3

Substituting (5) into the Schrödinger equation (2) and using (1) yields a differential equation for the radial part of the wave function $R_{nl}(r)$:

$$R''_{nl} + \left(\frac{2mE_{nl}}{\hbar^2} - \frac{m^2\omega^2}{\hbar^2} r^2 - \frac{l(l+1)}{r^2} \right) R_{nl}(r) = 0 \ . \tag{6}$$

Using the abbreviations in (7.5), this differential equation becomes identical with (7.6), except for the angular-momentum term $l(l+1)/r^2$, usually termed the *angular-momentum barrier*,

$$R''_{nl} + \left(k^2 - \lambda^2 r^2 - \frac{l(l+1)}{r^2} \right) R_{nl} = 0 \ . \tag{7}$$

This differential equation can be transformed by a suitable substitution for R_{nl} to the same standard form (7.8) as previously (7.6).

This substitution will differ from (7.10) because of the angular-momentum barrier. As before, in the case of the linear oscillator, we try to split off the asymptotic behaviour of the wave function. If $r \to \infty$, we may neglect the angular-momentum term $l(l+1)/r^2$. Then the solution of the differential equation has to behave like

$$R_{nl}(r) \underset{r\to\infty}{\longrightarrow} \sim \exp[-(\lambda/2)r^2] \ . \tag{8}$$

At $r = 0$, the angular-momentum term becomes dominant, independent of the potential. Thus we try an expansion in a power series

$$R_{nl} = r^{\alpha} \sum_{i=0}^{\infty} a_i r^i \ .$$

Substitution into the asymptotic differential equation

$$R''_{nl} - \frac{l(l+1)}{r^2} R_{nl} = 0 \tag{9}$$

yields

$$\alpha(\alpha - 1) - l(l+1) = 0 \ ,$$

with solutions $\alpha_1 = -l$, $\alpha_2 = l+1$. Hence, we get

$$R_{nl}(r) \underset{r\to\infty}{\longrightarrow} \sim r^{l+1} \quad \text{or} \quad \sim r^{-l} \ . \tag{10}$$

The first possibility in (10) suggests the substitution

$$R_{nl}(r) = r^{l+1} \exp[-(\lambda/2)r^2] v(r) \ . \tag{11}$$

Notice that, quite analogously, we could continue the calculation using

$$R_{nl}(r) = r^{-l} \exp[-(\lambda/2)r^2] u(r) \ .$$

Example 7.3

However, this leads to exactly the same solutions as (11). This is not immediately obvious, but becomes clear by repeating the following steps, employing the substitution (11).[6] With (11), (7) changes to

$$v'' + 2\left(\frac{l+1}{r} - \lambda r\right)v' - (\lambda(2l+3) - k^2)v = 0 \ . \tag{12}$$

By substituting the variable

$$t = \lambda r^2 \ , \tag{13}$$

(12) transforms into a Kummer differential equation [see (17) of Exercise 7.1]:

$$t\frac{\mathrm{d}^2 v}{\mathrm{d}t^2} + \left(l + \frac{3}{2} - t\right)\frac{\mathrm{d}v}{\mathrm{d}t} - \left[\frac{1}{2}\left(l + \frac{3}{2}\right) - \frac{\kappa}{2}\right]v = 0 \ , \tag{14}$$

with $\kappa = k^2/2\lambda = \hbar k^2/2m\omega = E/\hbar\omega$ [see (7.9)]. It has the solutions

$$\begin{aligned} v(r) &= C\,{}_1F_1\left[\tfrac{1}{2}(l + \tfrac{3}{2} - \kappa), l + \tfrac{3}{2}; \lambda r^2\right] \\ &\quad + C_2 r^{-(2l+1)} \times {}_1F_1\left[\tfrac{1}{2}(-l + \tfrac{1}{2} - \kappa), -l + \tfrac{1}{2}, \lambda r^2\right] \ . \end{aligned} \tag{15}$$

When $l \neq 0$, the second particular solution cannot be normalized, since it diverges too strongly at $r = 0$. Therefore we set $C_2 = 0$. The same is also valid for $l = 0$. To prove this is not trivial, so we will now demonstrate the verification in detail. (In the case of the linear oscillator, the second particular solution would make sense from a physical point of view. The difference between the cases originates from the fact that, earlier, the normalization integral was one-dimensional, whereas, now, it is three-dimensional; i.e. it has to be integrated with a different volume element.)

We start by requiring that the momentum operator $-\mathrm{i}\hbar\nabla$ be self-adjoint, since it represents a physical quantity and therefore should have real eigenvalues:

$$\int u_n^*(-\mathrm{i}\hbar\nabla)u_m\,\mathrm{d}\tau = \int(-\mathrm{i}\hbar\nabla u_n)^* u_m\,\mathrm{d}\tau \ , \tag{16}$$

where u_n, u_m are elements of a complete orthonormal set of solutions that belong to a certain Hamiltonian, e.g. to the Hamiltonian (1):

$$\int u_n^* u_m\,\mathrm{d}\tau = \delta_{nm} \ . \tag{17}$$

Since the u_n form a complete set, we may expand the components of $-\mathrm{i}\hbar\nabla u_m$ in terms of the u_n:

$$-\mathrm{i}\hbar\nabla_i u_m = \sum_k \alpha_{ik} u_k \ . \tag{18}$$

[6] See J.M. Eisenberg, W. Greiner: *Nuclear Theory 1, Nuclear Models*, 3rd ed. (North-Holland, Amsterdam 1987) pp. 145–146.

Example 7.3

Then from (16) we can obtain

$$\int u_n^*(-\mathrm{i}\hbar\nabla)(-\mathrm{i}\hbar\nabla u_m)\,\mathrm{d}\tau = \int(-\mathrm{i}\hbar\nabla u_n)^*(\mathrm{i}\hbar\nabla u_m)\,\mathrm{d}\tau \ , \tag{19}$$

or

$$\int u_n^*\Delta u_m\,\mathrm{d}\tau = -\int \nabla u_n^*\nabla u_m\,\mathrm{d}\tau \ , \tag{20}$$

employing ***Green***'s theorem, which requires that the surface integral vanish. If we multiply (20) by $-\hbar^2/2m$ and set $n = m$, the left-hand side becomes the mean value of kinetic energy of the state u_n:

$$\langle E_{\mathrm{kin}}\rangle_n = \frac{\hbar^2}{2m}\int |\nabla u_n|^2\,\mathrm{d}\tau \ . \tag{21}$$

The second particular solution in (15) with $l = 0$ behaves like

$$\frac{R(r)}{r} \sim \frac{1}{r}\,{}_1F_1\left(\frac{1}{4}-\frac{\kappa}{2},\frac{1}{2};\ \lambda r^2\right)\mathrm{e}^{-(\lambda/2)r^2} \ , \tag{22}$$

which means that the integral (21) diverges, whereas the expectation value of the potential energy $m\omega^2 r^2/2$ remains finite (this straightforward calculation is left to the reader as an exercise). The reason for the divergence of the integral of kinetic energy is the divergence of the wave function at the origin $r = 0$. There it holds that

$$\nabla\frac{R}{r} \sim \frac{1}{r^2}$$

and therefore

$$\left(\nabla\frac{R}{r}\right)2r^2\,\mathrm{d}r \sim \mathrm{d}r \ .$$

This term obviously diverges. Consequently, we also have to rule out solution (22).

We now return to (15). Employing the same arguments as in the case of the linear oscillator, we conclude that, since the solutions are required to be regular at infinity, the hypergeometric function has to break off, leading this time to the condition

$$\tfrac{1}{2}(l+\tfrac{3}{2}-\kappa) = -n \ , \quad (n \in \mathbb{N}_0) \ ,$$

i.e. to a quantization of energy:

$$E_{nl} = \hbar\omega(2n+l+\tfrac{3}{2}) \ . \tag{23}$$

The term $(3/2)\hbar\omega$ represents the *zero-point energy of the three-dimensional harmonic oscillator*. Since there are now zero-point oscillations along the x, y and z axes, the zero-point energy is three times as large as in the one-dimensional case.

Example 7.3

Spectrum of the three-dimensional harmonic oscillator

$N=2n+l$	n	l	Energy	Number of degenerate states
0	0	0(s)	$\frac{3}{2}\hbar\omega$	1
1	0	1(p)	$\frac{5}{2}\hbar\omega$	3
2	1 0	0(s) 2(d)	$\left.\frac{7}{2}\hbar\omega\right\}$	1 6 5
3	1 0	1(s) 3(f)	$\left.\frac{9}{2}\hbar\omega\right\}$	3 10 7
4	2 1 0	0(s) 2(d) 4(g)	$\left.\frac{11}{2}\hbar\omega\right\}$	1 5 15 9

The complete, not yet normalized, eigenfunction belonging to the eigenvalue E_{nl} is then

$$\psi_{nlm}(r,\vartheta,\varphi) = r^l \mathrm{e}^{-(\lambda/2)r^2}\,{}_1F_1(-n, l+\tfrac{3}{2}; \lambda r^2) Y_{lm}(\vartheta,\varphi) \ . \tag{24}$$

The $2l+1$ eigenstates with the same (n, l) but different magnetic quantum numbers m are degenerate. Furthermore, the states with constant $N \equiv 2n + l$ are degenerate, too. Therefore N is sometimes called the *principal quantum number*. The table shows the degeneracy of the lowest eigenstates of the three-dimensional oscillator.

In spectroscopy, it is common to use the notation s, p, d, f, ... for the angular momenta $l = 0, 1, 2, 3, \ldots$. For example, a 5p state is characterized by the principal quantum number $N = 5$ and angular momentum $l = 1$ (see table).

The diagram illustrates the spectrum of the three-dimensional harmonic oscillator. The three-dimensional oscillator is of fundamental importance for

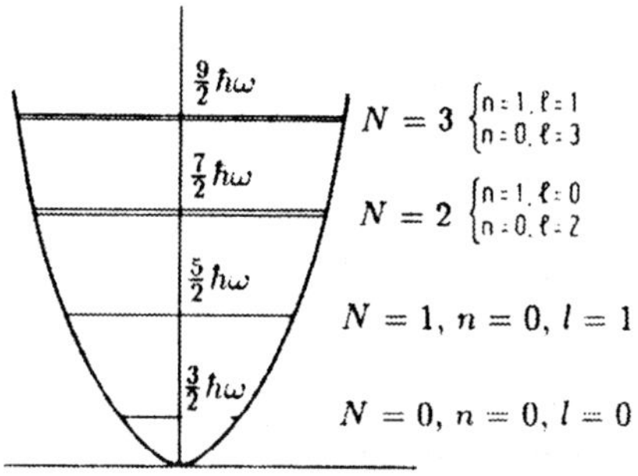

The level structure of a particle in a three-dimensional oscillator potential. Notice the degeneracy of the higher states

Example 7.3

nuclear physics in the so-called shell model for nuclei, when establishing the corresponding Hamiltonian. In fact, it is an essential part of it: the shell model of the nucleus is based on the assumption that the individual nucleons move in an average potential that is generated by all the nucleons together.

This average potential is often approximated by the three-dimensional oscillator.[7] In addition, there is a so-called *spin–orbit interaction*, with which we will become familiar when discussing relativistic quantum mechanics.

7.6 Biographical Notes

WEBER, Heinrich, German mathematician, *5.3.1842 Heidelberg, † 17.5.1913 Strasbourg. W. studied in Heidelberg, Leipzig and Königsberg. In 1873 he was professor in Berlin, 1884 in Marburg, 1892 in Göttingen and from 1895 on, in Strasbourg. W. made significant contributions to mathematical physics, number theory and algebra. He was also co-author of excellent text books.

GAUSS, Carl Friedrich, German mathematician, astronomer and physicist, *30.4.1777 Braunschweig (Brunswick), † 23.2.1855 Göttingen. G. was a day-labourer's son and attracted attention very early on because of his extraordinary mathematical talents. Beginning in 1791, he was educated at the expense of the Duke of Braunschweig (Brunswick). G. studied 1795–98 in Göttingen and took his doctorate in 1799 in Helmstedt. From 1807 on, G. was director of the Göttingen observatory and professor at the University of Göttingen. He refused all offers to go elsewhere, for example, to join the academy in Berlin. G. started his scientific activities in 1791 with investigations concerning the geometric-arithmetic mean, the distribution of prime numbers, and in 1792, the *foundations of geometry*. By 1794 G. had discovered the *method of least squares*, and in 1795 he began to work intensely on the theory of numbers, for example, on the quadratic reciprocity law. In 1796, G. published his first work; in it, he furnished proof that equilateral n-gons, in particular the 17-gons, can be constructed with a compass and straightedge, if n is a Fermat-type prime number. In his doctoral thesis (1799), G. furnished proof of the *fundamental theorem of algebra*, which he followed with further proofs. From his bequest we know that in the same year G. had already developed the foundations of the theory of elliptic and modul functions. His first extensive work, published in 1801, is his famous *Disquisitiones arithmeticae*, which marks as the beginning of modern number theory. It contains the theory of quadratic congruences and the first proof of the quadratic reciprocity law, the *theorema aureum*, as well as the *science of circle partitions*. From about 1801 on, G. became interested in astronomy. The results of his studies were the calculation of the planetoid Ceres's orbit (1801), research into secular perturbations (1809 and 1818), and the attraction of the universal ellipsoid (1813). In 1812, G. published his treatise on the *hypergeometric series*, which is first

[7] See J.M. Eisenberg, W. Greiner: *Nuclear Theory 1, Nuclear Models*, 3rd ed. (North-Holland, Amsterdam 1987).

correct and systematic investigation of convergence. From 1820 on, G. devoted himself increasingly to geodesy. His most outstanding theoretical achievement was the *theory of surfaces*, which contains the *Theorema egregium* (1827). G. practised geodesy, too, e.g. he performed comprehensive surveys in the years 1821–25. In spite of such extensive achievements, in 1825 and in 1831 his publications on *biquadratic remainders* also appeared. The second of these treatises offers a description of complex numbers in the plane and a novel theory of prime numbers. In his last years G. also enjoyed physical problems. His most important contributions are the invention of the electric telegraph, made in 1833/34 together with W. Weber, and in 1839/40 the *potential theory*, which became a new branch of mathematics. Many of G.'s important results only became known from his diary and letters; for example, G. had discovered non-Euclidean geometry by 1816. G.'s reluctance to publish important results originated in the extraordinarily high standards he set for the presentation of his research, and in his efforts to avoid unnecessary argument.

KUMMER, Ernst Eduard, German mathematician, *29.1.1810 Sorau (Zary), † 14.5.1893 Berlin. K. was a teacher at the grammar school in Liegnitz from 1832 to 1842; later he was at the University of Breslau (Wrocław) until 1856. In the following years, until 1883, he was a professor at the University of Berlin. His major mathematical achievements are the differential geometry of congruences and the introduction of the ideal numbers to the theory of algebraic number fields.

LAGUERRE, Edmond Nicolas, French mathematician, *9.4.1834, † 13.8.1886 Bar-le-Duc. L. was one of the founders of modern geometry. He became a member of the Académic Française in 1885. Besides geometrical problems (especially the interpretation of imaginary and real geometry), L. particularly furthered the theories of algebraic equations and continued fractions.

GREEN, George, English mathematician, *14.7.1793 at Nottingham, † 31.3.1841 at Sneinton near Nottingham. Besides his activities as successor to his father who was a baker and miller, G. followed closely all discoveries concerning electricity and read the works of Laplace. After further studies at Cambridge, he worked there at Caius college. His main work "Essays on the Application of Mathematical Analysis to Theories of Electricity and Magnetism" (1828) represents the first attempt at a mathematical description of the phenomena of electricity, and marks, along with the work of Gauss, the beginning of potential theory.

8. The Transition from Classical to Quantum Mechanics

8.1 Motion of the Mean Values

We consider a Hermitian operator $\hat{L}$. The mean value of the operator is, as we know, defined as

$$\overline{L} = \int \psi^* \hat{L} \psi \, \mathrm{d}V \, . \tag{8.1}$$

Since both the operator $\hat{L}$ and the wave function ψ can be time dependent, the mean value $\overline{L}$ will in general be time dependent, too. When we evaluate the temporal variation of $\hat{L}$, we can exchange differentiation and integration. This yields

$$\frac{\mathrm{d}}{\mathrm{d}t}\overline{L} = \int \psi^* \frac{\partial \hat{L}}{\partial t} \psi \, \mathrm{d}V + \int \left(\frac{\partial \psi^*}{\partial t} \hat{L} \psi + \psi^* \hat{L} \frac{\partial \psi}{\partial t} \right) \mathrm{d}V \, . \tag{8.2}$$

The first integral represents the mean value of the partial temporal derivative of the operator $\hat{L}$. The second integral can be simplified with the aid of the time-dependent Schrödinger equation:

$$\frac{\partial \psi}{\partial t} = -\frac{\mathrm{i}}{\hbar} \hat{H} \psi \, , \qquad \frac{\partial \psi^*}{\partial t} = \frac{\mathrm{i}}{\hbar} \hat{H}^* \psi^* \, . \tag{8.3}$$

If we make use of the Hermiticity of $\hat{H}$ we obtain

$$\frac{\mathrm{d}}{\mathrm{d}t}\overline{L} = \int \psi^* \frac{\partial \hat{L}}{\partial t} \psi \, \mathrm{d}V + \frac{\mathrm{i}}{\hbar} \int \psi^* [\hat{H}, \hat{L}]_- \psi \, \mathrm{d}V \tag{8.4}$$

or, more simply,

$$\frac{\mathrm{d}}{\mathrm{d}t}\overline{L} = \overline{\frac{\partial \hat{L}}{\partial t}} + \frac{\mathrm{i}}{\hbar} \overline{[\hat{H}, \hat{L}]_-} \, . \tag{8.5}$$

Taking (8.5) as a basis, we can neatly define the total temporal derivative of the operator $\mathrm{d}\hat{L}/\mathrm{d}t$:

$$\frac{\mathrm{d}\hat{L}}{\mathrm{d}t} = \frac{\partial \hat{L}}{\partial t} + \frac{\mathrm{i}}{\hbar} [\hat{H}, \hat{L}]_- \, . \tag{8.6}$$

From this definition, we see that the temporal derivative of the mean value $\hat{L}$ is equal to the mean value of $\mathrm{d}\hat{L}/\mathrm{d}t$. The sum and product obey the usual differentiation rules:

$$\frac{\mathrm{d}}{\mathrm{d}t}(\hat{A}+\hat{B}) = \frac{\mathrm{d}\hat{A}}{\mathrm{d}t} + \frac{\mathrm{d}\hat{B}}{\mathrm{d}t} \,, \tag{8.7}$$

$$\frac{\mathrm{d}}{\mathrm{d}t}(\hat{A}\hat{B}) = \frac{\mathrm{d}\hat{A}}{\mathrm{d}t}\hat{B} + \hat{A}\frac{\mathrm{d}\hat{B}}{\mathrm{d}t} \,, \tag{8.8}$$

which can be seen by applying (8.6). The proof of (8.8) is as follows:

$$\begin{aligned}
\frac{\mathrm{d}}{\mathrm{d}t}(\hat{A}\hat{B}) &= \frac{\partial}{\partial t}(\hat{A}\hat{B}) + \frac{\mathrm{i}}{\hbar}[\hat{H},\hat{A}\hat{B}]_- \\
&= \frac{\partial A}{\partial t}\hat{B} + \hat{A}\frac{\partial \hat{B}}{\partial t} + \frac{\mathrm{i}}{\hbar}(\hat{H}\hat{A}\hat{B} - \hat{A}\hat{H}\hat{B} + \hat{A}\hat{H}\hat{B} - \hat{A}\hat{B}\hat{H}) \\
&= \frac{\partial \hat{A}}{\partial t}\hat{B} + \hat{A}\frac{\partial \hat{B}}{\partial t} + \frac{\mathrm{i}}{\hbar}([\hat{H},\hat{A}]_-\hat{B} + \hat{A}[\hat{H},\hat{B}]_-) \\
&= \frac{\mathrm{d}\hat{A}}{\mathrm{d}t}\hat{B} + \hat{A}\frac{\mathrm{d}\hat{B}}{\mathrm{d}t} \,.
\end{aligned} \tag{8.9}$$

8.2 Ehrenfest's Theorem

We consider the time derivative of the coordinate (or momentum) operator. Neither operator is explicitly time dependent; hence, the following is valid for the x components:

$$\frac{\mathrm{d}\hat{x}}{\mathrm{d}t} = \frac{\mathrm{i}}{\hbar}[\hat{H},\hat{x}]_- \,, \tag{8.10}$$

$$\frac{\mathrm{d}\hat{p}_x}{\mathrm{d}t} = \frac{\mathrm{i}}{\hbar}[\hat{H},\hat{p}_x]_- \,. \tag{8.11}$$

For the other components, analogous expressions are valid. To evaluate the commutators, let us have a look at the Hamiltonian of a particle in the potential:

$$\hat{H} = \frac{1}{2m}(\hat{p}_x^2 + \hat{p}_y^2 + \hat{p}_z^2) + \hat{V}(x,y,z) \,. \tag{8.12}$$

The operator $\hat{x}$ commutes with $\hat{p}_z^2$, $\hat{p}_y^2$ and the potential, which is supposed to be an exclusively space-dependent function. Thus

$$\begin{aligned}
[\hat{H},x]_- &= \frac{1}{2m}[\hat{p}_x^2,x]_- = \frac{1}{2m}[\hat{p}_x(x\hat{p}_x - \mathrm{i}\hbar) - x\hat{p}_x^2] \\
&= \frac{1}{2m}[(-\mathrm{i}\hbar + x\hat{p}_x)\hat{p}_x - \mathrm{i}\hbar\hat{p}_x - x\hat{p}_x^2] = \frac{\hbar}{\mathrm{i}}\frac{\hat{p}_x}{m} \,.
\end{aligned} \tag{8.13}$$

The commutator with the component of momentum $\hat{p}_x$ yields

$$[\hat{H}, \hat{p}_x]_- = [\hat{V}(x, y, z), \hat{p}_x]_- = -\frac{\hbar}{\mathrm{i}}\frac{\partial \hat{V}}{\partial x} \ . \tag{8.14}$$

Hence, from (8.10) and (8.11) we get

$$\frac{\mathrm{d}\hat{x}}{\mathrm{d}t} = \frac{\hat{p}_x}{m} \ , \tag{8.15}$$

$$\frac{\mathrm{d}\hat{p}_x}{\mathrm{d}t} = -\frac{\partial \hat{V}}{\partial x} \ . \tag{8.16}$$

Thus the same relations exist between the operators of position and momentum as between the coordinates of position and momentum in classical mechanics:

$$\frac{\mathrm{d}x}{\mathrm{d}t} = \frac{p_x}{m} = v_x \ , \qquad \frac{\mathrm{d}p_x}{\mathrm{d}t} = -\frac{\partial V}{\partial x} = F_x \ . \tag{8.17}$$

Evaluating the mean values of (8.15) and (8.16) and considering $\overline{\mathrm{d}x/\mathrm{d}t} = \mathrm{d}\bar{x}/\mathrm{d}t$, both of ***Ehrenfest***'s theorems follow

$$\frac{\mathrm{d}}{\mathrm{d}t}\int \psi^* \hat{x} \psi \,\mathrm{d}x = \frac{1}{m}\int \psi^* \hat{p}_x \psi \,\mathrm{d}x \ ,$$

$$\frac{\mathrm{d}}{\mathrm{d}t}\int \psi^* \hat{p}_x \psi \,\mathrm{d}x = -\int \psi^* \frac{\partial V}{\partial x} \psi \,\mathrm{d}x \ . \tag{8.18}$$

This statement is summed up in *Ehrenfest's Theorem* (1927): The mean values of quantum-mechanical quantities move according to classical equations.

8.3 Constants of Motion, Laws of Conservation

A time-dependent operator is a constant of motion if the operator commutes with the Hamiltonian. Indeed, for the case of time independence we have

$$\frac{\mathrm{d}\hat{L}}{\mathrm{d}t} = \frac{\partial \hat{L}}{\partial t} + \frac{\mathrm{i}}{\hbar}[\hat{H}, \hat{L}]_- = 0 \ . \tag{8.19}$$

If the operator itself is not explicitly time dependent, i.e. $\partial \hat{L}/\partial t = 0$, it follows that $[\hat{H}, \hat{L}] = 0$. Thus only those operators $\hat{L}$ represent constants of motion which (1) are not explicitly time dependent and (2) commute with the Hamiltonian. This fact will be very important in our further studies of quantum mechanics.

The operator $\hat{H}$ of the total energy obviously commutes with itself. It is therefore exactly a constant of motion when it is not explicitly time dependent. This is just the law of *conservation of energy*.

The momentum $\hat{p}$ is not explicitly time dependent. With respect to (8.14), it immediately follows that $\hat{p}_x =$ constant if $\partial V/\partial x = 0$. Thus the law of *conservation of momentum* is valid in quantum mechanics, too.

For central forces, the associated potential $V(r)$ is only a function of the radius r. The angular-momentum operator $\hat{L}^2 = -\hbar^2\nabla^2_{\vartheta,\varphi}$ [see (4.75)] thus commutes with $V(r)$. The complete Hamiltonian is [see (4.82a)]

$$\hat{H} = \hat{T}_r + \hat{L}^2/2mr^2 + \hat{V}(r) \; ; \tag{8.20}$$

hence,

$$[\hat{H}, \hat{L}^2] = 0 \; . \tag{8.21}$$

Thereby, the law of *conservation of angular momentum* is valid (Kepler's second law: the area theorem). The same consideration is true for the z component of angular momentum, because $[\hat{L}^2, \hat{L}_z] = 0$ and $[\hat{H}, \hat{L}_z] = 0$.

EXERCISE

8.1 Commutation Relations

Problem. Show by application of the canonical commutation relation

$$[\hat{p}_i, \hat{q}_k]_- = \frac{\hbar}{\mathrm{i}}\delta_{ik} \tag{1}$$

that the commutation relations

$$[\hat{H}, \hat{p}_i]_- = -\frac{\hbar}{\mathrm{i}}\frac{\partial \hat{H}}{\partial \hat{q}_i} \; , \quad [\hat{H}, \hat{q}_i]_- = \frac{\hbar}{\mathrm{i}}\frac{\partial \hat{H}}{\partial \hat{p}_i} \tag{2}$$

are valid for Hamiltonians of the form

$$\hat{H} = \hat{q}_i, \hat{q}_i^2, \ldots, \hat{q}_i^n, \hat{p}_i^n \quad \text{and} \tag{3}$$

$$\hat{H} = \sum_k C_{mn}\hat{p}_k^m \hat{q}_k^n \; . \tag{4}$$

Solution. The proof is given by complete induction. The equation

$$[\hat{H}, \hat{p}_i]_- = -\frac{\hbar}{\mathrm{i}}\frac{\partial \hat{H}}{\partial \hat{q}_i} \tag{5}$$

is obviously true for $\hat{H} = \hat{q}_i$. We assume it also to be true for $\hat{H} = \hat{q}_i^n$. Then, for $\hat{H} = \hat{q}_i^{n+1}$, the following is valid:

Exercise 8.1

$$[\hat{q}_i^{n+1},\hat{p}_i]_- = \hat{q}_i^{n+1}\hat{p}_i - \hat{p}_i\hat{q}_i^{n+1} = \hat{q}_i(\hat{q}_i^n\hat{p}_i - \hat{p}_i\hat{q}_i^n) + (\hat{q}_i\hat{p}_i - \hat{p}_i\hat{q}_i)\hat{q}_i^n$$
$$= \hat{q}_i\left(-\frac{\hbar}{\mathrm{i}}n\hat{q}_i^{n-1}\right) + \left(-\frac{\hbar}{\mathrm{i}}\right)\hat{q}_i^n$$
$$= -\frac{\hbar}{\mathrm{i}}(n+1)\hat{q}_i^n = -\frac{\hbar}{\mathrm{i}}\frac{\partial \hat{H}}{\partial \hat{q}_i}\ ; \tag{6}$$

hence, the relation is valid for all n.

The proof for $[\hat{H},\hat{q}_i]_-$ with $\hat{H} = \hat{p}_i^n$ is performed analogously.

If $\hat{H} = \sum C_{mn}\hat{p}_i^m\hat{q}_i^n$, then we have

$$[\hat{H},\hat{p}_i]_- = \sum C_{mn}\hat{p}_i^m[\hat{q}_i^n,\hat{p}_i] = \sum C_{mn}\hat{p}_i^m\left(-\frac{\hbar}{\mathrm{i}}\frac{\partial \hat{q}_i^n}{\partial \hat{q}_i}\right)$$
$$= \sum C_{mn}\hat{p}_i^m\left(-\frac{\hbar}{\mathrm{i}}n\hat{q}_i^{n-1}\right) = -\frac{\hbar}{\mathrm{i}}\frac{\partial \hat{H}}{\partial \hat{q}_i}\ . \tag{7}$$

The proof for $[\hat{H},\hat{q}_i]_-$ is performed analogously.

EXERCISE

8.2 The Virial Theorem

The *virial theorem* formulates a general relation between the mean value of the kinetic energy $\langle|\hat{T}|\rangle$ and the potential V:

$$2\langle|\hat{T}|\rangle = \langle|\hat{\boldsymbol{r}}\cdot\nabla V(\hat{\boldsymbol{r}})|\rangle\ . \tag{1}$$

It is valid both in classical mechanics and quantum mechanics and can be proved similarly in both disciplines. In classical mechanics, we start with the *temporal mean value* of the time derivative of the quantity $\hat{\boldsymbol{r}}\cdot\boldsymbol{p}$, i.e. $\mathrm{d}(\hat{\boldsymbol{r}}\cdot\boldsymbol{p})/\mathrm{d}t$, which vanishes for periodic motions. Correspondingly, in quantum mechanics we consider the expectation value of $\mathrm{d}(\hat{\boldsymbol{r}}\cdot\hat{\boldsymbol{p}})/\mathrm{d}t$ and get

$$\left\langle\left|\frac{\mathrm{d}}{\mathrm{d}t}\hat{\boldsymbol{r}}\cdot\hat{\boldsymbol{p}}\right|\right\rangle = \frac{\mathrm{d}}{\mathrm{d}t}\langle|\hat{\boldsymbol{r}}\cdot\hat{\boldsymbol{p}}|\rangle = \frac{1}{\mathrm{i}\hbar}\langle|[\hat{\boldsymbol{r}}\cdot\hat{\boldsymbol{p}},\hat{H}]_-|\rangle = 0\ . \tag{2}$$

The last identity can be verified easily in energy representation:

$$\langle\psi_E|[\hat{\boldsymbol{r}}\cdot\hat{\boldsymbol{p}},\hat{H}]_-|\psi_E\rangle = \langle\psi_E|\hat{\boldsymbol{r}}\cdot\hat{\boldsymbol{p}}E - E\hat{\boldsymbol{r}}\cdot\hat{\boldsymbol{p}}|\psi_E\rangle$$
$$= (E-E)\langle\psi_E|\hat{\boldsymbol{r}}\cdot\hat{\boldsymbol{p}}|\psi_E\rangle = 0\ . \tag{3}$$

The last step is based on the Hermiticity of $\hat{H}$ and the reality of E. On the other hand, the commutator $[\hat{\boldsymbol{r}}\cdot\hat{\boldsymbol{p}},\hat{H}]_-$ can be easily computed for all $\hat{H}$ of the form

Exercise 8.2

$\hat{H} = \hat{p}^2/2m + V(r)$:

$$
\begin{aligned}
[\hat{\boldsymbol{r}}\cdot\hat{\boldsymbol{p}}, \hat{H}]_- &= \left[x\hat{p}_x + y\hat{p}_y + z\hat{p}_z, \frac{\hat{p}_x^2+\hat{p}_y^2+\hat{p}_z^2}{2m} + V(x,y,z) \right]_- \\
&= \frac{\mathrm{i}\hbar}{m}(\hat{p}_x^2+\hat{p}_y^2+\hat{p}_z^2) - \mathrm{i}\hbar\left(x\frac{\partial V}{\partial x} + y\frac{\partial V}{\partial y} + z\frac{\partial V}{\partial z} \right) \\
&= 2\mathrm{i}\hbar\hat{T} - \mathrm{i}\hbar(\hat{\boldsymbol{r}}\cdot\nabla V) \ . \qquad (4)
\end{aligned}
$$

Thus we have $2\langle|\hat{T}|\rangle = \langle|\hat{\boldsymbol{r}}\nabla V(r)|\rangle$. We note that it does not matter for the proof whether we start from $\hat{\boldsymbol{r}}\cdot\hat{\boldsymbol{p}}$ or $\hat{\boldsymbol{p}}\cdot\hat{\boldsymbol{r}}$, because the difference between the terms is a constant which obviously commutates with $\hat{H}$. If V is a spherically symmetric potential, e.g. $V(r) \sim r^n$, the virial theorem yields $2\langle|\hat{T}|\rangle = n\langle|V|\rangle$. This is valid for all n, where the expectation value $\langle|V|\rangle$ must exist, of course.

8.4 Quantization in Curvilinear Coordinates

Equation (8.6), which gives us the total time derivative of an operator $\hat{F}$

$$
\frac{\mathrm{d}\hat{F}}{\mathrm{d}t} = \frac{\partial \hat{F}}{\partial t} + \frac{1}{\mathrm{i}\hbar}[\hat{F}, \hat{H}]_- \ , \qquad (8.22)
$$

has a formal analogue in classical mechanics, the ***Poisson** bracket*. The total temporal derivative of a function $F(p_i, q_i, t)$ is given by

$$
\frac{\mathrm{d}F}{\mathrm{d}t} = \frac{\partial F}{\partial t} + \sum_{i=1}^{f}\left(\frac{\partial F}{\partial q_i}\dot{q}_i + \frac{\partial F}{\partial p_i}\dot{p}_i \right) \ . \qquad (8.23)
$$

Here, p_i, q_i are the generalized momenta and coordinates, respectively, and f is the number of degrees of freedom. By using Hamilton's equations, the second term on the right-hand side of (8.23) transforms to

$$
\sum_{i=1}^{f}\left(\frac{\partial F}{\partial q_i}\dot{q}_i + \frac{\partial F}{\partial p_i}\dot{p}_i \right) = \sum_{i=1}^{f}\left(\frac{\partial F}{\partial q_i}\frac{\partial H}{\partial p_i} - \frac{\partial F}{\partial p_i}\frac{\partial H}{\partial q_i} \right) \equiv \{F, H\} \ ,
$$

and, hence, (8.23) becomes

$$
\frac{\mathrm{d}F}{\mathrm{d}t} = \frac{\partial F}{\partial t} + \{F, H\} \ . \qquad (8.24)
$$

The analogy between the classical equation (8.24) and the quantum-mechanical equation (8.22) is obvious. The so-defined term $\{F, H\}$, involving the Hamiltonian function H, is called the Poisson bracket. The transition from classical mechanics to quantum mechanics can obviously be performed by the transition to operators and the replacement of the Poisson bracket $\{\,,\,\}$ by the commutator $(1/\mathrm{i}\hbar)[\,,\,]_-$. The operator $(1/\mathrm{i}\hbar)[\hat{F}, \hat{H}]_-$ is also called the *quantum-mechanical Poisson bracket.*

These considerations of analogy can be continued. In classical mechanics we work with canonical variables and say a transformation from q_i, p_i to Q_i, P_i is canonical if Q_i and P_i again fulfil Hamilton's equations. This means the transition

$$H(p_i, q_i) \rightarrow \mathcal{H}(P_i, Q_i) \ , \tag{8.25}$$

where $\mathcal{H}$ stands for the new Hamiltonian function depending on the coordinates P_i and Q_i. The same statement can be expressed by the Poisson bracket; indeed

$$\{q_i, p_i\} = \delta_{ij} \ , \tag{8.26}$$

because the latter is equivalent to

$$\sum_{\sigma=1}^{f} \left(\frac{\partial q_i}{\partial q_\sigma} \frac{\partial p_j}{\partial p_\sigma} - \frac{\partial q_i}{\partial p_\sigma} \frac{\partial p_j}{\partial q_\sigma} \right) = \delta_{ij} \ . \tag{8.27}$$

The term $(\partial q_i/\partial p_\sigma)(\partial p_j/\partial q_\sigma)$ always vanishes and $(\partial q_i/\partial q_\sigma)(\partial p_j/\partial p_\sigma)$ gives the identity only for $i = j$, hence, δ_{ij}. If we transform to Q_i, P_i, the transformation is then only canonical if

$$\{Q_i, P_j\} = \delta_{ij} \tag{8.28}$$

is valid. (This will be further explained in Exercise 8.4.) Furthermore, the following equations also hold:

$$\{Q_i, Q_j\} = 0 \quad \text{and} \quad \{P_i, P_j\} = 0 \ . \tag{8.29}$$

Proceeding to quantum mechanics, we get the same relation, provided that we set $\hat{p}_j = -\mathrm{i}\hbar\partial/\partial x_j$ in the above-defined quantum-mechanical Poisson bracket:

$$-\frac{\mathrm{i}\hbar}{\mathrm{i}\hbar} \left[x_i, \frac{\partial}{\partial x_j} \right]_- = \delta_{ij} \ . \tag{8.30}$$

Thus the momentum is replaced by the operator. Likewise, both of the above-considered relations (8.29) are valid in quantum mechanics:

$$[x_i, x_j]_- = 0 \quad \text{and} \quad \left[\frac{\partial}{\partial x_i}, \frac{\partial}{\partial x_j} \right]_- = 0 \ . \tag{8.31}$$

Moreover, for the classical Poisson bracket, the following relations are valid:

$$\begin{aligned}
&\{A, B\} = -\{B, A\} \ , \quad \{A, C\} = 0 \quad \text{for} \quad C = \text{const} \ , \\
&\{A_1 A_2, B\} = \{A_1, B\} A_2 + A_1 \{A_2, B\} \ , \\
&\{A_1 + A_2, B\} = \{A_1, B\} + \{A_2, B\} \ , \\
&\{A, \{B, C\}\} + \{B, \{C, A\}\} + \{C, \{A, B\}\} = 0 \quad \text{(the Jacobi identity)} \ . \qquad (8.32)
\end{aligned}$$

It can easily be checked that the quantum-mechanical commutator fulfils the same algebraic relations. This fact was first noted by P.A.M. Dirac, who used it to show the formal analogy between quantum and Hamiltonian mechanics.

The transition from classical to quantum mechanics can formally be achieved by a special canonical transformation with the commutators

$$[\hat{p}_i, x_j]_- = \mathrm{i}\hbar\delta_{ij} \ , \quad [x_i, x_j]_- = 0 \ , \quad [\hat{p}_i, \hat{p}_j]_- = 0 \ . \qquad (8.33)$$

A short example shows, nevertheless, that one must take care with this transition, now that the momentum is an operator. The terms

$$p^2 \ , \quad \frac{1}{x} pxp \ , \quad \frac{1}{x^2} pxpx \ , \quad \text{etc.} \qquad (8.34)$$

are equivalent in classical mechanics. If, however, we replace the momentum by the operator $\hat{p} = -\mathrm{i}\hbar\partial/\partial x$, we get different terms in quantum mechanics.

Similar difficulties arise if we use *curvilinear coordinates*. The special nature of Cartesian coordinates becomes clear when considering kinetic energy. The kinetic energy in generalized coordinates is of the form

$$T = \sum_{i,k=1}^{3} m_{ik}(q_1, q_2, q_3) p_i p_k \ . \qquad (8.35)$$

The mass coefficients m_{ik} are generally functions of space. Hence, if the momentum is measured, they cannot simultaneously be exactly determined. Here, this means that the kinetic energy cannot be determined by measuring only the momenta. In Cartesian coordinates for a particle of mass m, the coefficients obey the relation $m_{ik} = (2m)^{-1}\delta_{ik}$; thus

$$T = \frac{p_x^2 + p_y^2 + p_z^2}{2m} \ , \qquad (8.36)$$

and the kinetic energy is determined exclusively by the momenta. Of course, the form of the kinetic energy is essential in determining the Hamiltonian operator (Hamiltonian). To obtain the Hamiltonian operator from the Hamiltonian function, it is always necessary to transform the function into Cartesian coordinates before putting in the operators. This is the safest way to pass from the classical to the corresponding quantum-mechanical system.

We now show the different outcomes of the two procedures, taking a central potential as an example.

Correct Way

We consider a central force problem (e.g. the hydrogen atom) with the potential $V(r) = -e^2/r$. The Hamiltonian function in Cartesian coordinates takes the form $H = \boldsymbol{p}^2/2m + V(r)$. We replace the momentum by the operator $-\mathrm{i}\hbar\nabla$ and obtain the Hamiltonian

$$\hat{H} = -\frac{\hbar^2\boldsymbol{\nabla}^2}{2m} + V(\boldsymbol{r}) = -\frac{\hbar^2\Delta}{2m} + V(\boldsymbol{r}) \ . \tag{8.37}$$

Now we must transform the Δ operator into the spherical coordinates r, ϑ, φ, which are appropriate to the problem. The outcome of this computation is well known and leads to the Schrödinger equation

$$\begin{aligned}\hat{H}\psi =& -\frac{\hbar^2}{2m}\left[\frac{1}{r^2}\frac{\partial}{\partial r}r^2\frac{\partial}{\partial r}\psi + \frac{1}{r^2\sin\vartheta}\frac{\partial}{\partial\vartheta}\left(\sin\vartheta\frac{\partial}{\partial\vartheta}\psi\right)\right.\\ &\left.+\frac{1}{r^2\sin^2\vartheta}\frac{\partial^2}{\partial\varphi^2}\psi\right] + V(\boldsymbol{r})\psi = E\psi \ .\end{aligned} \tag{8.38}$$

Incorrect Way

We start from the classical Hamiltonian function and transform it from Cartesian to spherical coordinates. Thus we obtain the Hamiltonian function:

$$H = \frac{1}{2m}\left(p_r^2 + \frac{1}{r^2}p_\vartheta^2 + \frac{1}{r^2\sin^2\vartheta}p_\varphi^2\right) + V(\boldsymbol{r}). \tag{8.39}$$

Now, performing the transition to quantum mechanics with the transformation equations:

$$\begin{aligned}&[\hat{p}_r, r] = -\mathrm{i}\hbar \ , \quad \hat{p}_r = -\mathrm{i}\hbar\frac{\partial}{\partial r} \ ;\\ &[\hat{p}_\vartheta, \vartheta] = -\mathrm{i}\hbar \ , \quad \hat{p}_\vartheta = -\mathrm{i}\hbar\frac{\partial}{\partial\vartheta} \ ;\\ &[\hat{p}_\varphi, \varphi] = -\mathrm{i}\hbar \ , \quad \hat{p}_\varphi = -\mathrm{i}\hbar\frac{\partial}{\partial\varphi} \ ,\end{aligned} \tag{8.40}$$

and noting the term for the quantum-mechanical Hamiltonian,

$$\hat{H} = \frac{1}{2m}\left(\frac{\partial^2}{\partial r^2} + \frac{1}{r^2}\frac{\partial^2}{\partial\vartheta^2} + \frac{1}{r^2\sin^2\vartheta}\frac{\partial^2}{\partial\varphi^2}\right) + V(\boldsymbol{r}) \ , \tag{8.41}$$

we recognize that the outcomes of the two procedures differ; in the second case, we have lost some terms during the transformation.

The transformations of kinetic energy operators from Cartesian to curvilinear coordinates seem to be very awkward, and there may be cases in which it is impossible to write down the kinetic energy in Cartesian coordinates. Thus the question of how to proceed in such a case inevitably arises.

In the following, we explain a method by which, starting from an arbitrary system of coordinates, we can obtain the correct term of the quantum-mechanical Hamiltonian.

We consider a system of N particles with $3N$ degrees of freedom. The Cartesian coordinates of the particles are $x_1, x_2, \dots, x_{3N}$. The associated Cartesian momenta are denoted by $p_1, p_2, \dots, p_{3N}$. Thus the classical Hamiltonian function is

$$H = \sum_{k=1}^{3N} \frac{p_k^2}{2m} + V(x_1, \dots, x_{3N}) \; . \tag{8.42}$$

The kinetic energy occuring therein is of Cartesian structure, i.e. it is diagonal in the momenta and has constant mass factors $1/2m$. On the basis of our experience with the single-particle system [e.g. the hydrogen atom – see (8.36) and (8.38)], we are able to transform this kinetic energy into its quantum-mechanical form. The operator of the kinetic energy is

$$\hat{T} = -\frac{\hbar^2}{2m}\Delta_{3N} = -\frac{\hbar^2}{2m}\left(\frac{\partial^2}{\partial x_1^2} + \dots + \frac{\partial^2}{\partial x_{3N}^2}\right) , \tag{8.43}$$

where the $3N$-dimensional Laplace operator (Laplacian) is written in brackets. It should be made clear that this is a model which is indeed justified by our experience.

Analogously, we now consider curvilinear coordinates $(u_1, u_2, \dots, u_{3N})$. The square of a length element in $3N$-dimensional space takes the form

$$\mathrm{d}s^2 = \sum_{i,k=1}^{3N} g_{ik}\,\mathrm{d}u_i\,\mathrm{d}u_k \; . \tag{8.44}$$

The coefficients $g_{ik}(u_j)$ are the elements of the *metric tensor*; in general they depend on the coordinates u_j. The kinetic energy yields through

$$T = \frac{m}{2}\left(\frac{\mathrm{d}s}{\mathrm{d}t}\right)^2 \tag{8.45}$$

the relation

$$T = \frac{m}{2}\sum_{i,k=1}^{3N} g_{ik}\frac{\mathrm{d}u_i}{\mathrm{d}t}\frac{\mathrm{d}u_k}{\mathrm{d}t} \; . \tag{8.46}$$

Thus the $g_{ik}(u_j)$ are a kind of mass coefficient. We call the determinant of the matrix (g_{ik}) $\det(g_{ik}) = g$, and the inverse of the matrix $(g_{ik})^{-1} = (g^{ik})$.

In curvilinear coordinates, Laplace's operator Δ_{3N} is[1]

$$\Delta = \frac{1}{\sqrt{g}}\sum_{i,k=1}^{3N}\frac{\partial}{\partial u_i}\left(\sqrt{g}g^{ik}\frac{\partial}{\partial u_k}\right) \; . \tag{8.47}$$

[1] See, e.g. M.R. Spiegel: *Vector Analysis* (Schaum, New York 1959).

Using this form of the Laplace operator we always obtain the correct operator of the kinetic energy in the transition to quantum mechanics, namely

$$\hat{T} = -\frac{\hbar^2}{2m}\frac{1}{\sqrt{g}}\sum_{i,k}\frac{\partial}{\partial u_i}\sqrt{g}g^{ik}\frac{\partial}{\partial u_k} \ . \tag{8.48}$$

This general method of quantization is especially important in nuclear physics for the quantization of collective phenomena.[2] In the case of a vibrating nucleus, for example, we are dealing with a system the mass of which (*vibrating mass*) depends on the amplitude of the oscillation. The greater the oscillation, the more nucleons participate in the motion (cf. Fig. 8.1). The metric tensor (*mass tensor*) becomes coordinate dependent and the quantization (8.48) is of decisive importance.

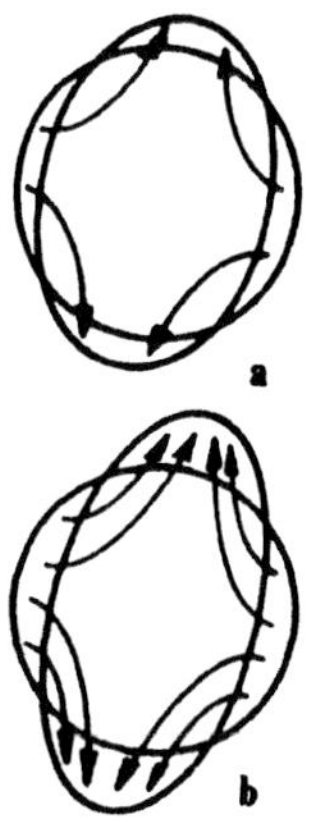

Fig. 8.1a,b. Surface oscillations of a nucleus. At (**a**), small amplitudes, fewer nucleons participate in the motion than for (**b**), large amplitudes. The vibrating mass thus becomes amplitude dependent (coordinate dependent)

EXAMPLE

8.3 The Kinetic-Energy Operator in Spherical Coordinates

Using the example of spherical coordinates, we once again derive the kinetic-energy operator (8.38), but now from (8.48), which applies to all coordinates. The length element in spherical coordinates is

$$ds^2 = dr^2 + r^2\,d\vartheta^2 + r^2\sin^2\vartheta\,d\varphi^2 = \sum_i g_{ii}(du_i)^2 \ . \tag{1}$$

Thus we obtain for the single elements of the metric tensor

$$g_{11} = 1 \ , \quad g_{22} = r^2 \ , \quad g_{33} = r^2\sin^2\vartheta \ , \tag{2}$$

forming the diagonal matrix

$$(g_{ik}) = \begin{pmatrix} 1 & 0 & 0 \\ 0 & r^2 & 0 \\ 0 & 0 & r^2\sin^2\vartheta \end{pmatrix} \ . \tag{3}$$

The determinant and the inverse matrix are easily evaluated:

$$\det(g_{ik}) \equiv g = r^4\sin^2\vartheta \quad \text{and}$$
$$(g^{ik}) = \begin{pmatrix} 1 & 0 & 0 \\ 0 & 1/r^2 & 0 \\ 0 & 0 & 1/r^2\sin^2\vartheta \end{pmatrix} \ . \tag{4}$$

[2] See J.M. Eisenberg and W. Greiner: *Nuclear Theory 1: Nuclear Models*, 3rd ed. (North-Holland, Amsterdam 1987).

As all $g_{ik} = 0$ for $i \neq k$, the kinetic energy operator consists of only three terms:

$$\begin{aligned}\hat{T} &= -\frac{\hbar^2}{2m}\frac{1}{\sqrt{g}}\left(\frac{\partial}{\partial r}\sqrt{g}\frac{\partial}{\partial r} + \frac{\partial}{\partial\vartheta}\frac{\sqrt{g}}{r^2}\frac{\partial}{\partial\vartheta} + \frac{\partial}{\partial\varphi}\frac{\sqrt{g}}{r^2\sin^2\vartheta}\frac{\partial}{\partial\varphi}\right) \\ &= -\frac{\hbar^2}{2m}\left(\frac{\partial^2}{\partial r^2} + \frac{2}{r}\frac{\partial}{\partial r} + \frac{1}{r^2\sin\vartheta}\frac{\partial}{\partial\vartheta}\sin\vartheta\frac{\partial}{\partial\vartheta} + \frac{1}{r^2\sin^2\vartheta}\frac{\partial^2}{\partial\varphi^2}\right) . \end{aligned} \quad (5)$$

This result agrees with (8.38), which was obtained by first making the transition from classical to quantum mechanics in Cartesian coordinates and then performing the transformation to spherical coordinates.

EXERCISE

8.4 Review of Some Useful Relations of Classical Mechanics: Lagrange and Poisson Brackets

Consider a transformation from a set of space and momentum coordinates q_i, p_i to a new set Q_i, P_i, where

$$Q_i = Q_i(q_j, p_j, t) , \quad P_i = P_i(q_j, p_j, t) . \quad (1)$$

This transformation is called *canonical* if a function $\mathcal{H}(Q_i, P_i, t)$ (Hamiltonian) exists, such that

$$H(q_i, p_i) \to \mathcal{H}(Q_i, P_i) . \quad (2)$$

$H = \sum_i p_i q_i - L$ and $\mathcal{H} = \sum_i P_i Q_i - L'$ are the relations between the Lagrangian L and H, and L' and $\mathcal{H}$, respectively. In short,

$$\dot{Q}_i = \frac{\partial\mathcal{H}}{\partial P_i} , \quad \dot{P}_i = -\frac{\partial\mathcal{H}}{\partial Q_i} . \quad (3)$$

Poincaré's Theorem states that the following surface integral is invariant under canonical transformations:

$$J_1 = \iint_A \sum_i \mathrm{d}q_i\,\mathrm{d}p_i .$$

Hence, it easily follows that the ***Lagrange*** *bracket* is invariant under canonical transformation.

$$\{\{u, v\}\} = \sum_i \left(\frac{\partial q_i}{\partial u}\frac{\partial p_i}{\partial v} - \frac{\partial p_i}{\partial u}\frac{\partial q_i}{\partial v}\right) . \quad (4)$$

The *Poisson bracket* is defined by

$$\{u, v\} = \sum_i \left(\frac{\partial u}{\partial q_i}\frac{\partial v}{\partial p_i} - \frac{\partial u}{\partial p_i}\frac{\partial v}{\partial q_i}\right) . \quad (5)$$

Exercise 8.4

Problem. (a) Prove ***Poincaré***'s Theorem.

(b) Show the invariance of the Lagrange bracket under canonical transformations.

(c) Show the invariance of the so-called fundamental Poisson brackets under arbitrary canonical transformations:

$$\begin{aligned}
&\{p_i, p_j\} = 0 \ , \\
&\{q_i, q_j\} = 0 \ . \\
&\{q_i, p_j\} = \delta_{ij} \ .
\end{aligned} \tag{6}$$

(d) Verify the relation

$$\{F, G\}_{q,p} = \{F, G\}_{Q,P} \tag{7}$$

for two arbitrary functions F and G, i.e. the invariance of the Poisson bracket under arbitrary canonical transformations.

Solution. (a) The position of a point on a two-dimensional area A in phase space is completely determined by two parameters, u and v. On this area we can express the coordinates q_i and p_i as functions of u and v, since $q_i = q_i(u, v)$, $p_i = p_i(u, v)$. With the aid of the Jacobi determinant, the area elements $\mathrm{d}u\,\mathrm{d}v$ and $\mathrm{d}q_i\,\mathrm{d}p_i$ can be transformed into each other; the Jacobi determinant being

$$\frac{\partial(q_i, p_i)}{\partial(u, v)} = \begin{vmatrix} \dfrac{\partial q_i}{\partial u} & \dfrac{\partial p_i}{\partial u} \\ \dfrac{\partial q_i}{\partial v} & \dfrac{\partial p_i}{\partial v} \end{vmatrix} \ . \tag{8}$$

The two surface elements are related by

$$\mathrm{d}q_i\,\mathrm{d}p_i = \frac{\partial(q_i, p_i)}{\partial(u, v)}\,\mathrm{d}u\,\mathrm{d}v \ , \tag{9}$$

i.e. the statement that J_1 has the same value for all canonical transformations,

$$\iint\limits_A \sum_i \mathrm{d}q_i\,\mathrm{d}p_i = J_1 = \iint\limits_A \sum_k \mathrm{d}Q_k\,\mathrm{d}P_k \ , \tag{10}$$

can be expressed with the aid of (9) as

$$\iint\limits_A \sum_i \frac{\partial(q_i, p_i)}{\partial(u, v)}\,\mathrm{d}u\,\mathrm{d}v = \iint\limits_A \sum_k \frac{\partial(Q_k, P_k)}{\partial(u, v)}\,\mathrm{d}u\,\mathrm{d}v \ . \tag{11}$$

Thus the proof is reduced to the statement

$$\sum_i \frac{\partial(q_i, p_i)}{\partial(u, v)} = \sum_k \frac{\partial(Q_k, P_k)}{\partial(u, v)} \ , \tag{12}$$

Exercise 8.4

i.e. both Jacobi determinants are identical. Now we consider the canonical transformation $q, p \to Q, P$ to be achieved by the generating function $F_2(q, P, t)$, for which

$$p_i = \frac{\partial F_2}{\partial q_i}\,, \quad Q_i = \frac{\partial F_2}{\partial P_i}\,, \quad \mathcal{H} = H + \frac{\partial F_2}{\partial t} \tag{13}$$

is valid.

For the calculation of the functional determinant we first need

$$\frac{\partial p_i}{\partial u} = \frac{\partial}{\partial u}\left(\frac{\partial F_2}{\partial q_i}\right)\,, \tag{14}$$

where we have used (13). Owing to its definition, F_2 depends only on q_k and P_k; here, time acts as a parameter, not a coordinate. With the aid of the total differential for F_2, we get

$$\frac{\partial p_i}{\partial u} = \sum_k \frac{\partial^2 F_2}{\partial q_i \partial P_k}\frac{\partial P_k}{\partial u} + \sum_k \frac{\partial^2 F_2}{\partial q_i \partial q_k}\frac{\partial q_k}{\partial u}\,. \tag{15}$$

An analogous result is obtained for $\partial p_i/\partial v$, so that the functional determinant assumes the form

$$\sum_i \frac{\partial(q_i, p_i)}{\partial(u, v)} = \sum_i \begin{vmatrix} \dfrac{\partial q_i}{\partial u}\,, & \displaystyle\sum_k \frac{\partial^2 F_2}{\partial q_i \partial P_k}\frac{\partial P_k}{\partial u} + \sum_k \frac{\partial^2 F_2}{\partial q_i \partial q_k}\frac{\partial q_k}{\partial u} \\ \dfrac{\partial q_i}{\partial v}\,, & \displaystyle\sum_k \frac{\partial^2 F_2}{\partial q_i \partial P_k}\frac{\partial P_k}{\partial v} + \sum_k \frac{\partial^2 F_2}{\partial q_i \partial q_k}\frac{\partial q_k}{\partial v} \end{vmatrix}\,. \tag{16}$$

After application of the rules of calculation for addition and multiplication of constant factors with determinants, this is transformed to

$$\sum_i \frac{\partial(q_i, p_i)}{\partial(u, v)} = \sum_{i,k} \frac{\partial^2 F_2}{\partial q_i \partial q_k} \begin{vmatrix} \dfrac{\partial q_i}{\partial u} & \dfrac{\partial q_k}{\partial u} \\ \dfrac{\partial q_i}{\partial v} & \dfrac{\partial q_k}{\partial v} \end{vmatrix} + \sum_{i,k} \frac{\partial^2 F_2}{\partial q_i \partial P_k} \begin{vmatrix} \dfrac{\partial q_i}{\partial u} & \dfrac{\partial P_k}{\partial u} \\ \dfrac{\partial q_i}{\partial v} & \dfrac{\partial P_k}{\partial v} \end{vmatrix}\,. \tag{17}$$

It is apparent that the first term is antisymmetric with respect to changing the indices of summation i and k, because this only involves changing two columns. Thus the first term vanishes and can be replaced by another vanishing term:

$$\sum_i \frac{\partial(q_i, p_i)}{\partial(u, v)} = \sum_{i,k} \frac{\partial^2 F_2}{\partial P_i \partial P_k} \begin{vmatrix} \dfrac{\partial P_i}{\partial u} & \dfrac{\partial P_k}{\partial u} \\ \dfrac{\partial P_i}{\partial v} & \dfrac{\partial P_k}{\partial v} \end{vmatrix} + \sum_{i,k} \frac{\partial^2 F_2}{\partial q_i \partial P_k} \begin{vmatrix} \dfrac{\partial q_i}{\partial u} & \dfrac{\partial P_k}{\partial u} \\ \dfrac{\partial q_i}{\partial v} & \dfrac{\partial P_k}{\partial v} \end{vmatrix}\,. \tag{18}$$

If (18) is transformed into the form of (16), the element a_{11} reads

$$\sum_i \frac{\partial^2 F_2}{\partial P_i \partial P_k}\frac{\partial P_i}{\partial u} + \sum_i \frac{\partial^2 F_2}{\partial q_i \partial P_k}\frac{\partial q_i}{\partial u} = \frac{\partial}{\partial u}\frac{\partial F_2}{\partial P_k}\,. \tag{19}$$

Exercise 8.4

Here, contrary to (16), we have moved the sum over k in front of the determinant and the sum over i into the determinant. Because of (13), it holds in (1) that

$$\frac{\partial F_2}{\partial P_k} = Q_k \; ; \quad \text{hence,}$$

$$\sum_i \frac{\partial(q_i, p_i)}{\partial(u, v)} = \sum_k \begin{vmatrix} \frac{\partial Q_k}{\partial u} & \frac{\partial P_k}{\partial u} \\ \frac{\partial Q_k}{\partial v} & \frac{\partial P_k}{\partial v} \end{vmatrix} = \sum_k \frac{\partial(Q_k, P_k)}{\partial(u, v)} \; . \tag{20}$$

Thus, together with (12), ***Poincaré***'s statement is proved.

(b) Since we have already verified Poincaré's statement in part (a), the statement in (4) holds:

$$\sum_i \frac{\partial(q_i, p_i)}{\partial(u, v)} = \sum_i \frac{\partial(Q_i, P_i)}{\partial(u, v)} \; . \tag{21}$$

It is identical with the just proved relation (20) for the Jacobi determinants. Indeed, rewriting this expression as

$$\Leftrightarrow \sum_i \left(\frac{\partial q_i}{\partial u}\frac{\partial p_i}{\partial v} - \frac{\partial p_i}{\partial u}\frac{\partial q_i}{\partial v} \right) = \sum_i \left(\frac{\partial Q_i}{\partial u}\frac{\partial P_i}{\partial v} - \frac{\partial P_i}{\partial u}\frac{\partial Q_i}{\partial v} \right) \; , \tag{22}$$

we see that it is equivalent to the invariance of the Lagrange bracket:

$$\{\{u, v\}\}_{p,q} = \{\{u, v\}\}_{P,Q} \; . \tag{23}$$

(c) First we show a useful relation. Let u_l, $l = 1, 2, \dots, 2n$ be a set of $2n$ independent functions such that every u_l is a function of the $2n$ coordinates $q_1, q_2, \dots, q_n, p_1, p_2, \dots, p_n$. Then the relation

$$\sum_{l=1}^{2n} \{\{u_l, u_i\}\}\{u_l, u_j\} = \delta_{ij} \tag{24}$$

is always valid. According to the definition of the Lagrange and Poisson brackets, it follows that

$$\begin{aligned} \sum_l^{2n} \{\{u_l, u_i\}\}\{u_l, u_j\} = \sum_l^{2n}\sum_k^{n}\sum_m^{n} & \left(\frac{\partial q_k}{\partial u_l}\frac{\partial p_k}{\partial u_i} - \frac{\partial q_k}{\partial u_i}\frac{\partial p_k}{\partial u_l} \right) \\ \times & \left(\frac{\partial u_l}{\partial q_m}\frac{\partial u_j}{\partial p_m} - \frac{\partial u_j}{\partial q_m}\frac{\partial u_l}{\partial p_m} \right) \; . \end{aligned} \tag{25}$$

The first term can be transformed to

$$\begin{aligned} \sum_{k,m}^{n} \frac{\partial p_k}{\partial u_i}\frac{\partial u_j}{\partial p_m} \sum_l^{2n} \frac{\partial q_k}{\partial u_l}\frac{\partial u_l}{\partial q_m} &= \sum_{k,m} \frac{\partial p_k}{\partial u_i}\frac{\partial u_j}{\partial p_m}\frac{\partial q_k}{\partial q_m} \\ &= \sum_{k,m} \frac{\partial p_k}{\partial u_i}\frac{\partial u_j}{\partial p_m}\delta_{km} = \sum_k^{n} \frac{\partial u_j}{\partial p_k}\frac{\partial p_k}{\partial u_i} \; . \end{aligned} \tag{26}$$

Exercise 8.4

The last term in (25) can be evaluated in the same way:

$$\left(\sum_{k,m}^{n} \frac{\partial q_k}{\partial u_i} \frac{\partial u_j}{\partial q_m}\right) \left(\sum_{l}^{2n} \frac{\partial p_k}{\partial u_l} \frac{\partial u_l}{\partial p_m}\right) = \sum_{k}^{n} \frac{\partial u_j}{\partial q_k} \frac{\partial q_k}{\partial u_i} \ . \tag{27}$$

So that the sum of the terms in (26) and (27) yields a total differential in u_j:

$$\sum_{k}^{n} \left(\frac{\partial u_j}{\partial p_k} \frac{\partial p_k}{\partial u_i} + \frac{\partial u_j}{\partial q_k} \frac{\partial q_k}{\partial u_i} \right) = \frac{\partial u_j}{\partial u_i} = \delta_{ij} \ . \tag{28}$$

The second and the third terms of (25) always vanish. We show this as an example for the second term:

$$\left(\sum_{k,m}^{n} \frac{\partial q_k}{\partial u_i} \frac{\partial u_j}{\partial p_m}\right) \left(\sum_{l}^{2n} \frac{\partial p_k}{\partial u_l} \frac{\partial u_l}{\partial q_m}\right) = 0 \ , \tag{29}$$

because

$$\sum_{l}^{2n} \frac{\partial p_k}{\partial q_m} = 0 \ . \tag{30}$$

Thus we have shown that

$$\sum_{l}^{2n} \{\{u_l, u_i\}\}\{u_l, u_j\} = \frac{\partial u_j}{\partial u_i} = \delta_{ij} \ . \tag{31}$$

We note that up to now, the choice of the coordinate system has been irrelevant. Hence, (31) is valid for all coordinate transformations, not only for canonical ones. The latter quality serves to evaluate several Poisson brackets, without committing ourselves to a certain system of coordinates.

For the $2n$ independent functions u_l, we choose the set $q_1, \dots, q_n$, $p_1, \dots, p_n$ and consider in particular the case $u_i = q_i$, $u_j = p_j$.

Thus (24) or (31) (which are the same) yields

$$\sum_{l}^{n} \{\{p_l, q_i\}\}\{p_l, p_j\} + \sum_{l}^{n} \{\{q_l, q_i\}\}\{q_l, p_j\} = 0 \ , \tag{32}$$

because $\partial u_i/\partial u_j = \partial q_i/\partial p_j = 0$ for all i, j. As was shown in part (b), the Lagrange bracket is invariant under canonical transformations. We will use this property in order to show the invariance of the Poisson brackets appearing in (32).

The following expressions are thus invariant:

$$\{\{p_l, q_i\}\} = -\delta_{il} \quad \text{and} \quad \{\{q_l, q_i\}\} = 0 \ . \tag{33}$$

Exercise 8.4

Inserting these into (32) the second term vanishes and we are left with

$$\{p_i, p_j\} = 0 \ . \tag{34}$$

Since (32) is valid for all transformations, the same applies for the Poisson bracket above.

Choosing $u_i = p_i$ and $u_j = q_i$ in (31) we obtain

$$\{q_i, q_j\} = 0 \ , \tag{35}$$

which is again valid for all transformations. Choosing $u_i = q_i$, $u_i = q_j$ in (31) yields

$$\sum_l^n \{\{q_l, q_i\}\}\{q_l, q_j\} + \sum_l^n \{\{p_l, q_i\}\}\{p_l, q_i\} = \delta_{ij} \ . \tag{36}$$

Since the first term vanishes, the second expression has to satisfy

$$-\sum_l \delta_{il}\{\{p_l, q_j\}\} = \delta_{ij} \ , \tag{37}$$

which is only possible if

$$\{q_i, p_j\} = \delta_{ij} \ . \tag{38}$$

Thus the invariance of the fundamental Poisson bracket under arbitrary canonical transformations is shown by use of the invariance properties of the Lagrange bracket. In particular

$$\{q_i, p_j\} = \{Q_i, P_j\} = \delta_{ij} \ , \quad \text{i.e.}$$
$$\left(\sum_{l=1}^n \frac{\partial q_i}{\partial q_l}\frac{\partial p_j}{\partial p_l} - \frac{\partial q_i}{\partial p_l}\frac{\partial p_j}{\partial q_l}\right) = \left(\sum_{l=1}^n \frac{\partial Q_i}{\partial Q_l}\frac{\partial P_j}{\partial P_l} - \frac{\partial Q_i}{\partial P_l}\frac{\partial P_j}{\partial Q_l}\right) = \delta_{ij} \ . \tag{39}$$

(d) For two arbitrary functions F and G, the Poisson bracket is defined with respect to the set q, p in the following way:

$$\{F, G\}_{q,p} = \sum_j \left(\frac{\partial F}{\partial q_j}\frac{\partial G}{\partial p_j} - \frac{\partial F}{\partial p_j}\frac{\partial G}{\partial q_j}\right) \ . \tag{40}$$

The q_j and p_k are functions of the new variables Q_j and P_k, respectively, and vice versa. Hence, the function G can also be expressed in terms of Q_i, P_k. This possibility is now used by transforming (40) into

$$\begin{aligned}\{F, G\}_{q,p} = \sum_{j,k} \Bigg[&\frac{\partial F}{\partial q_j}\left(\frac{\partial G}{\partial Q_k}\frac{\partial Q_k}{\partial p_j} + \frac{\partial G}{\partial P_k}\frac{\partial P_k}{\partial p_j}\right) \\ &- \frac{\partial F}{\partial p_j}\left(\frac{\partial G}{\partial Q_k}\frac{\partial Q_k}{\partial q_j} + \frac{\partial G}{\partial P_k}\frac{\partial P_k}{\partial q_j}\right)\Bigg] \ .\end{aligned} \tag{41}$$

Exercise 8.4

A clever reshuffling of several terms leads to

$$\{F,G\}_{q,p}=\sum_k\left(\frac{\partial G}{\partial Q_k}\{F,Q_k\}_{q,p}+\frac{\partial G}{\partial P_k}\{F,P_k\}_{q,p}\right) . \tag{42}$$

Replacing F by Q_j and G by F similarly yields

$$\{Q_k,F\}_{q,p}=\sum_j\frac{\partial F}{\partial Q_j}\{Q_k,Q_j\}_{q,p}+\sum_j\frac{\partial F}{\partial P_j}\{Q_k,P_j\}_{q,p} . \tag{43}$$

As we have only invariant Poisson brackets left [cf. (a)], we can start to evaluate

$$\{Q_k,F\}_{q,p}=\sum_j\frac{\partial F}{\partial P_j}\delta_{jk}=\frac{\partial F}{\partial P_k} . \tag{44}$$

Here the analogous relations to (35) and (38) for Q, P are used (the invariance of the fundamental Poisson brackets!). Substituting P_j for F and F for G, on the other hand, yields

$$\{P_k,F\}_{q,p}=\sum_j\frac{\partial F}{\partial Q_j}\{P_k,Q_j\}+\sum_j\frac{\partial F}{\partial P_j}\{P_k,P_j\} ;$$

thus

$$\{F,P_k\}_{q,p}=\frac{\partial F}{\partial Q_k} , \quad \{F,Q_k\}=-\frac{\partial F}{\partial P_k} . \tag{45}$$

Substitution of the results (44) and (45) into (42) finally gives

$$\{F,G\}_{q,p}=\sum_k\left(\frac{\partial F}{\partial Q_k}\frac{\partial G}{\partial P_k}-\frac{\partial F}{\partial P_k}\frac{\partial G}{\partial Q_k}\right)=\{F,G\}_{Q,P} . \tag{46}$$

Thus we have demonstrated the invariance of the general Poisson bracket under canonical transformations.

8.5 Biographical Notes

EHRENFEST, Paul, Austrian physicist, *Vienna 18.1.1880, †Leiden 25.9.1933. E. was a professor in Leiden (Netherlands) from 1912 on. He contributed to atomic physics with his hypothesis of adiabatic invariants (BR).

POISSON, Siméon Denis, French mathematician, *Pithiviers 21.6.1781, †Paris 25.4.1840. P. was a student of the École Polytechnique and he was employed there after completing his studies, being a professor from 1802. P. was a member of the bureau of lengths and of the Académie des Sciences. He was a French peer from 1837. P. worked in many fields, e.g. general mechanics, heat conduction, potential theory, differential equations and the calculus of probabilities.

POINCARÉ, Henri, French mathematician, *Nancy 29.4.1854, †Paris 17.7.1912. P. studied at the École Polytechnique and became professor at Caen in 1879, later, at Paris. He produced more than 30 books. At the turn of the century he was believed to be the outstanding mathematician of his age. P.'s greatest contribution to mathematical physics was a paper on the dynamics of the electron (1906) in which he obtained, independently of Einstein, many of the results of special relativity. Einstein developed the theory from elementary considerations about light signalling, whereas P.'s treatment was based on the theory of electromagnetism and was thus restricted. P.'s writings on the philosophy of science were as important as his contributions to mathematics. He became a member of the Académie Française in 1908 (taken from Encyclopedia Britannica, 1960 edition).

LAGRANGE, Joseph Louis, French mathematician, *Torino 25.1.1736, †Paris 10.4.1813. L. came from a French-Italian family, and in 1755 became professor in Torino. In 1766 he went to Berlin as the director of the mathematical-physical class of the academy. In 1786, after the death of Friedrich II, he went to Paris, where he considerably supported the reform of the measuring system, and where he was professor at several universities. His very extensive work contains a new foundation of variational calculus (1760) and its application to dynamics, contributions to the three-body problem (1772), the application of the theory of chain fractions to the solution of equations (1767), number-theoretical problems, and an unsuccessful reduction of infinitesimal calculus to algebra. With his "Mécanique analytique" (1788), L. became the initiator of analytic mechanics. Important for function theory is his "Théorie des fonctions analytiques, contenant les principes du calcul différentiel" (1789), and for algebra his "Traité de la résolution des équations numériques de tous degrés" (1798).

9. Charged Particles in Magnetic Fields

9.1 Coupling to the Electromagnetic Field

If a charged particle of charge e moves in an electromagnetic field, the Lorentz force

$$F = e\left(E + \frac{v}{c} \times B\right) \tag{9.1}$$

acts on the particle. The electric and magnetic field strengths can be expressed by the corresponding potentials $A(r, t)$ and $\phi(r, t)$ according to

$$E = -\nabla\phi - \frac{1}{c}\frac{\partial A}{\partial t} , \qquad B = \nabla \times A . \tag{9.2}$$

Here, $A(r, t)$ is the vector potential and $\phi(r, t)$ the Coulomb potential. In classical mechanics, this motion is described by the Hamiltonian function

$$H = \frac{1}{2m}\left(p - \frac{e}{c}A\right)^2 + e\phi , \tag{9.3}$$

which will be shown in Exercise 9.1. This indicates the simplest way of coupling the electric field to the motion of the particle. The momentum p is replaced by the term $p - (e/c)A$. The substitution $p - (e/c)A$ is gauge invariant and is called the *minimal coupling*. Hamilton's *canonical momentum* p is the sum of the *kinetic momentum* mv and the term $(e/c)A$, which is determined by the vector potential. Thus

$$p = mv + \frac{e}{c}A . \tag{9.4}$$

The transition to quantum mechanics is obtained by replacing the canonical momentum p by $(\hbar/\mathrm{i})\nabla$, according to the rules of quantization in the coordinate representation (see Chap. 8). Thus we obtain the Hamiltonian

$$\hat{H} = \frac{1}{2m}\left(\frac{\hbar}{\mathrm{i}}\nabla - \frac{e}{c}A\right)^2 + e\phi . \tag{9.5}$$

Calculating the square, it should be noted that, in general, the gradient and vector potentials do not commute. We get

$$\hat{H} = -\frac{\hbar^2}{2m}\Delta - \frac{e\hbar}{2\mathrm{i}mc}(\nabla\cdot A + A\cdot\nabla) + \frac{e^2}{2mc^2}A^2 + e\phi \ ,$$

$$\hat{H} = -\frac{\hbar^2}{2m}\Delta + \frac{\mathrm{i}e\hbar}{mc}A\cdot\nabla + \frac{\mathrm{i}e\hbar}{2mc}(\nabla\cdot A) + \frac{e^2}{2mc^2}A^2 + e\phi \ . \tag{9.6}$$

It is well known that the electromagnetic potentials A and ϕ are not unique, but are gauge dependent. Particularly in the Coulomb gauge, it holds that $\nabla \cdot A = 0$; thus the third term vanishes. If we change the order of the terms and use, for the sake of clarity, the momentum operator $\hat{p}$, we obtain

$$\hat{H} = \frac{\hat{p}^2}{2m} + e\phi - \frac{e}{mc}A\cdot\hat{p} + \frac{e^2}{2mc^2}A^2 \ ,$$

$$\hat{H} = \hat{H}_0 - \frac{e}{mc}A\cdot\hat{p} + \frac{e^2}{2mc^2}A^2 \ . \tag{9.7}$$

Here, the operator $\hat{H}_0$ represents the motion of the particle without a magnetic field; the coupling of the motion of the particle to the magnetic field is given by the product $A\cdot\hat{p}$. The third term depends only on the A field; for normal field strengths of small magnitude, it can be dropped. If the vector potential A describes a plane electromagnetic wave, the coupling terms in (9.7) lead to radiative transitions (emission and absorption). The states of the particle in an electromagnetic field are given as solutions of the Schrödinger equation with the Hamiltonian derived above in (9.5):

$$\left\{\frac{[\hat{p}-(e/c)A]^2}{2m} + e\phi\right\}\psi = \mathrm{i}\hbar\frac{\partial}{\partial t}\psi \ . \tag{9.8}$$

We can check that Ehrenfest's theorem is also valid for this Schrödinger equation, for which we shall now prove the gauge invariance. Gauge invariance means that the solutions of the Schrödinger equation describe the same physical states if we apply to the potentials the transformations

$$A' = A + \nabla f(r,t) \quad \text{and} \quad \phi' = \phi - \frac{1}{c}\frac{\partial f(r,t)}{\partial t} \tag{9.9}$$

with the arbitrary function $f(r, t)$. Using the four-component relativistic notation by introducing the four-vector A_μ, these transformations read

$$A'_\mu = A_\mu + \frac{\partial f}{\partial x^\mu} \ , \quad \text{with}$$

$$A_\mu = \{A, \mathrm{i}\phi\} \quad \text{and} \quad \mu = 1,2,3,4 \tag{9.10}$$

where $x^1 = x$, $x^2 = y$, $x^3 = z$, $x^4 = \mathrm{i}ct$.

If we denote the Hamiltonian with primed potentials by $\hat{H}'$, the corresponding Schrödinger equation becomes

$$\hat{H}'\psi' = \mathrm{i}\hbar\frac{\partial}{\partial t}\psi' \ . \tag{9.11}$$

Our statement now is that ψ and ψ' differ only by a phase factor. If so, the gauge transformation does not change the physical quantities, because, during their calculation, only products of the form $\psi^*\psi$ or matrix elements $\langle\psi|\dots|\psi\rangle$ in which the phase cancels occur. We start with

$$\psi' = \psi \exp\left(\frac{\mathrm{i}e}{\hbar c} f(\boldsymbol{r},t)\right) \tag{9.12}$$

and insert this in (9.11), which thus becomes

$$\frac{[\hat{\boldsymbol{p}} - (e/c)\boldsymbol{A} - (e/c)\nabla f]^2}{2m}\psi \exp\left(\frac{\mathrm{i}e}{\hbar c} f\right) + \left(e\phi - \frac{e}{c}\frac{\partial f}{\partial t}\right)\psi \exp\left(\frac{\mathrm{i}e}{\hbar c} f\right)$$
$$= \mathrm{i}\hbar\frac{\partial\psi}{\partial t}\exp\left(\frac{\mathrm{i}e}{\hbar c} f\right) - \frac{e}{c}\frac{\partial f}{\partial t}\psi \exp\left(\frac{\mathrm{i}e}{\hbar c} f\right) . \tag{9.13}$$

We can easily see that

$$\begin{aligned}\left(\hat{\boldsymbol{p}} - \frac{e}{c}\boldsymbol{A}'\right)\psi' &= \left(\frac{\hbar}{\mathrm{i}}\nabla - \frac{e}{c}\boldsymbol{A} - \frac{e}{c}\nabla f\right)\psi \exp\left(\frac{\mathrm{i}e}{\hbar c} f\right) \\ &= \exp\left(\frac{\mathrm{i}e}{\hbar c} f\right)\left(\frac{\hbar}{\mathrm{i}}\nabla + \frac{e}{c}\nabla f - \frac{e}{c}\boldsymbol{A} - \frac{e}{c}\nabla f\right)\psi \\ &= \exp\left(\frac{\mathrm{i}e}{\hbar c} f\right)\left(\frac{\hbar}{\mathrm{i}}\nabla - \frac{e}{c}\boldsymbol{A}\right)\psi .\end{aligned} \tag{9.14}$$

Applying the operator $[\hat{\boldsymbol{p}} - (e/c)\boldsymbol{A}']$ once again, we obtain the equation

$$\hat{H}\psi = \mathrm{i}\hbar\frac{\partial\psi}{\partial t} . \tag{9.15}$$

In other words, (9.15) follows from (9.11) by using (9.12). This outcome shows us that the solutions of the Schrödinger equation (9.8) still describe the same physical states, even after gauge transformation. The states ψ_n and ψ_n' differ only by a unique (i.e. state-independent) phase factor $\exp[(\mathrm{i}e/\hbar c) f(\boldsymbol{r},t)]$. The physical observables are not affected by this as mentioned above. It is clear that it is not the canonical momentum $p \to -\mathrm{i}\hbar\nabla$ (the expectation value of which is not gauge invariant), but the genuine kinetic momentum $m\boldsymbol{v} \leftrightarrow \mathrm{i}\hbar\nabla - (e/c)\boldsymbol{A}$ (which is gauge invariant), that represents a measurable quantity.

Hence, if in a physical problem the momentum operator $\hat{\boldsymbol{p}}$ appears, the operator $\hat{\boldsymbol{p}}$ must always be replaced by $\hat{\boldsymbol{p}} - (e/c)\boldsymbol{A}$ if electromagnetic fields are present. This is the only way to guarantee gauge invariance in quantum theory; otherwise, certain potentials $\boldsymbol{A}$ and ϕ could be determined in quantum mechanics, and this should not be possible!

We shall now summarize once more the principal idea of gauge invariance in quantum mechanics, in relativistic notation. The gauge transformation for the electromagnetic fields $A_\mu(x_\nu)$ is

$$A'_\mu = A_\mu + \frac{\partial f}{\partial x_\mu} , \quad \text{with}$$
$$A_\mu = \{\boldsymbol{A}, \mathrm{i}\phi\} \quad \text{and} \quad x_\mu = \{\boldsymbol{x}, \mathrm{i}ct\} . \tag{9.16}$$

This leaves the electromagnetic observables, i.e. the field strengths $\boldsymbol{E}$ and $\boldsymbol{B}$, unchanged. The four-momentum operator is given by

$$\hat{p}_\mu = -\mathrm{i}\hbar \left\{ \frac{\partial}{\partial x_1}, \frac{\partial}{\partial x_2}, \frac{\partial}{\partial x_3}, \frac{\partial}{\partial \mathrm{i}ct} \right\} = \left\{ \hat{\boldsymbol{p}}, \frac{\mathrm{i}\hat{E}}{c} \right\} , \tag{9.17}$$

and minimal coupling is achieved through the replacement

$$\hat{p}_\mu \to \hat{p}_\mu - \frac{e}{c} A_\mu . \tag{9.18}$$

In quantum mechanics, the gauge transformation (9.16) must be supplemented by the phase transformation of the wave function

$$\psi'(x_r) = \psi(x_\mu) \exp\left(\frac{\mathrm{i}e}{\hbar c} f(x_\mu) \right) , \tag{9.19}$$

so that

$$\left(\hat{p}_\mu - \frac{e}{c} A'_\mu \right) \psi' = \left(\hat{p}_\mu - \frac{e}{c} A_\mu - \frac{e}{c} \frac{\partial f}{\partial x_\mu} \right) \exp\left(\frac{\mathrm{i}e}{\hbar c} f(x_\mu) \right) \psi(x_\mu)$$
$$= \exp\left(\frac{\mathrm{i}e}{\hbar c} f(x_\mu) \right) \left(\hat{p}_\mu - \frac{e}{c} A_\mu \right) \psi(x_\mu) \tag{9.20}$$

holds. Then we can be certain that observables of the type

$$\left\langle \psi'_f(x_\mu) \left| V(x_\mu) \right| \psi'_i(x_\mu) \right\rangle = \left\langle \psi_f(x_\mu) \left| V(x_\mu) \right| \psi_i(x_\mu) \right\rangle \quad \text{and} \tag{9.21}$$
$$\left\langle \psi'_f(x_\mu) \left| F\left(\hat{p}_\mu - \frac{e}{c} A'_\mu \right) \right| \psi'_i(x_\mu) \right\rangle = \left\langle \psi_f(x_\mu) \left| F\left(\hat{p}_\mu - \frac{e}{c} A_\mu \right) \right| \psi_i(x_\mu) \right\rangle$$

are unchanged by (i.e. are invariant under) gauge transformations. The equations in (9.20) are exactly the right-hand side of the former equations (9.13) for $\mu = 4$ and (9.14) for $\mu = 1, 2, 3$, respectively. The following examples and exercises will further clarify this discussion.

EXAMPLE

9.1 The Hamilton Equations in an Electromagnetic Field

Let $q_1, q_2, \dots, q_s, \dots, q_f$ be the generalized position coordinates determining the configuration of the system, and $p_1, p_2, \dots, p_s, \dots, p_f$ the canonical conjugated momenta. The Hamiltonian (in classical mechanics we prefer to call it Hamiltonian function) H is a function of those position coordinates and momenta, and, in general, of time t.

Hamilton's equations are, as we know,

$$\frac{\mathrm{d}p_s}{\mathrm{d}t} = -\frac{\partial H}{\partial q_s} , \quad \frac{\mathrm{d}q_s}{\mathrm{d}t} = \frac{\partial H}{\partial p_s} . \tag{1}$$

The derivative of any function $F(q_i, p_j, t)$ of the generalized coordinates, momenta and time, with respect to time, is

$$\frac{\mathrm{d}F}{\mathrm{d}t} = \frac{\partial F}{\partial t} + \sum_{s=1}^{f} \frac{\partial F}{\partial q_s}\frac{\mathrm{d}q_s}{\mathrm{d}t} + \sum_{s=1}^{f} \frac{\partial F}{\partial p_s}\frac{\mathrm{d}p_s}{\mathrm{d}t} . \tag{2}$$

Using Hamilton's equation (1) we can transform (2) into the following form:

$$\frac{\mathrm{d}F}{\mathrm{d}t} = \frac{\partial F}{\partial t} + \{H, F\} , \tag{3}$$

where $\{H, F\}$ is

$$\{H, F\} = \sum_{s=1}^{f} \left\{ \frac{\partial F}{\partial q_s}\frac{\partial H}{\partial p_s} - \frac{\partial H}{\partial q_s}\frac{\partial F}{\partial p_s} \right\} , \tag{4}$$

and is the so-called Poisson bracket [cf. (8.14)].

Obviously, Hamilton's equations (1) can now be written

$$\frac{\mathrm{d}p_s}{\mathrm{d}t} = \{H, p_s\} , \quad \frac{\mathrm{d}q_s}{\mathrm{d}t} = \{H, q_s\} , \quad s = 1, 2, \dots, f \tag{5}$$

[we have only to set $F = p_s$ and $F = q_s$ in (3)].

In Chap. 8 we learned that in quantum mechanics, equations of motion are written in an analogous way. In the special case of a Cartesian system and of a particle in a field derivable from a potential function $V(x, y, z, t)$, we have

$$H = \frac{p_x^2 + p_y^2 + p_z^2}{2m} + V(x, y, z, t) , \tag{6}$$

where $q_1 = x$, $q_2 = y$, $q_3 = z$, and $p_1 = p_x$, $p_2 = p_y$, and $p_3 = p_z$. With (5) we obtain

$$\begin{aligned} \frac{\mathrm{d}p_x}{\mathrm{d}t} &= \{H, p_x\} = -\frac{\partial H}{\partial x} = -\frac{\partial V}{\partial x} , \\ \frac{\mathrm{d}x}{\mathrm{d}t} &= \{H, x\} = \frac{\partial H}{\partial p_x} = \frac{p_x}{m} . \end{aligned} \tag{7}$$

Example 9.1

The equations of the other coordinates and momenta can be obtained in the same way. From (7) we get

$$m\frac{\mathrm{d}^2x}{\mathrm{d}t^2} = -\frac{\partial V}{\partial x} \ , \tag{8}$$

i.e. *Newton's equation of motion.*

Consider now the motion of a charged particle with charge e and mass m in an electromagnetic field described by a potential $\phi = (1/e)V(\boldsymbol{r}, t)$ and a vector potential $\boldsymbol{A}$, so that

$$\boldsymbol{E} = -\nabla\phi - \frac{1}{c}\frac{\partial \boldsymbol{A}}{\partial t} \ , \tag{9}$$

$$\boldsymbol{B} = \operatorname{curl} \boldsymbol{A} \ , \tag{10}$$

where $\boldsymbol{E}$ and $\boldsymbol{B}$ are the electric and magnetic fields. In this case, the Hamiltonian function can be written as

$$H = \frac{1}{2m}\left(\boldsymbol{p} - \frac{e}{c}\boldsymbol{A}\right)^2 + e\phi \ . \tag{11}$$

Indeed, we will show that the Hamilton equations that emerge from this function,

$$\frac{\mathrm{d}p_x}{\mathrm{d}t} = -\frac{\partial H}{\partial x} \ , \quad \frac{\mathrm{d}p_y}{\mathrm{d}t} = -\frac{\partial H}{\partial y} \ , \quad \frac{\mathrm{d}p_z}{\mathrm{d}t} = -\frac{\partial H}{\partial z} \ , \tag{12}$$

$$\frac{\mathrm{d}x}{\mathrm{d}t} = \frac{\partial H}{\partial p_x} \ , \quad \frac{\mathrm{d}y}{\mathrm{d}t} = \frac{\partial H}{\partial p_y} \ , \quad \frac{\mathrm{d}z}{\mathrm{d}t} = \frac{\partial H}{\partial p_z} \ , \tag{13}$$

are equivalent to Newton's equations for the same particle under influence of the Lorentz force:

$$\begin{aligned} m\frac{\mathrm{d}^2\boldsymbol{r}}{\mathrm{d}t^2} &= e\left(\boldsymbol{E} + \frac{1}{c}\boldsymbol{v}\times\boldsymbol{B}\right) \ , \quad \text{or} \\ m\frac{\mathrm{d}^2x}{\mathrm{d}t^2} &= e\left[E_x + \frac{1}{c}\left(\frac{\mathrm{d}y}{\mathrm{d}t}B_z - \frac{\mathrm{d}z}{\mathrm{d}t}B_y\right)\right] \ , \\ m\frac{\mathrm{d}^2y}{\mathrm{d}t^2} &= e\left[E_y + \frac{1}{c}\left(\frac{\mathrm{d}z}{\mathrm{d}t}B_x - \frac{\mathrm{d}x}{\mathrm{d}t}B_z\right)\right] \ , \\ m\frac{\mathrm{d}^2z}{\mathrm{d}t^2} &= e\left[E_z + \frac{1}{c}\left(\frac{\mathrm{d}x}{\mathrm{d}t}B_y - \frac{\mathrm{d}y}{\mathrm{d}t}B_x\right)\right] \ . \end{aligned} \tag{14}$$

Inserting H from (11) into (12) and (13), we can write, after derivation,

$$\begin{aligned} \frac{\mathrm{d}p_x}{\mathrm{d}t} = \frac{e}{mc}\bigg[&\left(p_x - \frac{e}{c}A_x\right)\frac{\partial A_x}{\partial x} + \left(p_y - \frac{e}{c}A_y\right)\frac{\partial A_y}{\partial x} \\ &+ \left(p_z - \frac{e}{c}A_z\right)\frac{\partial A_z}{\partial x}\bigg] - e\frac{\partial \phi}{\partial x} \ . \end{aligned} \tag{15}$$

Example 9.1

From (13) we get

$$\frac{dx}{dt} = \frac{1}{m}\left(p_x - \frac{e}{c}A_x\right) ,$$
$$\frac{dy}{dt} = \frac{1}{m}\left(p_y - \frac{e}{c}A_y\right) ,$$
$$\frac{dz}{dt} = \frac{1}{m}\left(p_z - \frac{e}{c}A_z\right) . \tag{16}$$

This implies that

$$\frac{dp_x}{dt} = m\frac{d^2x}{dt^2} + \frac{e}{c}\frac{dA_x}{dt} . \tag{17}$$

Thus (15) can now be written in the following form:

$$m\frac{d^2x}{dt^2} + \frac{e}{c}\frac{dA_x}{dt} = \frac{e}{c}\frac{dx}{dt}\frac{\partial A_x}{\partial x} + \frac{e}{c}\frac{dy}{dt}\frac{\partial A_y}{\partial x} + \frac{e}{c}\frac{dz}{dt}\frac{\partial A_z}{\partial x} - e\frac{\partial \phi}{\partial x} . \tag{18}$$

Since the value of the vector potential $\boldsymbol{A}$ is obtained at the position of the charge e, the total derivative of A_x with respect to time is

$$\frac{dA_x}{dt} = \frac{\partial A_z}{\partial t} + \frac{\partial A_x}{\partial x}\frac{dx}{dt} + \frac{\partial A_x}{\partial y}\frac{dy}{dt} + \frac{\partial A_x}{\partial z}\frac{dz}{dt} . \tag{19}$$

After inserting into (15) and (16) the values of $[p_x - (e/c)A_x]$, $[p_y - (e/c)A_y]$ and $[p_z - (e/c)A_z]$ and of dp_x/dt from (17), we find with the help of (19) that

$$m\frac{d^2x}{dt^2} = -\frac{e}{c}\frac{\partial A_x}{\partial t} - e\frac{\partial \phi}{\partial x} + \frac{e}{c}\left[\frac{dy}{dt}\left(\frac{\partial A_y}{\partial x} - \frac{\partial A_x}{\partial y}\right) + \frac{dz}{dt}\left(\frac{\partial A_z}{\partial x} - \frac{\partial A_x}{\partial z}\right)\right] . \tag{20}$$

Here we can use the formulae (9) and (10), which connect fields and potential, to get

$$m\frac{d^2x}{dt^2} = eE_x + \frac{e}{c}\left(\frac{dy}{dt}B_z - \frac{dz}{dt}B_y\right) . \tag{21}$$

This is the first of the equations in (14); we can derive the other two relations in the same way. We thus see that the Hamilton equations (12) and (13), resulting from the Hamiltonian function (11), are equivalent to Newton's equations (14). The potentials $\boldsymbol{A}$ and ϕ can be chosen at will, if only formulae (9) and (10) lead to the required electromagnetic field. Using $\boldsymbol{A}'$ and ϕ' instead of $\boldsymbol{A}$ and ϕ, where

$$\boldsymbol{A}' = \boldsymbol{A} + \nabla f \quad \text{and} \quad \phi' = \phi - \frac{1}{c}\frac{\partial f}{\partial t} , \tag{22}$$

and f is an arbitrary function of the position coordinates and time, we get $\boldsymbol{E}' = \boldsymbol{E}$ and $\boldsymbol{B}' = \boldsymbol{B}$. When replacing $\boldsymbol{A}$ and ϕ in the Hamiltonian function (11) by $\boldsymbol{A}'$

Example 9.1

and ϕ', we get the equation of motion (20), with A and ϕ replaced by A' and ϕ', i.e. with the *same* equations (14). Thus, using (22), we have shown that equations (14) are independent of the choice of the potentials. This property of Hamilton's equations is known as *gauge invariance.*

Note that the Hamiltonian function H is changed by the transformation (22) in contrast to the equations (14). For instance, the motion in a homogeneous constant electric field oriented along the x axis can be described by the potentials $A = 0$ and $\phi = -Ex$ as well as, for example, $A' = (-cEt, 0, 0)$ and $\phi' = 0$, according to (22). It can easily be verified that both choices lead to Newton's equation of uniformly accelerated motion, but in the first case, the Hamiltonian function represents the total energy of the particle, and in the second, it represents the kinetic energy.

EXERCISE

9.2 The Lagrangian and Hamiltonian of a Charged Particle

Problem. Determine the Lagrangian and the Hamiltonian of a charged particle in an electromagnetic field. Use vector calculus as far as possible.

Solution. The effect of the electromagnetic field on a charged particle can be described by a velocity-dependent generalized potential.

Starting with the Lorentz force, we determine this potential, and the Lagrangian and Hamiltonian functions. The Lorentz force takes the form

$$F = e\left(E + \frac{v}{c} \times B\right) . \tag{1}$$

We can express the electric and magnetic fields by the potentials

$$E = -\nabla\phi - \frac{1}{c}\frac{\partial A}{\partial t} , \qquad B = \nabla \times A . \tag{2}$$

Insertion of (2) into the Lorentz force (1) yields

$$F = e\left(-\nabla\phi - \frac{1}{c}\frac{\partial A}{\partial t} + \frac{1}{c} v \times (\nabla \times A)\right) . \tag{3}$$

We can use the relation

$$B \times (\nabla \times C) = \nabla(B \cdot C) - (B \cdot \nabla)C - (C \cdot \nabla)B - C \times (\nabla \times B)$$

to transform the triple vector product

$$v \times (\nabla \times A) = \nabla(v \cdot A) - (v \cdot \nabla)A , \tag{4}$$

since the velocity v is not an explicit function of the position.

The total derivative of the vector potential with respect to time is given by

Exercise 9.2

$$\frac{\mathrm{d}\boldsymbol{A}}{\mathrm{d}t} = \frac{\partial \boldsymbol{A}}{\partial t} + (\boldsymbol{v}\cdot\nabla)\boldsymbol{A} \ . \tag{5}$$

The first term is the explicit change of the vector potential with time; the second term stems from the fact that the position at which the value of the potential is obtained changes because of the particle's motion.

We now replace the vector product (3) by the relations (4) and (5) and get

$$\boldsymbol{F} = e\left[-\nabla\phi + \frac{1}{c}\nabla(\boldsymbol{v}\cdot\boldsymbol{A}) - \frac{1}{c}\frac{\mathrm{d}\boldsymbol{A}}{\mathrm{d}t}\right] \ . \tag{6}$$

To deduce the generalized forces Q_i from a velocity-dependent potential $U(q_i, \dot{q}_i)$, we rely on the Lagrangian formalism, where the relation

$$Q_i = -\frac{\partial U}{\partial q_i} + \frac{\mathrm{d}}{\mathrm{d}t}\left(\frac{\partial U}{\partial \dot{q}_i}\right) \tag{7}$$

is valid. For comparison with (7), we transform

$$\frac{\mathrm{d}\boldsymbol{A}}{\mathrm{d}t} = \frac{\mathrm{d}}{\mathrm{d}t}\nabla_{\boldsymbol{v}}(\boldsymbol{A}\cdot\boldsymbol{v}) \ , \tag{8}$$

where $\nabla_{\boldsymbol{v}}$ signifies the derivative (gradient) with respect to the three components of the velocity. We take, for instance, the x component and compare (6) and (7) using relation (8):

$$F_x = -\frac{\partial}{\partial x}\left(e\phi - \frac{e}{c}\boldsymbol{v}\cdot\boldsymbol{A}\right) + \frac{\mathrm{d}}{\mathrm{d}t}\frac{\partial}{\partial v_x}\left(e\phi - \frac{e}{c}\boldsymbol{v}\cdot\boldsymbol{A}\right) \ . \tag{9}$$

Since the electrostatic potential $\phi(\boldsymbol{r}, t)$ is independent of velocity, we were able to add it to the last term. Hence, we get the generalized potential,

$$U = e\phi - \frac{e}{c}\boldsymbol{v}\cdot\boldsymbol{A} \ . \tag{10}$$

Using $L = T - U$ yields the Lagrangian

$$L = \frac{1}{2}m\boldsymbol{v}^2 - e\phi + \frac{e}{c}\boldsymbol{v}\cdot\boldsymbol{A} \ , \tag{11}$$

and, in the form of generalized coordinates, we have

$$L = \frac{1}{2}m\sum_i\left(\dot{q}_i^2 - e\phi(q_i)\right) + \frac{e}{c}\sum_i \dot{q}_i A_i \ . \tag{12}$$

The *canonical momentum* is given by

$$p_i = \frac{\partial L}{\partial \dot{q}_i} = m\dot{q}_i + \frac{e}{c}A_i \tag{13}$$

Exercise 9.2

or, in vector form,

$$p = mv + \frac{e}{c}A \ . \tag{14}$$

Now the Hamiltonian can be derived from the Lagrange function L by

$$H = \sum_i p_i \dot{q}_i - L \ . \tag{15}$$

It has the form

$$H = \frac{1}{2m}\left(p - \frac{e}{c}A\right)^2 + e\phi \ , \tag{16}$$

where the velocity is replaced using (14).

EXERCISE

9.3 Landau States

Problem. (a) What is the Schrödinger equation for the motion of charged particles in a constant magnetic field $\boldsymbol{B} = Be_z$? Choose the following vector potential:

$$\boldsymbol{A} = (-By, 0, 0) \quad \text{and} \quad \phi = 0 \ .$$

(b) Show that the following separation of variables,

$$\psi(x, y, z) = \mathrm{e}^{\mathrm{i}(\alpha x + \beta z)}\varphi(y) \ ,$$

used after substituting $y = y' - \hbar\alpha c/eB$, leads to the equation of a harmonic oscillator.

(c) What are the energy eigenvalues?

Solution. (a) It can be easily verified that the chosen vector potential in fact leads to the magnetic field $\boldsymbol{B} = Be_z$.

As has already been shown [see (9.8)], the Schrödinger equation (in its stationary form) is

$$\frac{1}{2m}\left(\hat{\boldsymbol{p}} - \frac{e}{c}\boldsymbol{A}\right)^2 \psi(\boldsymbol{r}) = E\psi(\boldsymbol{r}) \ . \tag{1}$$

Calculation of the product $[\hat{\boldsymbol{p}} - (e/c)\boldsymbol{A}]^2$ and insertion of $\boldsymbol{A}$ yields

$$\frac{1}{2m}\left[-\hbar^2\frac{\partial^2}{\partial x^2} - \frac{e}{c}\left(\mathrm{i}\hbar\frac{\partial}{\partial x}By + By\mathrm{i}\hbar\frac{\partial}{\partial x}\right) + \frac{e^2}{c^2}B^2y^2 - \hbar^2\frac{\partial^2}{\partial y^2} - \hbar^2\frac{\partial^2}{\partial z^2}\right]\psi = E\psi \tag{2}$$

Exercise 9.3

or

$$\left(-\frac{\hbar^2}{2m}\Delta - \mathrm{i}\frac{\hbar eB}{mc}y\frac{\partial}{\partial x} + \frac{e^2B^2}{2mc^2}y^2\right)\psi = E\psi \ . \tag{3}$$

(b) The expression $\psi(x, y, z) = \exp(\mathrm{i}\alpha x + \mathrm{i}\beta z)\varphi(y)$ with two constants α and β leads to

$$\left[-\frac{\hbar^2}{2m}(-\alpha^2 - \beta^2) - \frac{\hbar^2}{2m}\frac{\partial^2}{\partial y^2} + \frac{\hbar eB\alpha}{mc}y + \frac{e^2B^2}{2mc^2}y^2\right] \times \mathrm{e}^{\mathrm{i}(\alpha x+\beta z)}\varphi(y) = E\,\mathrm{e}^{\mathrm{i}(\alpha x+\beta z)}\varphi(y) \tag{4}$$

and thus

$$\left(-\frac{\hbar^2}{2m}\frac{\mathrm{d}^2}{\mathrm{d}y^2} + \frac{\hbar eB\alpha}{mc}y + \frac{e^2B^2}{2mc^2}y^2\right)\varphi(y) = \left(E - \frac{\hbar^2}{2m}\alpha^2 - \frac{\hbar^2}{2m}\beta^2\right)\varphi(y) \ . \tag{5}$$

That $\psi \propto \exp(\mathrm{i}\alpha x + \mathrm{i}\beta z)\varphi(y)$ seems to imply that the particle is free to move in the x and z directions ($\perp \boldsymbol{B}$ and $\parallel \boldsymbol{B}$), and that this motion is related to the kinetic energies $(\hbar^2/2m)\alpha^2$ and $(\hbar^2/2m)\beta^2$, respectively.

We will soon come back to this point. Now, substituting

$$y = y' - \frac{\hbar c\alpha}{eB} = y' - \frac{\hbar\alpha}{m\omega_0} \tag{6}$$

and setting

$$\omega_0 = \frac{eB}{mc} \quad \text{and} \quad \varepsilon = E - \frac{\hbar^2}{2m}\beta^2 \ , \tag{7}$$

the Schrödinger equation becomes

$$\left[-\frac{\hbar^2}{2m}\frac{\mathrm{d}^2}{\mathrm{d}y'^2} + \hbar\omega_0\alpha\left(y' - \frac{\hbar\alpha}{m\omega_0}\right) + \frac{m}{2}\omega_0^2\left(y' - \frac{\hbar\alpha}{m\omega_0}\right)^2\right]\varphi'(y') = \left(\varepsilon - \frac{\hbar^2}{2m}\alpha^2\right)\varphi'(y') \ . \tag{8}$$

This can be simplified to

$$\left(-\frac{\hbar^2}{2m}\frac{\mathrm{d}^2}{\mathrm{d}y'^2} + \frac{m}{2}\omega_0^2 y'^2\right)\varphi'(y') = \varepsilon\varphi'(y') \ . \tag{9}$$

Now we have, once again, the equation of a harmonic oscillator. Note that the "kinetic energy" in the x direction $(\hbar^2/2m)\alpha^2$ has now been absorbed into the y' degree of freedom.

Exercise 9.3

(c) From the above we can immediately write down the energy eigenvalues, namely

$$\varepsilon_n = \hbar\omega_0(n+\frac{1}{2}) \ , \quad n = 0, 1, 2, \dots \ . \tag{10}$$

The functions $\varphi'(y')$ are related to the Hermite polynomials and are located around

$$y' = 0 \ , \quad \text{i.e.} \quad y_0 = -(\hbar c/eB)\alpha \ .$$

The total energy is

$$E_n(\beta) = \frac{\hbar^2}{2m}\beta^2 + \hbar\omega_0\left(n+\frac{1}{2}\right) \ . \tag{11}$$

Neglecting the motion in the z direction ($\beta = 0$), the energy $E_n(0)$ is quantized.

For a given α, the wave function

$$\psi(x, y, z) = \exp(\mathrm{i}\alpha x + \mathrm{i}\beta z)\varphi(y)$$

is localized in the y direction, but not in the x direction. This result is unexpected, for both directions should be equally represented. However, as we have seen above, the energy is independent of α, so that we have infinite degeneracy.

Thus wave packets of the form

$$\psi_{n\beta}(x, y, z) = \int_{-\infty}^{+\infty} c(\alpha)\,\mathrm{e}^{\mathrm{i}(\alpha x+\beta z)}\varphi_\alpha(y)\,\mathrm{d}\alpha \ , \tag{12}$$

where $c(\alpha)$ can be (nearly) chosen at will, are also solutions of the Schrödinger equation (2). Therefore we can choose $c(\alpha)$ so that the solution is located in the x direction, too. Such bound states in the x–y plane are unrestricted in the z direction, i.e. along the direction of the magnetic field $\boldsymbol{B}$. They correspond classically to electrons orbiting perpendicular to $\boldsymbol{B}$, but moving with constant velocity (momentum) along $\boldsymbol{B}$ and are called ***Landau** states*; the energy levels (11) are *Landau levels*.

9.2 The Hydrogen Atom

The most important example of the motion of a particle in a potential field is the hydrogen atom.

Electrons and protons attract each other with the force e^2/r^2, corresponding to the potential $-e^2/r$. Here, r is the coordinate of relative motion, and is all that interests us for the moment. We choose the proton to be the centre of our coordinate system; the mass m used in the following context is then the reduced mass of the electron:

$$m = \frac{m_e}{1+m_e/m_p} \approx m_e\left(1-\frac{1}{1836}\right) . \tag{9.22}$$

Since we have a central potential, we use spherical coordinates. The stationary Schrödinger equation is then

$$\hat{H}\psi = E\psi = \left(\frac{\hat{p}^2}{2m} - \frac{e^2}{r}\right)\psi . \tag{9.23}$$

The squared momentum operator

$$\hat{p}^2 = -\hbar^2 \Delta = -\hbar^2 \left(\frac{1}{r^2}\frac{\partial}{\partial r} r^2 \frac{\partial}{\partial r} + \frac{1}{r^2}\Delta_{\vartheta,\varphi}\right)$$

can be divided with the aid of $\hat{\boldsymbol{L}}^2 = -\hbar^2 \Delta_{\vartheta,\varphi}$ into a radial part and a rotational part containing the angular-momentum operator $\hat{\boldsymbol{L}}$ (see Example 4.9). Consequently, the Schrödinger equation takes the form

$$\left(\frac{1}{r^2}\frac{\partial}{\partial r} r^2 \frac{\partial}{\partial r} - \frac{\hat{\boldsymbol{L}}^2}{\hbar^2 r^2}\right)\psi + \frac{2m}{\hbar^2}\left(E+\frac{e^2}{r}\right)\psi = 0 . \tag{9.24}$$

In the Schrödinger equation a centrifugal term $-\hat{\boldsymbol{L}}^2/2mr^2$ appears, similar to that in Kepler's problem in classical mechanics.

With the following separation of variables,

$$\psi(r,\vartheta,\varphi) = \frac{R(r)}{r} Y(\vartheta,\varphi) , \tag{9.25}$$

it is possible to separate (9.24) into a radial and an angular part. We start with

$$\frac{1}{r^2}\frac{\partial}{\partial r} r^2 \frac{\partial}{\partial r}\frac{R(r)}{r} = \frac{1}{r}\frac{\partial^2 R(r)}{\partial r^2} \tag{9.26}$$

and introduce the separation constant $l(l+1)$ to get

$$\begin{aligned} \frac{r^2}{R(r)}\frac{\partial^2 R(r)}{\partial r^2} + r^2 \frac{2m}{\hbar^2}\left(E+\frac{e^2}{r}\right) &= \frac{1}{\hbar^2}\frac{1}{Y(\vartheta,\varphi)}\hat{\boldsymbol{L}}^2 Y(\vartheta,\varphi) \\ &= l(l+1) . \end{aligned} \tag{9.27}$$

Hence, we have the two equations

$$\frac{\partial^2 R_l}{\partial r^2} + \left[\frac{2m}{\hbar^2}\left(E + \frac{e^2}{r}\right) - \frac{l(l+1)}{r^2}\right] R_l(r) = 0 \quad \text{and} \tag{9.28}$$

$$\hat{L}^2 Y_{lm}(\vartheta, \varphi) = \hbar^2 l(l+1) Y_{lm}(\vartheta, \varphi) \;, \quad \text{with}$$

$$l = 0, 1, 2, \dots \quad \text{and} \quad -l \le m \le +l \; . \tag{9.29}$$

The solutions of the angular differential equation (9.29) are the already familiar spherical harmonics $Y_{lm}(\vartheta, \varphi)$ (see Examples 4.8 and 4.9). The separation constant is the quantum number of the square of the angular momentum $\hat{L}^2 = \hbar^2 l(l+1)$. The additional quantum number m appearing in (9.29) characterizes the z component of the angular momentum $\hat{L}_z \ge m\hbar$. [The solution of (9.29) will be discussed once more, in detail, in Exercise 9.4.] The radial function $R_l(r)$ depends on the total angular-momentum quantum number l, as can be seen in (9.28). We will soon see that the condition of square-integrability for the wave function (normalization) calls for another quantum number, the so-called *radial quantum number* n_r.

To find the energy spectrum, it is sufficient to deal with the radial part, because the energy E appears only in (9.28). Indeed, since the problem is spherically symmetric, the energy can only depend on the radial part $R(r)$ of the wave function. (In the classical Kepler problem, the energy depends on the distance between the particles.)

$$\int \psi\psi^* \, \mathrm{d}V = 1 \tag{9.30}$$

leads to

$$\int_0^\infty R_l(r) R_l^*(r) \, \mathrm{d}r = 1 \tag{9.31}$$

because of the separation (9.25) and the orthonormality of the spherical harmonics. Here, we only determine the (discrete) bound states which are characterized by negative energy eigenvalues.[1]

To find a suitable substitution for solving the differential equation (9.28), it is useful to consider first the limits $r \to 0$ and $r \to \infty$. For $r \to 0$, the term containing the angular momentum is dominant and one gets the equation

$$\frac{\mathrm{d}^2 R_l}{\mathrm{d}r^2} - \frac{l(l+1)}{r^2} R_l = 0 \; . \tag{9.32}$$

[1] A discussion of the solutions of the continuum ($E > 0$) can be found e.g. in A.S. Davydov: *Quantum Mechanics* (Pergamon Press, Oxford 1965).

By trying a power series $R_l = r^\alpha(1 + a_1 r + a_2 r^2 + \dots)$ and neglecting the higher-order terms, only the following lowest-order term remains:

$$\alpha(\alpha-1)r^{\alpha-2} - l(l+1)r^{\alpha-2} = 0 \ . \tag{9.33}$$

The solutions for α are then $\alpha = l+1$ and $\alpha = -l$. The case $\alpha = -l$ leads, as does the three-dimensional oscillator (cf. Exercise 7.2), to the same solutions as the case $\alpha = l+1$.

In the other asymptotic limit ($r \to \infty$), we can approximate (9.28) by

$$\frac{d^2 R_l}{dr^2} + \frac{2m}{\hbar^2} E R_l = 0 \ . \tag{9.34}$$

The abbreviation

$$\gamma^2 = -\frac{2m}{\hbar^2} E \tag{9.35}$$

is often used and logically chosen, since the energies of the bound states should be negative. In this case the solution of (9.34) is

$$U_l = A\,e^{-\gamma r} + B\,e^{\gamma r} \ , \tag{9.36}$$

where we have to exclude the second term, because it becomes infinite as $r \to \infty$. With the solutions of both extreme ("asymptotic") cases, (9.32) and (9.34), we try the substitution

$$R_l(r) = r^{l+1} e^{-\gamma r} F(r) \ . \tag{9.37}$$

After inserting this into (9.28) and writing

$$z = 2\gamma r \quad \text{and} \quad k = \frac{me^2}{\gamma \hbar^2} \ , \tag{9.38}$$

we get

$$z\frac{d^2 F}{dz^2} + (2l + 2 - z)\frac{dF}{dz} - (l + 1 - k)F = 0 \ . \tag{9.39}$$

Recalling the mathematical discussion in Chap. 7, we recognize this as Kummer's differential equation. The solution is given in (18) of Exercise 7.1. We exclude the second term of the total solution, as it behaves like $r^{-2l-1} (r \to 0)$, i.e. $R_l \sim r^{-l}$ is always divergent.

Thus we obtain

$$F = C\,{}_1F_1(l + 1 - k, 2l + 2; 2\gamma r) \ . \tag{9.40}$$

To be normalizable, the confluent series should end at a certain term; this requirement leads to the quantization of the energy.

Setting

$$l+1-k=-n_r\ ,\quad n_r=0,1,2,\dots\ , \tag{9.41}$$

we rearrange the terms, getting

$$k=n_r+l+1=n\ . \tag{9.42}$$

The number n is the *principal quantum number* ($n=1,2,\dots$) and is determined by the radial quantum number n_r ($n_r=0,1,2,\dots$) and the angular-momentum quantum number l ($l=0,1,2,\dots$).

The definitions (9.35) and (9.38) allow us to determine the binding energy:

$$E_n=-\frac{me^4}{2\hbar^2}\frac{1}{n^2}\equiv-\frac{1}{2}\frac{e^2}{a_0n^2}\ , \tag{9.43}$$

where $a_0=\hbar^2/me^2=0.53$ Å is called the ***Bohr** radius.*

When we set $n=1$, we obtain the *binding energy of the hydrogen atom* in the ground state,

$$E_0=-\frac{1}{2}\frac{e^2}{a_0}=-13.6\,\mathrm{eV}\ . \tag{9.44}$$

The wave functions of the hydrogen atom are

$$\begin{aligned}\psi_{nlm}(\boldsymbol{r})&=N_{nl}r^l\mathrm{e}^{-\gamma_nr}\,{}_1F_1(-n_r,2l+2,2\gamma_nr)Y_{lm}(\vartheta,\varphi)\\&=N_{nl}\frac{R_{nl}(r)}{r}Y_{lm}(\vartheta,\varphi)\ ,\end{aligned} \tag{9.45}$$

where $\gamma_n=me^2/\hbar^2n=1/na_0$ and with the normalization constant

$$N_{nl}=\frac{1}{(2l+1)!}\sqrt{\frac{(n+l)!}{2n(n-l-1)!}}(2\gamma_n)^{l+3/2}a_0^{3/2}\ ,\quad n=n_r+l+1\ . \tag{9.46}$$

The radial part of the wave function $R_{nl}(r)$ obviously depends on two quantum numbers, n and l (or n_r and l). The dependence of l results from the separation of variables in (9.25), by which the rotational term $l(l+1)/r^2$ was introduced into the differential equation (9.28), whereas the dependence on n is caused by the eigenvalue equation, which originated from the requirement that the wave function be square-integrable [normalization condition (9.31)].

The ψ_{nlm} are eigenfunctions of the Schrödinger equation (9.24) belonging to the energy eigenvalues E_n. Equations (9.41) and (9.42) allow the quantum numbers l and m to have the values $0\le l\le(n-1)$ and $-l\le m\le l$. Counting all possible states of the same energy, we see that every eigenvalue is degenerate n^2 times:

$$\sum_{l=0}^{n-1}\sum_{m=-l}^{l}m=\sum_{l=0}^{n-1}(2l+1)=n^2\ . \tag{9.47}$$

Tables 9.1 and 9.2 present the normalized wave functions for the lowest states of the hydrogen atom. In the second table, the wave functions are separated into the radial ($R_{nl}(r)/r$) and angular ($Y_{lm}(\vartheta, \varphi)$) parts. The energies ($E_n$) depend only on the principal quantum number n and are shown in the last column. The energy units are $-e^2/2a_0 = -13.6\,\text{eV}$, i.e. the ground-state energy, and $\gamma_n = 1/na_0$, $a_0 = \hbar^2/me^2 = 0.52\,\text{Å}$. Every state with the eigenfunction ψ_{nlm} characterized

Table 9.1. The wave functions ψ_{nlm} for the lowest states of the Schrödinger hydrogen atom

n	l	m			$\psi_{nlm}(r, \vartheta, \varphi)$			E_n
1	0	0	$\frac{1}{\sqrt{\pi}}$	$\times\gamma^{3/2}$		$\times e^{-\gamma r}$		1
2	0	0	$\frac{1}{\sqrt{\pi}}$	$\times\gamma_2^{3/2}$	$\times(1-\gamma r)$	$\times e^{-\gamma_2 r}$		$\frac{1}{4}$
2	1	0	$\frac{1}{\sqrt{\pi}}$	$\times\gamma_2^{5/2}$	$\times r$	$\times e^{-\gamma_2 r}$	$\times\cos\vartheta$	$\frac{1}{4}$
2	1	±1	$\frac{1}{\sqrt{2\pi}}$	$\times\gamma_2^{5/2}$	$\times r$	$\times e^{-\gamma_2 r}$	$\times\sin\vartheta\, e^{\pm i\varphi}$	$\frac{1}{4}$
3	0	0	$\frac{1}{3\sqrt{\pi}}$	$\times\gamma_3^{3/2}$	$\times(3-6\gamma r+2\gamma^2 r^2)$	$\times e^{-\gamma_3 r}$		$\frac{1}{9}$
3	1	0	$\frac{2}{\sqrt{3\pi}}$	$\times\gamma_3^{5/2}$	$\times(2-\gamma r)r$	$\times e^{-\gamma_3 r}$	$\times\cos\vartheta$	$\frac{1}{9}$
3	1	±1	$\frac{1}{\sqrt{3\pi}}$	$\times\gamma_3^{5/2}$	$\times(2-\gamma r)r$	$\times e^{-\gamma_3 r}$	$\times\sin\vartheta\, e^{\pm i\varphi}$	$\frac{1}{9}$
3	2	0	$\frac{1}{3\sqrt{2\pi}}$	$\times\gamma_3^{7/2}$	$\times r^2$	$\times e^{-\gamma_3 r}$	$\times(3\cos^2\vartheta-1)$	$\frac{1}{9}$
3	2	±1	$\frac{1}{\sqrt{3\pi}}$	$\times\gamma_3^{7/2}$	$\times r^2$	$\times e^{-\gamma_3 r}$	$\times\sin\vartheta\cos\vartheta\, e^{\pm i\varphi}$	$\frac{1}{9}$
3	2	±2	$\frac{1}{2\sqrt{3\pi}}$	$\times\gamma_3^{7/2}$	$\times r^2$	$\times e^{-\gamma_3 r}$	$\times\sin^2\vartheta\, e^{\pm 2i\varphi}$	$\frac{1}{9}$

Table 9.2. The wave functions in Table 9.1, separated into radial and angular parts

n	l	m			$R_{nl}(r)$		$Y_{lm}(\vartheta, \varphi)$	E_n
1	0	0	2	$\times\gamma^{3/2}$	$\times$	$\times e^{-\gamma r}\times\frac{1}{\sqrt{4\pi}}$		1
2	0	0	2	$\times\gamma_2^{3/2}$	$\times(1-\gamma r)$	$\times e^{-\gamma_2 r}\times\frac{1}{\sqrt{4\pi}}$		$\frac{1}{4}$
2	1	0	$\frac{2}{\sqrt{3}}$	$\times\gamma_2^{5/2}$	$\times r$	$\times e^{-\gamma_2 r}\times\sqrt{\frac{3}{4\pi}}\cos\vartheta$		$\frac{1}{4}$
2	1	±1	$\frac{2}{\sqrt{3}}$	$\times\gamma_2^{5/2}$	$\times r$	$\times e^{-\gamma_2 r}\times\sqrt{\frac{3}{8\pi}}\sin\vartheta$	$\times e^{\pm i\varphi}$	$\frac{1}{4}$
3	0	0	$\frac{2}{3}$	$\times\gamma_3^{3/2}$	$\times(3-6\gamma r+2\gamma^2 r^2)$	$\times e^{-\gamma_3 r}\times\frac{1}{\sqrt{4\pi}}$		$\frac{1}{9}$
3	1	0	$\frac{\sqrt{8}}{3}$	$\times\gamma_3^{5/2}$	$\times(2-\gamma r)$	$\times e^{-\gamma_3 r}\times\sqrt{\frac{3}{4\pi}}\cos\vartheta$		$\frac{1}{9}$
3	1	±1	$\frac{\sqrt{8}}{3}$	$\times\gamma_3^{5/2}$	$\times(2-\gamma r)$	$\times e^{-\gamma_3 r}\times\sqrt{\frac{3}{8\pi}}\sin\vartheta$	$\times e^{\pm i\varphi}$	$\frac{1}{9}$
3	2	0	$\sqrt{\frac{8}{45}}$	$\times\gamma_3^{7/2}$	$\times r^2$	$\times e^{-\gamma_3 r}\times\sqrt{\frac{5}{4\pi}}\left(\frac{3}{2}\cos^2\vartheta-\frac{1}{2}\right)$		$\frac{1}{9}$
3	2	±2	$\sqrt{\frac{8}{45}}$	$\times\gamma_3^{7/2}$	$\times r^2$	$\times e^{-\gamma_3 r}\times\sqrt{\frac{5}{24\pi}}3\sin\vartheta\cos\vartheta$	$\times e^{\pm i\varphi}$	$\frac{1}{9}$
3	2	±1	$\sqrt{\frac{8}{45}}$	$\times\gamma_3^{7/2}$	$\times r^2$	$\times e^{-\gamma_3 r}\times\sqrt{\frac{5}{96\pi}}3\sin^2\vartheta$	$\times e^{\pm 2i\varphi}$	$\frac{1}{9}$

by the three quantum numbers n, l and m is an eigenstate of three simultaneously measurable quantities:

(1) the energy $E_n = (-me^4/2\hbar^2)(1/n^2)$,

(2) the squared angular momentum $\hat{L}^2$ and

(3) the projection of the angular momentum on the z axis $\hat{L}_z$.

The principal quantum number n characterizes the energy level E_n; the *(azimuthal) quantum number l* indicates the magnitude of the angular momentum $\hat{L}^2$; and the *magnetic quantum number m* gives the size of the z component of angular momentum, $\hat{L}_z$. Thus the eigenvalues of the three quantities E_n, $\hat{L}^2$ and $\hat{L}_z$ are sufficient to determine the wave function $\psi_{nlm}(r, \vartheta, \varphi)$.

The probability of finding an electron with the wave function $\psi_{nlm}(r, \vartheta, \varphi)$ in the volume element $\mathrm{d}V = r^2 \sin\vartheta \, \mathrm{d}\vartheta \, \mathrm{d}\varphi \mathrm{d}r$ is

$$w_{nlm}(r, \vartheta, \varphi) \mathrm{d}V = |\psi_{nlm}(r, \vartheta, \varphi)|^2 \mathrm{d}V \ . \tag{9.48}$$

If we insert

$$\psi_{nlm}(r, \vartheta, \varphi) = \frac{R_{nl}(r)}{r} Y_{lm}(\vartheta, \varphi) \ ,$$

we can write the probability in the following way:

$$w_{nlm}(r, \vartheta, \varphi) r^2 \mathrm{d}r \, \mathrm{d}\Omega = R_{nl}^2(r) \mathrm{d}r |Y_{lm}(\vartheta, \varphi)|^2 \mathrm{d}\Omega \ . \tag{9.49}$$

Integration over $\mathrm{d}\Omega$ yields the probability $w_{nl}(r) \mathrm{d}r$ of having an electron between two spherical surfaces of the radii r and $r + \mathrm{d}r$:

$$w_{nl} \mathrm{d}r = w_{nlm}(r) r^2 \mathrm{d}r = R_{nl}^2(r) \mathrm{d}r \ . \tag{9.50}$$

For example, in the state ψ_{100}, the probability is

$$w_{10}(r) \mathrm{d}r = N_{10}^2 \mathrm{e}^{-2r/a_0} r^2 \mathrm{d}r \ , \tag{9.51}$$

where N_{10} is the normalization constant (9.46). (Plots of the probability w against r are shown in the lower part of Fig. 9.1.)

The wave functions of the hydrogen atom also describe the states of ions with only one electron, such as He^+, Li^{++}, ... The only difference lies in replacing the charge e^2 by Ze^2 (see Sect. 9.7).

If we deal with atoms with a nuclear charge number Z greater than 1, we have to replace a_0 by a_0/Z and the maximum of the probability approaches the nucleus like $1/Z$, i.e. the electron is forced by the stronger Coulomb forces into an orbit closer to the nucleus.

The maximum of the function $R_{nl}^2(r)$, i.e. the most probable distance of the electron, is given for the state ψ_{100} by

$$r_0 = a_0 \ , \quad a_0 = \frac{\hbar^2}{me^2} = 0.53 \, \text{Å} \ . \tag{9.52}$$

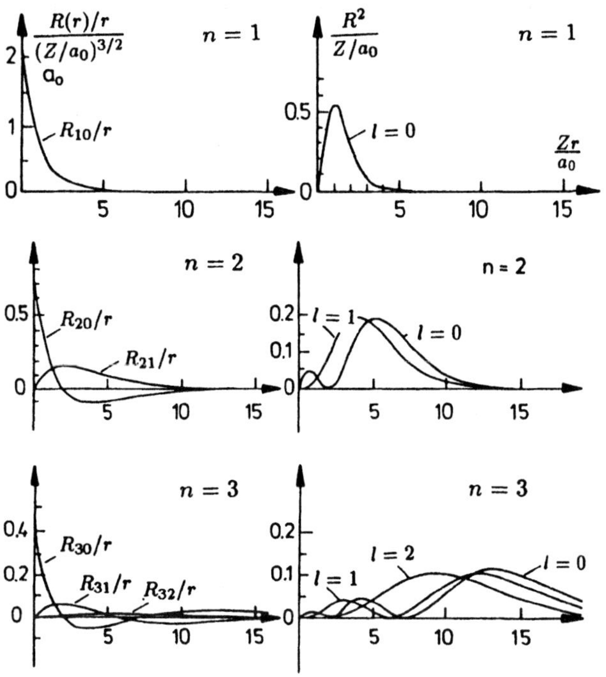

Fig. 9.1. The normalized radial functions R_{nl}/r (*left-hand side*) and the normalized probability densities w_{nl} (*right-hand side*) of the hydrogen atom for principal quantum numbers $n = 1, 2$ and 3

This is the classical *Bohr radius*, since according to classical theory, the electron should move around the nucleus on a circle with radius a_0.

With increasing principle quantum number n, the maximum of the charge distribution shifts away from the nucleus; the electron is less tightly bound.

According to the radial quantum number n_r, there are in general several maxima, a *principle maximum* and some *supplementary maxima* (see Fig. 9.1).

9.3 Three-Dimensional Electron Densities

Looking at the sketches in Fig. 9.2, the question arises why there are nonsymmetric states in the spherically symmetric Coulomb potential. Of course, the nonspherical symmetry of the wave function is immediately acceptable if a weak magnetic field is applied. The distributions shown are cylindrically symmetric to the z axis. The prominence of the z axis originates from the choice (orientation) of the spherical coordinates. Physically the z axis can be fixed, for example, by a (weak) magnetic field. The complete solution of the Schrödinger equation (9.23, 9.24) corresponding to the energy eigenvalue E_n is a linear combination of all ψ_{nlm}, since the wave function ψ_n is n^2 times degenerate. Thus, in the absence

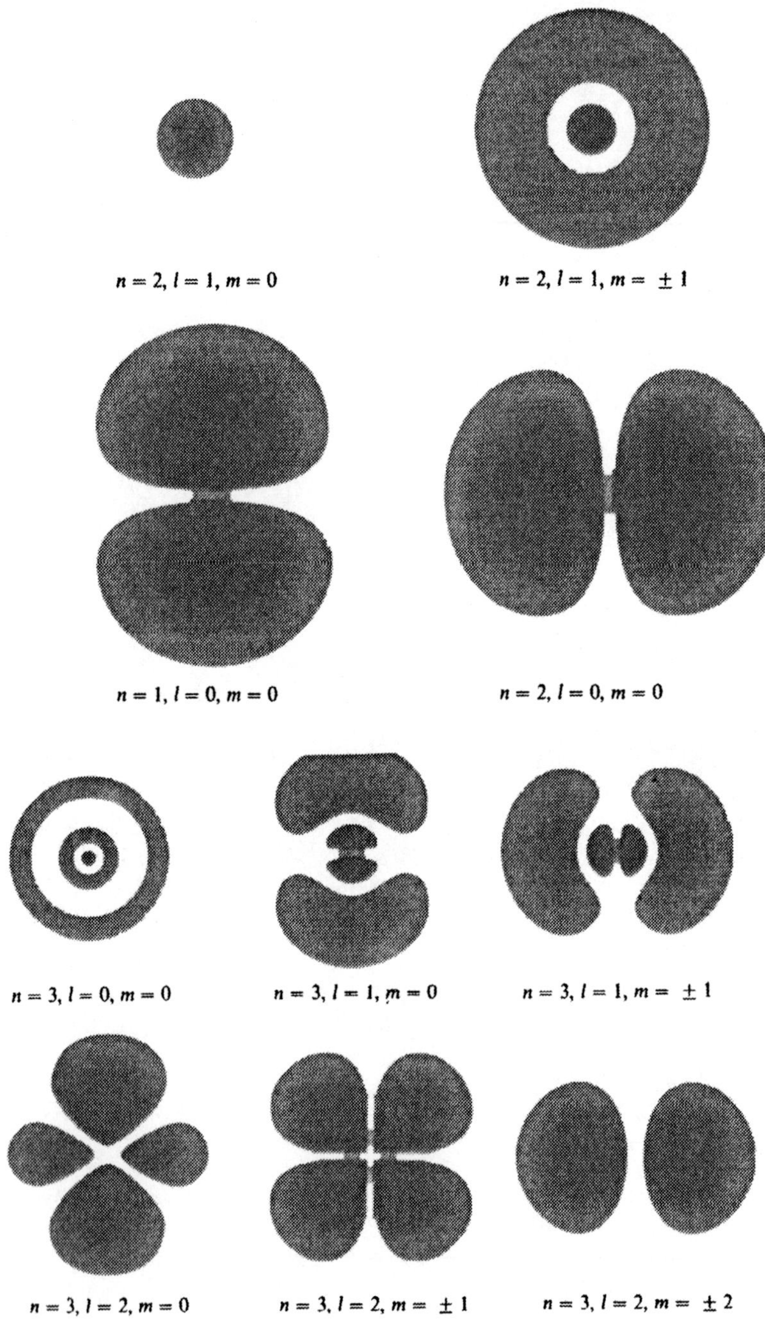

Fig. 9.2a. Section through the electronic density distribution $|\Psi|^2$ of several states of the hydrogen atom. The density of the hatching corresponds to the probability density of the electrons

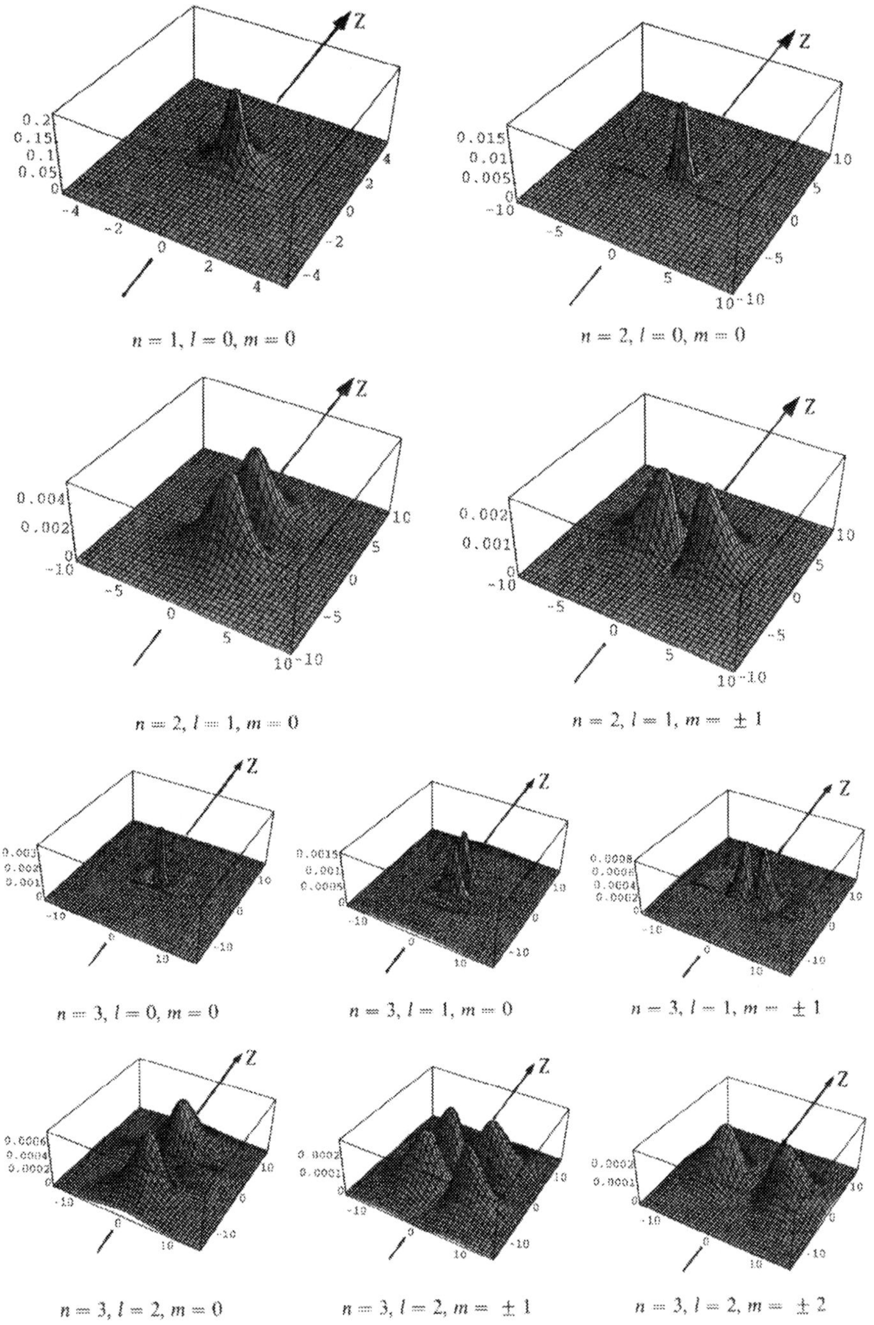

Fig. 9.2b. Three-dimensional plot of the electronic density distribution $|\Psi|^2$ of the lowest states of the hydrogen atom. The radial variables are in units of the Bohr radius (0.53 Å)

of a magnetic field, we will generally have

$$\psi_n = \sum_{l=0}^{n-1} \sum_{m=-l}^{l} a_{nlm} \psi_{nlm} , \tag{9.53}$$

with arbitrary coefficients a_{nlm}. In particular, we can construct states ψ_n containing the ψ_{nlm} with equal probability. Since the wave functions ψ_{nlm} are orthonormalized, the squares of the expansion coefficients are in the latter case the reciprocal of the degeneracy factor

$$|a_{nlm}|^2 = \frac{1}{n^2} . \tag{9.54}$$

This is true, as already indicated, if, for some physical reason (e.g. a magnetic field), none of the components is dominant.

The superposition of all ψ_{nlm} in (9.53) is in fact a spherically symmetric state, which is easy to verify. If a particular direction is selected by external fields, degeneracy ceases, and the electron density is anisotropic (e.g. the Stark and Zeeman effects).

9.4 The Spectrum of Hydrogen Atoms

The energy values of (9.43) characterize the energy levels of the hydrogen atom:

$$E_n = -\frac{me^4}{2\hbar^2}\frac{1}{n^2} = -\frac{1}{2}\frac{e^2}{a_0}\frac{1}{n^2} . \tag{9.55}$$

During the transition of an electron from the level E_n to another level $E_{n'}$, the atom emits a photon of energy

$$\hbar\omega_{nn'} = E_n - E_{n'} \quad \text{(Bohr's frequency condition)} . \tag{9.56}$$

Inserting E_n (or $E_{n'}$) we get

$$\omega_{nn'} = \frac{e^4 m}{2\hbar^3}\left(\frac{1}{n'^2} - \frac{1}{n^2}\right) , \quad n' < n , \tag{9.57}$$

and the frequency

$$\nu_{nn'} = \mathrm{R}\left(\frac{1}{n'^2} - \frac{1}{n^2}\right) , \tag{9.58}$$

where $R = me^4/4\pi\hbar^3 = 3.27 \times 10^{+15}\,s^{-1}$ is the ***Rydberg*** *constant*. The quantity $E_n/\hbar$ is called the *spectral term*. The differences between these terms determine $\omega_{nn'}$.

Figure 9.3 shows the energy-level diagram of the hydrogen atom and the most important transitions. We see that with increasing principal quantum number n, the differences between energy levels decrease, i.e.

$$\lim_{n\to\infty} E_n = 0 \quad \text{and} \quad \lim_{n\to\infty}(E_n - E_{n-1}) = 0 \ . \tag{9.59}$$

If the energies are positive, the values are arbitrarily close together. This continuum describes an ionized atom. The *ionization energy* is the *negative binding energy*.

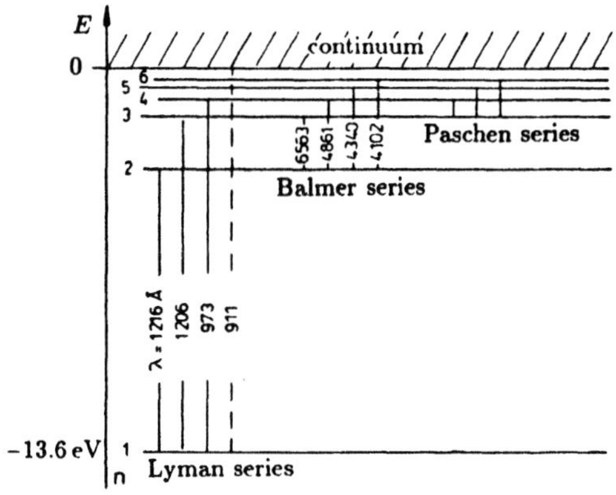

Fig. 9.3. Energy levels and spectral series of the hydrogen atom

All frequencies involved in transitions which end on the same lower state form a *spectral series*. The transitions to the ground states $n' = 1$ constitute the ***Lyman*** *series*. The frequencies are

$$\nu = R\left(\frac{1}{1^2} - \frac{1}{n^2}\right) , \quad n = 2, 3, \ldots \ . \tag{9.60}$$

The transitions to states with $n' = 2, 3, 4$ and 5 comprise the ***Balmer***, ***Ritz-Paschen***, ***Brackett*** and ***Pfund*** series. Recently, hydrogen-like atoms in highly excited levels up to $n = 100$ have been observed; they are called *Rydberg atoms*. Their diameter is around 10^5 times larger than the diameter of the ground state.[2]

[2] See also M.L. Littman et al.: Phys. Rev. **20**, 2251 (1979).

9.5 Currents in the Hydrogen Atom

The operator of the current density j was introduced in Chap. 6 as

$$j = \frac{\mathrm{i}\hbar}{2\mu}(\psi\nabla\psi^* - \psi^*\nabla\psi) \ . \tag{9.61}$$

Here, μ denotes the mass of the electron. This letter is used to distinguish it from the magnetic quantum number m.

The eigenfunction of the hydrogen atom (9.45) is written as

$$\psi_{nlm} = N_{nl}\frac{R_{nl}(r)}{r}P_l^{|m|}(\vartheta)\,\mathrm{e}^{\mathrm{i}m\varphi} \ ,$$

where $R_{nl}(r)$ is the radial part and N_{nl} the normalization constant [see (9.45) and (9.46)]. We use spherical coordinates to facilitate calculation. Then ∇ reads

$$\nabla = \left\{\frac{\partial}{\partial r}, \frac{1}{r}\frac{\partial}{\partial\vartheta}, \frac{1}{r\sin\vartheta}\frac{\partial}{\partial\varphi}\right\} \ . \tag{9.62}$$

The components of the current density are now

$$\begin{aligned}
j_r^{(nlm)} &= \frac{\mathrm{i}\hbar}{2\mu}\left(\psi_{nlm}\frac{\partial}{\partial r}\psi_{nlm}^* - \psi_{nlm}^*\frac{\partial}{\partial r}\psi_{nlm}\right) \ , \\
j_\vartheta^{(nlm)} &= \frac{\mathrm{i}\hbar}{2\mu}\left(\psi_{nlm}\frac{1}{r}\frac{\partial}{\partial\vartheta}\psi_{nlm}^* - \psi_{nlm}^*\frac{1}{r}\frac{\partial}{\partial\vartheta}\psi_{nlm}\right) \ , \\
j_\varphi^{(nlm)} &= \frac{\mathrm{i}\hbar}{2\mu}\left(\psi_{nlm}\frac{1}{r\sin\vartheta}\frac{\partial}{\partial\varphi}\psi_{nlm}^* - \psi_{nlm}^*\frac{1}{r\sin\vartheta}\frac{\partial}{\partial\varphi}\psi_{nlm}\right) \ .
\end{aligned} \tag{9.63}$$

Then we get

$$\begin{aligned}
\psi_{nlm}\frac{\partial}{\partial r}\psi_{nlm}^* &= N_{nl}^2\left(\frac{R_{nl}(r)}{r}P_l^{|m|}(\vartheta)\,\mathrm{e}^{\mathrm{i}m\varphi}\right)\frac{\partial}{\partial r}\left(\frac{R_{nl}^*(r)}{r}P_l^{*|m|}(\vartheta)\,\mathrm{e}^{-\mathrm{i}m\varphi}\right) \\
&= N_{nl}^2\left(P_l^{|m|}(\vartheta)\right)^2\frac{R_{nl}(r)}{r}\frac{\partial}{\partial r}\left(\frac{R_{nl}(r)}{r}\right) = \psi_{nlm}^*\frac{\partial}{\partial r}\psi_{nlm}
\end{aligned} \tag{9.64}$$

as well as

$$\psi_{nlm}\frac{1}{r}\frac{\partial}{\partial\vartheta}\psi_{nlm}^* = \psi_{nlm}^*\frac{1}{r}\frac{\partial}{\partial\vartheta}\psi_{nlm} \ ; \tag{9.65}$$

$R_{nl}(r)$ and $P_l^{|m|}(\vartheta)$ are real functions. It follows immediately that

$$j_r = j_\vartheta = 0 \ . \tag{9.66}$$

This is quite reasonable, because a current in a radial direction would cause the entire charge either to be collected in the nucleus or to be emitted from the atom after a certain period of time.

The only nonvanishing component of the current is the φ component, as this is where the only derivative of a complex part of the function enters, according to the last equation in (9.63).

The current density in the φ direction is

$$j_\varphi = \frac{i\hbar}{2\mu}\left(\psi_{nlm}\frac{1}{r\sin\vartheta}(-im)\psi^*_{nlm} - \psi^*_{nlm}\frac{1}{r\sin\vartheta}(im)\psi_{nlm}\right)$$
$$= \frac{\hbar m}{\mu r\sin\vartheta}|\psi_{nlm}|^2 \ , \quad (9.67)$$

meaning that the azimuthal current is mainly determined by the azimuthal quantum number m, certainly a very reasonable result. The notion of the electron circling the nucleus seems intuitively correct and is based on the Bohr model.

9.6 The Magnetic Moment

If $\mathrm{d}\sigma$ is an area vertical to the current direction (see Fig. 9.4), the current $\mathrm{d}I_\varphi$ passing through this area is

$$\mathrm{d}I_\varphi = j_\varphi \mathrm{d}\sigma \ . \quad (9.68)$$

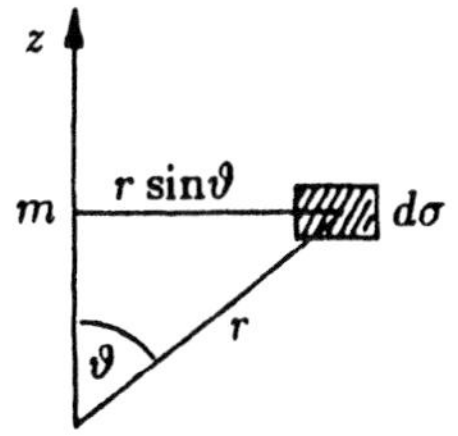

Fig. 9.4. Calculation of the magnetic moment

In standard texts on electrodynamics it is shown that the electric current $\mathrm{d}I$, circling around a plane area F, causes a magnetic moment $\mathrm{d}M = (F/c)\,\mathrm{d}I$.[3] Its z component in an atom is therefore

$$\mathrm{d}M_z = \frac{F}{c}\mathrm{d}I_\varphi = \frac{1}{c}j_\varphi F\,\mathrm{d}\sigma \ . \quad (9.69)$$

We have to multiply the particle current density by the charge $-e$ to get the electric current density, which is now needed.

Since $F = \pi r^2\sin^2\vartheta$, the magnetic moment becomes

$$\mathrm{d}M_z = \frac{-e\hbar m}{cr\sin\vartheta\mu}|\psi_{nlm}|^2\pi r^2\sin^2\vartheta\,\mathrm{d}\sigma \ , \quad (9.70)$$

i.e.

$$\mathrm{d}M_z = -\frac{e\hbar m}{c\mu}|\psi_{nlm}|^2\pi r\sin\vartheta\,\mathrm{d}\sigma \ ,$$

and finally

$$M_z = -\frac{e\hbar m}{2c\mu} \ , \quad (9.71)$$

[3] See J.D. Jackson: *Classical Electrodynamics*, 2nd ed. (Wiley, New York 1975) and W. Greiner: *Classical Electrodynamics* (Springer, New York 1998).

since $dV = 2\pi r \sin\vartheta \, d\sigma$ is the volume of the current element through $d\sigma$ and the integration over the normalized wave function is 1. As there are no supplementary currents in the atom, the magnetic moment is

$$M = M_z = -\mu_B m \ , \tag{9.72}$$

where $\mu_B = -e\hbar/2\mu c$ is the so-called *Bohr magneton*. The absolute value of the maximal magnetic moment is $\mu_B l$; that of the minimal moment is zero.

Noting that the z component of the angular momentum has the value $L_z = m\hbar$, we get the *gyromagnetic factor* (g factor for short) of the electron with respect to the angular momentum. It is defined as the ratio of the absolute value of the magnetic moment $|M_z|$ divided by the angular momentum in units of $\hbar$, i.e.

$$g = \frac{|M_z|/\mu_B}{|L_z|/\hbar} \ , \tag{9.73}$$

and therefore $g = 1$. By definition, the magnetic moment is measured in units of μ_B and the angular momentum is measured in units of $\hbar$, which explains the numerator $|M_z|/\mu_B$ and the denominator $|L_z|/\hbar$ in (9.73).

Since the electron has another angular momentum, the spin, we can define another g factor with respect to the spin. This will be done in Chaps. 12 and 13.

As we see, there indeed exist real electric currents in the atom, similar to those which Bohr assumed to be caused by circling electrons. In quantum mechanics as well, the semi-classical Bohr model yields a vivid picture of the states described in an accurate manner by quantum mechanics.

9.7 Hydrogen-like Atoms

Ions or atoms having only one valence electron in the outermost shell can be described as hydrogen-like atoms. In Chap. 14 we will see that every electron state can be occupied by only one electron because of Pauli's principle. Furthermore, in our case, we have not taken into account up to now the spin in the wave function. As we shall see, the electron spin can take two values: spin up and spin down, with respect to the z axis. Thus a state ψ_{nlm} can be occupied by two electrons, as it should be for complete wave functions, instead of one. In addition, we have to consider the fact that the inner electrons screen the nuclear potential, this effect can be described by an *effective nuclear charge number* Z_{eff}. This number is Z (the number of protons in the nucleus) reduced by the integral over the electron density in a sphere of radius r:

$$
\begin{aligned}
Z_{\text{eff}}(r) &= Z - \frac{4\pi}{e}\int_0^r \varrho r'^2\,\mathrm{d}r' \\
&= Z - 4\pi \times 2\sum_{nlm}\int_0^r |\psi_{nlm}(r')|^2 r'^2\,\mathrm{d}r' \; .
\end{aligned}
\tag{9.74}
$$

Here, ϱ is the total spatial charge density and the sum is extended over all totally occupied shells. In practice, the value of the effective charge number is determined (fitted) by the experiment. This procedure provides a useful possibility for describing the spectra of the alkali atoms.

Recently, with the development of heavy-ion accelerators, which allow the production of, e.g., high-energy uranium ions (up to 1 GeV/nucleon), it has become possible to produce "bare" heavy nuclei. For example, uranium nuclei without any electrons, or with only one or two electrons, have been observed. Clearly, the electron wave function of a 91-fold ionized uranium atom is a hydrogen-type wave function. However, relativistic and quantum-electrodynamical effects become important for these large-Z atoms, and this opens up a new field of research.

An iterative method of calculation is offered by the *Hartree method*. Here, the potential of an electron i is the superposition of the central Coulomb potential $-Ze^2/r_i$ and the potential derived from the remaining electrons. This leads to a stationary Schrödinger equation of the form

$$
\left(\frac{\hat{p}_i^2}{2m} - \frac{Ze^2}{r_i} + 2e^2\sum_{\substack{j=1\\ j\neq i}}^{Z}\int \frac{\psi_j^*\psi_j(\boldsymbol{r}_j)}{|\boldsymbol{r}_i - \boldsymbol{r}_j|}\,\mathrm{d}V_j\right)\psi_i = E_i\psi_i \; , \quad i = 1, 2, \dots, Z \; .
\tag{9.75}
$$

The terms in brackets indicate, respectively, the kinetic energy, the Coulomb interaction energy of the electron i with the nucleus and the Coulomb interaction energies between electron i and all other electrons.

Thus we get Z coupled differential equations for the various wave functions $\psi_i(r_i)$. Moreover, because of the quadratic terms in ψ_j, these equations are nonlinear. They can be iteratively solved, starting with the hydrogen wave functions. The Hartree method does not lead to very accurate results because a quantum-mechanical interaction between two identical particles, the so-called *exchange interaction*, is neglected. This supplementary effect (which we will treat for the two-electron atom in Chap. 15) is considered in the ***Hartree–Fock** method*.

EXERCISE

9.4 The Angular-Dependent Part of the Hydrogen Wave Function

We have already introduced the Schrödinger equation for the hydrogen atom and separated it into a radial and an angular-dependent part (with the separation constant C). The differential equation for the angular-dependent part of the wave function [see (9.27) and (9.29)] is of the form

$$\Delta_{\vartheta,\varphi} Y + CY = 0 \ . \tag{1}$$

We shall now determine the solutions and the corresponding quantum numbers of the differential equation.

Even though we have already treated the spherical harmonics in Example 4.9, it will be educational to derive them once more in a somewhat different way.

The angular-dependent part of the Laplacian is

$$\Delta_{\vartheta,\varphi} = \frac{1}{\sin\vartheta}\frac{\partial}{\partial\vartheta}\left(\sin\vartheta\frac{\partial}{\partial\vartheta}\right) + \frac{1}{\sin^2\vartheta}\frac{\partial^2}{\partial\varphi^2} \ . \tag{2}$$

To solve (1), we insert this Laplacian:

$$\frac{1}{\sin\vartheta}\frac{\partial}{\partial\vartheta}\left(\sin\vartheta\frac{\partial Y}{\partial\vartheta}\right) + \frac{1}{\sin^2\vartheta}\frac{\partial^2 Y}{\partial\varphi^2} + CY = 0 \ . \tag{3}$$

The variables ϑ and φ are separated as follows:

$$Y(\vartheta,\varphi) = \Theta(\vartheta)\phi(\varphi) \ . \tag{4}$$

Multiplied by $\sin^2\vartheta/(\Theta(\vartheta)\phi(\varphi))$, (3) takes the form

$$\frac{\sin^2\vartheta}{\Theta(\vartheta)\sin\vartheta}\frac{\partial}{\partial\vartheta}\left(\sin\vartheta\frac{\partial\Theta(\vartheta)}{\partial\vartheta}\right) + C\sin^2\vartheta = -\frac{1}{\phi(\varphi)}\frac{\partial^2\phi(\varphi)}{\partial\varphi^2} \ .$$

The left-hand side of this equation depends only on ϑ; the right-hand side, only on φ. So we set both sides equal to a constant K and obtain the different equations

$$\frac{1}{\sin\vartheta}\frac{\mathrm{d}}{\mathrm{d}\vartheta}\left(\sin\vartheta\frac{\mathrm{d}\Theta(\vartheta)}{\mathrm{d}\vartheta}\right) + C\Theta(\vartheta) - K\frac{\Theta(\vartheta)}{\sin^2\vartheta} = 0 \ , \tag{5}$$

$$\frac{\mathrm{d}^2\phi(\varphi)}{c\varphi^2} + K\phi(\varphi) = 0 \ . \tag{6}$$

The solution of (6) is

$$\phi(\varphi) = \mathrm{e}^{\pm\mathrm{i}\sqrt{K}\varphi} \ .$$

We require that the wave function be single valued. This means

$$\mathrm{e}^{\pm\mathrm{i}\sqrt{K}\varphi} = \mathrm{e}^{\pm\mathrm{i}\sqrt{K}(\varphi+2\pi)} = \mathrm{e}^{\pm\mathrm{i}\sqrt{K}\varphi\pm2\mathrm{i}\sqrt{K}\pi} \ .$$

Exercise 9.4

From this equation we get $K = m^2$ and $m = 0, 1, 2, 3, \dots$, where m is the magnetic quantum number (quantum number of the magnetic moment), as we already know. Thus the integer values of m follow from the uniqueness of the wave function.

With the substitution

$$t = \cos\vartheta \;, \quad \sin\vartheta = \sqrt{1-t^2} \;,$$
$$\mathrm{d}\vartheta = -\frac{\mathrm{d}t}{\sqrt{1-t^2}} \;,$$

(5) takes the form

$$\frac{\mathrm{d}}{\mathrm{d}t}\left[(1-t^2)\frac{\mathrm{d}\Theta}{\mathrm{d}t}\right] + \left(C - \frac{m^2}{1-t^2}\right)\Theta = 0 \;. \tag{7}$$

To solve (7), we try

$$\Theta = (1-t^2)^{m/2} v_m(t) \;.$$

The equation for $v_m(t)$ then reads

$$\frac{\mathrm{d}}{\mathrm{d}t}\left\{(1-t^2)\frac{\mathrm{d}}{\mathrm{d}t}[(1-t^2)^{m/2} v_m(t)]\right\} + \left(C - \frac{m^2}{1-t^2}\right)(1-t^2)^{m/2} v_m(t) = 0 \;,$$

from which we obtain, after performing the differentiation and reordering,

$$(1-t^2)v_m''(t) - 2(m+1)t v_m'(t) + [C - m(m+1)]v_m(t) = 0 \;. \tag{8}$$

Differentiation of (8) yields the same differential equation for $v_m'(t)$, with the coefficients m replaced by $m+1$:

$$(1-t^2)(v_m')'' - 2(m+2)t(v_m')' + [C - (m+1)(m+2)](v_m') = 0 \;.$$

Thus solutions $v_m' = v_{m+1}$ are possible or, represented by a function v_0,

$$v_m(t) = \frac{\mathrm{d}^m v_0(t)}{\mathrm{d}t^m} \;. \tag{9}$$

In the case $m = 0$, the differential equation (8) reads:

$$(1-t^2)v_0'' - 2t v_0' + C v_0 = 0 \;; \tag{10}$$

this is *Legendre's differential equation* (see Example 4.8). We try to solve it in the vicinity of $t = 0$ by inserting a power series

$$v_0(t) = a_0 + a_1 t + a_2 t^2 + a_3 t^3 + \dots \;, \tag{11}$$

where $v_0(t=0) = a_0$ and $v_0'(t=0) = a_1$.

Exercise 9.4

To determine the coefficients of the series, we differentiate this twice, term by term:

$$v_0' = a_1 + 2a_2t + 3a_3t^2 + 4a_4t^3 + \dots \ , \tag{12}$$

$$v_0'' = 2a_2 + 3\times 2a_3t + 4\times 3a_4t^2 + \dots \ . \tag{13}$$

When we insert (11), (12) and (13) into Legendre's differential equation (10), we find

$$\sum_{\nu=2}^{\infty}\nu(\nu-1)a_\nu t^{\nu-2} - \sum_{\nu=2}^{\infty}\nu(\nu-1)a_\nu t^\nu - \sum_{\nu=1}^{\infty}2\nu a_\nu t^\nu + C\sum_{\nu=0}^{\infty}a_\nu t^\nu = 0 \ . \tag{14}$$

Since the factor $\nu(\nu-1)$ vanishes for $\nu=0$ and $\nu=1$, we can change the summation and write

$$\sum_{\nu=2}^{\infty}\nu(\nu-1)a_\nu t^\nu = \sum_{\nu=0}^{\infty}\nu(\nu-1)a_\nu t^\nu \ ,$$

or

$$\sum_{\nu=1}^{\infty}2\nu a_\nu t^\nu = \sum_{\nu=0}^{\infty}2\nu a_\nu t^\nu \ .$$

This simplifies (14) to

$$\sum_{\nu=2}^{\infty}\nu(\nu-1)a_\nu t^{\nu-2} = \sum_{\nu=0}^{\infty}[\nu(\nu+1)-C]a_\nu t^\nu \ .$$

To compare the coefficients of the power t^l, we have to set $\nu = l+2$ on the left-hand side and $\nu = l$ on the right-hand side. We then obtain

$$(l+2)(l+1)a_{l+2} = [l(l+1)-C]a_l \ . \tag{15}$$

With the recursion formula (15) we are able to evaluate all coefficients from a_0 and a_1 because

$$a_{l+2} = \frac{l(l+1)-C}{(l+1)(l+2)}a_l \quad (l\geq 0) \ .$$

By successive insertion of this equation, it can be shown that the coefficients satisfy the general relations

$$a_{2k} = (-1)^k C(C-2\times 3)\dots[C-(2k-2)(2k-1)]\frac{a_0}{(2k)!} \ ,$$

$$a_{2k+1} = (-1)^k (C-1\times 2)\dots[C-(2k-1)(2k)]\frac{a_1}{(2k+1)!} \ .$$

The complete solution of the Legendre differential equation is then given by the sum of the two power series:

Exercise 9.4

$$
\begin{aligned}
v_0(t) = a_0 \Big\{ & 1 - C\frac{t^2}{2!} + C(C-2\times 3)\frac{t^4}{4!} - C(C-2\times 3)(C-4\times 5)\frac{t^6}{6!} + \dots \\
& + (-1)^k C(C-2\times 3)\dots[C-(2k-2)(2k-1)]\frac{t^{2k}}{(2k)!} + \dots \Big\} \\
+ a_1 \Big\{ & t - (C-1\times 2)\frac{t^3}{3!} + (C-1\times 2)(C-3\times 4)\frac{t^5}{5!} + \dots \\
& + (-1)^k (C-1\times 2)(C-3\times 4)\dots[C-(2k-1)2k]\,\frac{t^{2k+1}}{(2k+1)!} \\
& + \dots \Big\} \,.
\end{aligned}
$$

Each of these series diverges if it does not terminate at a certain point. The series can be forced to converge by setting either $a_0 = 0$ and $a_1 \neq 0$ or $a_0 \neq 0$ and $a_1 = 0$.

Furthermore, we choose $C = l(l+1)$, where $l = 0, 2, 4, \dots$ in the first case, and $l = 1, 3, 5, \dots$ in the second case. Then only a finite number of coefficients

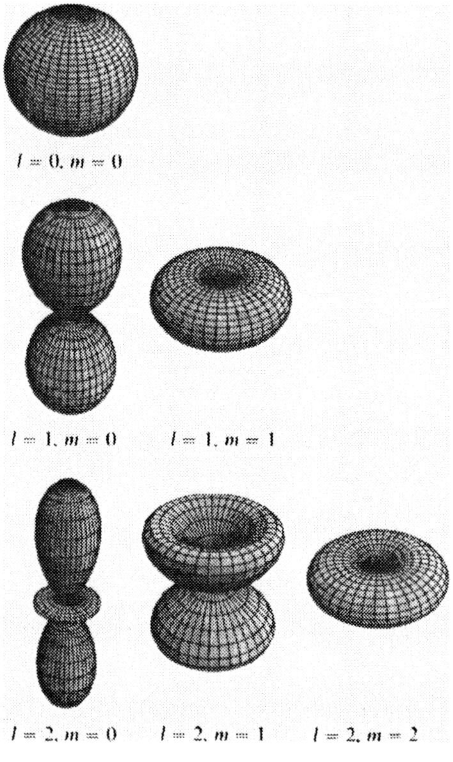

The spherical harmonics $Y_{lm}(\theta, \phi)$ for the lowest values of l and m. We have plotted the function $|Y_{lm}(\theta, \phi)|^2$ in spherical coordinates. For a given direction of θ and ϕ in the coordinate system, the distance of the surface to the origin equals the squared absolute value of the spherical harmonics

Exercise 9.4

are nonzero, and the power series converges, i.e. we obtain polynomials. These polynomials are the only solutions which are regular at $|t| = 1$, and which can be considered as solutions of the physical problem. They are known as *Legendre polynomials* $P_l(t)$ $(l = 0, 1, 2, \dots)$ (see also Example 4.8). They are normalized so that $P_l(1) = 1$, and they satisfy the following orthogonality relation:

$$\int_{-1}^{+1} P_l(t) P_{l'}(t)\,\mathrm{d}t = \frac{2}{2l+1}\delta_{ll'} \ .$$

Reversing the various substitutions, the complete solutions of (1) read

$$Y(\vartheta, \varphi) = Y_{lm}(\vartheta, \varphi) \equiv \mathrm{e}^{\pm\mathrm{i}m\varphi} \sin^m \vartheta \frac{\mathrm{d}^m P_l(\cos\vartheta)}{\mathrm{d}(\cos\vartheta)^m} \ .$$

We see, again, that the angular-dependent part of the hydrogen wave function is represented by the spherical harmonics Y_{lm}.

EXAMPLE

9.5 Spectrum of a Diatomic Molecule

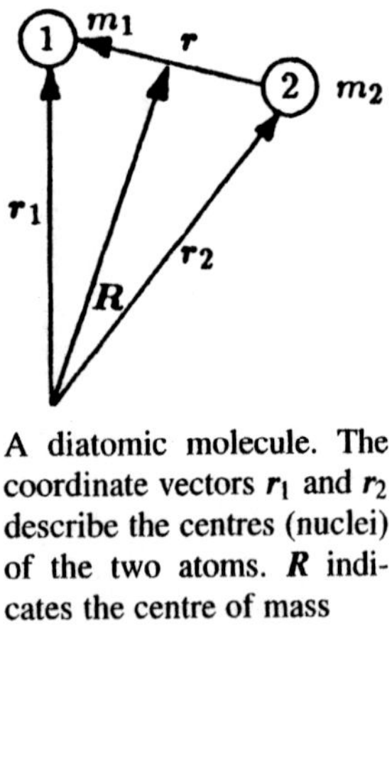

A diatomic molecule. The coordinate vectors $\boldsymbol{r}_1$ and $\boldsymbol{r}_2$ describe the centres (nuclei) of the two atoms. $\boldsymbol{R}$ indicates the centre of mass

With the techniques we have developed so far, we now want to determine the spectrum of a diatomic molecule in a qualitative way. The potential between the atoms is assumed to be local and not explicitly time dependent, it is given as a function of the distance between the atoms (see figure),

$$V = V(\boldsymbol{r}_1, \boldsymbol{r}_2).$$

The Laplacian appearing in the Schrödinger equation has to be applied to the coordinates of both atoms:

$$\Delta = \Delta_1 + \Delta_2 \ , \quad \text{with}$$

$$\Delta_1 = \frac{\partial^2}{\partial x_1^2} + \frac{\partial^2}{\partial y_1^2} + \frac{\partial^2}{\partial z_1^2} \quad \text{and}$$

$$\Delta_2 = \frac{\partial^2}{\partial x_2^2} + \frac{\partial^2}{\partial y_2^2} + \frac{\partial^2}{\partial z_2^2} \ .$$

Hence, the stationary Schrödinger equation becomes

$$\left(\frac{-\hbar^2}{2m_1}\Delta_1 + \frac{-\hbar^2}{2m_2}\Delta_2\right)\psi(\boldsymbol{r}_1, \boldsymbol{r}_2) + V(\boldsymbol{r}_1, \boldsymbol{r}_2)\psi(\boldsymbol{r}_1, \boldsymbol{r}_2) = E\psi(\boldsymbol{r}_1, \boldsymbol{r}_2) \ . \tag{1}$$

Example 9.5

Introducing the centre-of-mass coordinate $\boldsymbol{R}$ and the relative coordinate $\boldsymbol{r}$, the two-body problem can be reduced to an equivalent one-body problem. The following relations hold:

$$\begin{aligned} M\boldsymbol{R} &= m_1\boldsymbol{r}_1 + m_2\boldsymbol{r}_2 \quad \text{with the total mass} \\ M &= m_1 + m_2 \ , \quad \text{and} \qquad (2) \\ \boldsymbol{r} &= \boldsymbol{r}_1 - \boldsymbol{r}_2 \ . \qquad (3) \end{aligned}$$

We also have to express the Laplacian in terms of the new coordinates. Taking the x coordinate as an example, we obtain, with the definitions (2) and (3),

$$X = \frac{m_1x_1 + m_2x_2}{M} \ , \quad x = x_1 - x_2 \ .$$

Thus, for the derivatives with respect to x_1 and x_2, we get

$$\begin{aligned} \frac{\partial}{\partial x_1} &= \frac{m_1}{M}\frac{\partial}{\partial X} + \frac{\partial}{\partial x} \quad \text{and} \\ \frac{\partial}{\partial x_2} &= \frac{m_2}{M}\frac{\partial}{\partial X} - \frac{\partial}{\partial x} \ . \end{aligned}$$

Hence,

$$-\frac{\hbar^2}{2m_1}\frac{\partial^2}{\partial x_1^2} - \frac{\hbar^2}{2m_2}\frac{\partial^2}{\partial x_2^2} = -\frac{\hbar^2}{2M}\frac{\partial^2}{\partial X^2} - \frac{\hbar^2}{2\mu}\frac{\partial^2}{\partial x^2} \ ,$$

where $1/\mu = 1/m_1 + 1/m_2$ is the *reduced mass*. Analogous results follow for the other components, and the Schrödinger equation (1) takes the form

$$-\frac{\hbar^2}{2M}\Delta_R\psi(\boldsymbol{r}, \boldsymbol{R}) - \frac{\hbar^2}{2\mu}\Delta_r\psi(\boldsymbol{r}, \boldsymbol{R}) + V(r)\psi(\boldsymbol{r}, \boldsymbol{R}) = E\psi(\boldsymbol{r}, \boldsymbol{R}) \ .$$

Using $\psi(\boldsymbol{r}, \boldsymbol{R}) = f(\boldsymbol{r})F(\boldsymbol{R})$ and splitting up the energy into $E = E_r + E_R$, we separate the differential equation into the *centre-of-mass motion*

$$-\frac{\hbar^2}{2M}\Delta_R F(\boldsymbol{R}) = E_R F(\boldsymbol{R}) \ , \qquad (4)$$

and the *relative motion*

$$-\frac{\hbar^2}{2\mu}\Delta_r f(\boldsymbol{r}) + V(r)f(\boldsymbol{r}) = E_r f(\boldsymbol{r}) \ . \qquad (5)$$

Equation (4) no longer contains the potential; the motion of the centre of mass is free and described by a plane wave:

$$F(\boldsymbol{R}) = C\exp\left(-\frac{\mathrm{i}}{\hbar}\boldsymbol{P}\cdot\boldsymbol{R}\right) \ ,$$

where $\boldsymbol{P}^2 = 2ME_R$. This is quite reasonable, since we expect the molecule to move, as a whole, freely in space.

Example 9.5

In the equation for the relative motion we perform the usual separation of variables for a central potential:

$$f(\boldsymbol{r}) = f(r, \vartheta, \varphi) = \frac{R(r)}{r} Y_{lm}(\vartheta, \varphi) \ .$$

This leads to the radial equation [see e.g. (9.28)]

$$-\frac{\hbar^2}{2\mu}\frac{\partial^2 R}{\partial r^2} + W_l(r) = E_R R \ , \tag{6}$$

with the *effective potential*

$$W_l(r) = V(r) + \frac{\hbar^2}{2\mu}\frac{l(l+1)}{r^2} \ , \tag{7}$$

which is the sum of the true potential $V(r)$ and the rotational energy $L^2/2\mu r^2$, as in classical mechanics. To get a qualitative survey of the energy eigenvalues E_n which constitute the spectrum of the molecule, we assume an acceptable form for the potential.

As illustrated by the upper figure, the potential should be repulsive if the two atoms are too close together. At a value r_0, it should have a minimum and for large distances r, it should be attractive and tend towards zero. The repulsion for values $r < r_0$ is produced by the practically "naked" (i.e. without electrons) nuclei facing each other. The minimum at $r = r_0$ is caused by the molecular electrons, which move around both centres. [We shall explain this later on in greater detail (see hydrogen molecule in Example 15.2).]

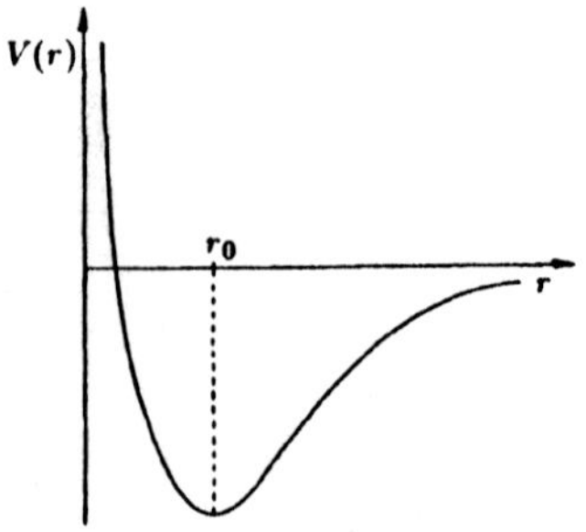

Qualitative form of the potential $V(r)$ between two atoms

If the molecule has an angular momentum, the repulsive centrifugal potential has to be added. Therefore the minimum becomes less marked and shifts towards larger distances, as shown in the lower figure.

In the case of small oscillations, the energy eigenvalues can be calculated by replacing the potential about the minimum by a parabola. Since the location of the minimum depends on the angular momentum, we call it r_l. Now we expand $W_l(r)$ around the point r_l:

$$W_l(r) = W_l(r_l) + \left.\frac{\mathrm{d}W_l(r)}{\mathrm{d}r}\right|_{r=r_l}(r-r_l) + \frac{1}{2}\left.\frac{\mathrm{d}^2 W_l(r)}{\mathrm{d}r^2}\right|_{r=r_l}(r-r_l)^2 + \dots \ . \tag{8}$$

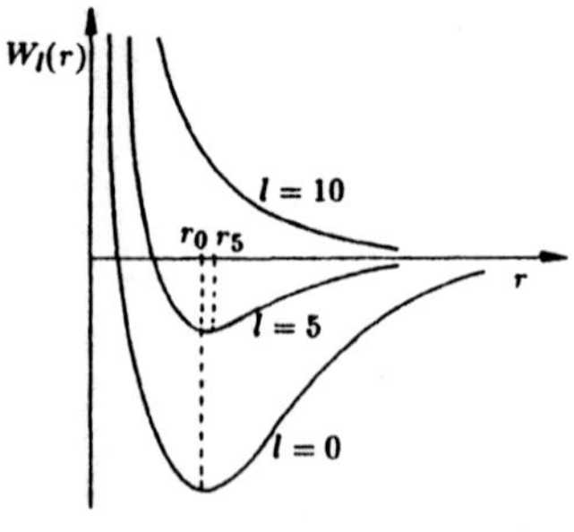

Qualitative behaviour of the effective potential $W_l(r)$ between two nuclei. r_0, r_5 (or in general r_l) indicate the positions of the minima of the l-dependent potential

The higher terms are neglected, because we consider small oscillations about the equilibrium position $|r - r_l| \ll r_l$ only. The second derivative can be written as

$$\frac{\mathrm{d}^2 W_l(r_l)}{\mathrm{d}r^2} \equiv \mu\omega_l^2 \ . \tag{9}$$

This gives us approximately a parabolic potential. Since the first derivative vanishes at the point of equilibrium $r = r_l$, (8) becomes, using the abbreviation $x = r - r_l$,

$$W_l(r) = V(r_l) + \frac{\hbar^2 l(l+1)}{2\mu r_l^2} + \frac{1}{2}\mu\omega_l^2 x^2 \ .$$

Example 9.5

With the moment of inertia $\Theta_l = \mu r_l^2$, the Schrödinger equation (6) changes into

$$-\frac{\hbar^2}{2\mu}\frac{d^2R}{dx^2}+\left[V(r_l)+\frac{\hbar^2 l(l+1)}{2\Theta_l}+\frac{1}{2}\mu\omega_l^2 x^2\right]R = E_r R \ .$$

Accordingly, our approximation leads to the linear harmonic-oscillator equation. This can be easily seen by substituting

$$E' = E - V(r_l) - \frac{\hbar^2 l(l+1)}{2\Theta_l} \ ,$$

which yields

$$-\frac{\hbar^2}{2\mu}\frac{d^2R}{dx^2}+\frac{1}{2}\mu\omega_l^2 x^2 R = E'R \ .$$

As we already know, the eigenvalues of the linear harmonic oscillator are given by (see Chap. 7)

$$E_n' = \hbar\omega_l\left(n+\frac{1}{2}\right) \ , \quad n = 0, 1, 2, \dots \ ,$$

and thus the whole spectrum results as

$$E = E_{nl} = V(r_l) + \hbar\omega_l\left(n+\frac{1}{2}\right)+\frac{\hbar^2 l(l+1)}{2\Theta_l} \ . \tag{10}$$

The energy obviously consists of a *rotational* part $\hbar^2 l(l+1)/2\Theta_l$, and a *vibrational* part $\hbar\omega_l(n+1/2)$. In addition, the vibrational frequency ω_l is determined by the rotation; ω_l depends on l [see (9)]. Because of our approximation, the solution (10) is valid for small quantum numbers n and l only.

The rotations are observed in the far part and the vibrations in the nearer part of the infra-red spectrum. This means that the level density of the rotation at a given vibration energy exceeds the density of levels for different quantum numbers n. In other words, the rotational states can be classified according to the vibration states. The levels of vibration are equidistant, which we see from (10), while the ratios between the rotation levels are $1:2:3:4:\dots$, if the moment of inertia Θ_l remains constant. The scheme of a rotation–vibration spectrum is shown in the following picture:

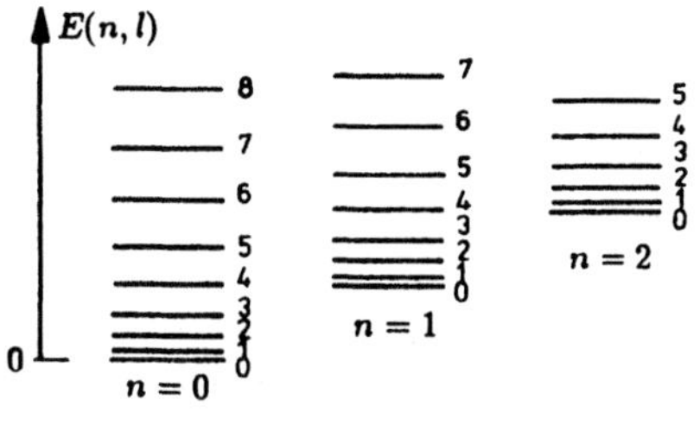

Rotation–vibration spectrum of a diatomic molecule

Example 9.5

Such rotation–vibration spectra also exist for nuclei and probably even for elementary particles; for such problems we refer to the literature.[4]

EXAMPLE

9.6 Jacobi Coordinates

The Jacobi coordinates are a generalization of the relative and centre-of-mass coordinates used in Example 9.5. The latter are suitable for describing a two-body system. But what does the treatment of an N-body problem look like? Jacobi coordinates give the answer.

First we take the particles 1 and 2 and treat them in the usual way:

$$\xi_1 = \frac{m_1 x_1}{m_1} - x_2 = x_1 - x_2 \ ,$$
$$\eta_1 = \frac{m_1 y_1}{m_1} - y_2 = y_1 - y_2 \ ,$$
$$\zeta_1 = \frac{m_1 z_1}{m_1} - z_2 = z_1 - z_2 \ . \tag{1}$$

The first Jacobian vector $\boldsymbol{\xi}_1 = \{\xi_1, \eta_1, \zeta_1\}$ is the relative vector between particle 1 and particle 2. The second Jacobian coordinate is now defined as the relative vector between the centre of mass of the first two particles and the third one. The third Jacobian vector connects the fourth particle with the centre of mass of the first three particles and so on (cf. Sect. 14.2). Therefore we have

$$\begin{aligned}
\xi_1 &= \frac{m_1 x_1}{m_1} - x_2 \ , \\
\xi_2 &= \frac{m_1 x_1 + m_2 x_2}{m_1 + m_2} - x_3 \ , \\
&\ \vdots \\
\xi_j &= \frac{\sum_{k=1}^{j} m_k x_k}{\sum_{k=1}^{j} m_k} - x_{j+1} \ , \\
&\ \vdots \\
\xi_N &= \frac{1}{M} \sum_{k=1}^{N} m_k x_k \equiv X \ ,
\end{aligned} \tag{2}$$

and analogous equations are valid for the η_i and the ζ_i components. This is illustrated in the figure below.

[4] See, e.g., J.M. Eisenberg and W. Greiner: *Nuclear Theory 1, Nuclear Models*, 3rd ed. (North-Holland, Amsterdam 1987).

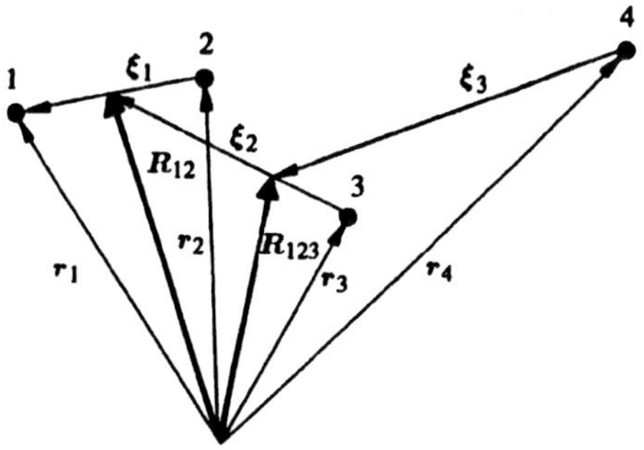

The Jacobi coordinates for four particles, numbered 1, 2, 3, 4. $\boldsymbol{r}$ are their position vectors. $\boldsymbol{R}_{12}$, $\boldsymbol{R}_{123}$ are the vectors for the centre of mass of the first two and three particles, respectively. The Jacobi-coordinate vectors ξ_i point from the particle $i+1$ to the centre of mass of the first i particles

Evidently, $\boldsymbol{\xi}_N$ is the centre-of-mass vector of the whole system. Now, we want to transform the kinetic energy operator

$$\hat{T} = \frac{-\hbar^2}{2}\sum_{k=1}^{N}\frac{1}{m_k}\left(\frac{\partial^2}{\partial x_k^2}+\frac{\partial^2}{\partial y_k^2}+\frac{\partial^2}{\partial z_k^2}\right) \tag{3}$$

to Jacobi coordinates. First we note that from the transformation formulas (2) it follows that

$$\frac{\partial \xi_j}{\partial x_k} = \frac{m_k}{M_j}\ , \quad k \le j\ ; \quad \frac{\partial \xi_j}{\partial x_k} = -1\ , \quad k = j+1\ ;$$
$$\frac{\partial \xi_j}{\partial x_k} = 0\ , \quad k > j+1\ , \qquad \text{where} \tag{4}$$

$$M_j = \sum_{k=1}^{j} m_k \tag{5}$$

is the total mass of the first j particles. With the help of (4) and (5) we find:

$$\sum_{k=1}^{N}\frac{\partial\psi}{\partial x_k} = \sum_{k=1}^{N}\sum_{j=1}^{N}\frac{\partial\psi}{\partial\xi_j}\frac{\partial\xi_j}{\partial x_k} = \sum_{j=1}^{N}\frac{\partial\psi}{\partial\xi_j}\sum_{k=1}^{N}\frac{\partial\xi_j}{\partial x_k}$$
$$= \sum_{j=1}^{N}\frac{\partial\psi}{\partial\xi_j}\left(\sum_{k=1}^{j}\frac{m_k}{M_j}+\frac{\partial\xi_j}{\partial x_{j+1}}\right) = \frac{\partial\psi}{\partial\xi_N}\ . \tag{6}$$

Note that $(\sum_{k=1}^{j} m_k/M_j - 1) = 0$ for $j < N$. Only in the case $j = N$ does the expression in parentheses become nonzero and attain the value 1.

The kinetic-energy operator is calculated in an analogous way. It is sufficient to evaluate the operator:

$$\hat{D}_x\psi = \sum_{k=1}^{N}\frac{1}{m_k}\frac{\partial^2\psi}{\partial x_k^2} = \sum_{k=1}^{N}\frac{1}{m_k}\sum_{j=1}^{N}\sum_{j'=1}^{N}\frac{\partial^2\psi}{\partial\xi_j\partial\xi_{j'}}\frac{\partial\xi_j}{\partial x_k}\frac{\partial\xi_{j'}}{\partial x_k}\ . \tag{7}$$

Example 9.6

Then we get, by using (4) and (5),

$$
\begin{aligned}
\hat{D}_z\psi = &\sum_{k=1}^{N}\frac{1}{m_k}\left(\sum_{j=k}^{N}\sum_{j'=k}^{N}\frac{m_k^2}{M_jM_{j'}}\frac{\partial^2\psi}{\partial\xi_j\partial\xi_{j'}}\right)\\
&-2\sum_{k=2}^{N}\sum_{j=k}^{N}\frac{1}{m_k}\frac{m_k}{M_j}\frac{\partial^2\psi}{\partial\xi_j\partial\xi_{k-1}}+\sum_{k=2}^{N}\frac{1}{m_k}\frac{\partial^2\psi}{\partial\xi_{k-1}^2}\\
= &\sum_{k=1}^{N}\frac{1}{m_k}\left(2\sum_{j'=k}^{N}\sum_{j>j'}^{N}\frac{m_k^2}{M_jM_{j'}}\frac{\partial^2\psi}{\partial\xi_j\partial\xi_{j'}}-2\sum_{j=k}^{N}\frac{m_k}{M_j}\frac{\partial^2\psi}{\partial\xi_j\partial\xi_{k-1}}\right)\\
&+\sum_{k=1}^{N}\frac{1}{m_k}\left(\sum_{j=k}^{N}\frac{m_k^2}{M_j^2}\frac{\partial^2\psi}{\partial\xi_j^2}+\frac{\partial^2\psi}{\partial\xi_{k-1}^2}\right)\\
=&2\left(\sum_{k=1}^{N}\sum_{j'=k}^{N}\sum_{j>j'}^{N}\frac{m_k}{M_jM_{j'}}\frac{\partial^2\psi}{\partial\xi_j\partial\xi_{j'}}-\sum_{k=1}^{N}\sum_{j=k}^{N}\frac{1}{M_j}\frac{\partial^2\psi}{\partial\xi_j\partial\xi_{k-1}}\right)\\
&+\sum_{k=1}^{N}\left(\sum_{j=k}^{N}\frac{m_k}{M_j^2}\frac{\partial^2\psi}{\partial\xi_j^2}+\frac{1}{m_k}\frac{\partial^2\psi}{\partial\xi_{k-1}^2}\right)\\
\overset{(*)}{=}&2\left(\sum_{j'=1}^{N}\sum_{j>j'}^{N}\sum_{k=1}^{j'}\frac{m_k}{M_jM_{j'}}\frac{\partial^2\psi}{\partial\xi_j\partial\xi_{j'}}-\sum_{j=1}^{N}\sum_{k=1}^{j}\frac{1}{M_j}\frac{\partial^2\psi}{\partial\xi_j\partial\xi_{k-1}}\right)\\
&+\sum_{k=1}^{N}\left(\sum_{j=k}^{N}\frac{m_k}{M_j^2}\frac{\partial^2\psi}{\partial\xi_j^2}+\frac{1}{m_k}\frac{\partial^2\psi}{\partial\xi_{k-1}^2}\right)\\
=&2\left(\sum_{j'=1}^{N}\sum_{j>j'}^{N}\frac{1}{M_j}\frac{\partial^2\psi}{\partial\xi_j\partial\xi_{j'}}-\sum_{j=1}^{N}\sum_{k'=0}^{j-1}\frac{1}{M_j}\frac{\partial^2\psi}{\partial\xi_j\partial\xi_{k'}}\right)\qquad\text{[because of (5)]}\\
&+\sum_{k=1}^{N}\left(\sum_{j=k}^{N}\frac{m_k}{M_j^2}\frac{\partial^2\psi}{\partial\xi_j^2}+\frac{1}{m_k}\frac{\partial^2\psi}{\partial\xi_{k-1}^2}\right)\\
=&2\left(\sum_{j=1}^{N}\sum_{j'=1}^{j-1}\frac{1}{M_j}\frac{\partial^2\psi}{\partial\xi_j\partial\xi_{j'}}-\sum_{j=1}^{N}\sum_{j'=1}^{j-1}\frac{1}{M_j}\frac{\partial^2\psi}{\partial\xi_j\partial\xi_{j'}}\right)\\
&+\sum_{k=1}^{N}\left(\sum_{j=k}^{N}\frac{m_k}{M_j^2}\frac{\partial^2\psi}{\partial\xi_j^2}+\frac{1}{m_k}\frac{\partial^2\psi}{\partial\xi_{k-1}^2}\right)\\
=&\sum_{k=1}^{N}\left(\sum_{j=k}^{N}\frac{m_k}{M_j^2}\frac{\partial^2\psi}{\partial\xi_j^2}+\frac{1}{m_k}\frac{\partial^2\psi}{\partial\xi_{k-1}^2}\right)\ .
\end{aligned}
\tag{8}
$$

Example 9.6

Thereby we have changed the order of summation over k, j and j' at $(*)$. The last sum transforms into

$$
\begin{aligned}
&\sum_{k=1}^{N}\frac{1}{m_k}\left(\sum_{j=k}^{N}\frac{m_k^2}{M_j^2}\frac{\partial^2\psi}{\partial\xi_j^2}+\frac{\partial^2\psi}{\partial\xi_{k-1}^2}\right)\\
&=\sum_{j=1}^{N}\sum_{k=1}^{j}\frac{m_k}{M_j^2}\frac{\partial^2\psi}{\partial\xi_j^2}+\sum_{k=1}^{N-1}\frac{1}{m_{k+1}}\frac{\partial^2\psi}{\partial\xi_k^2}\\
&=\sum_{j=1}^{N}\frac{1}{M_j}\frac{\partial^2\psi}{\partial\xi_j^2}+\sum_{j=1}^{N-1}\frac{1}{m_{j+1}}\frac{\partial^2\psi}{\partial\xi_j^2}\\
&=\frac{1}{M}\frac{\partial^2\psi}{\partial\xi_N^2}+\sum_{j=1}^{N-1}\left(\frac{1}{M_j}+\frac{1}{m_{j+1}}\right)\frac{\partial^2\psi}{\partial\xi_j^2}\ ,
\end{aligned}
\tag{9}
$$

which means that

$$
\hat{D}_x\psi=\frac{1}{M}\frac{\partial^2\psi}{\partial\xi_N^2}+\sum_{j=1}^{N-1}\frac{1}{\mu_j}\frac{\partial^2\psi}{\partial\xi_j^2}\ ,
\tag{10}
$$

where μ_j is the reduced mass of the centre of mass of the first j particles and the particle with the number $j+1$:

$$
\frac{1}{\mu_j}=\frac{1}{M_j}+\frac{1}{m_{j+1}}\ .
\tag{11}
$$

Taking into account that

$$
\hat{D}\psi=(\hat{D}_x+\hat{D}_y+\hat{D}_z)\psi\ ,
\tag{12}
$$

we get from (10) the relation

$$
\hat{D}\psi=\frac{1}{M}\nabla_N^2\psi+\sum_{j=1}^{N-1}\frac{1}{\mu_j}\nabla_j^2\psi\ .
\tag{13}
$$

This is the kinetic-energy operator (up to the factor $-\hbar^2/2$), expressed in Jacobi coordinates. As expected, the centre-of-mass motion separates. It is described by the first term on the right-hand side in (13).

9.8 Biographical Notes

BOHR, Niels Hendrik David, Danish physicist, *Copenhagen 7.10.1885, † Copenhagen 18.11.1962. A professor from 1916, in 1920 B. became director of the Institute of Theoretical Physics at the University of Copenhagen. In 1913 he succeeded in applying Planck's quantum hypothesis (1900) to Rutherford's planetary atomic model. This Bohr model was the first to explain theoretically the spectral series of hydrogen. B. generalized this model to include the description of other elements and developed a theory of the periodic system of the elements. His *correspondence principle*, which was named after him, established a relation between the classical and the new quantum theory. In 1920 he received the Nobel Prize in physics. Together with the young W. Heisenberg, B. developed the "Copenhagen interpretation" of quantum mechanics in 1927, the now prevalent physical interpretation of quantum-theoretical formalism, based on the Heisenberg uncertainty principle and the well-known duality of particles and waves. Later on, he worked on problems of nuclear and elementary particle physics. From 1933 to 1936 he used the "sandbag model" for describing nuclear reactions of collisions. His interpretation of the nuclear fission of uranium was important for its later technical use. From 1943 to 1945, B. worked on the development of the atomic bomb in Los Alamos.

LANDAU, Lew Dawidowitsch, Soviet physicist, *22.1.1908 Baku, † 1.4.1968 Moscow, director of the Institute for Theoretical Physics of the Soviet Academy of Sciences. L. investigated, in particular, diamagnetism and low-temperature physics. In 1962 he received the Nobel Prize for his explanation of superfluidity especially as it appears in He II.

RYDBERG, Janne (John) Robert, Swedish physicist, *Halmstad 8.11.1854, † Lund 28.12.1919. From 1901 on, a professor in Lund, R. worked on the periodic system of the elements and series spectra. In 1889 he submitted his "Recherches sur la constitution des spectres d'émission des éléments" to the Swedish Academy of Sciences. The *Rydberg constant* and, recently, *Rydberg atoms* were named after him. In 1913 he published his papers "Elektron, der erste Grundstoff" and "Untersuchungen über das System der Grundstoffe".

LYMAN, Theodore, Amer. physicist, *Boston 23.11.1874, † Cambridge (Mass.) 11.10.1954. From 1910 to 1947, L. was director of the Jefferson Physical Laboratory at Harvard University. He was a pioneer in the field of UV spectroscopy, and in 1906 he discovered a series of the hydrogen atom in the UV range, which was named after him.

BALMER, Johann Jakob, Swiss mathematician, *Lausanne (Basel) 1.5.1825, † Basel 12.3.1898. B. taught at the Basel Lady's College and from 1865 until 1890 was also lecturer at the University of Basel. In 1885 he was the first to construct a formula describing those parts of the hydrogen spectrum known at that time (Balmer series). Later he recognized the relation between the constant h of his formula and the Rydberg constant and its significance as the limit of the series.

PASCHEN, Friedrich, German physicist, *Schwerin 2.1.1865, † Potsdam 25.2.1947. P. was professor in Tübingen and in Bonn. In 1924, he became president of the Physikalisch Technische Reichsanstalt Berlin and a professor in Berlin. He constructed very sensitive galvanometers and quadrant electrometers and worked together with

C. Runge, mainly in spectroscopic experiments. In 1908 he extended the Balmer formula to the IR lines of the hydrogen spectrum (*Paschen series*). In 1912/13, together with Back, he discovered the *Paschen-Back effect*: the splitting of lines in strong magnetic fields. The *Paschen Law* (1889) states that the ignition voltage of a discharge in a gas depends only on the distance between the electrodes and on the gas pressure.

BRACKETT, F. P. Brackett, American astronomer, *1865, † 1953; his most important achievement was the discovery of the Brackett-series in the spectrum of the hydrogen atom. It describes the radiation due to eclectron transitions from excited states into the shell with the principal quantum number $n = 4$.

HARTREE, Douglas Rayner, British physicist and mathematician, *Cambridge 27.3.1897, † Cambridge 12.2.1958. From 1929–37, H. was a professor of applied mathematics, and then of theoretical physics, at the University of Manchester; from 1946–58, he was a professor in Cambridge. H. became a member of the Royal Society in 1932. His most important achievement was the development of approximation methods for the calculation of quantum-mechanical wave functions of many-electron systems. Furthermore, he worked on problems of digital calculators, ballistics and atmospheric physics. His main publications were Numerical Analysis (1952) and The Calculation of Atomic Structures (1957).

FOCK, Wladimir Alexandrowitsch, Soviet physicist, *Petersburg 22.12.1898. F. became a professor at the University of Leningrad in 1932. In 1939, he joined the Academy of Sciences of the USSR and collaborated at several institutes. F.'s main field of work has been quantum mechanics and quantum electrodynamics, he is one of the founders of the quantum theory of many-particle systems. In 1926 he set up a relativistic equation for particles without spin in magnetic fields, independent of O. Klein. In 1928, together with M. Born, he demonstrated the validity of the adiabatic principle in quantum mechanics and developed an approximation method for wave equations of many-particle system (the *Hartree-Fock* method) in 1930. From 1932 to 1934, he generalized the Schrödinger equation for systems of variable-particle number in so-called "Fock space". He also worked on the interpretation of quantum theory, general relativity ("Theory of space, time and gravitation" 1955), the theory of elasticity, and the theory of refraction and propagation of radio waves.

10. The Mathematical Foundations of Quantum Mechanics II

10.1 Representation Theory

The state of a particle is completely described by the normalized wave function $\psi(\boldsymbol{r}, t)$, which we have used until now. In the Schrödinger equation,

$$\left(\frac{\hat{\boldsymbol{p}}^2}{2m} + V(\boldsymbol{r})\right) \psi(\boldsymbol{r}, t) = \mathrm{i}\hbar \frac{\partial}{\partial t} \psi(\boldsymbol{r}, t) \ , \tag{10.1}$$

which gives us the evolution in time of the state, we expressed the momentum operator by the differential operator, i.e.

$$\hat{\boldsymbol{p}} = -\mathrm{i}\hbar \nabla \ . \tag{10.2}$$

This representation $\psi(\boldsymbol{r}, t)$ of a particle state is called the *coordinate representation*. Because of Heisenberg's uncertainty principle, the momentum $\boldsymbol{p}$ of a particle is not exactly known if its position $\boldsymbol{r}$ is fixed. According to (3.50), the average momentum is

$$\langle \hat{\boldsymbol{p}} \rangle = \int \psi^*(\boldsymbol{r}, t)(-\mathrm{i}\hbar \nabla)\psi(\boldsymbol{r}, t)\, \mathrm{d}V \ . \tag{10.3}$$

We can extract information about the momentum of a particle from the wave function $\psi(\boldsymbol{r}, t)$ if we expand it in terms of eigenfunctions of the momentum operator; this is simply a Fourier transformation. The Fourier integral reads

$$\begin{aligned} \psi(\boldsymbol{r}, t) &= \frac{1}{(2\pi\hbar)^{3/2}} \int a(\boldsymbol{p}, t) \exp\left(\frac{\mathrm{i}}{\hbar} \boldsymbol{p} \cdot \boldsymbol{r}\right) \mathrm{d}^3 p \\ &= \int a(\boldsymbol{p}, t) \psi_{\boldsymbol{p}}(\boldsymbol{r})\, \mathrm{d}^3 p \ . \end{aligned} \tag{10.4}$$

The integration is extended over all of momentum space; the function $a(\boldsymbol{p}, t)$ is the Fourier transform of $\psi(\boldsymbol{r}, t)$ at time t. The plane waves $\psi_{\boldsymbol{p}}(\boldsymbol{r})$ are eigenfunctions of the momentum (see Example 4.4). Indeed, we have

$$\psi_{\boldsymbol{p}} = \frac{1}{(2\pi\hbar)^{3/2}} \exp\left(\frac{\mathrm{i}}{\hbar} \boldsymbol{p} \cdot \boldsymbol{r}\right) \ , \quad \text{with} \quad \hat{\boldsymbol{p}}\psi_{\boldsymbol{p}} = \frac{\hbar}{\mathrm{i}} \nabla \exp\left(\frac{\mathrm{i}}{\hbar} \boldsymbol{p} \cdot \boldsymbol{r}\right) = \boldsymbol{p}\psi_{\boldsymbol{p}} \ . \tag{10.5}$$

Now, by inspection of (10.4) it becomes evident that the function $a(\boldsymbol{p},t)$ describes the particle state as completely as the function $\psi(\boldsymbol{r},t)$. We call $a(\boldsymbol{p},t)$ the *momentum representation* of the state of the particle. With the reciprocity of the Fourier transformation, it follows from (10.4) that

$$\begin{aligned} a(\boldsymbol{p},t) &= \frac{1}{(2\pi\hbar)^{3/2}} \int \psi(\boldsymbol{r},t) \exp\left(\frac{\mathrm{i}}{\hbar}\boldsymbol{p}\cdot\boldsymbol{r}\right) \mathrm{d}^3r \\ &= \int \psi(\boldsymbol{r},t)\psi_p^*(\boldsymbol{r})\,\mathrm{d}^3r \ . \end{aligned} \tag{10.6}$$

Hence, if $\psi(\boldsymbol{r},t)$ is known, we can construct $a(\boldsymbol{p},t)$ according to (10.6); and, vice versa, if $a(\boldsymbol{p},t)$ is known, we are able to construct $\psi(\boldsymbol{r},t)$ through (10.4). Analogously, the equivalence of the normalization can easily be shown:

$$\int |\psi(\boldsymbol{r},t)|^2\,\mathrm{d}^3r = \int |a(\boldsymbol{p},t)|^2\,\mathrm{d}^3p \ . \tag{10.7}$$

Indeed, (3.41) expresses this fact for particles within a box. The relation corresponding to (10.3) for the average of the position operator reads

$$\langle\hat{\boldsymbol{r}}\rangle = \int a^*(\boldsymbol{p},t)(\mathrm{i}\hbar\nabla_p)a(\boldsymbol{p},t)\,\mathrm{d}^3p \ , \tag{10.8}$$

where $\nabla_p = (\partial/\partial p_x, \partial/\partial p_y, \partial/\partial p_z)$ is the *nabla* or *del operator in momentum space*. Indeed, we can easily calculate with (10.4)

$$\begin{aligned} \langle\boldsymbol{r}\rangle &= \int \psi^*(\boldsymbol{r},t)\,\boldsymbol{r}\,\psi(\boldsymbol{r},t)\,\mathrm{d}^3r \\ &= \int \mathrm{d}^3r\,\mathrm{d}^3p\,\mathrm{d}^3p'\,a^*(\boldsymbol{p},t)\psi_p^*(\boldsymbol{r})\,\boldsymbol{r}\,a(\boldsymbol{p}',t)\psi_{p'}(\boldsymbol{r}) \\ &= \int \mathrm{d}^3p\,\mathrm{d}^3p'\,a^*(\boldsymbol{p},t)a(\boldsymbol{p}',t) \int \mathrm{d}^3r\,\psi_p^*(\boldsymbol{r})\boldsymbol{r}\psi_{p'}(\boldsymbol{r}) \ . \end{aligned} \tag{10.9}$$

Now, by use of the first equation in (10.5), we can replace the vector $\boldsymbol{r}$ in the space integral by

$$\begin{aligned} \int \mathrm{d}^3r\,\psi_p^*(\boldsymbol{r})\,\boldsymbol{r}\,\psi_{p'}(\boldsymbol{r}) &= \int \psi_p^*(\boldsymbol{r})(-\mathrm{i}\hbar\nabla_{p'})\psi_{p'}(\boldsymbol{r})\,\mathrm{d}^3r \\ &= -\mathrm{i}\hbar\nabla_{p'} \int \psi_p^*(\boldsymbol{r})\psi_{p'}(\boldsymbol{r})\,\mathrm{d}^3r \\ &= -\mathrm{i}\hbar\nabla_{p'}\delta^3(\boldsymbol{p}-\boldsymbol{p}') \ , \end{aligned} \tag{10.10}$$

so that (10.9) becomes

$$\begin{aligned} \langle\boldsymbol{r}\rangle &= \int \mathrm{d}^3p\,\mathrm{d}^3p'\,a^*(\boldsymbol{p},t)a(\boldsymbol{p}',t)(-\mathrm{i}\hbar\nabla_{p'})\delta^3(\boldsymbol{p}-\boldsymbol{p}') \\ &= \int \mathrm{d}^3p\,a^*(\boldsymbol{p},t)\Big[a(\boldsymbol{p}',t)(-\mathrm{i}\hbar)\delta^3(\boldsymbol{p}-\boldsymbol{p}')\Big|_{-\infty}^{+\infty} \\ &\quad - \int \mathrm{d}^3p'(-\mathrm{i}\hbar\nabla_{p'})a(\boldsymbol{p}',t)\delta^3(\boldsymbol{p}-\boldsymbol{p}')\Big] \\ &= \int \mathrm{d}^3p\,a^*(\boldsymbol{p},t)(\mathrm{i}\hbar\nabla_p)a(\boldsymbol{p},t) \ . \end{aligned}$$

The function $a(\boldsymbol{p}, t)$ represents the momentum distribution of the particle state $\psi(\boldsymbol{r}, t)$. The absolute square $|a(\boldsymbol{p}, t)|^2$ gives the probability of finding the particle with definite momentum $\boldsymbol{p}$, i.e. with the wave function

$$\psi_p(\boldsymbol{r}) = \frac{1}{(2\pi\hbar)^{3/2}} \exp\left(\frac{\mathrm{i}}{\hbar}\boldsymbol{p}\cdot\boldsymbol{r}\right)$$

in the state $\psi(\boldsymbol{r}, t)$. Hence, $|a(\boldsymbol{p}, t)|^2$ is the probability density in momentum space.

Up to now we have based our considerations on the physical point of view that the coordinate wave function $\psi(\boldsymbol{r}, t)$ of a particle is determined by measuring its spatial distribution. The momentum distribution follows by Fourier transformation. But often in physics we must adopt a reverse approach; for example, in electron scattering experiments, momentum distributions (form factors) are measured. Then the (spatial) charge distribution of a nucleus follows from a Fourier analysis (see, e.g., Example 11.8).

Coordinate representation and momentum representation are equally suited to describing the state of a particle. Equations (10.4) and (10.6) allow transition from one type of representation to the other.

Let us now briefly consider the *energy representation.* For simplicity we assume the particle to have a discrete energy spectrum with eigenvalues $E_1, E_2, \dots, E_n, \dots$ and a corresponding system of orthonormal eigenfunctions $\psi_1, \psi_2, \dots, \psi_n, \dots$. The expansion of the general wave function $\psi(\boldsymbol{r}, t)$ in terms of energy eigenfunctions reads

$$\psi(\boldsymbol{r}, t) = \sum_n a_n(t)\psi_n(\boldsymbol{r}) \ , \tag{10.11}$$

where the index n indicates the energy dependence. We can get the expansion coefficients from (10.11) by multiplying it by ψ_m^* and integrating over the whole space:

$$a_m(t) = \int \psi_m^*(\boldsymbol{r})\psi(\boldsymbol{r}, t)\,\mathrm{d}^3r \ . \tag{10.12}$$

It is clear that the state of the particle is completely determined by the set of a_n, i.e. the energy representation. Indeed, $\psi(\boldsymbol{r}, t)$ and the $a_n(t)$ follow from one another; the transformations are given by (10.11) and (10.12). This is completely analogous to the former situation in which we were able to evaluate $\psi(\boldsymbol{r}, t)$ from $a(\boldsymbol{p}, t)$ with (10.4) or $a(\boldsymbol{p}, t)$ from $\psi(\boldsymbol{r}, t)$ with (10.6).

EXAMPLE

10.1 Momentum Distribution of the Hydrogen Ground State

To apply our formalism we evaluate the momentum distribution of an electron in the ground state of a hydrogen atom. The normalized wave function of this state is given by [$\gamma_n = 1/na_0$; see (9.45), (9.46) and Table 9.1]

$$\psi(\boldsymbol{r}, t) = \frac{\mathrm{e}^{-\mathrm{i}\omega t}}{\sqrt{\pi a_0^3}} \exp\left(-\frac{r}{a_0}\right) , \tag{1}$$

where ω denotes the frequency and a_0, Bohr's radius. The momentum representation is given by

$$a(\boldsymbol{p}, t) = \frac{1}{(2\pi\hbar)^{3/2}} \int \psi(\boldsymbol{r}, t) \exp\left(-\frac{\mathrm{i}}{\hbar}\boldsymbol{p}\cdot\boldsymbol{r}\right) \mathrm{d}V . \tag{2}$$

Inserting (1), we get

$$a(\boldsymbol{p}, t) = \frac{\mathrm{e}^{-\mathrm{i}\omega t}}{\pi^2(2\hbar a_0)^{3/2}} \int_{-\infty}^{\infty} \exp\left(-\frac{r}{a_0}\right) \exp\left(-\frac{\mathrm{i}}{\hbar}\boldsymbol{p}\cdot\boldsymbol{r}\right) \mathrm{d}V . \tag{3}$$

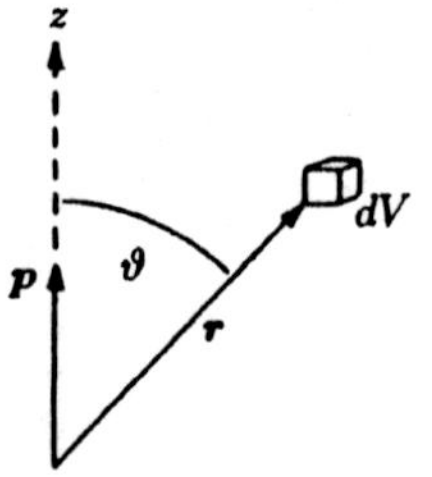

Special choice of the coordinates to evaluate the integral

To simplify our integral we choose the z axis parallel to the momentum and get in spherical coordinates (see figure)

$$a(\boldsymbol{p}, t) = \frac{\mathrm{e}^{-\mathrm{i}\omega t}}{\pi^2(2a_0\hbar)^{3/2}} \int \exp\left(-\frac{r}{a_0}\right) \exp\left(-\frac{\mathrm{i}}{\hbar} pr\cos\vartheta\right) r^2 \mathrm{d}r \sin\vartheta\, \mathrm{d}\vartheta\, \mathrm{d}\varphi ,$$

$$a(\boldsymbol{p}, t) = \frac{\mathrm{e}^{-\mathrm{i}\omega t}}{\pi\sqrt{2a_0^3\hbar^3}} \int_0^{\infty} \left[\exp\left(-\frac{r}{a_0}\right) \int_{-1}^{1} \exp\left(-\frac{\mathrm{i}}{\hbar} pr\cos\vartheta\right) \mathrm{d}\cos\vartheta\right] r^2 \mathrm{d}r . \tag{4}$$

Angular integration yields

$$a(\boldsymbol{p}, t) = \frac{\mathrm{i}\mathrm{e}^{-\mathrm{i}\omega t}}{\pi p\sqrt{2a_0^3\hbar}} \int_0^{\infty} \left\{\exp\left[-r\left(\frac{1}{a_0} + \mathrm{i}\frac{p}{\hbar}\right)\right] - \exp\left[-r\left(\frac{1}{a_0} - \mathrm{i}\frac{p}{\hbar}\right)\right]\right\} r\,\mathrm{d}r , \tag{5}$$

and from this equation, it follows directly that

$$a(\boldsymbol{p}, t) = \frac{1}{\pi}\left(\frac{2a_0}{\hbar}\right)^{3/2} \frac{\mathrm{e}^{-\mathrm{i}\omega t}}{[1 + (p^2/\hbar^2)a_0^2]^2} . \tag{6}$$

The probability density in configuration space is obtained from (1) as

$$|\psi(\boldsymbol{r},t)|^2 = \frac{\exp[-(r/a_0)^2]}{\pi a_0^3} ,$$

while we obtain the density in momentum space from (6) as

$$|a(\boldsymbol{p},t)|^2 = \frac{8a_0^3}{\pi\hbar^3[1+(p^2a_0^2/\hbar^2)]^4} . \tag{7}$$

The form of both densities can be shown as in the figures.

Integration over the momentum density also yields the correct value of one:

$$\int\limits_{-\infty}^{\infty} |a(\boldsymbol{p})|^2 \,\mathrm{d}^3p = \int |a(\boldsymbol{p})|^2 4\pi p^2 \,\mathrm{d}p = \frac{32}{\pi}\int \frac{x^2\,\mathrm{d}x}{(1+x^2)^4} = 1 . \tag{8}$$

Here we have substituted $pa_0/\hbar = x$. The momentum distribution (7) can be verified experimentally by observing photoelectrons in ionization experiments or by measuring inelastic electron scattering. Relation (7) has been confirmed in such experiments.

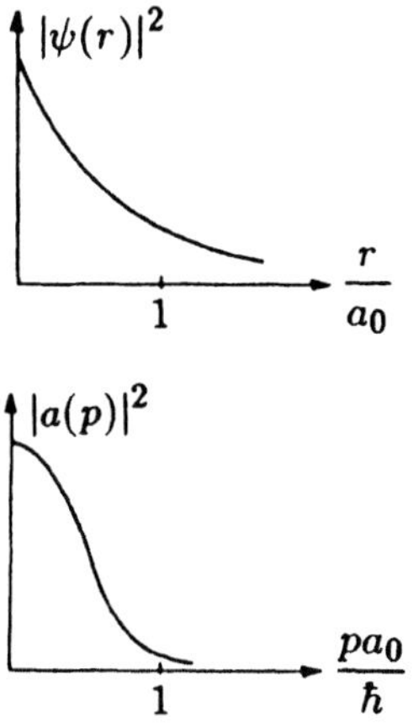

Probability distribution in configuration (*top*) and in momentum space (*bottom*) for the hydrogen ground state

10.2 Representation of Operators

The operator equation

$$\varphi = \hat{L}\psi \tag{10.13}$$

transforms a function ψ to another function φ. For an explicit calculation we have to choose a certain form of representation. Until now we have used the coordinates $\boldsymbol{r}$ to express an operator; i.e. we have been working in *coordinate representation*. In this case, the operator $\hat{L}$ generally takes the form

$$\hat{L} = \hat{L}(\hat{\boldsymbol{p}},\boldsymbol{r}) = \hat{L}\left(\frac{\hbar}{\mathrm{i}}\nabla,\boldsymbol{r}\right) . \tag{10.14}$$

If we change the representation of the wave function, we have to transform the operator accordingly.

Let us consider first the *energy representation*. We expand the wave functions $\psi(\boldsymbol{r})$ and $\varphi(\boldsymbol{r})$ in (10.13) in terms of eigenfunctions of the energy, i.e. of the Hamiltonian ($\hat{H}\psi_n = E_n\psi_n$). Thus we have

$$\psi(\boldsymbol{r}) = \sum_n a_n\psi_n(\boldsymbol{r}) \quad \text{and} \quad \varphi(\boldsymbol{r}) = \sum_n b_n\psi_n(\boldsymbol{r}) . \tag{10.15}$$

The energy representation of the functions ψ and φ is then given, according to the previous section, by the set of coefficients a_n and b_n, respectively.

To get the energy representation of the operator $\hat{L}$, we insert the expansions (10.15) into (10.13), yielding

$$\sum_n b_n \psi_n = \hat{L} \sum_n a_n \psi_n = \sum a_n \hat{L} \psi_n \ .$$

After multiplication by ψ_m^* and integration, it follows from this equation that

$$\sum_n b_n \delta_{mn} = \sum_n a_n \int \psi_m^* \hat{L} \psi_n \, dV \ . \tag{10.16}$$

This suggests introducing the *matrix element*

$$L_{mn} = \int \psi_m^* \hat{L} \psi_n \, dV \tag{10.17}$$

as an abbreviation, so that we can now write for (10.16):

$$b_m = \sum_n L_{mn} a_n \ . \tag{10.18}$$

This equation is the *energy representation* of (10.13). The total set of the L_{mn}, i.e. the matrix L_{mn}, constitutes the *energy representation* of $\hat{L}$. As already indicated, because of the two indices, the L_{mn} are combined into a matrix. Since both indices run over the same set of numbers and because the number of energy eigenvalues is infinite, the matrix L_{mn} is a *quadratic infinite* matrix.

To give an example of an operator with continuous eigenvalues, we shall now compute the *momentum representation* of (10.13). The problem is simplified by considering the one-dimensional case ($\boldsymbol{r} \to x$, $\boldsymbol{p} \to p_x \equiv p$).

Since we are looking for the momentum representation, we expand in terms of momentum eigenfunctions, i.e.

$$\psi_p = \frac{1}{\sqrt{2\pi\hbar}} \exp\left(\frac{\mathrm{i}}{\hbar} px\right) \ ,$$

but now we do not write them explicitly, and have

$$\psi(x) = \int a(p) \psi_p(x) \, dp \ , \quad \varphi(x) = \int b(p) \psi_p(x) \, dp \ . \tag{10.19}$$

The functions $a(p)$ and $b(p)$ are the momentum representations of ψ and φ, respectively. Inserting (10.19) into (10.13) yields

$$\int b(p) \psi_p(x) \, dp = \hat{L} \int a(p) \psi_p(x) \, dp = \int a(p) \hat{L} \psi_p(x) \, dp \ . \tag{10.20}$$

Since the operator $\hat{L}$ is assumed to be given in coordinate representation (10.14), it depends on x and not on p. We can therefore write it under the integral. Multiplication by $\psi^*_{p'}(x)$ and integration lead to

$$\int b(p)\,\mathrm{d}p \int \psi^*_{p'}(x)\psi_p(x)\,\mathrm{d}x = \int a(p)\,\mathrm{d}p \int \psi^*_{p'}(x)\hat{L}\psi_p(x)\,\mathrm{d}x \ . \qquad (10.21)$$

With the orthogonality relation

$$\int \psi^*_{p'}(x)\psi_p(x)\,\mathrm{d}x = \delta(p'-p) \qquad (10.22)$$

and the abbreviation

$$L_{p'p} = \int \psi^*_{p'}(x)\hat{L}\psi_p(x)\,\mathrm{d}x \ , \qquad (10.23)$$

we get the *momentum representation* of (10.13):

$$b(p') = \int L_{p'p}a(p)\,\mathrm{d}p \ . \qquad (10.24)$$

The indices p and p' are continuous and hence the matrix element

$$L_{p'p} = L(p', p)$$

is a function of the variables p and p'. But the term "matrix element" is also used in this case. The *infinite matrix* $(L_{pp'})$ *is the momentum representation of* $\hat{L}$. For explicit calculations see Examples 10.2 and 10.3.

Now we want to review some laws of matrix calculus and show their validity for matrix elements of operators.

A matrix $L = (L_{mn}\delta_{mn})$ is called *diagonal*; especially for $L_{nn} = 1$ we have the *unit matrix* $E = (\delta_{mn})$. The matrix which is called the *complex conjugate matrix* of L is defined by

$$L^* = (L^*)_{mn} = (L)^*_{mn} \ . \qquad (10.25)$$

The *transposed* matrix $\tilde{L}$ of $L = (L_{mn})$ is

$$\tilde{L} = (\tilde{L})_{mn} = (L)_{nm} \ . \qquad (10.26)$$

It is obtained from the original matrix by transposing the indices, i.e. mirroring the matrix elements on the principal diagonal. The elements of the *adjoint matrix* L^+ fulfil the relation

$$(L^+)_{mn} = (\tilde{L})_{mn} = (L^*)_{nm} \ . \qquad (10.27)$$

In the case $L = L^+$, we call the matrix L *self-adjoint or Hermitian*. Now we show that a Hermitian operator is represented by a Hermitian matrix. Indeed,

$$L_{mn} = \int \psi_m^* \hat{L} \psi_n \,\mathrm{d}x = \int \psi_n \hat{L}^* \psi_m^* \,\mathrm{d}x$$
$$= \left(\int \psi_n^* \hat{L} \psi_m \,\mathrm{d}x \right)^* = L_{nm}^* \ . \tag{10.28}$$

Two matrices are added component by component:

$$C_{nm} = A_{nm} + B_{nm} \ . \tag{10.29}$$

Let $\hat{C}$ be the sum of the operators $\hat{A}$ and $\hat{B}$. We can show that the matrix corresponding to $\hat{C}$ is the sum of the matrices corresponding to $\hat{A}$ and $\hat{B}$:

$$C_{mn} = \int \psi_m^* \hat{C} \psi_n \,\mathrm{d}x = \int \psi_n^* (\hat{A} + \hat{B}) \psi_n \,\mathrm{d}x$$
$$= \int \psi_m^* \hat{A} \psi_n \,\mathrm{d}x + \int \psi_m^* \hat{B} \psi_n \,\mathrm{d}x = A_{mn} + B_{mn} \ . \tag{10.30}$$

The multiplication of matrices is defined as

$$C_{mn} = \sum_k A_{mk} B_{kn} \ . \tag{10.31}$$

Let us prove that the matrices of operators fulfil the same relation. If $\hat{C} = \hat{A}\hat{B}$, then obviously

$$C_{mn} = \int \psi_m^* \hat{C} \psi_n \,\mathrm{d}x = \int \psi_m^* \hat{A} \hat{B} \psi_n \,\mathrm{d}x = \int \psi_m^* \hat{A} (\hat{B} \psi_n) \,\mathrm{d}x \ . \tag{10.32}$$

We set $(\hat{B}\psi_n) = \varphi_n(x)$ and expand this function in terms of orthogonal functions $\psi_k(x)$

$$\varphi_n(x) = \hat{B}\psi_n = \sum_k b_{kn} \psi_k(x) \ , \tag{10.33}$$

with coefficients b_{kn}:

$$b_{kn} = \int \psi_k^* \hat{B} \psi_n \,\mathrm{d}x = B_{kn} \ . \tag{10.34}$$

By inserting (10.34) into (10.33) and the result into (10.32), we obtain for C_{mn}

$$C_{mn} = \int \psi_m^* \hat{A} \left(\sum_k b_{kn} \psi_k \right) \mathrm{d}x = \int \psi_m^* \sum_k b_{kn} \hat{A} \psi_k \,\mathrm{d}x$$
$$= \sum_k B_{kn} \int \psi_m^* \hat{A} \psi_k \,\mathrm{d}x \ . \tag{10.35}$$

In analogy to (10.34), we define

$$\int \psi_m^*(x) \hat{A} \psi_k(x) \,\mathrm{d}x = A_{mk} \ , \tag{10.36}$$

and verify with

$$C_{mn} = \sum_k B_{kn} A_{mk} = \sum_k A_{mk} B_{kn} \tag{10.37}$$

that the multiplication rule (10.31) is also valid for matrices which belong to operators.

In the following, the wave functions $\varphi(x)$ and $\psi(x)$ are represented by the numbers b_n and a_n, respectively. We want to transform the equation

$$\varphi = \hat{L}\psi \tag{10.38}$$

into matrix form. Therefore we use the column vectors

$$(a_n) = \begin{pmatrix} a_1 \\ a_2 \\ \vdots \end{pmatrix} , \quad (b_n) = \begin{pmatrix} b_1 \\ b_2 \\ \vdots \end{pmatrix} . \tag{10.39}$$

Now, we can replace the equation $\varphi = \hat{L}\psi$, which is equivalent to $b_m = \sum_n L_{mn} a_n$, by the matrix equation

$$(b_n) = (L_{mn})(a_n) , \tag{10.40}$$

where (L_{mn}) corresponds to the operator $\hat{L}$. Explicitly this equation reads

$$\begin{pmatrix} b_1 \\ b_2 \\ \vdots \end{pmatrix} = \begin{pmatrix} L_{11} & L_{12} \dots & \\ L_{21} & L_{22} \dots & \\ \vdots & & \dots \end{pmatrix} \begin{pmatrix} a_1 \\ a_2 \\ \vdots \end{pmatrix} . \tag{10.41}$$

We call (a_n) and (b_n) the *representations* of the wave functions ψ and φ, respectively in the chosen basis ψ_k.

The expectation value $\langle\psi|\hat{L}|\psi\rangle$ of an operator $\hat{L}$ in the state $\psi(x)$ is easily determined in matrix representation. We have

$$\begin{aligned} \bar{L} &= \langle\psi(x)|\,\hat{L}\,|\psi(x)\rangle = \int \psi^*(x)\hat{L}\psi(x)\,\mathrm{d}x \\ &= \int \mathrm{d}x \sum_{n,m} a_n^* \psi_n^*(x)\hat{L}(x) a_m \psi_m(x) \\ &= \sum_{n,m} a_n^* a_m \int \mathrm{d}x\, \psi_n^*(x)\hat{L}(x)\psi_m(x) = \sum_{n,m} a_n^* L_{nm} a_m \\ &= (a_1^*, a_2^*, \dots) \begin{pmatrix} L_{11} & L_{12}\dots \\ \vdots & \end{pmatrix} \begin{pmatrix} a_1 \\ a_2 \\ \vdots \end{pmatrix} . \end{aligned} \tag{10.42}$$

The results of this section are quite important. We have learned that in addition to coordinate representation, a whole variety of representations exist for expressing quantum-mechanical relations. Later on in Chap. 12, this fact will prove useful for describing spin.

EXAMPLE

10.2 Momentum Representation of the Operator *r*

Let us transform the operator of the x coordinate $\hat{x} = x$ into momentum representation. According to (10.23) it is

$$x_{p_x' p_x} = \int \psi^*_{p_x'}(x)\, x\, \psi_{p_x}(x)\, \mathrm{d}x \;. \tag{1}$$

For momentum eigenfunctions we take again the plane waves

$$\psi_{p_x}(x) = \frac{\exp[\mathrm{i}(p_x/\hbar)x]}{\sqrt{2\pi\hbar}}$$

and get

$$x_{p_x p_x'} = \frac{1}{2\pi\hbar} \int \exp\left(-\mathrm{i}\frac{p_x}{\hbar}x\right) x \exp\left(\mathrm{i}\frac{p_x'}{\hbar}x\right) \mathrm{d}x \;. \tag{2}$$

This can be written as a partial derivative with respect to p_x':

$$\begin{aligned} x_{p_x p_x'} &= \frac{1}{2\pi\hbar} \int \exp\left(-\mathrm{i}\frac{p_x}{\hbar}x\right) \frac{\hbar}{\mathrm{i}} \frac{\partial}{\partial p_x'} \exp\left(\mathrm{i}\frac{p_x'}{\hbar}x\right) \mathrm{d}x \\ &= \frac{\hbar}{\mathrm{i}} \frac{\partial}{\partial p_x'} \frac{1}{2\pi\hbar} \int \exp\left(\frac{\mathrm{i}}{\hbar}(p_x' - p_x)x\right) \mathrm{d}x \\ &= \frac{\hbar}{\mathrm{i}} \frac{\partial}{\partial p_x'} \delta(p_x' - p_x) \;. \end{aligned} \tag{3}$$

This is the *momentum representation of the operator of the x coordinate in matrix representation*. With the momentum representations $b(p_x)$ and $a(p_x)$ of the functions $\varphi(x)$ and $\psi(x)$, respectively, the equation $\varphi(x) = x\psi(x)$ becomes

$$\begin{aligned} b(p_x) &= \int x_{p_x p_x'} a(p_x')\, \mathrm{d}p_x' \\ &= -\int \mathrm{i}\hbar \left(\frac{\partial}{\partial p_x'} \delta(p_x' - p_x)\right) a(p_x')\, \mathrm{d}p_x' \;. \end{aligned} \tag{4}$$

Partial integration yields

$$\begin{aligned} b(p_x) &= -\mathrm{i}\hbar[\delta(p_x' - p_x)a(p_x')]_{-\infty}^{+\infty} + \mathrm{i}\hbar \int \delta(p_x' - p_x) \frac{\partial a(p_x')}{\partial p_x'} \mathrm{d}p_x' \\ &= \mathrm{i}\hbar \frac{\partial}{\partial p_x} a(p_x) \;. \end{aligned} \tag{5}$$

Comparing this equation with $\varphi(x) = \hat{x}\psi(x) = x\psi(x)$, which is in coordinate representation, we note that the coordinate x is replaced by the operator $\mathrm{i}\hbar\,\partial/\partial p_x$;

Example 10.2

the latter has to be interpreted as the momentum representation of the operator x. In the case of the other coordinates y and z, the derivation is similar, and so we obtain the following momentum representation of $\hat{\boldsymbol{r}}$:

$$\hat{\boldsymbol{r}} = \mathrm{i}\hbar \nabla_p \,, \tag{6}$$

where ∇_p is the nabla or del operator (gradient) in momentum space.

The following table shows the connection betweeen the momentum and the coordinate operators:

Representation	$\hat{\boldsymbol{r}}$	$\hat{\boldsymbol{p}}$
Configuration space (coordinate representation)	$\boldsymbol{r}$	$-\mathrm{i}\hbar\nabla$
Momentum space (momentum representation)	$\mathrm{i}\hbar\nabla_p$	$\boldsymbol{p}$

In coordinate representation, $\hat{\boldsymbol{r}} = \boldsymbol{r}$ is simply an ordinary vector whose components are numbers; $\hat{\boldsymbol{p}} = -\mathrm{i}\hbar\nabla$ is a vector whose components are differential operators with respect to x. In momentum representation, the situation is reversed: $\hat{\boldsymbol{r}} = \mathrm{i}\hbar\nabla_p$ is a vector whose components are differential operators with respect to $\boldsymbol{p}$, while $\hat{\boldsymbol{p}} = \boldsymbol{p}$ is an ordinary vector with numbers as components.

EXAMPLE

10.3 The Harmonic Oscillator in Momentum Space

Now we show that the solution of the one-dimensional quantum-mechanical oscillator yields the same eigenvalues in momentum space as in configuration space. To do this, we replace the Hamiltonian in x-space representation,

$$\hat{H} = \frac{\hat{p}^2}{2m} + \frac{m}{2}\omega^2 x^2 \,, \tag{1}$$

by the Hamiltonian in its momentum-space representation (see the table in Example 10.2):

$$\hat{H} = \frac{p^2}{2m} + \frac{m}{2}\omega^2 (\mathrm{i}\hbar)^2 \frac{\partial^2}{\partial p^2} \,. \tag{2}$$

The Schrödinger equation $\hat{H}\psi = E\psi$ now reads

$$\left(\frac{p^2}{2m} - \frac{m}{2}\omega^2\hbar^2 \frac{\partial^2}{\partial p^2}\right)\psi = E\psi \,, \quad \text{or} \tag{3}$$

$$\left(\frac{p^2}{2m} - \frac{m}{2}\omega^2\hbar^2 \frac{\partial^2}{\partial p^2} - E\right)\psi = 0 \,. \tag{4}$$

Example 10.3

We divide both sides of the equation by $m^2\omega^2$ and get, after reordering

$$\left(-\frac{\hbar^2}{2m}\frac{\partial^2}{\partial p^2}+\frac{p^2}{2m^3\omega^2}-\frac{E}{m^2\omega^2}\right)\psi=0\ . \tag{5}$$

With the substitutions

$$E'=\frac{E}{m^2\omega^2}\quad\text{and}\quad\omega'=\frac{1}{m^2\omega}\ , \tag{6}$$

the differential equation (5) becomes

$$\left(-\frac{\hbar^2}{2m}\frac{\partial^2}{\partial p^2}+\frac{m}{2}\omega'^2p^2-E'\right)\psi=0\ . \tag{7}$$

This equation takes the well-known form of the oscillator equation in configuration space [see (7.4)]. Hence, all the conclusions reached in Chap. 7 can be applied with the result that

$$E'=\hbar\omega'\left(n+\frac{1}{2}\right)\ . \tag{8}$$

By resubstituting E' and ω' we get

$$E=\hbar\omega\left(n+\frac{1}{2}\right)\ , \tag{9}$$

i.e. the same energy values as in our former calculation in configuration space (see Chap. 7). Furthermore, the wave functions are of the same form in both representations.

In the following we give the *matrix representation* (we shall also call it the *matrix form*) of several operators. This will be useful in our later studies.

Matrix of the Coordinate Operator $\hat{x}$ in Configuration Space

We show that the matrix form of the coordinate $\hat{x}$ in x representation is given by

$$x_{xx'}=x'\delta(x-x')\ . \tag{10.43}$$

Indeed, the laws of matrix multiplication give

$$\varphi(x')=\int\limits_{-\infty}^{\infty}x_{x'x}\psi(x)\,\mathrm{d}x=\int\limits_{-\infty}^{\infty}x'\delta(x-x')\psi(x)\,\mathrm{d}x=x'\psi(x')\ , \tag{10.44}$$

i.e. the matrix for $\hat{x}$ produces the right factor (eigenvalue) x' in the equation $\varphi(x')=x'\psi(x')$. Hence, we call $x_{xx'}=x'\delta(x-x')$ the *matrix form* of the coordinate x in x representation.

Matrix of $V(x)$ in Configuration Space

Let $V(x)$ be an arbitrary function of the x coordinate. As above, we insert $V_{x'x} = V(x')\delta(x-x')$ into the equation

$$\varphi(x) = V(x)\psi(x) \tag{10.45}$$

and get

$$\begin{aligned}\varphi(x') &= \int\limits_{-\infty}^{\infty} V_{x'x}\psi(x)\,\mathrm{d}x = \int\limits_{-\infty}^{\infty} V(x')\delta(x-x')\psi(x)\,\mathrm{d}x \\ &= V(x')\psi(x') \ .\end{aligned} \tag{10.46}$$

Obviously, (10.45) and (10.46) are identical. Hence, $V_{xx'} = V(x')\delta(x-x')$ is the matrix form of the potential in coordinate (x) representation.

Matrix Form of the Momentum Operator in Configuration Space (x Representation)

This matrix reads

$$\hat{p}_{x'x} = \mathrm{i}\hbar\frac{\partial}{\partial x}\delta(x'-x) \ . \tag{10.47}$$

By way of proof, we insert it into

$$\varphi(x) = \hat{p}\psi(x) \tag{10.48}$$

and get

$$\varphi(x') = \int\limits_{-\infty}^{\infty} \hat{p}_{x'x}\psi(x)\,\mathrm{d}x = \mathrm{i}\hbar\int\limits_{-\infty}^{\infty}\frac{\partial}{\partial x}\delta(x'-x)\psi(x)\,\mathrm{d}x \ . \tag{10.49}$$

Partial integration yields

$$\varphi(x') = \mathrm{i}\hbar[\delta(x'-x)\psi(x)|_{-\infty}^{+\infty} - \mathrm{i}\hbar\int\limits_{-\infty}^{\infty}\delta(x'-x)\frac{\partial}{\partial x}\psi(x)\,\mathrm{d}x \ . \tag{10.50}$$

The first term vanishes again and so

$$\varphi(x') = -\mathrm{i}\hbar\int\limits_{-\infty}^{\infty}\delta(x'-x)\frac{\partial}{\partial x}\psi(x)\,\mathrm{d}x = -\mathrm{i}\hbar\frac{\partial}{\partial x'}\psi(x') \ , \tag{10.51}$$

which is the standard form of (10.48) and verifies our assumption.

10.3 The Eigenvalue Problem

An important and frequently encountered problem in quantum mechanics involves finding eigenvalues and eigenfunctions of a given operator $\hat{A}$. If the operator $\hat{A}$ is given in the representation of its eigenfunctions, the diagonal elements of the corresponding matrix A_{mn} are just its eigenvalues. Let us develop methods for finding eigenvalues and eigenfunctions of the operator $\hat{A}$ if it is not given in its eigenrepresentation.

The eigenfunctions ψ_a of $\hat{A}$ fulfil the equation

$$\hat{A}\psi_a(x) = a\psi_a(x) \ . \tag{10.52}$$

We expand them in terms of functions φ_n which are not eigenfunctions of $\hat{A}$:

$$\psi_a(x) = \sum c_n^a \varphi_n(x) \ . \tag{10.53}$$

The combination of (10.52) and (10.53) yields

$$\hat{A}\sum c_n^a \varphi_n = a\sum c_n^a \varphi_n \ . \tag{10.54}$$

After multiplying by φ_k^* and integrating, we have

$$\sum_n c_n^a A_{kn} = ac_k^a \ , \tag{10.55}$$

where the abbreviation A_{kn} means

$$A_{kn} = \int \varphi_k^* \hat{A}\varphi_n \, \mathrm{d}V \ . \tag{10.56}$$

Let us now assume A_{kn} is given and the eigenvalues a and the eigenvectors $\{a_n^a\}$ in (10.55) for the given matrix (A_{kn}) are to be computed. If we know both, the eigenvalue problem is solved in any representation, since with $\{c_n^a\}$, we can construct via (10.53) the eigenfunctions of $\hat{A}$, i.e. $\psi_a(x)$, in x representation, too. To find the c_n^a, it is convenient to write (10.55) in the form

$$\sum_n (A_{kn} - a\delta_{kn})c_n^a = 0 \ . \tag{10.57}$$

Obviously (10.57) represents an infinite homogeneous system of equations for the coefficients c_n^a. Such a system has a nontrivial solution if the determinant of coefficients vanishes, i.e.

$$\det(A_{kn} - a\delta_{kn}) = 0 \ . \tag{10.58}$$

The problem is that, in general, this determinant is infinite. To solve (10.58), we consider secular determinants of Nth degree:

$$D_N(a) = \begin{vmatrix} A_{11}-a & A_{12} & \vdots & A_{1N} \\ A_{21} & A_{22}-a & \vdots & A_{2N} \\ \cdots & \cdots & \cdots & \cdots \\ A_{N1} & A_{N2} & \vdots & A_{NN}-a \end{vmatrix} = 0 \ . \tag{10.59}$$

This is a truncation of the expansion (10.53) at a certain value $n = N$. We check for convergence by increasing the parameter N. The equation $D_N(a) = 0$ is of Nth degree and therefore yields N solutions for a. These solutions

$$a_1^{(N)}, a_2^{(N)}, \dots, a_N^{(N)} \tag{10.60}$$

are all real, because $D_N(a)$ is the determinant of a Hermitian matrix (the operator $\hat{A}$ is assumed to be Hermitian, as all operators in quantum mechanics associated with observables should be).

Now we evaluate each eigenvalue a_i for a sequence of increasing determinants D_N and get a sequence of solutions:

$$a_i^{(1)}, a_i^{(2)}, \dots, a_i^{(N)} \to a_i \ . \tag{10.61}$$

The convergence of this sequence can be explained physically. The matrix elements A_{kn} measure the correlation between the states φ_k and φ_n. But in the case $n \gg k$, this connection will be negligible (for example, highly excited states hardly disturb the ground state). Then the A_{kn} usually get very small and contribute only very little to the first roots of the secular determinant.

We insert each of the so-calculated a_i into (10.57) and obtain the coefficients $c_n(a_i)$ and, with (10.53), the eigenfunctions

$$\psi_{a_i} = \sum_n c_n(a_i)\varphi_n(x) \ . \tag{10.62}$$

When the spectra and matrices of the operators are continuous, we get an integral instead of a sum in (10.57) and this equation becomes a ***Fredholm** integral equation* of the second kind:

$$\int A(\xi', \xi)c(\xi)\,\mathrm{d}\xi = ac(\xi') \ . \tag{10.63}$$

We shall have to deal with such continuum problems later on in quantum electrodynamics in the discussion of spontaneous vacuum decay. They also appear in the decay of bound states into several continua. At this point we shall not deal with them any further.

10.4 Unitary Transformations

An operator $\hat{A}$ can be represented by matrices in several ways. Indeed, for any complete set of wave functions $\psi_n(x)$, we can construct the corresponding representation of the operator $\hat{A}$ [see (10.17)]. Now we consider the transformation behaviour of these matrices when changing the representations.

An operator $\hat{A}$ may be given in a representation with a basis of functions $\psi_n(\boldsymbol{r})$, which are the eigenfunctions of an operator $\hat{L}$ (i.e. $\hat{L}\psi_n(\boldsymbol{r}) = L_n\psi_n(\boldsymbol{r})$). Then,

$$A_{mn} = \int \psi_m^*(\boldsymbol{r})\hat{A}\psi_n(\boldsymbol{r})\,\mathrm{d}V \tag{10.64}$$

is the *L representation* of the operator $\hat{A}$. On the other hand, a representation of $\hat{A}$ with eigenfunctions $\varphi_\mu(\boldsymbol{r})$ of $\hat{M}$ [i.e. $\hat{M}\varphi_\mu(\boldsymbol{r}) = M_\mu\varphi(\boldsymbol{r})$] is also possible:

$$A_{\mu\nu} = \int \varphi_\mu^*(\boldsymbol{r})\hat{A}\varphi_\nu(\boldsymbol{r})\,\mathrm{d}V \ . \tag{10.65}$$

To distinguish between these representations, we use Latin indices for the L representation and Greek indices for the M representation.

Now we want to determine the transformation matrix which connects (A_{mn}) with $(A_{\mu\nu})$. Therefore we expand the eigenfunctions of $\hat{M}$ in terms of eigenfunctions of $\hat{L}$:

$$\varphi_\mu = \sum_n S_{n\mu}\psi_n \ . \tag{10.66}$$

Multiplication by ψ_m^* and integration yield

$$\int \psi_m^*\varphi_\mu\,\mathrm{d}V = \sum_n S_{n\mu}\delta_{mn} = S_{m\mu} \ . \tag{10.67}$$

Obviously *the matrix element $S_{m\mu}$ is the projection of ψ_m onto the state φ_μ.* Replacing the $\hat{M}$ eigenfunctions (in 10.65) according to (10.66) leads to

$$A_{\mu\nu} = \int \sum_n S_{n\mu}^*\psi_n^*\hat{A}\sum_m S_{m\nu}\psi_m\,\mathrm{d}V = \sum_{n,m} S_{n\mu}^*S_{m\nu}\int \psi_n^*\hat{A}\psi_m\,\mathrm{d}V \ ,$$

$$A_{\mu\nu} = \sum_{nm} S_{n\mu}^*A_{nm}S_{m\nu} \ . \tag{10.68}$$

Now, using the elements of the adjoint matrix

$$(\tilde{S})_{n\mu}^* = (S_{\mu n})^+ \tag{10.69}$$

gives the transformation rule between the matrices of $\hat{A}$ in the two representations:

$$(A_{\mu\nu}) = \sum_{n,m} (S^+)_{\mu n}(A_{nm})(S_{m\nu}) \ , \tag{10.70}$$

or, denoting the matrices by capital letters only,

$$A_M = S^+ A_L S \ . \tag{10.71}$$

The indices M and L refer to the various representations of $\hat{A}$. The requirement that the ψ_n as well as the φ_μ be orthonormal wave functions implies the unitarity of S. This is shown in the following derivation:

$$\delta_{\mu\nu} = \int \varphi_\mu^* \varphi_\nu \,\mathrm{d}V = \int \sum_m S_{m\mu}^* \psi_m^* \sum_n S_{n\nu}\psi_n \,\mathrm{d}V = \sum_{n,m} S_{m\mu}^* S_{n\nu}\delta_{mn} \ ,$$

$$\delta_{\mu\nu} = \sum_m S_{m\mu}^* S_{m\nu} = (S^+ S)_{\mu\nu} \ ; \tag{10.72}$$

i.e. the product of S and its adjoint matrix S^+ is equal to the unit matrix:

$$S^+ S = \mathbb{1} \ . \tag{10.73}$$

Unitarity also means the equivalence of the adjoint S^+ and the reciprocal matrix S^{-1}. We note that a unitary matrix is not necessarily Hermitian:

$$S^+ = S^{-1} \neq S \ . \tag{10.74}$$

The physical meaning of the unitary transformation (10.66) is the conservation of probability: if a particle is in a state φ_u with probability 1, it can be found with the probability $|S_{\mu n}|^2$ in the states ψ_n. The set $|S_{\mu 1}|^2, \dots , |S_{\mu n}|^2, \dots$ then gives the probability distribution of the particle with respect to the states ψ_n. Therefore it must hold that

$$\sum_n |S_{\mu n}|^2 = \sum_n S_{\mu n}^* S_{\mu n} = 1 \ , \tag{10.75}$$

i.e., according to (10.72), S is unitary.

An important and frequently used theorem is the invariance of the trace of a matrix under unitary transformations. The trace of a matrix A is denoted by $\mathrm{tr}A$ and is defined as the sum of all diagonal elements. According to (10.68) and (10.72), we calculate

$$\begin{aligned}
\mathrm{tr}A_M &= \sum_{\mu} A_{\mu\mu} = \sum_{\mu}\sum_{n,m} S^*_{n\mu} A_{nm} S_{m\mu} \ , \\
&= \sum_{\mu,n,m} A_{nm} S_{m\mu} S^*_{n\mu} \ , \\
&= \sum_{n,m} A_{nm} (SS^+)_{mn} = \sum_{n,m} A_{nm}\delta_{mn} \ , \\
&= \sum_{n} A_{nn} = \mathrm{tr}A_L \ . \qquad (10.76)
\end{aligned}$$

Hence, $\mathrm{tr}A_M = \mathrm{tr}A_L$. Thus the trace of a matrix does not depend on the particular representation.

10.5 The S Matrix

The temporal evolution of a system can be described as a series of unitary transformations. The operator of this time-evolution transformation will be denoted by $\hat{S}$; the corresponding matrix is the *S matrix* (scattering matrix). We now derive the $\hat{S}$ operator and show some of its properties.

The operator in question has to transform a state at time $t = 0$ into the state at time t:

$$\psi(\boldsymbol{r}, t) = \hat{S}(t)\psi(\boldsymbol{r}, 0) \ . \qquad (10.77)$$

If we insert $\psi(\boldsymbol{r}, t)$ into the time-dependent Schrödinger equation, we can determine $\hat{S}$, namely

$$\begin{aligned}
&\left(\mathrm{i}\hbar\frac{\partial}{\partial t} - \hat{H}\right)\hat{S}(t)\psi(\boldsymbol{r}, 0) = \left(\mathrm{i}\hbar\frac{\partial \hat{S}}{\partial t} - \hat{H}\hat{S}\right)\psi(\boldsymbol{r}, 0) = 0 \ , \\
&\mathrm{i}\hbar\frac{\partial \hat{S}}{\partial t} - \hat{H}\hat{S} = 0 \ . \qquad (10.78)
\end{aligned}$$

In the case of $\hat{H}$ not being explicitly time dependent, the following solution results:

$$\hat{S} = \exp\left(-\frac{\mathrm{i}}{\hbar}\hat{H}t\right) \ . \qquad (10.79)$$

From (10.77), it follows that $\hat{S}(0) = 1$. Therefore the integration constant in (10.79) is set equal to 1. If we apply the operator $\hat{S}$ to a function as in (10.77), we expand the exponential function into a power series:

$$\hat{S} = \exp\left(-\frac{\mathrm{i}}{\hbar}\hat{H}t\right) = \sum_n \frac{1}{n!}\left(-\frac{\mathrm{i}}{\hbar}\hat{H}t\right)^n \ . \qquad (10.80)$$

We give special attention to the energy representation in which $\hat{H}$ is diagonal, i.e. $H\psi_n = E_n\psi_n(\boldsymbol{x})$. With

$$\psi(\boldsymbol{r},0) = \sum_n a_n\psi_n(\boldsymbol{r}) \ , \tag{10.81}$$

we get the temporal evolution of $\psi(\boldsymbol{r},0)$ by applying the operator $\hat{S}$ according to (10.77). This gives

$$\begin{aligned}
\psi(\boldsymbol{r},0) = \hat{S}\psi(\boldsymbol{r},0) &= \sum_n a_n\hat{S}\psi_n \\
&= \sum_n a_n \sum_k \frac{1}{k!}\left(-\frac{\mathrm{i}}{\hbar}\hat{H}t\right)^k \psi_n \\
&= \sum_n a_n \sum_k \frac{1}{k!}\left(-\frac{\mathrm{i}}{\hbar}E_n t\right)^k \psi_n \ ,
\end{aligned}$$

$$\hat{S}\psi(\boldsymbol{r},0) = \sum_n a_n \exp\left(-\frac{\mathrm{i}}{\hbar}E_n t\right)\psi_n(\boldsymbol{r}) \ . \tag{10.82}$$

Obviously the well-known time dependence for stationary states follows. In the energy representation, $\hat{S}$ is diagonal, as we can see from (10.82):

$$S_{mn} = \int \psi_m^*\hat{S}\psi_n \,\mathrm{d}V = \exp\left(-\frac{\mathrm{i}}{\hbar}E_n t\right)\delta_{mn} \ . \tag{10.83}$$

Equation (10.80) shows, too, that $\hat{S}$ is a unitary operator:

$$\hat{S}^+ = \left[\exp\left(-\frac{\mathrm{i}}{\hbar}\hat{H}t\right)\right]^+ = \exp\left(\frac{\mathrm{i}}{\hbar}\hat{H}^+t\right) = \exp\left(\frac{\mathrm{i}}{\hbar}\hat{H}t\right) = \hat{S}^{-1} \ , \tag{10.84}$$

because $\hat{H}$ is Hermitian. We now expand the wave function $\psi(r,t)$ with respect to the eigenfunctions φ_n of the operator $\hat{L}$:

$$\psi(\boldsymbol{r},t) = \sum_n b_n(t)\varphi_n(\boldsymbol{r}) \ . \tag{10.85}$$

If we again describe the temporal evolution by $\hat{S}$, according to (10.77) we obtain

$$\sum_n b_n(t)\varphi_n(\boldsymbol{r}) = \sum_n \hat{S}b_n(0)\varphi_n(\boldsymbol{r}) \ . \tag{10.86}$$

Multiplication by φ_m^* and subsequent integration yield the matrix equation

$$b_m(t) = \sum_n S_{mn}(t)b_n(0) \ , \tag{10.87}$$

where $S_{mn} = \int \varphi_m^*\hat{S}\varphi_n \,\mathrm{d}V$. Now let us consider the special case $b_n(0) = 1$. Then all other $b_{n'}(0)$, $n' \neq n$, are equal to zero because of normalization. This means that in the L representation, the particle at time $t = 0$ is completely in the state $\varphi_n(\boldsymbol{r})$. We can say that the system is prepared to be initially in state $\varphi_n(\boldsymbol{r})$. Consequently, (10.87) yields

$$b_m(t) = S_{mn}(t) \ . \tag{10.88}$$

This is an interesting result with a rather obvious physical interpretation.

The matrix element $S_{mn}(t)$ then yields the amplitude with which the system has passed over from state φ_n into the state φ_m after time t. Or in other words, the value

$$\omega(n \to m) = |S_{mn}(t)|^2 \tag{10.89}$$

gives us the *transition probability* from state φ_n into state φ_m under the influence of $\hat{H}$. This relation will play an important role in our subsequent calculations of transition probabilities of a quantum-mechanical system and in the calculation of quantum-electrodynamic scattering processes (transitions from an ingoing to an outgoing state).[1]

10.6 The Schrödinger Equation in Matrix Form

As an example of the formalism developed so far, we look at the solution of the Schrödinger equation

$$\mathrm{i}\hbar \frac{\partial \psi}{\partial t} = \hat{H}\psi \ , \tag{10.90}$$

and use the energy representation for the wave function, i.e. the eigenrepresentation of the not explicitly time-dependent Hamiltonian,

$$\hat{H}\psi_n = E_n \psi_n \ . \tag{10.91}$$

Expanding the wave function with respect to eigenfunctions of the Hamiltonian,

$$\psi(\boldsymbol{r}, t) = \sum_n a_n(t)\psi_n(\boldsymbol{r}) \ , \tag{10.92}$$

and inserting it into the Schrödinger equation (10.90) we get

$$\mathrm{i}\hbar \sum_n \frac{\partial a_n}{\partial t}\psi_n(\boldsymbol{r}) = \sum_n E_n a_n \psi_n(\boldsymbol{r}) \ . \tag{10.93}$$

Multiplication by ψ_m^* and integration yield

$$\mathrm{i}\hbar \frac{\partial a_m(t)}{\partial t} = E_m a_m(t) \ . \tag{10.94}$$

[1] See Sect. 11.4.

Only the fact that we have chosen the energy representation of $\hat{H}$ with $H_{mn} = E_n\delta_{mn}$ is responsible for the differential equations for $a_m(t)$ not being coupled. The solution of (10.94) is:

$$a_m(t) = a_m(0)\exp\left(-\frac{\mathrm{i}}{\hbar}E_m t\right) \ . \qquad (10.95)$$

The amplitudes of the stationary states are time dependent; the integration constants result from the initial conditions. If we use a representation other than the energy representation, then (10.99) below, in which we can see the coupling of the different amplitudes $a_m(t)$ by the matrix elements, is valid.

Now, in a similar way, we want to calculate the temporal change of the mean value of an operator $\hat{L}$. The mean value is given by

$$\langle\hat{L}\rangle = \int \psi^*\hat{L}\psi\,\mathrm{d}V \ . \qquad (10.96)$$

Inserting the expansion of the eigenfunctions (10.92) yields

$$\begin{aligned}\langle\hat{L}\rangle &= \int \sum_m a_m^*(t)\psi_m^*(\boldsymbol{r})\hat{L}\sum_n a_n(t)\psi_n(\boldsymbol{r})\,\mathrm{d}V \\ &= \sum_{nm} a_m^*(t)L_{mn}a_n(t) \ ,\end{aligned} \qquad (10.97)$$

according to definition (10.17) of the matrix element. Equation (10.97) gives the mean value of the operator in matrix representation as a function of time. We take the temporal derivation of the mean value and get

$$\frac{\mathrm{d}\langle\hat{L}\rangle}{\mathrm{d}t} = \sum_{m,n}\frac{\partial a_m^*}{\partial t}L_{mn}a_n + \sum_{n,m}a_m^*\frac{\partial L_{mn}}{\partial t}a_n + \sum_{n,m}a_m^*L_{mn}\frac{\partial a_n}{\partial t} \ . \qquad (10.98)$$

The temporal derivatives of the evolution coefficients $a_m(t)$ are now expressed according to (10.90) and (10.93) by the matrix elements H_{nk} of the Hamiltonian:

$$\mathrm{i}\hbar\frac{\partial a_n}{\partial t} = \sum_k H_{nk}a_k \ . \qquad (10.99)$$

This is the Schrödinger equation (10.90) in matrix representation;[2] it is valid in any representation. In the energy representation (10.91), especially, it reduces to

[2] Heisenberg [Z. Phys. **33**, 879 (1925)] introduced matrix elements as the quantum-mechanical analogue to the Fourier amplitudes of classical mechanics. Just as a classical quantity is determined by its Fourier amplitude, the corresponding quantum-mechanical quantity should be given by all these matrix elements. Heisenberg did not initially use the expression "matrix element"; Born and Jordan first ascertained [Z. Phys. **34**, 858 (1925)] that the multiplication law for quantum-mechanical quantities, given by Heisenberg, is identical to ordinary matrix multiplication. The entire

the simple decoupled equations (10.94). If we insert (10.99) and the complex-conjugated formula into $\mathrm{d}\langle\hat{L}\rangle/\mathrm{d}t$ of (10.98), then we get

$$\frac{\mathrm{d}\langle\hat{L}\rangle}{\mathrm{d}t} = -\frac{1}{\mathrm{i}\hbar}\sum_{m,n,k} a_k^* H_{mk}^* L_{mn} a_n + \sum_{m,n} a_m^* \frac{\partial L_{mn}}{\partial t} a_n + \frac{1}{\mathrm{i}\hbar}\sum_{m,n,k} a_m^* L_{mn} a_k H_{nk} \ . \tag{10.100}$$

As the Hamiltonian is Hermitian, we have

$$H_{mk}^* = H_{km} \ , \tag{10.101}$$

and with a change of the indices of the first and third terms [in the first term we substitute $(m, n, k) \to (n, k, m)$], we are able to summarize as follows:

$$\frac{\mathrm{d}\langle\hat{L}\rangle}{\mathrm{d}t} = \sum_{m,n} a_m^* \frac{\partial L_{mn}}{\partial t} a_n + \frac{1}{\mathrm{i}\hbar}\sum_{m,k} a_m^* \sum_n (L_{mn}H_{nk} - L_{nk}H_{mn}) a_k \ . \tag{10.102}$$

According to the rules of matrix multiplication and with the introduction of operator products, a further simplification is possible, namely

$$\frac{\mathrm{d}\langle\hat{L}\rangle}{\mathrm{d}t} = \sum_{m,k} a_m^* \frac{\partial L_{mk}}{\partial t} a_k + \frac{1}{\mathrm{i}\hbar}\sum_{m,k} a_m^* (\hat{L}\hat{H} - \hat{H}\hat{L})_{mk} a_k \ . \tag{10.103}$$

We combine the double sums and introduce the commutator $[\hat{H}, \hat{L}]_-$ so that

$$\frac{\mathrm{d}\langle\hat{L}\rangle}{\mathrm{d}t} = \sum_{m,k} a_m^* \left(\frac{\partial L_{mk}}{\partial t} + \frac{\mathrm{i}}{\hbar}([\hat{H}, \hat{L}]_-)_{mk}\right) a_k \tag{10.104}$$

results. Setting $\mathrm{d}\langle\hat{L}\rangle/\mathrm{d}t = \langle \mathrm{d}L/\mathrm{d}t\rangle$ and using (10.97), we get the matrix element of the temporal change of the operator:

$$\left(\frac{\mathrm{d}\hat{L}}{\mathrm{d}t}\right)_{mn} = \frac{\partial L_{mn}}{\partial t} + \frac{\mathrm{i}}{\hbar}[\hat{H}, \hat{L}]_{mn} \ . \tag{10.105}$$

We already deduced this result in Chap. 8, and used it to derive Ehrenfest's theorems; now we have it in matrix representation.

---

theory was further developed and became firmly established with the help of matrix calculus [M. Born, W. Heisenberg, P. Jordan. Z. Phys. **35**, 557 (1926)]. The papers referred to here can be considered fundamental to the development of quantum mechanics.

10.7 The Schrödinger Representation

In our previous description of the dynamical evolution of a physical system we used *time-dependent state functions* $\psi(\boldsymbol{r}, t)$. The physical quantities, at least the not explicitly time-dependent ones, are described by *time-independent operators*. We call this type of description the *Schrödinger representation* or *Schrödinger picture*.

10.8 The Heisenberg Representation

In the *Heisenberg representation (Heisenberg picture)*, the situation is reversed: *the wave functions are time independent* and the dynamical evolution is described by *time-dependent operators*.

The two representations are completely equivalent in describing a system; they lead to the same expectation values, the same spectra, etc. The transition from one representation to another one is given by a unitary time-dependent transformation, as we will see below.

To explain the different types of representation (which are sometimes also called "pictures", i.e. Schrödinger picture, Heisenberg picture etc.) we look at a matrix element of an operator $\hat{L}$:

$$L_{mn} = \int \psi_m^*(\boldsymbol{r}, t)\hat{L}\psi_n(\boldsymbol{r}, t)\,\mathrm{d}V \ . \tag{10.106}$$

For the wave function, we write in energy representation:

$$\psi_m(\boldsymbol{r}, t) = \psi_m(\boldsymbol{r}) \exp\left(-\frac{\mathrm{i}}{\hbar}E_m t\right) \ . \tag{10.107}$$

The time dependence of the stationary state is given by an exponential factor. Inserting this into the integral (10.106) yields

$$\begin{aligned} L_{mn}(t) &= \int \psi_m^*(\boldsymbol{r}) \exp\left(\frac{\mathrm{i}}{\hbar}E_m t\right) \hat{L}\psi_n(\boldsymbol{r}) \exp\left(-\frac{\mathrm{i}}{\hbar}E_n t\right) \mathrm{d}V \\ &= \int \psi_m^*(\boldsymbol{r})\hat{L} \exp\left(\frac{\mathrm{i}}{\hbar}(E_m - E_n)t\right) \psi_n(\boldsymbol{r})\,\mathrm{d}V \ , \\ L_{mn} &= \int \psi_m^*(\boldsymbol{r})\hat{L}_{\mathrm{H}}(t)\psi_n(\boldsymbol{r})\,\mathrm{d}V \ . \end{aligned} \tag{10.108}$$

Of course, the matrix element has not changed during our manipulations. The equations (10.106) and (10.108) differ only in that the time dependence is in one case [(10.106)] in the wave function $\psi(\boldsymbol{r}, t)$; in the other case [(10.108)] in the operator $\hat{L}_{\mathrm{H}}(t)$. The operator in the Heisenberg representation is thus

$$\hat{L} \to \hat{L}_{\mathrm{H}} = \hat{L} \exp\left(\frac{\mathrm{i}}{\hbar}(E_m - E_n)\right) t \ . \tag{10.109}$$

This is true if the operator is not explicitly time dependent. In the general case we can describe the transition from Schrödinger to Heisenberg picture by a unitary transformation. With the operator

$$\hat{S} = \exp\left(-\frac{\mathrm{i}}{\hbar}\hat{H}t\right) \tag{10.110}$$

we get

$$\psi_{\mathrm{H}}(\boldsymbol{r}) = \hat{S}^{-1}\psi_{\mathrm{S}}(\boldsymbol{r},t) \tag{10.111}$$

for the wave functions, and for the operators

$$\hat{L}_{\mathrm{H}}(t) = \hat{S}^{-1}(t)\hat{L}_{\mathrm{S}}\hat{S}(t) \ , \tag{10.112}$$

where the index H *stands for Heisenberg* and S *for Schrödinger*. A comparison with (10.108) and (10.109) shows the validity of transformation (10.112) in the energy representation.

10.9 The Interaction Representation

If we have a system whose Hamiltonian splits into an $\hat{H}_0$ part and into an additional interaction $\hat{V}$,

$$\hat{H} = \hat{H}_0 + \hat{V} \ , \tag{10.113}$$

we describe it by the so-called *interaction representation* (or interaction picture). In this description, both the state functions and the operators are time dependent. It follows from the unitary transformation

$$\hat{S}_{\mathrm{I}} = \exp\left(\frac{\mathrm{i}}{\hbar}\hat{H}_0 t\right) \tag{10.114}$$

in the Schrödinger representation. Equation (10.114) is analogous to (10.110). As with (10.111), we get the wave function with

$$\psi_{\mathrm{I}}(\boldsymbol{r},t) = \hat{S}_{\mathrm{I}}^{-1}\psi_{\mathrm{S}}(\boldsymbol{r},t) \ . \tag{10.115}$$

The operator in the interaction picture is obtained as

$$\hat{L}_{\mathrm{I}}(t) = \hat{S}_{\mathrm{I}}^{-1}(t)\hat{L}_{\mathrm{S}}\hat{S}_{\mathrm{I}}(t) \ , \tag{10.116}$$

which is analogous to (10.112).

10.10 Biographical Notes

FREDHOLM, Erik Ivar, Swedish mathematician, *Stockholm 7.4.1866, †Mörby 17.8.1927. His famous work on integral equations was published in 1903, where he established the fundamentals of the modern theory of this topic. He was awarded with the Wallmark prize of the Swedish Academy of Science and the French Academy's prize. He was appointed professor of theoretical physics in Stockholm in 1906.

11. Perturbation Theory

An exact solution of the Schrödinger equation exists only for a few idealized problems; normally it has to be solved using an approximation method. Perturbation theory is applied to those cases in which the real system can be described by a small change in an easily solvable, idealized system. The Hamiltonian of the system is then of the form

$$\hat{H} = \hat{H}_0 + \varepsilon \hat{W} \ , \tag{11.1}$$

where $\hat{H}$ and $\hat{H}_0$ do not differ very much from each other. $\hat{H}_0$ is called the *Hamiltonian of the unperturbed system*; the perturbation $\varepsilon \hat{W}$ (i.e. the adaptation to the real system) *has to be very small*; and ε is a real parameter which allows the expansion of wave functions and energies into a power series in ε. The parameter ε is also called the *smallness* (or *perturbation*) *parameter*.

In this form we can describe a great number of problems encountered in atomic physics, in which the nucleus provides the strong central potential for the electrons; further interactions of less strength are described by the perturbation. Examples of these additional interactions are: the magnetic interaction (spin–orbit coupling), the electrostatic repulsion of electrons and the influence of external fields. For the present, we restrict ourselves to perturbations constant in time and a Hamiltonian $\hat{H}_0$, the spectrum of which is discrete and not degenerate.

11.1 Stationary Perturbation Theory

We assume that the Hamiltonian is split up according to (11.1), and that the eigenvalues and eigenfunctions of the unperturbed Hamiltonian $\hat{H}_0$ are known:

$$\hat{H}_0 \psi_n^0 = E_n^0 \psi_n^0 \ . \tag{11.2}$$

We are seeking the eigenvalues and eigenfunctions of the complete Hamiltonian $\hat{H}$, i.e.

$$\hat{H}\Psi = E\Psi \ , \tag{11.3}$$

$$(\hat{H}_0 + \varepsilon \hat{W})\Psi = E\Psi \ . \tag{11.4}$$

The desired exact wave function Ψ is expanded in terms of the known solutions ψ_n^0 of the unperturbed system:

$$\Psi(\boldsymbol{r}) = \sum_n a_n \psi_n^0(\boldsymbol{r}) \ . \tag{11.5}$$

Inserting this into (11.4) and using (11.2) yields

$$\sum_n a_n (E_n^0 - E + \varepsilon \hat{W}) \psi_n^0 = 0 \ .$$

Multiplication by ψ_m^{0*} and subsequent integration gives

$$\sum_n a_n [(E_n^0 - E)\delta_{mn} + \varepsilon W_{mn}] = 0 \ . \tag{11.6}$$

We have used the fact that the eigenfunctions are orthonormal:

$$\int \psi_m^{0*} \psi_n^0 \, \mathrm{d}V = \delta_{mn} \ .$$

The matrix element W_{mn} stands for

$$W_{mn} = \int \psi_m^{0*} \hat{W} \psi_n^0 \, \mathrm{d}V \ . \tag{11.7}$$

Equation (11.6) can be transformed into

$$a_m (E_m^0 - E + \varepsilon W_{mm}) + \varepsilon \sum_{n \neq m} a_n W_{mn} = 0 \ . \tag{11.6a}$$

For $\varepsilon = 0$, we have only the idealized state, with $a_m^0 = 1$ and $E^0 = E_m^0$, so that, according to (11.5), $\psi = \psi_m^0$. Now if $\varepsilon \neq 0$, the wave function will change and other neighbouring states ψ_n^0 with $n \neq m$ will be admixed (see Fig. 11.1).

To calculate this, we use the fact that the perturbation is small. We expand both the desired expansion coefficients a_m and the energy eigenvalues E_k in powers of the perturbation parameter ε:

$$\begin{aligned} a_m &= a_m^{(0)} + \varepsilon a_m^{(1)} + \varepsilon^2 a_m^{(2)} + \dots \ , \\ E &= E_k = E^{(0)} + \varepsilon E^{(1)} + \varepsilon^2 E^{(2)} + \dots \ . \end{aligned} \tag{11.8}$$

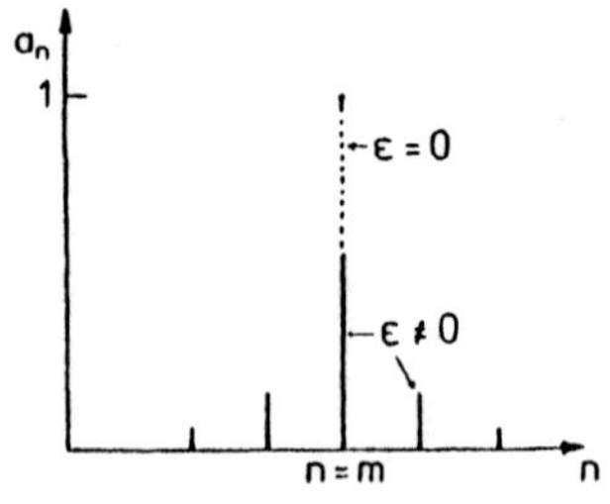

Fig. 11.1. Effect of perturbation: for $\varepsilon \neq 0$ other states ψ_n^0 are mixed with amplitudes a_n to the unperturbed state ψ_m^0. The latter is fully retained for $\varepsilon = 0$. The states in the vicinity of ψ_m^0 are more strongly admixed than those further away

The numbers in the brackets show us the degree of the approximation, e.g. $a_m^{(2)}$ means that this coefficient is small in second order in ε. We now insert the series (11.8) into (11.6a) and order in powers of ε:

$$\begin{aligned} &(E_m^0 - E^{(0)}) a_m^{(0)} + \varepsilon \left[(W_{mm} - E^{(1)}) a_m^{(0)} + (E_m^0 - E^{(0)}) a_m^{(1)} + \sum_{n \neq m} W_{mn} a_n^{(0)} \right] \\ &+ \varepsilon^2 \left[(W_{mm} - E^{(1)}) a_m^{(1)} + (E_m^0 - E^{(0)}) a_m^{(2)} + \sum_{n \neq m} W_{mn} a_n^{(1)} - E^{(2)} a_m^{(0)} \right] \\ &+ \varepsilon^3 [\dots] + \dots = 0 \ . \end{aligned} \tag{11.9}$$

From this formula we can determine the energy values and the expansion coefficients in the various orders of approximation, which we shall now study systematically.

0th Approximation

If we set $\varepsilon = 0$, there is no perturbation and (11.9) yields

$$(E_m^0 - E^{(0)})a_m^{(0)} = 0 \ . \qquad (11.10)$$

m runs over all levels, $m = 1, 2, 3, \ldots$. Let us focus on the level $m = k$ and look for the change of its energy and wave function. Equation (11.10) then yields

$$E^{(0)} = E_k^0 \ , \quad a_m^{(0)} = \delta_{mk} \ . \qquad (11.11)$$

1st Approximation

Inserting these values into (11.9) and taking into consideration only terms up to the first order in ε, we have

$$(E_m^0 - E_k^0)\delta_{mk} + \varepsilon\Bigg[(W_{mm} - E^{(1)})\delta_{mk} + (E_m^0 - E_k^0)a_m^{(1)} + \sum_{n \neq m} \delta_{nk} W_{mn}\Bigg] = 0 \ . \qquad (11.12)$$

The first term does not contribute at all because of the solution of the 0th approximation. For $m = k$, we get the energy shift of the k level in a first approximation as

$$E^{(1)} = W_{kk} \ . \qquad (11.13)$$

The admixture amplitudes for the other states follow from (11.12) for $m \neq k$:

$$(E_m^0 - E_k^0)a_m^{(1)} + W_{mk} = 0 \quad \text{and} \quad a_m^{(1)} = \frac{W_{mk}}{E_k^0 - E_m^0} \ , \quad m \neq k \ . \qquad (11.14)$$

In the case of $m = k$, we obviously do not obtain a condition for the $a_{m=k}^{(1)}$ from (11.12). Hence, we have to determine $a_k^{(1)}$ in a different manner, namely by the normalization of the wave function ψ_k. Indeed for ψ_k, we get, according to (11.8), in first-order perturbation theory

$$\psi = \sum_n a_n \psi_n^0 = \sum_n \left(\sum_{i=0,1} \varepsilon^i a_n^{(i)} \right) \psi_n^0 = \psi_k^0 + \varepsilon \left(a_k^{(1)} \psi_k^0 + \sum_{n \neq k} a_n^{(1)} \psi_n^0 \right)$$

$$= \psi_k^0 + \varepsilon \left(a_k^{(1)} \psi_k^0 + \sum_{n \neq k} \frac{W_{nk}}{E_k^0 - E_n^0} \psi_n^0 \right) . \tag{11.15}$$

Since the ψ_k should also span an orthonormal system of wave functions, we get

$$\langle \psi_k | \psi_k \rangle = 1 = \left\langle \psi_k^0 \middle| \psi_k^0 \right\rangle + \left\langle \psi_k^0 \middle| \varepsilon a_k^{(1)} \psi_k^0 \right\rangle + \left\langle \varepsilon a_k^{(1)} \psi_n^0 \middle| \psi_k^0 \right\rangle$$
$$+ \varepsilon^2 \left\langle a_k^{(1)} \psi_k^0 \middle| a_k^{(1)} \psi_k^0 \right\rangle = 1 + \varepsilon (a_k^{(1)} + a_k^{(1)*}) + \varepsilon^2 |a_k^{(1)}|^2 . \tag{11.16}$$

Neglecting the term proportional to ε^2 (because we are calculating up to first order only), we have

$$0 = \varepsilon (a_k^{(1)} + a_k^{(1)*}) . \tag{11.17}$$

As the wave function is determined only up to a phase factor, we can choose the $a_m^{(1)}$ to be real. Then, obviously, $a_k^{(1)} = 0$ results.

2nd Approximation

If we insert the values of the first approximation into (11.9) for $m = k$, only parts of the coefficient of ε^2 remain and for the energy it follows that

$$E^{(2)} = \sum_{n \neq k} \frac{W_{kn} W_{nk}}{E_k^0 - E_n^0} , \tag{11.18}$$

and analogously for the amplitudes with $m \neq k$,

$$a_m^{(2)} = -\frac{W_{kk} W_{mk}}{(E_m^0 - E_k^0)^2} + \sum_{n \neq k} \frac{W_{mn} W_{nk}}{(E_k^0 - E_n^0)(E_k^0 - E_m^0)} , \qquad m, n \neq k . \tag{11.19}$$

Again, we do not obtain a condition for $a_m^{(2)}$ in the case $m = k$, so we have to use the normalization condition of the wave function once more. This procedure can be continued so that the perturbation effects can be determined in any degree of approximation. According to (11.8) we got for the energy of state k in the 2nd approximation

$$E_k = E_k^0 + \varepsilon W_{kk} - \varepsilon^2 \sum_{n \neq k} \frac{W_{kn} W_{nk}}{E_n^0 - E_k^0} + \ldots . \tag{11.20}$$

This contains the interesting result that, in first order, the correction of the energy is simply the expectation value of the perturbation W, which is quite reasonable. If k denotes the ground state of a system, $E_k^0 < E_n^0$, and the effect of a second-order approximation is always negative, regardless of the sign of the perturbation, because

$$W_{nk}W_{kn} = |W_{kn}|^2 \quad \text{and} \quad E_n^0 - E_k^0 > 0$$

are always positive. This is an important fact, which we can use in many problems, particularly in those cases in which the first-order correction W_{kk} vanishes for one reason or another.

For the application of perturbation theory, we assumed the perturbation to be small, i.e. the energy levels and their differences are not changed significantly. We can express this in the following way:

$$\left| \frac{\varepsilon W_{mn}}{E_m^0 - E_n^0} \right| \ll 1 \quad \text{for} \quad m \neq n \ . \tag{11.21}$$

Since the energies E_m^0 and E_n^0 are very close to each other for large quantum numbers in the Coulomb field [see (9.43)], perturbation theory can only be applied to the case of strongly bound states. We have required during the derivation of the perturbation formulas that the nature of the spectra not be changed qualitatively. The perturbed states ψ_k should continuously emerge from the unperturbed states $\psi_k^{(0)}$ if the perturbation $\hat{W}$ is turned on.

11.2 Degeneracy

Now we shall briefly discuss the application of perturbation theory to a spectrum with degenerate states. Up to now we have talked about states without any degeneracy. Indeed, for any energy E_k^0, we have assumed that only one definite state ψ_k^0 exists; in a system in which degeneracy occurs, this is no longer the case. For a given level of energy E_n^0, a series of eigenfunctions $\psi_{n\beta}^0$, $\beta = 1, 2, \ldots, f_n$ might exist. (Here, β stands for one or more quantum numbers.) The energy eigenvalues are independent of β. Such a level is called *f_n-fold degenerate*.

If we go back to (11.6a), we will now have to write it in the following form:

$$a_{m\alpha}(E_m^0 + \varepsilon W_{m\alpha m\alpha} - E) + \varepsilon \sum_{n\beta \neq m\alpha} a_{n\beta} W_{m\alpha n\beta} = 0 \ , \tag{11.22}$$

where, according to (11.7), the matrix elements are given by

$$W_{m\alpha n\beta} = \int \psi_{m\alpha}^{0*} \hat{W} \psi_{n\beta}^0 \, \mathrm{d}V \ . \tag{11.23}$$

The energy eigenvalue E_n^0 of the unperturbed state contains no additional index. It is independent of α because of degeneracy; this is precisely the peculiarity of degeneracy.

If we look at the 0th approximation, we can see the effect of perturbation on the degenerate state quite clearly. From (11.10), we get for the 0th approximation for the level $m = k$

$$a^{(0)}_{k\alpha}(E^0_k - E^{(0)}) = 0\ . \tag{11.24}$$

Obviously,

$$E^{(0)} = E^0_k \quad \text{and} \quad a^{(0)}_{k\alpha} = a^0_{k\alpha} \neq 0 \quad \text{for} \quad \alpha = 1 \dots f_k$$
$$\text{and} \quad a^0_m = 0 \quad \text{for} \quad m \neq k\ .$$

The double sum over n and β reduces to a single sum over β for the 0th approximation (because $n = k$ only) and we get for the kth level

$$(E^0_k + \varepsilon W_{k\alpha k\alpha} - E)a^0_{k\alpha} + \varepsilon \sum_{\beta\neq\alpha}^{f_k} a^0_{k\beta} W_{k\alpha k\beta} = 0\ . \tag{11.25}$$

The index α runs from 1 to f_k. Therefore (11.25) represents a system of f_k linear equations for the $a^{(0)}_{k\alpha}$. The determinant of the system is of dimension f_k. It has to vanish if the linear system of equations is to give nontrivial solutions $a^0_{k\alpha}$, i.e. solutions not equal to zero. Hence,

$$D_k = \begin{vmatrix} E^0 + \varepsilon W_{11} - E & \varepsilon W_{12} \cdots & \varepsilon W_{1f_k} \\ \varepsilon W_{21} & E^0 + \varepsilon W_{22} - E \cdots & \varepsilon W_{2f_k} \\ \vdots & & \vdots \\ \varepsilon W_{f_k 1} & \cdots & E^0 + \varepsilon W_{f_k f_k} - E \end{vmatrix} = 0\ . \tag{11.26}$$

We have deleted the index k in the determinant because it always appears in the same way. Equation (11.26) is called a *secular equation.* It is an equation of degree f_k for the determination of the energy E and it thus has in general f_k solutions $E_{k\alpha}$ for E. As the perturbation $\varepsilon\hat{W}$ is small, the solutions are close together. *In general the degeneracy of a level is lifted under the influence of perturbation and the formerly f_k-fold degenerate state splits up energetically into f_k close-lying states* with energies $E_{k\alpha}$ $\alpha = 1, \dots, f_k$.

The appearance of a degeneracy can always be traced back to a symmetry of the system. For example, the $(2l+1)$-fold angular-momentum degeneracy of the state of a particle in the central potential (see Exercise 7.2 and Chap. 9) is a result of the spherical symmetry (isotropy of space) of the potential. If the symmetry is broken by a perturbation (broken symmetry), the degenerate levels split into a series of neighbouring levels. Such a perturbation may be caused by an additional weak interaction (e.g. spin–orbit coupling causes so-called fine-structure splitting) or by applying an external field.

The eigenfunctions $\varphi_{k\alpha}$ of the energies $E_{k\alpha}$ are special linear combinations of the degenerate states $\psi^0_{k\beta}$. The corresponding amplitudes $a^0_{k\alpha\beta}$ are obtained by insertion of the solutions $E = E_{k\alpha}$ into (11.25), which can then be solved for $a^0_{k\alpha}$. The resulting eigenfunctions are then of the form

$$\varphi_{k\alpha} = \sum_{\beta=1}^{f_k} a^0_{k\alpha\beta}\psi^0_{k\beta} \,. \tag{11.27}$$

Since we are now again dealing with nondegenerate levels, the further approximation can be obtained in perturbation theory as before.

EXAMPLE

11.1 The Stark Effect

As an example of perturbation theory, we now calculate the level splitting of a hydrogen atom in a homogeneous electric field. As we shall see, the effect of the electric field on an atom is to split the spectral lines. This phenomenon was experimentally shown by ***Stark*** in 1913.

Experiments showed that the effect of an electric field on hydrogen or other atoms depends on the field strength; this is, of course, to be expected. But the effect is different for hydrogen than for other atoms. The energy levels of hydrogen (e.g. the *Balmer series*) for weak fields split proportionally to the first power of the field strength (the so-called *linear Stark effect*). The splitting of the energy levels of all other atoms is proportional to the second power in the field strength (*quadratic Stark effect*).

There was no explanation for the *Stark*[1] *effect* in classical theory; only quantum mechanics indicated how to understand this phenomenon. We shall now discuss the theory of the linear Stark effect in greater detail, restricting ourselves to the second level ($n = 2$) of hydrogen.

The applied external electric field $\boldsymbol{E}$ (in the experiments it was 10^4–10^5 V/cm) is much smaller than the inner atomic one, which is caused by the nucleus and is of the order $E_{\text{nucl}} = e/a_0^2 \approx 5 \times 10^9$ V/cm (a_0 is the radius of the first Bohr orbit). To solve the problem we use perturbation theory of the degenerate case. *The potential energy of the electron in the external electric field, $\hat{V}$,* is treated as the perturbation.

The first level ψ_{100} in the hydrogen atom is not degenerate. Therefore, in the simplest case, we start from the splitting of the second level. As we know, the

[1] A general discussion of symmetry problems in quantum mechanics can be found in W. Greiner and B. Müller: *Quantum Mechanics – Symmetries*, 2nd ed. (Springer, Berlin, Heidelberg 1994)

Example 11.1

hydrogen levels are n^2-fold degenerate; i.e. four eigenfunctions belong to the energy $E_n^0 = E_2^0$ of the unperturbed hydrogen atom. These wave functions are (see Table 9.1)

$$\varphi_1 = \Psi_{200} = \frac{(1 - r/2a_0)}{\sqrt{2a_0^3}} \mathrm{e}^{-r/2a_0} Y_{00} \quad \text{(2s state)} , \tag{1a}$$

$$\varphi_2 = \Psi_{210} = \frac{r/2a_0}{\sqrt{6a_0^3}} \mathrm{e}^{-r/2a_0} Y_{10} \quad \text{(2p states)} , \tag{1b}$$

$$\varphi_{3,4} = \Psi_{21\pm1} = \frac{r/2a_0}{\sqrt{6a_0^3}} \mathrm{e}^{-r/2a_0} Y_{1\pm1} . \tag{1c}$$

The four-fold degeneracy is lifted by the appearance of an electric field. Now there is an additional potential energy for the electron in the homogeneous field $\boldsymbol{E}$ because of the adjustment of the dipole moment $e\boldsymbol{r}$ of the electron in the field. If we arrange the electric field in the z direction, the potential energy will be given by

$$V = -e\boldsymbol{r} \cdot \boldsymbol{E} = -ez\,|\boldsymbol{E}| = -e\,|\boldsymbol{E}|\,r\sqrt{\frac{4\pi}{3}} Y_{10} . \tag{2}$$

Let $\hat{H}_0$ denote the Hamiltonian of the unperturbed system. The perturbed Hamiltonian will then be

$$\hat{H} = \hat{H}_0 + \hat{V} .$$

We calculate the matrix elements of the perturbation according to (11.26) and use the functions φ_α introduced in (1). The matrix elements are of the form

$$V_{\alpha\beta} = \int_{-\infty}^{\infty} \varphi_\alpha^* \hat{V} \varphi_\beta \,\mathrm{d}V .$$

Most of the integrands are odd functions of the space coordinates; this can immediately be seen upon insertion of the perturbation (2) and functions φ_α of (1). After integration, only the matrix elements V_{12} and V_{21} turn out to be nonvanishing. In this case,

$$V_{12} = V_{21}$$
$$= \int_{-\infty}^{\infty} \frac{1 - r/2a_0}{\sqrt{2a_0^3}} \mathrm{e}^{-r/2a_0} (-e|\boldsymbol{E}|) \frac{r/2a_0}{\sqrt{6a_0^3}} \mathrm{e}^{-r/2a_0} r Y_{00}^* Y_{10} Y_{10} \sqrt{\frac{4\pi}{3}} \,\mathrm{d}V$$

Example 11.1

$$= -e\,|\boldsymbol{E}| \int_0^\infty (1 - r/2a_0)\frac{r}{12a_0^4}\,\mathrm{e}^{-r/a_0} r^3\,\mathrm{d}r \int \mathrm{d}\Omega |Y_{10}|^2(\vartheta,\varphi)$$

$$= -\frac{e|E|a_0}{12}\int_0^\infty \varrho^4\,\mathrm{d}\varrho(1-\varrho/2)\,\mathrm{e}^{-\varrho} \times 1 = 3e\,|\boldsymbol{E}|\,a_0\ . \quad (3)$$

Here, we have used the fact that $Y_{00} = (4\pi)^{-1/2}$. Because of the degeneracy of the system, the general solution of the Hamiltonian to the energy eigenvalue E_2^0 is given by a linear combination of the functions φ_α:

$$\Psi = \sum_{\alpha=1}^{4} a_\alpha \varphi_\alpha\ . \quad (4)$$

To determine the coefficients a_α, we use the system of equations (11.25), which in this case reads

$$(E_2^0 + V_{\alpha\alpha} - E)a_\alpha + \sum_{\beta\neq\alpha}^{4} a_\beta V_{\alpha\beta} = 0\ ,$$
$$\alpha = 1, 2, 3, 4\ . \quad (5)$$

Since all matrix elements vanish except for V_{12} and V_{21}, the system reduces to four equations, namely

$$(E_2^0 - E)a_1 + V_{12}a_2 = 0\ , \quad (6a)$$

$$V_{21}a_1 + (E_2^0 - E)a_2 = 0\ , \quad (6b)$$

$$(E_2^0 - E)a_3 = 0\ , \quad (6c)$$

$$(E_2^0 - E)a_4 = 0\ . \quad (6d)$$

To get a nontrivial solution the coefficient determinant has to vanish, i.e.

$$\begin{vmatrix} E_2^0 - E & V_{12} & 0 & 0 \\ V_{12} & E_2^0 - E & 0 & 0 \\ 0 & 0 & E_2^0 - E & 0 \\ 0 & 0 & 0 & E_2^0 - E \end{vmatrix} = 0\ .$$

From this we easily obtain the four values for the energies of the perturbed levels, which are, as we know, solutions of this determinant. The result is

$$E_a = E_b = E_2^0\ , \quad E_c = E_2^0 + V_{12}\ , \quad E_d = E_2^0 - V_{12}\ . \quad (7)$$

Example 11.1

Obviously, the superposition of the homogeneous field did not completely cancel degeneracy. This can be explained by the fact that we now have cylindrical symmetry instead of the spherical symmetry from before. In other words, no complete cancellation of symmetry occurs. The resulting splitting is shown in the following figure.

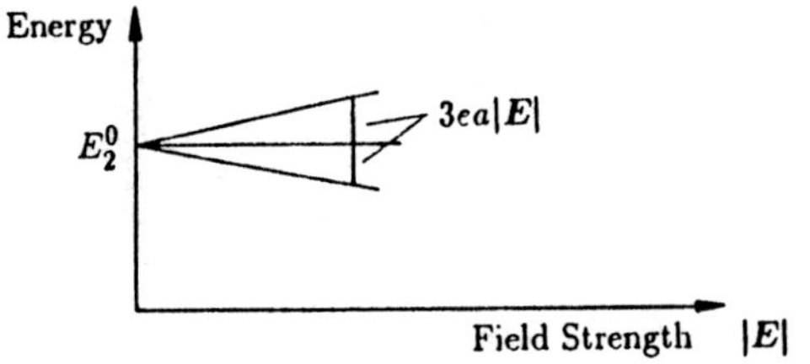

The four degenerate hydrogen levels belonging to the main quantum number $n = 2$, split up by the Stark effect into three levels. The medium energy, which corresponds to the unperturbed energy E_2^0, is still twice degenerate

The linear level splitting by the electric field appears only in hydrogen. It results from the linearity of V_{12} in E [see (3)] and because of (7).

There is no l degeneracy in many-electron systems; therefore, no average dipole moment exists, but the atom is polarized by the external field. The so-induced dipole moment is proportional to the field strength; therefore the energies of the atom change with $|\boldsymbol{E}|^2$. This phenomenon is called the *quadratic Stark effect*. To get the wave functions (4) corresponding to the energy values (7), we insert the energies into the system (6):

$$\begin{aligned}
&\text{For}\quad E_a = E_b = E_2^0\ , && \text{it follows that } a_1 = a_2 = 0\ , \quad a_3 \text{ and } a_4 \text{ arbitrary}\ .\\
&\text{For}\quad E_c = E_2^0 + V_{12}\ , && \text{it follows that } a_1 = a_2\ , \quad a_3 = a_4 = 0\ .\\
&\text{For}\quad E_d = E_2^0 - V_{12}\ , && \text{we get } a_1 = -a_2\ , \quad a_3 = a_4 = 0\ .
\end{aligned}$$

With the applied field $|\boldsymbol{E}|$ we thus obtain the following wave functions:

$$\text{for}\quad E = E_2^0:\quad \Psi_{\mathrm{III,IV}} = a_3\varphi_3 + a_4\varphi_4 = a_3\psi_{211} + a_4\Psi_{21-1}\ ,\quad \text{with}\quad a_3^2 + a_4^2 = 1\ ,$$

$$\begin{aligned}
\text{for}\quad E = E_2^0 + V_{12}:\quad \Psi_{\mathrm{II}} &= \frac{1}{\sqrt{2}}(\varphi_1 + \varphi_2)\\
&= \frac{1}{\sqrt{2}}(\psi_{200} + \psi_{210})\ ,\\
\text{for}\quad E = E_2^0 - V_{12}:\quad \Psi_{\mathrm{I}} &= \frac{1}{\sqrt{2}}(\varphi_1 - \varphi_2)\\
&= \frac{1}{\sqrt{2}}(\psi_{200} - \psi_{210})\ .
\end{aligned} \tag{8}$$

It is easy to show that the matrix, built by the functions $\Psi_{\mathrm{I,II,III,IV}}$, is diagonal.

The *Stark effect with $n = 2$* is *qualitatively* interpreted in the following way. As the motion of the characteristic wave function of the electron is *not spheri-*

cally symmetric (because of the degeneracy of the 2s and 2p states), the atom has an electric dipole moment $\boldsymbol{p}$. For this reason, an atom in the electric field

$$\boldsymbol{E} = (E_x = 0,\ E_y = 0,\ E_z = E)$$

gains the additional energy

$$V = -(\boldsymbol{p} \cdot \boldsymbol{E}) = -|\boldsymbol{p}|\, E \cos\gamma \ , \tag{9}$$

where γ is the angle between the direction of the electric dipole of the atom and the z direction.

If we compare this expression with (3) and (7), we see that the electric dipole moment of the atom is $|\boldsymbol{p}| = 3a_0 e$, where the solution Ψ_{I} corresponds to $\gamma = 0$, but Ψ_{II} corresponds to $\gamma = \pi$. For the third and fourth solutions we then have $\gamma = \pm\frac{\pi}{2}$. *The latter result is due to an electric dipole moment which is perpendicular to the electric field; for this reason, no additional energy is produced. In other words, the linear Stark effect in a hydrogen atom with $n = 2$ is caused by the characteristic electric dipole momentum $\boldsymbol{p}$.*

The results of these calculations, obtained by the use of quantum mechanics, agree with the experiments only for weak fields ($E \sim 10^3$ V/cm). For strong fields ($E > 10^4$ V/cm), additional splitting occurs (the quadratic Stark effect), caused by the fact that the degeneracy in the angular-momentum quantum number is broken. The Stark effect vanishes totally if the field strength is greater than 10^5 V/cm. This phenomenon is connected with the self-ionization of atoms in the electric field: electrons in an excited level lose their binding to the atom if strong homogeneous electric fields are superposed.

EXERCISE

11.2 Comparison of a Result of Perturbation Theory with an Exact Result

Consider a hydrogen-like atom with a central nuclear charge number Z, containing one $1s$ electron.

Problem. Calculate the change in energy by increasing the charge of the nucleus by one ($Z \to Z+1$). Use first-order perturbation theory and compare this with the exact result.

Solution. The wave function for the first electron reads

$$\psi_{1s} = \frac{1}{\sqrt{\pi}}\gamma^{3/2}\mathrm{e}^{-\gamma r} \ , \quad \gamma = \frac{me^2}{\hbar^2}Z \ . \tag{1}$$

The unperturbed energy is

$$E_{1s}(Z) = -\frac{me^4}{2\hbar^2}Z^2 \ . \tag{2}$$

Exercise 11.2

In first-order perturbation theory, the energy change is given by

$$\Delta E_{1s} = \langle\psi|\,\Delta\hat{H}\,|\psi\rangle \ , \tag{3}$$

i.e. by the expectation value of the perturbation operator $\Delta\hat{H}$. In our case, we have

$$\Delta\hat{H} = \Delta V = -\frac{e^2}{r} \ . \tag{4}$$

This yields

$$\begin{aligned}\Delta E_{1s} &= \frac{1}{\pi}\gamma^3 4\pi \int_0^\infty \mathrm{d}r\, r^2 \mathrm{e}^{-2\gamma r}\left(-\frac{e^2}{r}\right) \\ &= -4\gamma^3 e^2 \int_0^\infty \mathrm{d}r\, r\,\mathrm{e}^{-2\gamma r} = -4\gamma^3 e^2 \frac{1}{4\gamma^2} \\ &= -\gamma e^2 = -\frac{me^4}{\hbar^2} Z \ .\end{aligned} \tag{5}$$

In comparison with this, the exact result is

$$\begin{aligned}E_{1s}(Z+1) - E_{1s}(Z) &= -\frac{me^4}{2\hbar^2}[(Z+1)^2 - Z^2] \\ &= -\frac{me^4}{\hbar^2}\left(Z+\frac{1}{2}\right) \ .\end{aligned} \tag{6}$$

The first-order perturbation theory approximation is obviously quite good for large Z, as would be expected.

EXERCISE

11.3 Two-State Level Crossing

Let the Hamiltonian H_0 have two closely neighbouring levels with energies $E_1^{(0)} \approx E_2^{(0)}$ and eigenfunctions $\psi_1^{(0)}$, $\psi_2^{(0)}$. Let all other eigenvalues be very different from them, so that the two levels can be considered as rather isolated energetically.

Problem. Investigate the Hamiltonian

$$\hat{H} = \hat{H}_0 + \hat{V} \quad \text{and}$$

(a) show by use of perturbation theory that on the one hand, only these levels contribute to the correction of the energy eigenvalues and, on the other hand, that one cannot use perturbation theory any longer. What must ψ_1^0 and ψ_2^0 be like?

Exercise 11.3

(b) Show that perturbation theory can be improved by diagonalizing the two-level problem, which gives

$$E_{1,2} = \frac{1}{2}(H_{11}+H_{22}) \pm \frac{1}{2}\sqrt{(H_{11}-H_{22})^2+4\,|H_{12}|^2}\ ,$$

with

$$H_{ij} = \int \psi_i^{(0)*}\hat{H}\psi_j^{(0)}\,\mathrm{d}x\ .$$

(c) Plot $E_{1,2}$ as a function of $\Delta H = H_{11}-H_{22}$. What does $\Delta E = E_1 - E_2$ look like if, in the first approximation, a level crossing occurs, i.e. if the potential V is such that $H_{11} = H_{22}$?

Solution. (a) The first correction of the energy eigenvalues is

$$E_i^{(1)} = E_i^{(0)} + V_{ii} = H_{ii}\ , \tag{1}$$

and, in second order,

$$E_i^{(2)} = H_{ii} - \sum_{j\neq i} |V_{ij}|^2 \Big/ (E_j^{(0)} - E_i^{(0)})\ . \tag{2}$$

For the two levels we have $E_1^0 \approx E_2^0$, i.e. the denominator is going to be small and we can neglect all other terms if $V_{ij} \neq 0$ (this means that for a radially symmetric potential, the wave functions ψ_1 and ψ_2 must have the same angular-momentum quantum numbers). In this case we can neglect all states except 1 and 2. Nevertheless, because of the small energy denominators in (2) certain terms in all orders of perturbation theory are going to be very large and perturbation theory loses its meaning.

(b) Let the eigenfunctions of $\hat{H}_0$ be ψ_1^0, ψ_2^0 with

$$\hat{H}_0\psi_i^{(0)} = E_i^{(0)}\psi_i^{(0)}\ , \quad i = 1,2\ . \tag{3}$$

We diagonalize the total Hamiltonian $\hat{H} = \hat{H}_0 + V$ within this two-state basis, i.e. we search for

$$\hat{H}\psi = (\hat{H}_0 + \hat{V})\psi = E\psi\ , \tag{4}$$

with

$$\psi = a\psi_1^{(0)} + b\psi_2^{(0)}\ . \tag{5}$$

Multiplying from the left by ψ_1^{0*} and ψ_2^{0*}, and integrating over x, we get the following system of linear equations for a and b:

$$\begin{aligned} &(H_{11}-E)a + V_{12}b = 0\ ,\\ &V_{21}a + (H_{22}-E)b = 0\ , \end{aligned} \tag{6}$$

Exercise 11.3

which has a solution $(a,b) \neq (0,0)$ if

$$\det\begin{pmatrix} H_{11}-E & V_{12} \\ V_{21} & H_{22}-E \end{pmatrix} = (H_{11}-E)(H_{22}-E) - |V_{12}|^2$$
$$= E^2 - (H_{11}+H_{22})E + H_{11}H_{22} - |V_{12}|^2 = 0 \qquad (7)$$

or

$$E_{1,2} = \frac{1}{2}(H_{11}+H_{22}) \pm \frac{1}{2}\sqrt{(H_{11}-H_{22})^2 + 4\,|V_{12}|^2} \qquad (8)$$

is valid. If we can use ordinary perturbation theory, i.e. if

$$\left|E_1^{(0)} - E_2^{(0)}\right| \gg |V_{12}| \quad \text{and}$$
$$\left|E_1^{(0)} - E_2^{(0)}\right| \gg |V_{22} - V_{11}| \qquad (9)$$

is valid, we get from (8) the energies,

$$E_i = E_i^{(0)} + V_{ii} \pm |V_{12}|^2/(E_i^{(0)} - E_j^{(0)})$$
$$i = 1,2\,; \quad j = 2,1\,. \qquad (10)$$

This is exactly the result of second-order perturbation theory.

(c) Let us abbreviate

$$\Delta = H_{11} - H_{22}\,;$$

then

$$E_{1,2} = H_{11} - \frac{1}{2}\Delta \pm \frac{1}{2}\sqrt{\Delta^2 + 4\,|V_{12}|^2}\,. \qquad (11)$$

The shortest distance between the energy eigenvalues is $2|V_{12}|$, i.e. the level crossing is prevented by the interaction. This is related to the so-called *Landau–Zener effect*, which we shall not discuss here. It should only be mentioned that the Landau–Zener effect deals with the question of the transfer of a particle (e.g. an electron) occupying one of the two crossing levels. As we might intuitively imagine, the particle transfer to the other crossing state depends on how close the two levels come to each other and also on the velocity with which the level crossing is passed. Inserting the energies $E_{1,2}$ of (8) into (6), the new states can be easily calculated:

$$\psi_1 = a_1\psi_1^{(0)} + b_1\psi_2^{(0)}$$
$$= V_{12}\psi_1^{(0)} - \left(+\frac{1}{2}\Delta - \frac{1}{2}\sqrt{\Delta^2 + 4\,|V_{12}|^2}\right)\psi_2^{(0)}\,,$$
$$\psi_2 = a_2\psi_1^{(0)} + b_2\psi_2^{(0)}$$
$$= V_{12}\psi_1^{(0)} - \left(+\frac{1}{2}\Delta + \frac{1}{2}\sqrt{\Delta^2 + 4\,|V_{12}|^2}\right)\psi_2^{(0)}\,, \qquad (12)$$

Exercise 11.3

with the interesting result that the wave functions are intermixed strongly at the level crossing, but are practically unperturbed far away from the crossing point (large $|\Delta|$). Moreover, $\psi_1^{(\Delta)}$ approaches $\psi_2^{(0)}$ far to the left of the crossing point (large negative Δ) and it approaches $\psi_1^{(0)}$ far to the right of the crossing point (large positive Δ). For $\psi_2^{(\Delta)}$ the situation is reversed. This fact can be expressed as follows: the unperturbed wave functions do indeed cross and remain unchanged except for in the immediate vicinity of the crossing point. The energies $E_1^{(\Delta)}$ and $E_2^{(\Delta)}$ also change from one unperturbed value $E_1^{(0)}$ and $E_2^{(0)}$, respectively, to the other one, i.e. $E_2^{(0)}$ and $E_1^{(0)}$, respectively. The situation is illustrated in the figure below.

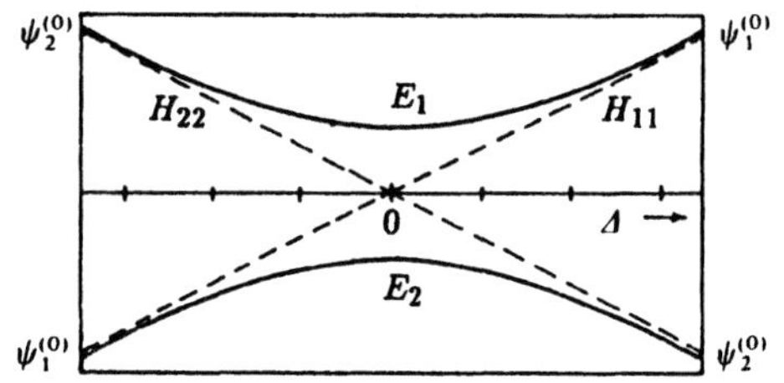

Level crossing: energy levels E_1, E_2, dependent on the energy difference Δ of the unperturbed system. The values of H_{11} and H_{22} are indicated by *dashed lines*

As an example of level crossing, we look at the electronic energy levels in a two-centered potential which is caused by two lead nuclei with a distance R between the two centres. This potential is no longer radially symmetric; it merely still has azimuthal symmetry, i.e. j_z is still a good quantum number. Furthermore, the Hamiltonian commutes with the parity operator. The solutions, shown below, were obtained not with the Schrödinger equation, but with the relativistic Dirac equation, but this is of no importance here. If we start from an $\hat{H}_0$ which belongs to a two-centre distance R and passes over to $R+\Delta R$, the potential changes by

$$\Delta V = V(R+\Delta R; \boldsymbol{r}) - V(R; \boldsymbol{r}) \ ,$$

$$V(R; \boldsymbol{r}) = Ze\left(\frac{1}{|\boldsymbol{r}-\frac{R}{2}\boldsymbol{e}_z|} + \frac{1}{|\boldsymbol{r}+\frac{R}{2}\boldsymbol{e}_z|}\right) \ . \tag{13}$$

This perturbation does not change the azimuthal and parity symmetries. The matrix elements V_{12} vanish if the states have different parity or different magnetic quantum numbers; otherwise, they are in general different from zero.

Now we can understand the figure shown in the margin.[2] (We shall not explain all symbols exactly.) As we can see, the $3s\sigma$ state crosses the $3d_{5/2}\pi$ state at $R \approx 650$ fm. This is possible because $V_{ij} = 0$ (different m quantum numbers, which are called here $\sigma(m=0)$ or $\pi(m=1)$). On the other hand, the $3s\sigma$ and the $3d_{5/2}\sigma$ states repel each other. These states have the same m quantum num-

[2] Figure appeared in G. Soff, W. Greiner, W. Betz, B. Müller: Phys. Rev. A **20**, 169 (1979).

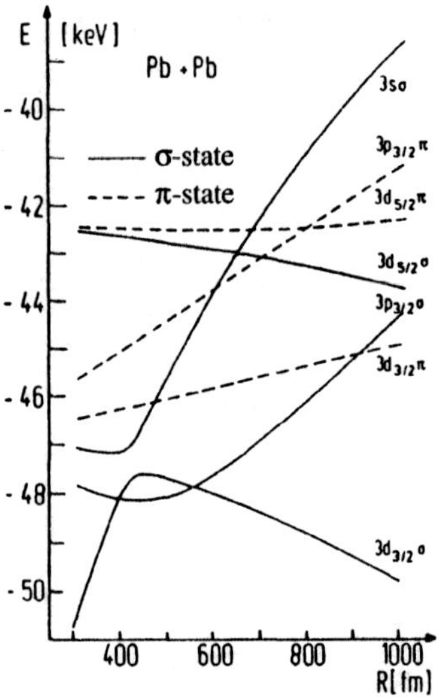

Example of level crossings in an electronic superheavy quasimolecule, as it is found in a heavy-ion collision of lead on lead. These crossings play an important role in the ionization process occurring during such atom–atom encounters

ber and the same parity. The $3p_{3/2}\sigma$ and the $3d_{3/2}\sigma$ states are able to cross each other, because they differ in parity ($\to V_{ij}=0$) while agreeing in the m quantum number.

EXERCISE

11.4 Harmonic Perturbation of a Harmonic Oscillator

Consider the harmonic oscillator Hamiltonian

$$\hat{H}=\hat{H}_0+\hat{W} \quad \text{with} \quad \hat{H}_0=-\frac{\hbar^2}{2m}\frac{\partial^2}{\partial x^2}+\frac{C_0}{2}x^2$$

and let $W=C_1x^2/2$ be an oscillator perturbation potential.

Problem. Calculate, using perturbation theory, the energy eigenvalues and compare them with the exact result.

Solution. In this case, the stationary solutions of the Schrödinger equation with the Hamiltonians $\hat{H}_0$ and $\hat{H}$ are known, namely those of the harmonic oscillator. The exact eigenvalues of $\hat{H}$ are

$$E_n=\hbar\omega(n+\tfrac{1}{2}) \ , \quad \omega=\sqrt{\frac{C_0+C_1}{m}} \ , \tag{1}$$

and those of $\hat{H}_0$ are

$$E_n^0=\hbar\omega_0(n+\tfrac{1}{2}) \ , \quad \omega_0=\sqrt{\frac{C_0}{m}} \ . \tag{2}$$

Nevertheless, we want to calculate the eigenvalues of $\hat{H}$ approximately by perturbation theory to test its effectiveness. It was shown in (11.20) that in second-order perturbation

$$E=E_l^0+W_{ll}+\sum_{n\neq l}\frac{W_{ln}W_{nl}}{E_l^0-E_n^0}+\dots \tag{3}$$

with

$$W_{ln}=\left\langle\phi_l\left|\hat{W}\right|\phi_n\right\rangle=W_{nl}^*=\left\langle\phi_l\left|\frac{C_1}{2}x^2\right|\phi_n\right\rangle \ . \tag{4}$$

To calculate W_{ln}, we need some of the relations already determined in Chap. 7, and again introduce the coordinate

$$\xi=\sqrt{\lambda}x=\sqrt{\frac{m\omega_0}{\hbar}}x \ .$$

The basis functions are [see (7.52)]

Exercise 11.4

$$\phi_n = \frac{(m\omega_0/\hbar)^{1/4}}{(\pi^{1/2}2^n n!)^{1/2}} \mathrm{e}^{-\xi^2/2} H_n(\xi) \ , \tag{5}$$

where $H_n(\xi)$ are the Hermitian polynomials. The ψ_n are the eigenfunctions of H_0. Then

$$\xi\phi_n = \sqrt{\frac{n}{2}}\phi_{n-1} + \sqrt{\frac{n+1}{2}}\phi_{n+1} \ . \tag{6}$$

We need, however,

$$\begin{aligned} \xi^2\phi_n &= \sqrt{\frac{n}{2}}\xi\phi_{n-1} + \sqrt{\frac{n+1}{2}}\xi\phi_{n+1} \\ &= \tfrac{1}{2}\sqrt{n(n-1)}\phi_{n-2} + \left(n+\tfrac{1}{2}\right)\phi_n + \frac{1}{2}\sqrt{(n+1)(n+2)}\,\phi_{n+2} \ . \end{aligned} \tag{7}$$

With this we can calculate

$$\begin{aligned} W_{ln} &= \left\langle \phi_l \left| \frac{C_1}{2}x^2 \right| \phi_n \right\rangle = \frac{C_1}{2}\frac{\hbar}{m\omega_0}\left\langle \phi_l \left| \xi^2 \right| \phi_n \right\rangle \\ &= \frac{C_1}{2}\frac{\hbar}{m\omega_0}\left[\frac{1}{2}\sqrt{n(n-1)}\delta_{l,n-2} + \left(n+\frac{1}{2}\right)\delta_{l,n} \right. \\ &\quad \left. + \frac{1}{2}\sqrt{(n+1)(n+2)}\delta_{l,n+2}\right] \ . \end{aligned} \tag{8}$$

So we get for the energy of the ground state ($l = 0$)

$$E = E^0_{l=0} + W_{00} + \sum_{n=1} \frac{W_{n0}W_{0n}}{E^0_{l=0} - E^0_n} + \dots \ . \tag{9}$$

As all $W_{0n} = 0$ except for $n = 2$, it follows that

$$\begin{aligned} E &= \frac{1}{2}\hbar\omega_0 + \left\langle \phi_0 \left| (C_1/2)x^2 \right| \phi_0 \right\rangle + \frac{|\langle\phi_0|(C_1/2)x^2|\phi_2\rangle|^2}{E^0_{l=0} - E^0_{n=2}} + \dots \\ &= \frac{1}{2}\hbar\omega_0 + \frac{C_1}{2}\frac{\hbar}{m\omega_0}\frac{1}{2} + \frac{(C_1^2/4)(\hbar^2/m^2\omega_0^2)(1/2)}{\frac{1}{2}\hbar\omega_0 - \frac{5}{2}\hbar\omega_0} + \dots \\ &= \frac{1}{2}\hbar\omega_0\left[1 + \frac{1}{2}\frac{C_1}{C_0} - \frac{1}{8}\left(\frac{C_1}{C_0}\right)^2 \pm \dots\right] \ , \end{aligned} \tag{10}$$

because $C_0 = m\omega_0^2$.

Exercise 11.4

In fact it is possible to take into consideration all orders of perturbation theory. The result will be

$$\begin{aligned} E &= \frac{1}{2}\hbar\omega_0\left[1+\frac{1}{2}\frac{C_1}{C_0}-\frac{1}{2\times 4}\left(\frac{C_1}{C_0}\right)^2\right. \\ &\quad \left.+\frac{1\times 3}{2\times 4\times 6}\left(\frac{C_1}{C_0}\right)^3-\frac{1\times 3\times 5}{2\times 4\times 6\times 8}\left(\frac{C_1}{C_0}\right)^4\pm\ldots\right] \\ &= \frac{1}{2}\hbar\omega_0\sqrt{1+\frac{C_1}{C_0}} \\ &= \frac{1}{2}\hbar\sqrt{\frac{C_0}{m}}\sqrt{1+\frac{C_1}{C_0}}=\frac{1}{2}\hbar\sqrt{\frac{C_0+C_1}{m}}=\frac{1}{2}\hbar\omega\ . \end{aligned} \tag{11}$$

With this we have found the exact result in perturbation theory of infinite order. It is also clear that second-order perturbation theory (10) yields a correction for the unperturbed oscillator towards the exact result for the modified oscillator (11).

EXERCISE

11.5 Harmonic Oscillator with Linear Perturbation

Consider the harmonic-oscillator Hamiltonian

$$\hat{H}=\hat{H}_0+W \quad \text{with} \quad \hat{H}_0=-\frac{\hbar^2}{2m}\frac{\partial^2}{\partial x^2}+\frac{C_0}{2}x^2$$

with $W=C_0ax$ as the perturbation potential.

Problem. Calculate, using perturbation theory, the energy eigenvalues and compare them with the exact result.

Solution. In this case, the exact stationary solutions are known, too:

$$-\frac{\hbar^2}{2m}\frac{\partial^2}{\partial x^2}\psi+\frac{C_0}{2}(x^2+2ax)\psi=E\psi\ . \tag{1}$$

We define $y=x+a$ and transform to

$$-\frac{\hbar^2}{2m}\frac{\partial^2}{\partial y^2}\psi+\frac{C_0}{2}(y^2-a^2)\psi=E\psi \tag{2}$$

or

$$\begin{aligned} &-\frac{\hbar^2}{2m}\frac{\partial^2}{\partial y^2}\psi+\frac{C_0}{2}y^2\psi=E'\psi\ ; \\ &E'=E+\frac{C_0}{2}a^2 \end{aligned} \tag{3}$$

Exercise 11.5

Equation (3) is the ordinary oscillator equation with the known eigenvalues

$$E'_n = \hbar\omega_0\left(n+\tfrac{1}{2}\right) \ , \quad n=0,1,\dots$$
$$\omega_0 = \sqrt{\frac{C_0}{m}} \tag{4}$$

or

$$E_n = \hbar\omega_0(n+\tfrac{1}{2}) - \frac{C_0}{2}a^2 \ . \tag{5}$$

Now we try to find this result by perturbation theory. In order to do so, we again need the matrix elements

$$W_{ln} = \langle\phi_l\,|C_0 ax|\,\phi_n\rangle = C_0 a\sqrt{\frac{\hbar}{m\omega_0}}\,\langle\phi_l\,|\xi|\,\phi_n\rangle \ ,$$
$$\omega_0 = \sqrt{\frac{C_0}{m}} \ . \tag{6}$$

For the energy of the ground state ($l=0$), we get, using

$$W_{0n} = C_0 a\sqrt{\frac{\hbar}{m\omega_0}}\,\langle\phi_0\,|\xi|\,\phi_n\rangle$$
$$= C_0 a\sqrt{\frac{\hbar}{m\omega_0}}\left(\sqrt{\frac{n}{2}}\delta_{0,n-1} + \sqrt{\frac{n+1}{2}}\delta_{0,n+1}\right) \ , \tag{7}$$

the equations

$$W_{00} = 0 \ , \quad W_{01} = C_0 a\sqrt{\frac{\hbar}{m\omega_0}}\sqrt{\frac{1}{2}} \ ;$$
$$W_{0n} = 0 \quad \text{for} \quad n \neq 0, 1 \ . \tag{8}$$

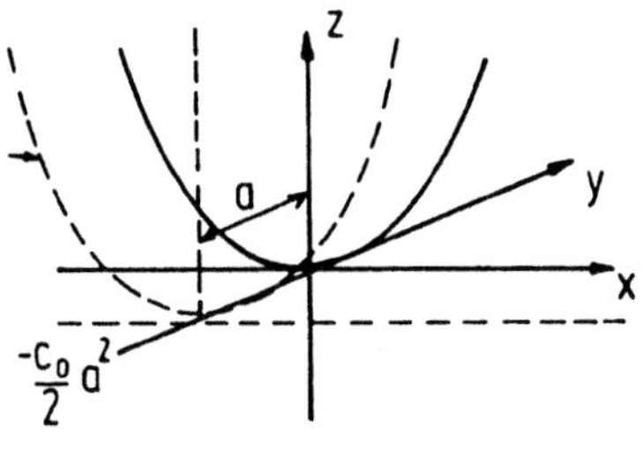

The superposition of an oscillator potential with a linear potential yields a shift of the original oscillator; otherwise everything remains the same. The dashed lines refer to the shifted coordinate system

Exercise 11.5

This yields

$$E_{n=0} = \frac{1}{2}\hbar\omega_0 + 0 + \frac{C_0^2 a^2 (\hbar/m\omega_0)\frac{1}{2}}{\frac{1}{2}\hbar\omega_0 - \frac{1}{2}\hbar\omega_0 3} + \dots$$
$$= \frac{1}{2}\hbar\omega_0 - \frac{C_0^2 a^2}{2m\omega_0^2} + \dots$$
$$= \frac{1}{2}\hbar\omega_0 - \frac{C_0}{2}a^2 \ . \qquad (9)$$

This agrees with the exact result above. Therefore we can conclude that for the ground state the higher orders of perturbation theory have to vanish identically.

11.3 The Ritz Variational Method

It is possible to determine the ground state without the explicit solution of the Schrödinger equation by requiring that its energy be the lowest of all possible energies for all possible wave functions. To understand this, we consider an arbitrary Hamiltonian $\hat{H}$ and demand that its spectrum have a *lower limit*. This means that it has a lowest, nondegenerate energy eigenvalue:

$$\hat{H}\psi_n = E_n\psi_n \ ; \quad (n = 0, 1, \dots) \ ; \qquad (11.28)$$

$$E_n > E_0 \ ; \quad (n \neq 0) \ . \qquad (11.29)$$

We can expand any arbitrary normalized wave function ψ into eigenfunctions of $\hat{H}$ and get

$$|\psi\rangle = \sum_n a_n\psi_n \ , \quad \sum_n |a_n|^2 = 1 \ . \qquad (11.30)$$

Then the mean energy of ψ is

$$\left\langle\psi\left|\hat{H}\right|\psi\right\rangle = \sum_{nm} a_n^*\left\langle\psi_n\left|\hat{H}\right|\psi_m\right\rangle a_m = \sum_{nm} a_n^* a_m E_n\delta_{nm}$$
$$= \sum_n E_n|a_n|^2 \geq E_0\sum_n |a_n|^2 = E_0 \ . \qquad (11.31)$$

This obviously means that every other state ψ, which differs from the true ground state ψ_0, has a higher energy than ψ_0. This result can also be written in the form

$$E_0 = \min_{\psi\in H}\left[\frac{\langle\psi|\hat{H}|\psi\rangle}{\langle\psi|\psi\rangle}\right] \ . \qquad (11.32)$$

Here, $\psi \in H$ indicates that ψ is an element of the Hilbert space H. ψ need not even be normalized in this equation. Finding the energy of the ground state has therefore become a *variational problem*. That the expression in brackets (in 11.32) is stationary we know from variational calculus to be a *necessary condition* for the extremum (minimum):

$$\delta(E_\psi) \equiv \delta \frac{\langle\psi|\hat{H}|\psi\rangle}{\langle\psi|\psi\rangle} = 0 \ . \tag{11.33}$$

With the well-known rule for differentiating the ratio of two functions, it follows that

$$\frac{(\delta\langle\psi|\hat{H}|\psi\rangle)\,\langle\psi|\psi\rangle - \langle\psi|\hat{H}|\psi\rangle(\delta\,\langle\psi|\psi\rangle)}{\langle\psi|\psi\rangle^2} = 0 \ , \tag{11.34}$$

where it is sufficient that the numerator vanishes. As ψ is a complex function, we can look at ψ and ψ^* as two independent functions. As in the case of Hamilton's principle in mechanics, we find

$$\frac{\delta}{\delta\psi^*(x)} \int \mathrm{d}^3x\, \psi^*(\hat{H}\psi) = \hat{H}\psi(x) \ , \tag{11.35a}$$

$$\frac{\delta}{\delta\psi^*(x)} \int \mathrm{d}^3x\, \psi^*\psi = \psi(x) \ . \tag{11.35b}$$

We then get, with (11.34), an eigenvalue equation for ψ:

$$\langle\psi|\psi\rangle\, \hat{H}\psi(x) - \langle\psi|\hat{H}|\psi\rangle\psi(x) = 0 \quad \text{or}$$

$$\hat{H}\psi(x) = \frac{\langle\psi|\hat{H}|\psi\rangle}{\langle\psi|\psi\rangle}\psi(x) \equiv E_\psi\psi(x) \ , \tag{11.36}$$

which is exactly the Schrödinger equation.

The variational principle (11.33), known as *Ritz's variational method*, is therefore equivalent to the Schrödinger formalism of quantum mechanics (for stationary states). Under the additional condition that E_ψ be the absolute minimum, we then get (11.32) and thus the ground state energy.

Ritz's method is used for many practical purposes; we proceed in the following way. The (test) wave function $\psi(x, \alpha_1, \dots, \alpha_n)$ is made to depend on the real parameters α_i; we then search according to (11.33) for the minimum of $E_\psi(\alpha_1, \dots, \alpha_n)$:

$$\frac{\partial}{\partial\alpha_i}(E_\psi(\alpha_1, \dots, \alpha_n)) = 0 \ . \tag{11.37}$$

In this way, we get an upper approximation for the energy in the ground state.

It is also possible to use Ritz's variational method for the lowest energy states of a special kind; for example, for the lowest state with angular momentum $l = 0$,

$l = 1, l = 2$ etc. We then find the lowest energy states for $l = 0$ or $l = 1$, or for $l = 2$, etc. The test wave functions have to be of the same special kind, i.e. they have to be angular-momentum functions for $l = 0, l = 1$, or $l = 2$ etc.

It is also possible to determine the second lowest state of the same kind of states if we demand that it be orthogonal to the lowest state. A great number of extensions are possible.

EXAMPLE

11.6 Application of the Ritz Variational Method: The Harmonic Oscillator

We are searching for the ground state of a particle in the oscillator potential

$$V(x) = \tfrac{1}{2}m\omega^2 x^2 \ . \tag{1}$$

As test wave function we use

$$|\varphi\rangle = A\exp\left(-\frac{\lambda^2}{2}x^2\right) , \tag{2}$$

with A and λ as free parameters. A describes, of course, the normalization of the wave function and is therefore trivial. The interesting parameter is λ. We get

$$\begin{aligned}\hat{H}\,|\varphi\rangle &= \left(-\frac{\hbar^2}{2m}\frac{\partial^2}{\partial x^2} + \frac{1}{2}m\omega^2 x^2\right) A\exp\left(-\frac{\lambda^2}{2}x^2\right) \\ &= A\left[\frac{\hbar^2}{2m}(\lambda^2 - \lambda^4 x^2) + \frac{1}{2}m\omega^2 x^2\right]\exp\left(-\frac{\lambda^2}{2}x^2\right)\end{aligned} \tag{3}$$

and consequently

$$\begin{aligned}\langle\varphi|\,\hat{H}\,|\varphi\rangle &= A^2\left[\frac{\hbar^2}{2m}\left(\lambda^2\frac{\sqrt{\pi}}{\lambda} - \lambda^4\frac{\sqrt{\pi}}{2\lambda^3}\right) + \frac{1}{2}m\omega^2\frac{\sqrt{\pi}}{2\lambda^3}\right] \\ &= \frac{A^2\sqrt{\pi}}{2\lambda^3}\left(\frac{\hbar^2\lambda^4}{2m} + \frac{1}{2}m\omega^2\right)\end{aligned} \tag{4}$$

and

$$\langle\varphi|\varphi\rangle = A^2\frac{\sqrt{\pi}}{\lambda} \ . \tag{5}$$

The energy as a function of λ is given by

$$E(\lambda) = \frac{\langle\varphi|\hat{H}|\varphi\rangle}{\langle\varphi|\varphi\rangle} = \frac{1}{2\lambda^2}\left(\frac{\hbar^2\lambda^4}{2m} + \frac{m\omega^2}{2}\right) \ . \tag{6}$$

Example 11.6

Hence, the Ritz procedure leads to

$$\frac{\partial E}{\partial \lambda} = \frac{\hbar^2 \lambda}{2m} - \frac{m\omega^2}{2\lambda^3} = 0 \ , \tag{7}$$

and thus

$$\lambda_0^4 = \frac{m^2\omega^2}{\hbar^2} \Rightarrow \lambda_0^2 = \frac{m\omega}{\hbar} \ . \tag{8}$$

Therefore the ground state energy is

$$\begin{aligned} E_0(\lambda_0) &= \frac{\hbar^2\lambda_0^2}{4m} + \frac{m\omega^2}{4\lambda_0^2} \\ &= \frac{\omega\hbar}{4} + \frac{\hbar\omega}{4} = \frac{1}{2}\hbar\omega \ . \end{aligned} \tag{9}$$

We see that in this special case, by using the variational method, we get exactly the ground state (see Chap. 7). The ground state wave function is then determined by inserting λ_0 from (8) into (2).

11.4 Time-Dependent Perturbation Theory

One of the main tasks of quantum mechanics is the *calculation of transition probabilities* from one state ψ_n to another state ψ_m. This occurs under the influence of a time-dependent perturbation $V(\boldsymbol{r}, t)$, which, so to say, "shakes" the system and so causes the transition. The question of the transition of a system from one state to another generally only makes sense if the cause of the transition, i.e. $V(\boldsymbol{r}, t)$, acts only within a finite *time period*, say from $t = 0$ to $t = T$. Except for this time period, the total energy is a *constant of motion*, which can be measured.

The change of the wave function while $V(\boldsymbol{r}, t)$ is acting is given by a Schrödinger equation. The solution of this equation, however, generally leads to great difficulties. General predictions can only be made if the transition is caused by *weak* influences, i.e. weak potentials $V(\boldsymbol{r}, t)$. These influences can be interpreted as *perturbations*.

If perturbations are already taken into account in the *Schrödinger equation*, it takes the following form

$$\mathrm{i}\hbar\frac{\partial\psi}{\partial t} = \hat{H}_0(\boldsymbol{r})\psi + V(\boldsymbol{r}, t)\psi \ . \tag{11.38}$$

Here, $\hat{H}_0(\boldsymbol{r})$ is the operator for the total energy of the system without perturbation; the index 0 stands for the time independence. $V(\boldsymbol{r}, t)$ is the perturbation (*perturbation potential*).

Fig. 11.2. General form of a perturbation in the time period $0 \le t \le T$. Such a perturbation can be caused, for example, by an external field, which is switched on during this period, or by a particle that is passing by. In the latter case, T is a measure of the collision (interaction) time

For the calculation of the transition probability $W_{mn}(t)$ from the energy level E_n to the energy level E_m of the unperturbed system [described by $\hat{H}_0(\boldsymbol{r})$] it is advisable to use the *E representation* (energy representation). But first we look for eigenvalues of the unperturbed problem, i.e.

$$\mathrm{i}\hbar\frac{\partial\tilde{\psi}}{\partial t} = \hat{H}_0(\boldsymbol{r})\tilde{\psi} \ . \tag{11.39}$$

If the stationary part of the normalized wave function satisfies the equation

$$\hat{H}_0(\boldsymbol{r})\psi_k(\boldsymbol{r}) = E_k\psi_k(\boldsymbol{r}) \ , \tag{11.40}$$

then the time-dependent functions

$$\tilde{\psi}_k(\boldsymbol{r},t) = \psi_k(\boldsymbol{r})\exp\left(-\frac{\mathrm{i}}{\hbar}E_kt\right) \tag{11.41}$$

are the solutions of the unperturbed system. They form a *complete set of functions* and the solution of the main problem (11.38) can be expanded in terms of these functions, i.e.

$$\psi(\boldsymbol{r},t) = \sum_k a_k(t)\psi_k(\boldsymbol{r})\exp\left(-\frac{\mathrm{i}}{\hbar}E_kt\right) = \sum_k a_k(t)\tilde{\psi}_k(\boldsymbol{r},t) \ . \tag{11.42}$$

Inserting this into the original equation (11.38) leads to

$$\mathrm{i}\hbar\sum_k\frac{\mathrm{d}a_k}{\mathrm{d}t}\tilde{\psi}_k + \sum_k a_k\mathrm{i}\hbar\frac{\partial\tilde{\psi}_k}{\partial t} = \sum_k a_k\hat{H}_0\tilde{\psi}_k + \sum_k a_kV\tilde{\psi}_k \tag{11.43}$$

or, because $\mathrm{i}\hbar\partial\tilde{\psi}_k/\partial t = \hat{H}_0\tilde{\psi}_k$,

$$\mathrm{i}\hbar\sum_k\frac{\mathrm{d}a_k}{\mathrm{d}t}\tilde{\psi}_k(\boldsymbol{r},t) = \sum_k a_k(t)V\tilde{\psi}_k(\boldsymbol{r},t) \ . \tag{11.44}$$

After multiplication by $\psi_m^*(\boldsymbol{r},t)$ this becomes

$$\begin{aligned}\mathrm{i}\hbar\sum_k\frac{\mathrm{d}a_k}{\mathrm{d}t}\psi_m^*\psi_k\exp\left[-\frac{\mathrm{i}}{\hbar}(E_k-E_m)t\right] \\ = \sum_k a_k(t)\psi_m^*\hat{V}\psi_k\exp\left[-\frac{\mathrm{i}}{\hbar}(E_k-E_m)t\right] \ .\end{aligned} \tag{11.45}$$

Considering the normalization of the wave functions ψ_k and the abbreviations

$$V_{mk}(t) \equiv \int \mathrm{d}^3x\,\psi_m^*V\psi_k \quad \text{and} \quad \omega_{km} \equiv \frac{E_k-E_m}{\hbar} \ , \tag{11.46}$$

after integration over $\mathrm{d}V$, (11.45) leads to

$$\mathrm{i}\hbar \sum_k \frac{\mathrm{d}a_k}{\mathrm{d}t} \delta_{mk}\, \mathrm{e}^{\mathrm{i}\omega_{mk}t} = \sum_k a_k(t) V_{mk}(t)\, \mathrm{e}^{\mathrm{i}\omega_{mk}t} \ . \tag{11.47}$$

With $\omega_{mm} = 0$, we finally get

$$\mathrm{i}\hbar \frac{\mathrm{d}a_m}{\mathrm{d}t} = \sum_k a_k(t) V_{mk}(t)\, \mathrm{e}^{\mathrm{i}\omega_{mk}t} \ . \tag{11.48}$$

The frequencies ω_{mk} are sometimes called *Bohr frequencies* for the transition $E_m \rightarrow E_k$.

We assume that at the beginning (i.e. before the perturbation sets in), the system is in the state E_n. So we have for $t = 0$

$$\psi(\boldsymbol{r}, 0) = \sum_k a_k(0) \tilde{\psi}_k(\boldsymbol{r}, 0) \stackrel{!}{=} \tilde{\psi}_n(\boldsymbol{r}, 0) = \psi_n(\boldsymbol{r}) \ . \tag{11.49}$$

This just means that

$$a_n(0) = 1 \quad \text{and} \quad a_k(0) = 0 \quad \text{for} \quad k \neq n \ , \tag{11.50}$$

and already suggests the interpretation of the $a_k(t)$.

To understand this even better, let us first look at the normalization of $\psi(\boldsymbol{r}, t)$. We find

$$\begin{aligned} 1 &= \int \mathrm{d}^3x\, \psi^*(\boldsymbol{r}, t)\psi(\boldsymbol{r}, t) \\ &= \sum_{k,k'} a_k^*(t) a_{k'}(t) \exp\left[\frac{\mathrm{i}}{\hbar}(E_{k'} - E_k)t\right] \int \mathrm{d}^3x\, \psi_k^*(\boldsymbol{r})\psi_{k'}(\boldsymbol{r}) \\ &= \sum_{k,k'} a_k^*(t) a_{k'}(t)\, \mathrm{e}^{\mathrm{i}\omega_{k'k}t} \delta_{kk'} \\ &= \sum_k |a_k(t)|^2 \ . \end{aligned} \tag{11.51}$$

The expansion coefficients $a_k(t)$ must obviously satisfy the normalization condition for all times t, especially in the interval of the perturbation ($0 \le t \le T$).

Now we want to discuss the meaning of the $a_k(t)$. At time t we can write the wave function $\psi(\boldsymbol{r}, t)$ as

$$\psi(\boldsymbol{r}, t) = \sum_k a_k(t) \psi_k(\boldsymbol{r}) \exp\left(-\frac{\mathrm{i}}{\hbar} E_k t\right) \ . \tag{11.52}$$

The matrix element

$$\begin{aligned} \langle \psi_m(\boldsymbol{r}) | \psi(\boldsymbol{r}, t) \rangle &= \left\langle \psi_m(\boldsymbol{r}) \middle| \sum_k a_k(t) \psi_k(\boldsymbol{r}) \exp\left(-\frac{\mathrm{i}}{\hbar} E_k t\right) \right\rangle \\ &= a_m(t) \exp\left(-\frac{\mathrm{i}}{\hbar} E_m t\right) \end{aligned} \tag{11.53}$$

describes the overlap between the time-dependent wave function $\psi(\boldsymbol{r}, t)$ and the stationary wave function $\psi_m(\boldsymbol{r})$. The probability of finding the state $\psi_m(\boldsymbol{r})$ in $\psi(\boldsymbol{r}, t)$ at the time t with the energy E_m is given, as is well known, by the square of this term, i.e. by

$$|\langle\psi_m(r)|\psi(r,t)\rangle|^2 = |a_m(t)|^2 \ . \tag{11.54}$$

Since, according to the initial conditions [see (11.50)], at $t=0$, $a_m(t=0)=\delta_{mn}$ holds and since, in general, $a_m(t)\neq 0$ (for all m) for $t>0$, it is evident that the quantities $|a_k(t)|^2$ give the probability of finding the system at time t in the state $\tilde{\psi}_k$ with the energy E_k. Taking the initial conditions into account, $|a_m(t)|^2$ is the probability for the transition from the state φ_n to φ_m in the period from $t=0$ to T:

$$W_{mn}(t) = |a_m(t)|^2 \ . \tag{11.55}$$

Now it is our task to calculate the amplitudes $a_m(t)$ from the coupled differential equations (11.48) and the initial conditions (11.50). So far, the problem is clearly and exactly formulated. The solution, however, can in general only be obtained approximately and successively. We consider the fact that $\hat{V}(\boldsymbol{r}, t)$ represents a small perturbation; in the absence of perturbation, the system remains unchanged in its initial state. Thus in *zeroth order* we can make the following approximation, considering only *small* perturbations:

$$a_k^{(0)}(t) = \delta_{nk} \ , \tag{11.56}$$

which means that we start the zero-order solution with the initial conditions (11.50). This approximation is used to calculate the next best approximation, as we insert this solution into the right-hand side of the differential equations (11.48) (*successive approximation*):

$$\mathrm{i}\hbar\frac{\mathrm{d}a_m^{(1)}}{\mathrm{d}t} = \sum_k a_k^{(0)}(t)V_{mk}(t)\,\mathrm{e}^{\mathrm{i}\omega_{mk}t} = V_{mn}(t)\,\mathrm{e}^{\mathrm{i}\omega_{mn}t} \ . \tag{11.57}$$

This procedure can be continued until we reach the precision desired or necessary. In general, the iteration procedure for the differential equations (11.48) can be formulated as

$$\mathrm{i}\hbar\frac{\mathrm{d}a_m^{(i+1)}}{\mathrm{d}t} = \sum_k a_k^{(i)}V_{mk}(t)\,\mathrm{e}^{\mathrm{i}\omega_{mk}t} \ . \tag{11.58}$$

We restrict ourselves to *first-order approximation* and find, after integration, that

$$a_m^{(1)}(t) = \frac{1}{\mathrm{i}\hbar}\int_0^t V_{mn}(\tau)\,\mathrm{e}^{\mathrm{i}\omega_{mn}\tau}\,\mathrm{d}\tau + \delta_{mn} \ . \tag{11.59}$$

Now the qualities of the perturbation, mentioned above, are used. Thus we assume that $\hat{V}(\boldsymbol{r},t)=0$ for $t<0$ and $t>T$. Further, we suppose that $V_{mn}(t)$ is so small that the first-order approximation holds even for $t=T$. Then we get for $t\geq T$

$$\begin{aligned} a_m^{(1)}(t) &= \frac{1}{\mathrm{i}\hbar}\int_0^T V_{mn}(\tau)\,\mathrm{e}^{\mathrm{i}\omega_{mn}\tau}\,\mathrm{d}\tau \\ &= \frac{1}{\mathrm{i}\hbar}\int_{-\infty}^{\infty} V_{mn}(\tau)\,\mathrm{e}^{\mathrm{i}\omega_{mn}\tau}\,\mathrm{d}\tau\ , \qquad m\neq n\ . \end{aligned} \tag{11.60}$$

This means that, in particular, $a_m^{(1)}(t)$ is constant in time for $t>T$. It becomes a constant of motion for $t>T$. The perturbation has ceased and the system has settled into a new state.

Let us now study the meaning of $a_m^{(1)}(t)$ in greater detail. For that purpose we note that the perturbation can be expanded in a *Fourier series*:

$$V(\boldsymbol{r},t)=\int_{-\infty}^{\infty} V(\boldsymbol{r},\omega)\,\mathrm{e}^{-\mathrm{i}\omega t}\,\mathrm{d}\omega\ . \tag{11.61}$$

According to the theorem of Fourier integrals, the *Fourier component* $V(\boldsymbol{r},\omega)$ is then

$$V(\boldsymbol{r},\omega)=\frac{1}{2\pi}\int_{-\infty}^{\infty} V(\boldsymbol{r},t)\,\mathrm{e}^{\mathrm{i}\omega t}\,\mathrm{d}t\ . \tag{11.62}$$

For the matrix element (11.46) we then find

$$\begin{aligned} V_{mn}(t) &= \int \mathrm{d}^3x\,\psi_m^*(\boldsymbol{r})V(\boldsymbol{r},t)\psi_n(\boldsymbol{r})\ , \\ &= \int_{-\infty}^{\infty} \mathrm{e}^{-\mathrm{i}\omega t}\,\mathrm{d}\omega \int \mathrm{d}^3x\,\psi_m^*(\boldsymbol{r})V(\boldsymbol{r},\omega)\psi_n(\boldsymbol{r})\ , \\ &= \int_{-\infty}^{\infty} \mathrm{e}^{-\mathrm{i}\omega t}\,V_{mn}(\omega)\,\mathrm{d}\omega\ , \end{aligned} \tag{11.63}$$

where $V_{mn}(\omega)$ is now the matrix element of the Fourier transform $V(\boldsymbol{r},\omega)$ because, according to the Fourier theorem and in analogy to equation (11.62),

$$V_{mn}(\omega)=\frac{1}{2\pi}\int_{-\infty}^{\infty} V_{mn}(t)\,\mathrm{e}^{\mathrm{i}\omega t}\,\mathrm{d}t\ . \tag{11.64}$$

Comparing this with the expression for $a_m^{(1)}(t)$ (11.60), we find the relation

$$a_m^{(1)}(t) = \frac{2\pi}{\mathrm{i}\hbar} V_{mn}(\omega_{mn}) \ . \tag{11.65}$$

Thus we obtain for the transition probability

$$W_{mn}(t) = \frac{4\pi^2}{\hbar^2} |V_{mn}(\omega_{mn})|^2 \ ; \quad t \geq T \ . \tag{11.66}$$

Hence, for times $t \geq T$, the transition probability W_{mn} is constant and – as we see – is only nonzero if $V_{mn}(\omega_{mn}) \neq 0$, too. This means the transition from the state ψ_n (level E_n) to the state ψ_m (level E_m) is only possible if the frequency $\omega_{mn} = (E_m - E_n)/\hbar$ is contained in the perturbation spectrum, i.e. in the Fourier spectrum $V_{mn}(\omega_{mn})$ of the perturbation [see (11.63)]. Thus the transition exhibits *resonance behaviour*.

Obviously, we have the same situation as with a system of oscillators with eigenfrequencies which are equal to the Bohr frequencies ω_{mn}. If an external perturbation occurs that varies in time, then only those oscillators are stimulated which have an eigenfrequency that is included in the Fourier spectrum of the perturbation.

11.5 Time-Independent Perturbation

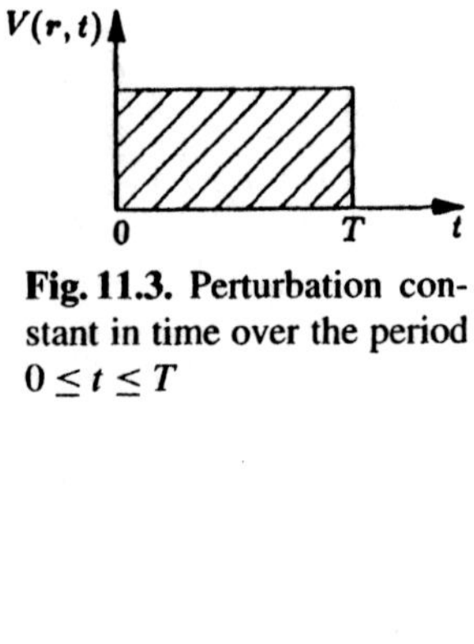

Fig. 11.3. Perturbation constant in time over the period $0 \leq t \leq T$

If

$$V(\boldsymbol{r},t) \begin{cases} = V(\boldsymbol{r}) & \text{for} \quad 0 \leq t \leq T \\ = 0 & \text{otherwise} \end{cases} \ , \tag{11.67}$$

i.e. if the perturbation is *not* time dependent while acting in the period $0 \leq t \leq T$ (see Fig. 11.3), then the integrals can be easily evaluated, and we find from (11.59) that

$$W_{mn}(t) = |a_m^{(1)}|^2 = \frac{1}{\hbar^2}|V_{mn}|^2 \underbrace{\left| \int_0^t d\tau \, \mathrm{e}^{\mathrm{i}\omega_{mn}\tau} \right|^2}_{f(t,\,\omega_{mn})} \ , \tag{11.68}$$

with

$$f(t,\omega) = \frac{1}{\omega^2}|\mathrm{e}^{\mathrm{i}\omega t} - 1|^2 = \frac{4}{\omega^2}\sin^2\frac{\omega}{2}t = \frac{2}{\omega^2}(1-\cos\omega t) \ . \tag{11.69}$$

As a function of ω, the quantity $f(t,\omega)$ takes the form shown in Fig. 11.4. It has a well-defined peak at $\omega = 0$ with the width $2\pi/t$, which becomes more distinct and sharper with increasing t.

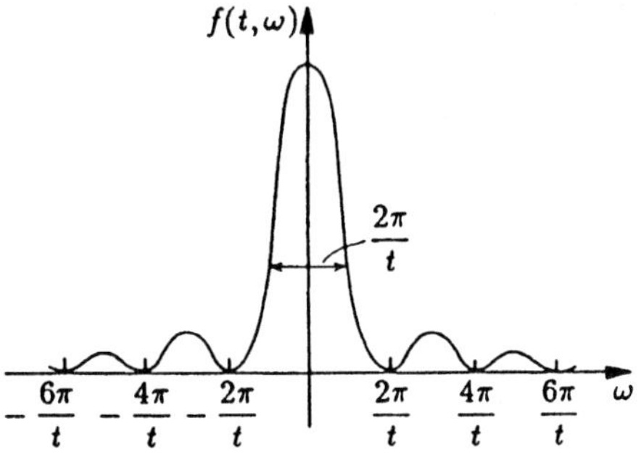

Fig. 11.4. The function $f(t, \omega)$ switches over to $2\pi t\delta(\omega)$ for $t \to \infty$, i.e. the maximum at $\omega = 0$ becomes increasingly sharp

The following relations are even exact (compare with Exercise 11.10)

$$\int_{-\infty}^{\infty} f(t, \omega)\, \mathrm{d}\omega = 2\pi t \tag{11.70}$$

and

$$\lim_{t\to\infty} \frac{f(t, \omega)}{t} = 2\pi\delta(\omega) \ . \tag{11.71}$$

For a fixed value of t, the probability W_{mn} in (11.68) depends in a simple way on the final state m. Up to a constant, it is the square of the perturbation matrix element $|V_{mn}|^2$ multiplied by the factor $f(t, \omega_{mn})$ that depends on the Bohr frequency ω_{mn} of this transition. Since this weighting factor $f(t, \omega_{mn})$ has a well-defined peak with the width $2\pi/t$ at $\omega_{mn} = 0$, transitions will mainly occur into such states that have energies in a band of the width $\delta E \simeq 2\pi\hbar/t$ around the energy of the initial state. This means: *the transitions conserve energy up to a value of the order* $\delta E \simeq 2\pi\hbar/t$. For $t \to \infty$ (and therefore $T \to \infty$) there are no transitions. This is intuitively clear, because a perturbation which is constant for all times cannot induce a transition; it does not "shake" the system. It is not surprising that here all frequencies occur, because the Fourier transform of a function constant in time over a certain period contains all frequencies except eventually those of a countable subset.

11.6 Transitions Between Continuum States

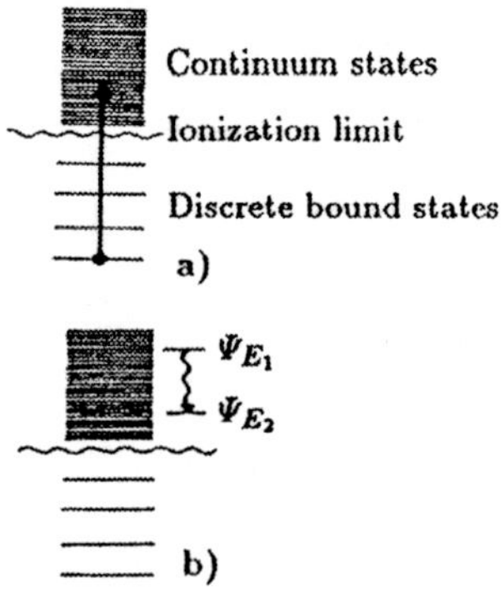

Fig. 11.5. (**a**) Transition from a discrete to a continuum state during ionization; (**b**) transition from the continuum state ψ_{E_1} to the continuum state ψ_{E_2} in the bremsstrahlung. At the same time, a photon with the energy $\hbar\omega = E_1 - E_2$ is emitted

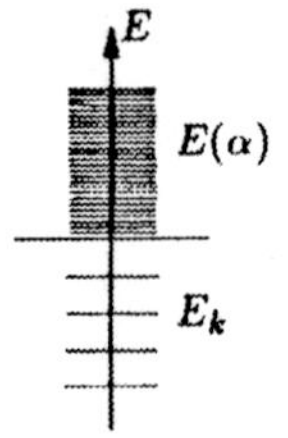

Fig. 11.6. Spectrum with a discrete (E_k) and a continuous $E(\alpha)$ part

So far we have considered an "unperturbed" operator $\hat{H}_0(\boldsymbol{r})$, which has *only a discrete spectrum.* We have also used a formalism which presumes that the states are *not degenerate.* By suitably changing this formalism, we can of course apply it to degenerate states. The generalization for a continuous spectrum is somewhat more complicated, but very often of practical importance, for example: the ionization of atoms (transition from a discrete bound state to a continuum state – see Fig. 11.5a) as a consequence of the perturbation field of a charged particle that is passing by, or the bremsstrahlung (continuum–continuum transition – see Fig. 11.5b) of charged particles as a result of acceleration or deceleration in the field of other particles. Let us now discuss this problem from a general point of view.

If the operator $\hat{H}_0(\boldsymbol{r})$ also has a continuous spectrum (see Fig. 11.5), we have as eigenfunctions

$$\hat{H}_0\psi_k(\boldsymbol{r}) = E_k\psi_k(\boldsymbol{r}) \quad \text{and} \quad \hat{H}_0\psi_\alpha(\boldsymbol{r}) = E(\alpha)\psi_\alpha(\boldsymbol{r}) \ . \tag{11.72}$$

Here, α is a continuous index that characterizes the continuum states of the spectrum. The stationary solutions, belonging to the time-dependent Schrödinger equation, are accordingly

$$\begin{aligned} \tilde{\psi}_k(\boldsymbol{r},t) &= \psi_k(\boldsymbol{r})\exp\left(-\frac{\mathrm{i}}{\hbar}E_k t\right) \ , \\ \tilde{\psi}_\alpha(\boldsymbol{r},t) &= \psi_\alpha(\boldsymbol{r})\exp\left(-\frac{\mathrm{i}}{\hbar}E(\alpha)t\right) \ . \end{aligned} \tag{11.73}$$

For the normalization of the eigenfunctions of the discrete states, we again have

$$\int \tilde{\psi}^*_{k'}(\boldsymbol{r},t)\tilde{\psi}_k(\boldsymbol{r},t)\,\mathrm{d}^3x = \delta_{k'k} \ . \tag{11.74}$$

For the overlap integrals between $\tilde{\psi}_k$ and $\tilde{\psi}_\alpha$, it holds that (because of the orthogonality of these states)

$$\int \tilde{\psi}^*_k(\boldsymbol{r},t)\tilde{\psi}_\alpha(\boldsymbol{r},t)\,\mathrm{d}^3x = 0 \ . \tag{11.75}$$

The normalization of the wave functions characterizing the continuous spectrum, however, is given by

$$\int \tilde{\psi}^*_{\alpha'}(\boldsymbol{r},t)\tilde{\psi}_\alpha(\boldsymbol{r},t)\,\mathrm{d}^3x = \frac{1}{n(\alpha)}\delta(\alpha-\alpha') \ , \tag{11.76}$$

where $n(\alpha)$ is a positive function of α. Obviously the functions $\tilde{\tilde{\psi}}_\alpha(\boldsymbol{r},t) = \sqrt{n(\alpha)}\tilde{\psi}_\alpha(\boldsymbol{r},t)$ are normalized to δ functions (see e.g. Chap. 5).

For the solution of the "disturbed" problem, we must use all eigenfunctions, i.e. the complete set, and so we obtain the linear combination

$$\psi(\boldsymbol{r},t)=\sum_k a_k(t)\tilde{\psi}_k(\boldsymbol{r},t)+\int a_\alpha(t)\tilde{\psi}_\alpha(\boldsymbol{r},t)\,\mathrm{d}\alpha\ . \tag{11.77}$$

Inserting this into the Schrödinger equation

$$\mathrm{i}\hbar\frac{\partial\psi}{\partial t}=(\hat{H}_0+V(\boldsymbol{r},t))\psi \tag{11.78}$$

yields

$$\begin{aligned}\mathrm{i}\hbar&\left(\sum_k\frac{\mathrm{d}a_k}{\mathrm{d}t}\tilde{\psi}_k+\int\frac{\mathrm{d}a_\alpha(t)}{\mathrm{d}t}\tilde{\psi}_\alpha\,\mathrm{d}\alpha\right)+\sum_k a_k\mathrm{i}\hbar\frac{\partial\tilde{\psi}_k}{\partial t}+\int a_\alpha(t)\mathrm{i}\hbar\frac{\partial\tilde{\psi}_\alpha}{\partial t}\,\mathrm{d}\alpha\\&=\sum_k a_k\hat{H}_0\tilde{\psi}_k+\int a_\alpha(t)\hat{H}_0\tilde{\psi}_\alpha\,\mathrm{d}\alpha+\sum_k a_kV\tilde{\psi}_k\\&\quad+\int a_\alpha(t)V\tilde{\psi}_\alpha\,\mathrm{d}\alpha\ ,\end{aligned} \tag{11.79}$$

so that

$$\begin{aligned}\mathrm{i}\hbar&\left(\sum_k\frac{\mathrm{d}a_k}{\mathrm{d}t}\tilde{\psi}_k+\int\frac{\mathrm{d}a_\alpha(t)}{\mathrm{d}t}\tilde{\psi}_\alpha\,\mathrm{d}\alpha\right)\\&=\sum_k a_kV\tilde{\psi}_k+\int a_\alpha(t)V\tilde{\psi}_\alpha\,\mathrm{d}\alpha\end{aligned} \tag{11.80}$$

remains. Now, proceeding as above [see (11.43)ff.], we find, after multiplication by the wave function $\tilde{\psi}_k^*$ or $\tilde{\psi}_{\alpha'}^*$, that

$$\begin{aligned}\mathrm{i}\hbar&\left\{\sum_k\frac{\mathrm{d}a_k}{\mathrm{d}t}\exp\left[\frac{\mathrm{i}}{\hbar}(E_{k'}-E_k)t\right]\psi_{k'}^*\psi_k\right.\\&\left.+\int\frac{\mathrm{d}a_\alpha(t)}{\mathrm{d}t}\exp\left[\frac{\mathrm{i}}{\hbar}(E_{k'}-E(\alpha))t\right]\psi_{k'}^*\psi_\alpha\,\mathrm{d}\alpha\right\}\\&=\sum_k a_k\exp\left[\frac{\mathrm{i}}{\hbar}(E_{k'}-E_k)t\right]\psi_{k'}^*V\psi_k\\&\quad+\int a_\alpha(t)\exp\left[\frac{\mathrm{i}}{\hbar}(E_{k'}-E(\alpha))t\right]\psi_{k'}^*V\psi_\alpha\,\mathrm{d}\alpha\end{aligned} \tag{11.81}$$

or

$$\begin{aligned}\mathrm{i}\hbar&\left\{\sum_k\frac{\mathrm{d}a_k}{\mathrm{d}t}\exp\left[\frac{\mathrm{i}}{\hbar}(E(\alpha')-E_k)t\right]\psi_{\alpha'}^*\psi_k\right.\\&\left.+\int\frac{\mathrm{d}a_\alpha(t)}{\mathrm{d}t}\exp\left[\frac{\mathrm{i}}{\hbar}(E(\alpha')-E(\alpha))t\right]\psi_{\alpha'}^*\psi_\alpha\,\mathrm{d}\alpha\right\}\end{aligned}$$

$$= \sum_k a_k \exp\left[\frac{\mathrm{i}}{\hbar}(E(\alpha') - E_k)t\right] \psi_{\alpha'}^* V \psi_k$$
$$+ \int a_\alpha(t) \exp\left[\frac{\mathrm{i}}{\hbar}(E(\alpha') - E(\alpha))t\right] \psi_{\alpha'}^* V \psi_\alpha \,\mathrm{d}\alpha \ . \tag{11.82}$$

Integration over the space coordinates, and taking into account the normalization and the overlap integrals, leads to

$$\mathrm{i}\hbar \sum_k \frac{\mathrm{d}a_k}{\mathrm{d}t} \exp\left[\frac{\mathrm{i}}{\hbar}(E_{k'} - E_k)t\right] \delta_{kk'}$$
$$= \sum_k a_k \exp\left[\frac{\mathrm{i}}{\hbar}(E_{k'} - E_k)t\right] \int \mathrm{d}^3x \, \psi_{k'}^* V \psi_k$$
$$+ \iint a_\alpha(t) \exp\left[\frac{\mathrm{i}}{\hbar}(E_{k'} - E(\alpha))t\right] \psi_{k'}^* V \psi_\alpha \,\mathrm{d}\alpha\,\mathrm{d}^3x \tag{11.83}$$

and

$$\mathrm{i}\hbar \int \frac{\mathrm{d}a_\alpha(t)}{\mathrm{d}t} \exp\left[\frac{\mathrm{i}}{\hbar}(E(\alpha') - E(\alpha))t\right] \frac{1}{n(\alpha)} \delta(\alpha - \alpha')\,\mathrm{d}\alpha$$
$$= \sum_k a_k \exp\left[\frac{\mathrm{i}}{\hbar}(E(\alpha') - E_k)t\right] \int \mathrm{d}^3x \, \psi_{\alpha'}^* V \psi_k$$
$$+ \int \mathrm{d}^3x \int a_\alpha(t) \exp\left[\frac{\mathrm{i}}{\hbar}(E(\alpha') - E(\alpha))t\right] \psi_{\alpha'}^* V \psi_\alpha \,\mathrm{d}\alpha \ . \tag{11.84}$$

Now, it is advisable to use for the interaction matrix element the same symbol as in (11.46), where the indices σ and τ can stand for both the continuous and the discrete spectrum. Thus

$$V_{\sigma\tau}(t) \equiv \int \mathrm{d}^3x \psi_\sigma^* V(r,t) \psi_\tau \ . \tag{11.85}$$

In the same way, we generalize the Bohr frequencies:

$$\omega_{\sigma\tau} = \frac{1}{\hbar}(E_\sigma - E_\tau) \ . \tag{11.86}$$

Here, either $E_\sigma = E(\alpha)$ or $E_\sigma = E_k$ is possible, depending on which part of the spectrum the index σ corresponds to. Then the coupled system of differential equations takes the form

$$\mathrm{i}\hbar \frac{\mathrm{d}a_{k'}}{\mathrm{d}t} = \sum_k a_k \mathrm{e}^{\mathrm{i}\omega_{k'k}t} V_{k'k}(t) + \int a_\alpha(t) \mathrm{e}^{\mathrm{i}\omega_{k'\alpha}} V_{k'\alpha}(t)\,\mathrm{d}\alpha \ , \tag{11.87a}$$

$$\frac{1}{n(\alpha')} \mathrm{i}\hbar \frac{\mathrm{d}a_{\alpha'}(t)}{\mathrm{d}t}$$
$$= \sum_k a_k \mathrm{e}^{\mathrm{i}\omega_{\alpha'k}t} V_{\alpha'k}(t) + \int a_\alpha(t) \mathrm{e}^{\mathrm{i}\omega_{\alpha'\alpha}t} V_{\alpha'\alpha}(t)\,\mathrm{d}\alpha \ . \tag{11.87b}$$

We already know the meaning of the a_k from above; for the $a_\alpha(t)$, the situation is slightly different. To determine their significance, we proceed as before in (11.51):

$$\begin{aligned}
1 &= \int d^3x\, \psi^*\psi = \sum_{kk'} a_k^* a_{k'}\, e^{i\omega_{kk'}t} \int d^3x\, \psi_k^*\psi_{k'} \\
&\quad + \sum_k \int d\alpha \left(a_k^* a_\alpha(t)\, e^{i\omega_{k\alpha}t} \int d^3x\, \psi_k^*\psi_\alpha + a_\alpha^*(t) a_k\, e^{i\omega_{\alpha k}t} \int d^3x\, \psi_\alpha^*\psi_k \right) \\
&\quad + \int d\alpha \int d\alpha' a_\alpha^*(t) a_{\alpha'}(t)\, e^{i\omega_{\alpha\alpha'}t} \int d^3x \psi_\alpha^*\psi_{\alpha'} \\
&= \sum_k |a_k|^2 + \int d\alpha \int d\alpha' a_\alpha^*(t) a_{\alpha'}(t)\, e^{i\omega_{\alpha\alpha'}t} \frac{1}{n(\alpha)} \delta(\alpha - \alpha') \\
&= \sum_k |a_k|^2 + \int d\alpha |a_\alpha(t)|^2 \frac{1}{n(\alpha)} \ .
\end{aligned} \tag{11.88}$$

Just as we expected for wave functions belonging to the continuous spectrum, only the expression

$$|a_\alpha(t)|^2 \frac{1}{n(\alpha)}\, d\alpha \tag{11.89}$$

has the meaning of a probability. More precisely, (11.89) gives the probability of finding the system in the range of states between α and $\alpha + d\alpha$.

Projectors

We denote by $B(\alpha)$ a small, connected range of values of the parameter α (this corresponds to a group of "neighbouring" states; see Fig. 11.7) so that the operator

$$\hat{P}_B = \int\limits_{B(\alpha)} d\alpha \tilde{\psi}_\alpha(\boldsymbol{r}, t) n(\alpha) \tilde{\psi}_\alpha^*(\boldsymbol{r}, t) \tag{11.90}$$

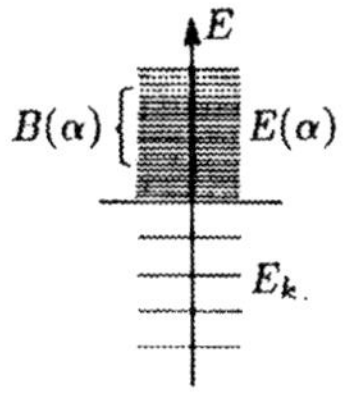

Fig. 11.7. The domain $B(\alpha)$ of the continuous spectrum

represents the *projector* (*projection operator*) onto those states contained in the interval and characterized by the values of the parameter α within the range of values $B(\alpha)$. The projector functions as follows:

$$\hat{P}_B \psi = \int\limits_{B(\alpha)} d\alpha \tilde{\psi}_\alpha(\boldsymbol{r}, t) n(\alpha) \int d^3x' \tilde{\psi}_\alpha^*(\boldsymbol{r}', t) \psi(\boldsymbol{r}', t) \ . \tag{11.91}$$

The action of $\hat{P}_B$ on a wave function $\psi(\boldsymbol{r}', t)$ is defined by the integral over $\boldsymbol{r}'$ in (11.91). The projector is somewhat similar to the Weyl packet (eigendifferential – see Sects. 4.4 and 5.1).

Now, if ψ is any wave function, we find from the expansion

$$\begin{aligned}\psi &= \sum_k a_k \tilde{\psi}_k(\boldsymbol{r}', t) + \int \mathrm{d}\alpha' a_{\alpha'}(t) \tilde{\psi}_{\alpha'}(\boldsymbol{r}', t) \\ &= \sum_k a_k \exp\left(-\frac{\mathrm{i}}{\hbar} E_k t\right) \psi_k(\boldsymbol{r}') \\ &\quad + \int \mathrm{d}\alpha' a_{\alpha'}(t) \exp\left(-\frac{\mathrm{i}}{\hbar} E(\alpha') t\right) \psi_{\alpha'}(\boldsymbol{r}') \end{aligned} \tag{11.92}$$

that

$$\begin{aligned}\hat{P}_B \psi &= \int\limits_{B(\alpha)} \mathrm{d}\alpha \tilde{\psi}_\alpha(\boldsymbol{r}', t) n(\alpha) \\ &\quad \left\{ \sum_k a_k \exp\left[\frac{\mathrm{i}}{\hbar}(E(\alpha) - E_k) t\right] \int \mathrm{d}^3 x' \psi_\alpha^*(\boldsymbol{r}') \psi_k(\boldsymbol{r}') \right. \\ &\quad \left. + \int \mathrm{d}\alpha' a_{\alpha'}(t) \exp\left[+\frac{\mathrm{i}}{\hbar}(E(\alpha) - E(\alpha'))t\right] \int \mathrm{d}^3 x' \psi_\alpha^*(\boldsymbol{r}') \psi_{\alpha'}(\boldsymbol{r}') \right\} \\ &= \int\limits_{B(\alpha)} \mathrm{d}\alpha \tilde{\psi}_\alpha(\boldsymbol{r}', t) n(\alpha) \int \mathrm{d}\alpha' a_{\alpha'}(t) \exp\left[+\frac{\mathrm{i}}{\hbar}(E(\alpha) - E(\alpha'))t\right] \\ &\quad \times \frac{1}{n(\alpha)} \delta(\alpha - \alpha') \\ &= \int\limits_{B(\alpha)} \mathrm{d}\alpha a_\alpha(t) \tilde{\psi}_\alpha(\boldsymbol{r}', t) \end{aligned} \tag{11.93}$$

holds. The application of the operator $\hat{P}_B$ thus *causes the projection of the wave function onto that domain of states $\tilde{\psi}_\alpha$ which is characterized by values α within the interval $B(\alpha)$*. This explains the name *projector*. If we now consider the energy $E(\alpha)$ as a new variable (i.e. we transform from α to the energy E), and if we name the corresponding range of values $B(E)$, we can also write the projector as

$$\hat{P}_B = \int\limits_{B(E)} \mathrm{d}E \tilde{\psi}_\alpha(\boldsymbol{r}', t) \varrho_\alpha(E) \tilde{\psi}_\alpha^*(\boldsymbol{r}', t) , \tag{11.94}$$

with

$$\varrho_\alpha(E) = n(\alpha) \frac{\mathrm{d}\alpha}{\mathrm{d}E} . \tag{11.95}$$

This quantity $\varrho_\alpha(E)$ is called the *density of states α at the energy E*. By inspection of the operator (11.94), we see that, indeed, *$\varrho_\alpha(E)$ is the number of states per unit of energy*. We shall now examine two properties of $\hat{P}_B$.

(a) Idempotency: $\hat{P}_B^2 = \hat{P}_B$

Indeed we find with $\psi = \sum a_k \tilde{\psi}_k + \int \mathrm{d}\alpha a_\alpha(t) \tilde{\psi}_\alpha$ that

$$
\begin{aligned}
\hat{P}_B^2 \psi &= \hat{P}_B(\hat{P}_B \psi) \\
&= \hat{P}_B \int\limits_{B(\alpha)} \mathrm{d}\alpha a_\alpha(t) \tilde{\psi}_\alpha(\boldsymbol{r}', t) \\
&= \int\limits_{B(\alpha')} \mathrm{d}\alpha' \tilde{\psi}_{\alpha'}(\boldsymbol{r}', t) n(\alpha') \int \mathrm{d}^3 x' \tilde{\psi}^*_{\alpha'}(\boldsymbol{r}', t) \int\limits_{B(\alpha)} \mathrm{d}\alpha a_\alpha(t) \tilde{\psi}_\alpha(\boldsymbol{r}', t) \\
&= \int\limits_{B(\alpha')} \mathrm{d}\alpha' \tilde{\psi}_{\alpha'}(\boldsymbol{r}', t) n(\alpha') \int\limits_{B(\alpha)} \mathrm{d}\alpha a_\alpha(t) \underbrace{\int \mathrm{d}^3 x' \tilde{\psi}^*_{\alpha'}(\boldsymbol{r}', t) \tilde{\psi}_\alpha(\boldsymbol{r}', t)}_{[1/n(\alpha)]\delta(\alpha - \alpha')} \\
&= \int\limits_{B(\alpha)} \mathrm{d}\alpha a_\alpha(t) \tilde{\psi}_\alpha(\boldsymbol{r}', t) = \hat{P}_B \psi \ .
\end{aligned}
\tag{11.96}
$$

Since this relation must be true for any ψ, we can conclude that the idempotency relation $\hat{P}_B^2 = \hat{P}_B$ holds generally.

(b) Hermiticity: $\hat{P}_B^+ = \hat{P}_B$

If $\phi = \sum b_k \tilde{\psi}_k + \int \mathrm{d}\alpha b_\alpha(t) \hat{\psi}_\alpha$, then it follows that

$$
\hat{P}_B \phi = \int\limits_{B(\alpha)} \mathrm{d}\alpha b_\alpha(t) \tilde{\psi}_\alpha(\boldsymbol{r}', t) \ . \tag{11.97}
$$

We investigate the action of the operator $\hat{P}_B^+$ for arbitrary ϕ and ψ by considering the integral

$$
\begin{aligned}
\int \mathrm{d}^3 x \phi^* \hat{P}_B^+ \psi &= \int \mathrm{d}^3 x (\hat{P}_B \phi)^* \psi \\
&= \int \mathrm{d}^3 x \left(\int\limits_{B(\alpha)} \mathrm{d}\alpha b_\alpha(t) \tilde{\psi}_\alpha(\boldsymbol{r}', t) \right)^* \left(\sum a_k \tilde{\psi}_k + \int \mathrm{d}\alpha' a_{\alpha'}(t) \tilde{\psi}_{\alpha'} \right) \\
&= \int\limits_{B(\alpha)} \mathrm{d}\alpha b^*_\alpha(t) \left(\underbrace{a_k \int \mathrm{d}^3 x \tilde{\psi}^*_\alpha \tilde{\psi}_k}_{=0} + \int a_{\alpha'}(t) \underbrace{\int \mathrm{d}^3 x \tilde{\psi}^*_\alpha \tilde{\psi}_{\alpha'}}_{[1/n(\alpha)]\delta(\alpha - \alpha')} \mathrm{d}\alpha' \right) \\
&= \int\limits_{B(\alpha)} \mathrm{d}\alpha b^*_\alpha(t) \int \mathrm{d}\alpha' a_{\alpha'}(t) \frac{1}{n(\alpha)} \delta(\alpha - \alpha') \\
&= \int\limits_{B(\alpha)} \mathrm{d}\alpha b^*_\alpha(t) a_\alpha(t) \frac{1}{n(\alpha)} \ .
\end{aligned}
\tag{11.98}
$$

On the other hand, we calculate

$$\begin{aligned}
&\int \mathrm{d}^3x\phi^* \hat{P}_B\psi \\
&= \int \mathrm{d}^3x \left(\sum b_k\tilde{\psi}_k + \int \mathrm{d}\alpha' b_{\alpha'}(t)\psi_{\alpha'}\right)^* \int\limits_{B(\alpha)} \mathrm{d}\alpha a_\alpha(t)\tilde{\psi}_\alpha(\boldsymbol{r}', t) \\
&= \int\limits_{B(\alpha)} \mathrm{d}\alpha a_\alpha(t) \left(\sum b_k^* \underbrace{\int \mathrm{d}^3x\tilde{\psi}_k^*\tilde{\psi}_\alpha}_{=0} + \int b_{\alpha'}^*(t) \underbrace{\int \mathrm{d}^3x\tilde{\psi}_\alpha^*\tilde{\psi}_{\alpha'}\,\mathrm{d}\alpha'}_{[1/n(\alpha)]\delta(\alpha-\alpha')}\right) \\
&= \int\limits_{B(\alpha)} \mathrm{d}\alpha a_\alpha(t) \int \mathrm{d}\alpha' b_{\alpha'}^*(t)\frac{1}{n(\alpha)}\delta(\alpha-\alpha') \\
&= \int\limits_{B(\alpha)} \mathrm{d}\alpha b_\alpha^*(t)a_\alpha(t)\frac{1}{n(\alpha)} = \int \mathrm{d}^3x\phi^*\hat{P}_B^+\psi \; .
\end{aligned} \tag{11.99}$$

As this relation is valid for any ϕ and ψ, it follows that $\hat{P}_B^+ = \hat{P}_B$. This means $\hat{P}_B$ is Hermitian.

Now we can return to the calculation of transition probabilities. Because $\hat{P}_B\psi$ yields that part of the wave function that lies in the range of neighbouring states ψ_α within the interval characterized by $B(\alpha)$, the absolute value $\int|\hat{P}_B\psi|\mathrm{d}^3x$ is just the probability W_B of finding the system in the states within $B(\alpha)$:

$$\begin{aligned}
W_B &\stackrel{\text{def}}{\equiv} \int |\hat{P}_B\psi|^2\mathrm{d}^3x = \int \mathrm{d}^3x(\hat{P}_B\psi)^*\hat{P}_B\psi(\boldsymbol{r}', t) \\
&= \int \mathrm{d}^3x\psi^*\hat{P}_B^+\hat{P}_B\psi = \int \mathrm{d}^3x\psi^*\hat{P}_B\hat{P}_B\psi = \int \mathrm{d}^3x\psi^*\hat{P}_B\psi(\boldsymbol{r}', t) \\
&= \int \mathrm{d}^3x\left(\sum a_k\tilde{\psi}_k(\boldsymbol{r}', t) + \int \mathrm{d}\alpha' a_{\alpha'}(t)\tilde{\psi}_{\alpha'}(\boldsymbol{r}', t)\right)^* \int\limits_{B(\alpha)} \mathrm{d}\alpha a_\alpha(t)\tilde{\psi}_\alpha(\boldsymbol{r}', t) \\
&= \int \mathrm{d}\alpha' a_{\alpha'}^*(t) \int\limits_{B(\alpha)} \mathrm{d}\alpha a_\alpha(t)\frac{1}{n(\alpha)}\delta(\alpha-\alpha') \\
&= \int\limits_{B(\alpha)} \mathrm{d}\alpha a_\alpha^*(t)a_\alpha\frac{1}{n(\alpha)} = \int\limits_{B(\alpha)} \mathrm{d}\alpha \left|\frac{a_\alpha(t)}{n(\alpha)}\right|^2 n(\alpha) \\
&= \int\limits_{B(E)} \left|\frac{a_\alpha(t)}{n(\alpha)}\right|^2 \varrho_\alpha(E)\,\mathrm{d}E \; .
\end{aligned} \tag{11.100}$$

It is now easy to determine the transition probability if the system is initially in a state n of the discrete spectrum, i.e. $a_k(0) = \delta_{nk}$, $a_\alpha(0) = 0$. Hence, in first order, we get from (11.87)

$$\mathrm{i}\hbar\frac{\mathrm{d}a_k^{(1)}}{\mathrm{d}t}=\mathrm{e}^{\mathrm{i}\omega_{kn}t}V_{kn}(t)\ ;\qquad \frac{1}{n(\alpha)}\mathrm{i}\hbar\frac{\mathrm{d}a_\alpha^{(1)}(t)}{\mathrm{d}t}=\mathrm{e}^{\mathrm{i}\omega_{\alpha n}t}V_{\alpha n}(t)\ . \qquad (11.101)$$

For the discrete part of the spectrum, there is no change compared with the earlier result [see (11.57)]. Therefore we shall further examine here only the second part, which describes transitions into the continuous spectrum. (Such problems arise, for example, in the case of ionization, when particles that pass by perturb the system.) After integration of (11.101) we find that

$$\frac{a_\alpha^{(1)}(t)}{n(\alpha)}=\frac{1}{\mathrm{i}\hbar}\int_0^t V_{\alpha n}(\tau)\,\mathrm{e}^{\mathrm{i}\omega_{\alpha n}\tau}\,\mathrm{d}\tau \qquad (11.102)$$

and

$$\begin{aligned}W_{n\to B}&=\int_{B(E)}\left|\frac{a_\alpha^{(1)}(t)}{n(\alpha)}\right|^2\varrho_\alpha(E)\,\mathrm{d}E\\&=\frac{1}{\hbar^2}\int_{B(E)}\left|\int_0^t V_{\alpha n}(\tau)\,\mathrm{e}^{\mathrm{i}\omega_{\alpha n}\tau}\,\mathrm{d}\tau\right|^2\varrho_\alpha(E)\,\mathrm{d}E\ ,\end{aligned} \qquad (11.103)$$

because if the transition occurs out of a particular state, then the transition probability is just the probability of finding the system in states $\tilde{\psi}_\alpha(E)$ of the range $B(E)$ at a later time. Note that $\alpha=\alpha(E)$ in (11.103)!

If the initial state (characterized by $\alpha=\beta$) belongs to the continuum, then $a_k^{(0)}=0$ and $a_\alpha^{(0)}=\delta(\beta-\alpha)$. The result is completely analogous, namely

$$\frac{a_\alpha(t)}{n(\alpha)}=\frac{1}{\mathrm{i}\hbar}\int_0^t V_{\alpha\beta}(\tau)\,\mathrm{e}^{\mathrm{i}\omega_{\alpha\beta}\tau}\,\mathrm{d}\tau\ . \qquad (11.104)$$

This leads to the transition probability $W_{\beta\to\alpha}$;

$$W_{\beta\to B}=\frac{1}{\hbar^2}\int_{B(E)}\left|\int_0^t V_{\alpha\beta}(\tau)\,\mathrm{e}^{\mathrm{i}\omega_{\alpha\beta}\tau}\,\mathrm{d}\tau\right|^2\varrho_\alpha(E)\,\mathrm{d}E\ . \qquad (11.105)$$

If the perturbation is constant in the interval $0<t<T$, i.e.

$$V(\boldsymbol{r},t)\begin{cases}=V(\boldsymbol{r}) & \text{for}\quad 0\le t\le T\\ =0 & \text{otherwise}\quad ,\end{cases}$$

then

$$V_{\alpha\beta}(\tau) \equiv \int \mathrm{d}^3 x \psi_\alpha^* \hat{V}(r) \psi_\beta = V_{\alpha\beta} \ , \tag{11.106}$$

so that

$$W_{\beta\to B} = \frac{1}{\hbar^2} \int\limits_{B(E)} |V_{\alpha\beta}|^2 f(t, \omega_{\alpha\beta}) \varrho_\alpha(E)\, \mathrm{d}E \ , \tag{11.107}$$

where again [see (11.69)]

$$f(t, \omega) = \left| \int\limits_0^t \mathrm{e}^{\mathrm{i}\omega\tau}\, \mathrm{d}\tau \right|^2 = \frac{2}{\omega^2}(1 - \cos \omega t) \ . \tag{11.108}$$

In (11.107), the squared matrix element $|V_{\alpha\beta}|^2$ cannot be taken out of the integral because it is related to the integration variable by $\alpha(E)$. Otherwise (11.107) is totally analogous to the earlier result (11.68), only that, here, the density of the final states $\varrho_\alpha(E)$ appears. If these final states are concentrated around a state with the energy E_1, i.e. $\varrho_\alpha(E) = \delta(E - E_1)$, then (11.68) can be immediately deduced from (11.107).

EXAMPLE

11.7 Transition Probability per Unit Time: Fermi's Golden Rule

Here we wish to examine the transition to states $\tilde{\psi}_\alpha$ within the energy interval $[E_1 - (\varepsilon/2), E_1 + (\varepsilon/2)]$. The width ε is chosen so small that in this interval $V_{\alpha\beta}$ and $\varrho_\alpha(E)$ can practically be considered to be constant, i.e. independent of α or E. Then we can write for the transition probability (11.107):

$$W_{\beta\to B} = \frac{1}{\hbar^2} |V_{\alpha\beta}|^2 \varrho_\alpha(E_1) \int\limits_{B(E)} f(t, \omega_{\alpha\beta})\, \mathrm{d}E \ , \tag{1}$$

where now $B(E) = [E|E_1 - (\varepsilon/2) < E < E_1 + (\varepsilon/2)]$.

If the time is chosen large enough, so that ε gets much larger than the oscillation frequencies contained in the function $f(t, \omega)$, i.e. so that

$$\varepsilon \gg \frac{2\pi\hbar}{t} \ , \tag{2}$$

[compare with Fig. 11.4 for $f(t, \omega)$], then the remaining integral in $W_{\alpha\to\beta}$ can be evaluated rather straightforwardly. Yet, two cases must be distinguished.

(a) The central peak of $f(t, \omega)$ lies outside the integration interval (see figure), i.e.

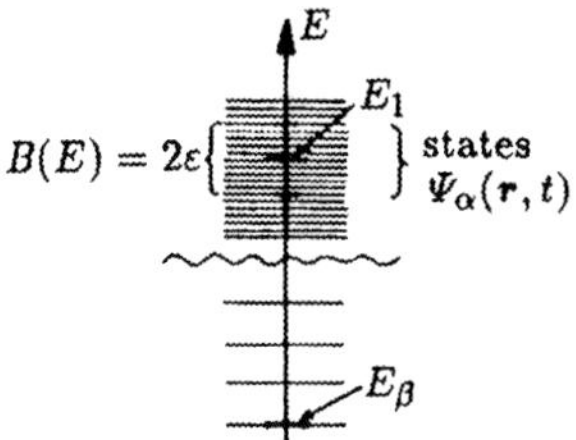

In case (a), transitions from the state ψ_β with the energy E_β into the continuum states $\tilde{\psi}_{\alpha(E)}$ in $B(E)$ around E_1 are considered

$$E_1 - E_\beta \gg \varepsilon \gg \frac{2\pi\hbar}{t} \ . \tag{3}$$

Obviously, in this case, the system changes its energy under the influence of the perturbation (*no conservation of energy*). With

$$f(t, \omega) = \frac{2}{\omega^2}(1 - \cos\omega t)$$

we obtain

$$\int\limits_{B(E)} f(t, \omega_{\alpha\beta})\,\mathrm{d}E = \int\limits_{E_1-\varepsilon/2}^{E_1+\varepsilon/2} \frac{2\hbar^2}{(E - E_\beta)^2}\,\mathrm{d}E - \int\limits_{(E_1-\varepsilon/2-E_\beta)/\hbar}^{(E_1+\varepsilon/2-E_\beta)/\hbar} \frac{2\hbar}{\omega^2}\cos\omega t\,\mathrm{d}\omega \ . \tag{4}$$

The second integral makes a negligible contribution because, according to the assumption (2) for ϱ, the integration interval contains many oscillations and $\cos\omega t$ repeatedly takes on all values between -1 and 1. Therefore

$$\int\limits_{B(E)} f(t, \omega_{\alpha\beta})\,\mathrm{d}E = -\frac{2\hbar^2}{E - E_\beta}\Bigg|_{E_1-\varepsilon/2}^{E_1+\varepsilon/2} = \frac{2\hbar^2\varepsilon}{(E_1 - E_\beta)^2 - \varepsilon^2/4} \approx \frac{2\hbar^2\varepsilon}{(E_1 - E_\beta)^2} \ , \tag{5}$$

and the *time-independent expression* for the transition probability results:

$$W_{\beta\to B(E_1)} = \frac{2\varepsilon|V_{\alpha\beta}|^2}{(E_1 - E_\beta)^2}\varrho(E_1) \ . \tag{6}$$

(b) We now consider the case

$$E_1 - E_\beta \approx \varepsilon \gg \frac{2\pi\hbar}{t} \ . \tag{7}$$

Here, the central peak lies within the integration interval, and the main contribution to the time integral comes from that part (see figure). We make only a small error by moving the boundaries of the integration to $\pm\infty$, i.e.

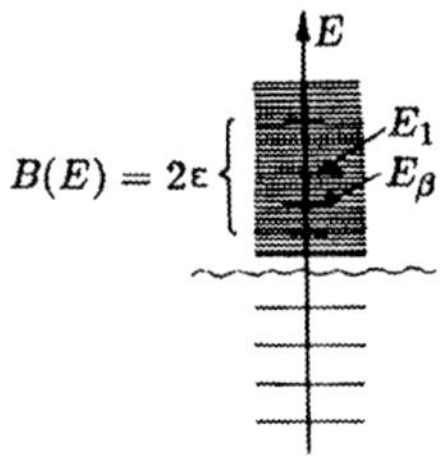

In case (b), E_β lies within the range $B(E)$

$$\int\limits_{B(E)} f(t, \omega_{\alpha\beta})\,\mathrm{d}E \approx 2\hbar \int\limits_{-\infty}^{\infty} \frac{1 - \cos\omega t}{\omega^2}\,\mathrm{d}\omega \ . \tag{8}$$

Example 11.7

The value of the integral is determined according to the residue theorem:

$$2\int_{-\infty}^{\infty}\frac{1-\cos\omega t}{\omega^2}\,\mathrm{d}\omega = 2\pi t\ , \quad \text{thus} \quad \int_{B(E)}^{-\infty} f(t,\omega_{\alpha\beta})\,\mathrm{d}E \approx 2\pi\hbar t \tag{9}$$

and with this

$$W_{\beta\to B(E)} = \frac{2\pi}{\hbar}|V_{\alpha\beta}(E)|^2\varrho_\alpha(E)t \tag{10}$$

results. Here, we have written E as the argument, because the main contribution results from $\omega = 0$, i.e. for $E_\alpha = E_\beta = E$. Since the final state has the same energy as the initial state, this transition *conserves energy*. Furthermore, because $\varepsilon \gg 2\pi\hbar/t$, the transition probability is larger than the sum of all other parts.

Now we introduce the *transition probability per unit time* $w_{\alpha\to\beta}$, which is naturally defined by

$$w_{\beta\to B(E)} \equiv \frac{\mathrm{d}W_{\beta\to B(E)}}{\mathrm{d}t}\ . \tag{11}$$

Since it was found in (6) that $W_{\alpha\to\beta}$ is time-independent for those transitions which do not conserve energy, $w_{\alpha\to\beta}$ vanishes. For energy-conserving transitions, however, we find the important formula

$$W_{\beta\to B(E)} = \frac{2\pi}{\hbar}|V_{\alpha\beta}(E)|^2\varrho_\alpha(E)\ . \tag{12}$$

It should be emphasized again that the matrix elements which appear and the density of states are assigned to states α, which have the same energy E_β as the initial state. The expression (12) is called *Fermi's golden rule.*

The conditions for the validity of this formula are evident from its derivation. Let us recapitulate the two most essential assumptions once more. It is necessary for the time t to be long enough to guarantee that $\varepsilon \gg 2\pi\hbar/t$; on the other hand, time t must be short enough to justify the first-order perturbation theory approximation, i.e. $w_{\alpha\to\beta}t \ll 1$.

EXAMPLE

11.8 Elastic Scattering of an Electron by an Atomic Nucleus

In this example, we give an application of Fermi's golden rule and simultaneously illustrate some concepts of scattering theory.

We examine the scattering of a high-velocity electron by a nucleus with the charge number Z. The electron-nucleus interaction – in our case the Coulomb energy – is treated as a perturbation:

$$V(\boldsymbol{r},\boldsymbol{R}) = \frac{Ze^2}{|\boldsymbol{r}-\boldsymbol{R}|}\exp\left(-\frac{|\boldsymbol{r}-\boldsymbol{R}|}{d}\right)\ . \tag{1}$$

Example 11.8

Let $\boldsymbol{r}$ mark the coordinates of the electron, $\boldsymbol{R}$ those of the nucleus. The exponential factor in (1) is introduced because of the screening effect of the electron cloud on the nuclear charge. We need it for mathematical reasons, too, since it helps to avoid divergencies when integrations are performed. The length d is a measure for the shielding distance. For $|\boldsymbol{r}-\boldsymbol{R}| \gg d$, the interaction disappears, because the charge of the nucleus is then completely shielded by the bound electrons.

Before the scattering, the whole system is described by the state $|\psi_\mathrm{i}\rangle$ and afterwards, by the state $|\psi_\mathrm{f}\rangle$. We want to calculate the probability for the transition $|\psi_\mathrm{i}\rangle \rightarrow |\psi_\mathrm{f}\rangle$ in order to learn something about the structure (charge distribution) of the nucleus when comparing it with experiments. The electron is in a state of motion with momentum $\boldsymbol{p}_0$ and energy E_0 before scattering occurs and is scattered by the Coulomb field (1) into a state with $\boldsymbol{p}$ and E. Since these are continuum states, the transition probability per unit time is given by (12) of Example 11.7:

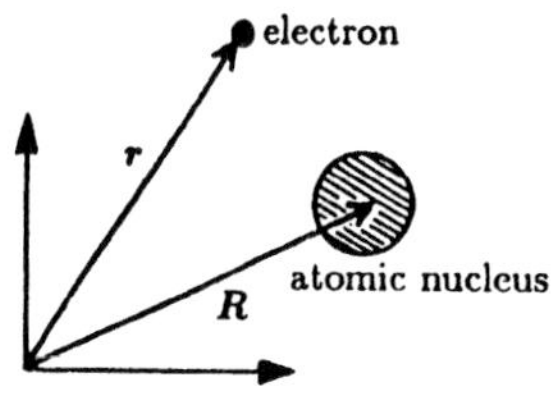

An electron is scattered by an atomic nucleus

$$w = \frac{2\pi}{\hbar}|\langle\psi_\mathrm{f}|V|\psi_\mathrm{i}\rangle|^2\varrho_\mathrm{f}(E) \ . \tag{2}$$

The total wave functions of the electron-nucleus system are products of the wave function of the electron and that of the nucleus. *We use plane waves as an approximation for the wave function of the electron.* This approximation is called the *Born approximation.* But it is only valid if the electron-nucleus interaction is small, i.e. the nuclear charge should not be too large and the velocity of the electron has to be great enough. These conditions are summarized by the relation

$$\frac{Z}{137}\frac{c}{v} \ll 1 \ . \tag{3}$$

Thus the wave functions are

$$\begin{aligned}\psi_\mathrm{i} &= \exp\left(\frac{\mathrm{i}}{\hbar}\boldsymbol{p}_0\cdot\boldsymbol{r}\right)\phi_\mathrm{i}(\boldsymbol{R}) \ ,\\ \psi_\mathrm{f} &= \exp\left(\frac{\mathrm{i}}{\hbar}\boldsymbol{p}\cdot\boldsymbol{r}\right)\phi_\mathrm{f}(\boldsymbol{R}) \ ,\end{aligned} \tag{4}$$

or, written in another way,

$$|\psi_\mathrm{i}\rangle = |\boldsymbol{k}_0\rangle\,|\mathrm{i}\rangle \ , \quad |\psi_\mathrm{f}\rangle = |\boldsymbol{k}_0\rangle\,|\mathrm{f}\rangle \ . \tag{5}$$

This can also be expressed as

$$\begin{aligned}\langle\boldsymbol{r}|\boldsymbol{k}_0\rangle &= \exp(\mathrm{i}\boldsymbol{k}_0\cdot\boldsymbol{r}) = \exp\left(\frac{\mathrm{i}}{\hbar}\boldsymbol{p}_0\cdot\boldsymbol{r}\right) \ ,\\ \langle\boldsymbol{r}|\boldsymbol{k}\rangle &= \exp(\mathrm{i}\boldsymbol{k}\cdot\boldsymbol{r}) = \exp\left(\frac{\mathrm{i}}{\hbar}\boldsymbol{p}\cdot\boldsymbol{r}\right) \ ,\\ \langle\boldsymbol{R}|\mathrm{i}\rangle &= \phi_\mathrm{i}(\boldsymbol{R}) \ ,\\ \langle\boldsymbol{R}|\mathrm{f}\rangle &= \phi_\mathrm{f}(\boldsymbol{R}) \ ,\end{aligned} \tag{6}$$

a form of notation used by some authors; we will also use it occasionally (see, e.g. Chap. 15).

Example 11.8

The $\phi(\boldsymbol{R})$ are the normalized wave functions of the nucleus. We did not normalize the plane waves (4) – based on (11.76) and (11.95) –, in order to define the density of the plane waves in a general way (i.e. also for plane waves which are not normalized on δ functions).

The ket vectors $|\boldsymbol{k}\rangle$ form a complete set of orthogonal functions:

$$\langle \boldsymbol{k}|\boldsymbol{k}'\rangle = (2\pi)^3\delta(\boldsymbol{k}-\boldsymbol{k}') \ , \quad \int |\boldsymbol{k}\rangle \frac{\mathrm{d}^3k}{(2\pi)^3} \langle \boldsymbol{k}| = \mathbb{1} \ . \tag{7}$$

These formulae will be necessary when calculating the density of states. But let us now evaluate the matrix elements. Since the wave function factorizes into an electronic and a nuclear part, the volume element $\mathrm{d}V$ also implies integration over both volumes $\mathrm{d}V_\mathrm{e}$ (electron space) and $\mathrm{d}V_\mathrm{n}$ (nucleus space). Thus

$$\begin{aligned}&\langle \psi_\mathrm{f}|V|\psi_\mathrm{i}\rangle \\ &\quad = Ze^2\int \phi_\mathrm{f}^*(\boldsymbol{R})\phi_\mathrm{i}(\boldsymbol{R})\int \frac{\exp(-|\boldsymbol{r}-\boldsymbol{R}|/d)}{|\boldsymbol{r}-\boldsymbol{R}|}\,\mathrm{e}^{\mathrm{i}(\boldsymbol{k}_0-\boldsymbol{k})\cdot\boldsymbol{r}}\,\mathrm{d}V_\mathrm{e}\,\mathrm{d}V_\mathrm{n} \ .\end{aligned} \tag{8}$$

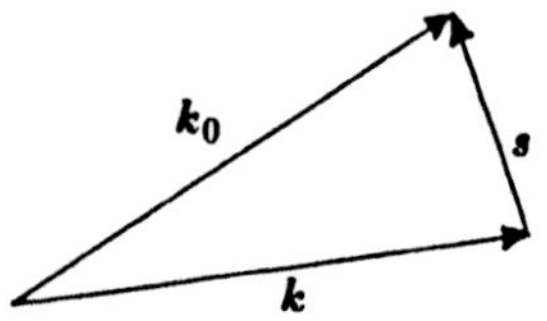

The vector $\hbar\boldsymbol{s}$ describes the momentum transferred from the electron to the nucleus $\boldsymbol{p}_0 = \hbar\boldsymbol{k}_0$ and $\boldsymbol{p} = \hbar\boldsymbol{k}$ are the momenta of the electron before and after the collision

The indices e and n stand for "electron" and "nucleus", respectively. We first calculate the integral over the electronic coordinates and abbreviate $\boldsymbol{k}_0-\boldsymbol{k}$ by the vector $\boldsymbol{s}$ (see upper figure):

$$\boldsymbol{s} = \boldsymbol{k}_0 - \boldsymbol{k} \ . \tag{9}$$

Therefore $\hbar\boldsymbol{s} = \boldsymbol{p}_0 - \boldsymbol{p}$ is the *momentum transfer* from the electron to the nucleus during the scattering process (see upper figure). The integral over $\mathrm{d}V_\mathrm{e}$ is evidently a function of R:

$$J_\mathrm{e}(\boldsymbol{R}) = \int \frac{\exp(-|\boldsymbol{r}-\boldsymbol{R}|/d)}{|\boldsymbol{r}-\boldsymbol{R}|}\,\mathrm{e}^{\mathrm{i}(\boldsymbol{k}_0-\boldsymbol{k})\cdot\boldsymbol{r}}\,\mathrm{d}V_\mathrm{e} \ . \tag{10}$$

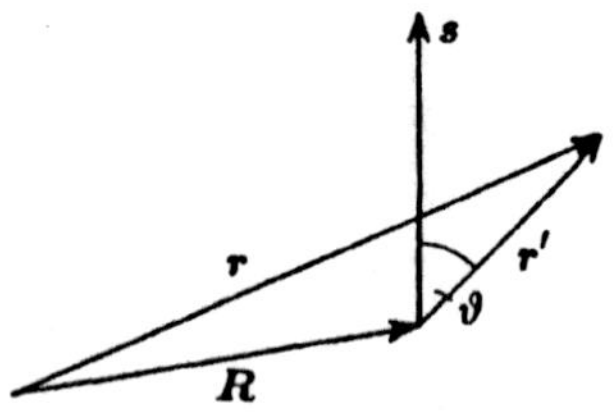

The new coordinates $\boldsymbol{r}'$. The vector $\boldsymbol{s}$ is taken as polar axis

Since we integrate over the whole space, we can change our system of coordinates and integrate over the whole space again (the integration boundaries need not be changed). We replace $\boldsymbol{r}$ by $\boldsymbol{r}' = \boldsymbol{r}-\boldsymbol{R}$ and introduce spherical coordinates with ϑ being the angle between $\boldsymbol{s}$ and $\boldsymbol{r}'$ (see lower figure).

In terms of these coordinates we have

$$\mathrm{d}V_\mathrm{e} = r'^2\,\mathrm{d}r'\,\sin\vartheta\,\mathrm{d}\vartheta\,\mathrm{d}\varphi \tag{11}$$

and, from $\boldsymbol{r} = \boldsymbol{r}' + \boldsymbol{R}$, we get

$$\begin{aligned}(\boldsymbol{k}_0-\boldsymbol{k})\cdot\boldsymbol{r} &= \boldsymbol{s}\cdot\boldsymbol{r} = \boldsymbol{s}\cdot\boldsymbol{r}' + \boldsymbol{s}\cdot\boldsymbol{R} \\ &= sr'\cos\vartheta + \boldsymbol{s}\cdot\boldsymbol{R} \ .\end{aligned} \tag{12}$$

Thus the integral becomes

Example 11.8

$$
\begin{aligned}
J_{\mathrm{e}}(\boldsymbol{R}) &= \int \frac{\mathrm{e}^{-r'/d}}{r'} \mathrm{e}^{\mathrm{i}(sr'\cos\vartheta + \boldsymbol{s}\cdot\boldsymbol{R})} r'^2 \sin\vartheta \, \mathrm{d}r' \, \mathrm{d}\vartheta \, \mathrm{d}\varphi \\
&= 2\pi \, \mathrm{e}^{\mathrm{i}\boldsymbol{s}\cdot\boldsymbol{R}} \int\limits_0^\infty \int\limits_0^\pi r' \, \mathrm{e}^{-r'/d} \mathrm{e}^{\mathrm{i}sr'\cos\vartheta} \sin\vartheta \, \mathrm{d}\vartheta \, \mathrm{d}r' \\
&= 2\pi \, \mathrm{e}^{\mathrm{i}\boldsymbol{s}\cdot\boldsymbol{R}} \int\limits_0^\infty r' \, \mathrm{e}^{-r'/d} \frac{1}{\mathrm{i}sr'} (\mathrm{e}^{\mathrm{i}sr'} - \mathrm{e}^{-\mathrm{i}sr'}) \, \mathrm{d}r' \\
&= 2\pi \, \mathrm{e}^{\mathrm{i}\boldsymbol{s}\cdot\boldsymbol{R}} \frac{1}{\mathrm{i}s} \int\limits_0^\infty (\mathrm{e}^{(\mathrm{i}s - 1/d)r'} - \mathrm{e}^{-(\mathrm{i}s + 1/d)r'}) \, \mathrm{d}r' \\
&= 2\pi \, \mathrm{e}^{\mathrm{i}\boldsymbol{s}\cdot\boldsymbol{R}} \frac{1}{\mathrm{i}s} \left(\frac{\mathrm{e}^{\mathrm{i}sr'}}{\mathrm{i}s - 1/d} + \frac{\mathrm{e}^{-\mathrm{i}sr'}}{\mathrm{i}s + 1/d} \right) \mathrm{e}^{-r'/d} \Bigg|_0^\infty \\
&= -2\pi \, \mathrm{e}^{\mathrm{i}\boldsymbol{s}\cdot\boldsymbol{R}} \frac{1}{\mathrm{i}s} \left(\frac{1}{\mathrm{i}s - 1/d} + \frac{1}{\mathrm{i}s + 1/d} \right) \\
&= \mathrm{e}^{\mathrm{i}\boldsymbol{s}\cdot\boldsymbol{R}} \frac{4\pi}{s^2 + 1/d^2} \; .
\end{aligned} \tag{13}
$$

This result can be simplified: the term $1/d^2$ in the denominator can be neglected if $s^2 d^2 \gg 1$. This means that the momentum transfer must not become too small. Hence,

$$
J_{\mathrm{e}}(\boldsymbol{R}) \approx \mathrm{e}^{\mathrm{i}\boldsymbol{s}\cdot\boldsymbol{R}} \frac{4\pi}{s^2} \; . \tag{14}
$$

Therefore the matrix element takes the form

$$
\langle \psi_{\mathrm{f}} | V | \psi_{\mathrm{i}} \rangle = Ze^2 \frac{4\pi}{s^2} \int \phi_{\mathrm{f}}^*(\boldsymbol{R}) \mathrm{e}^{\mathrm{i}\boldsymbol{s}\cdot\boldsymbol{R}} \phi_{\mathrm{i}}(\boldsymbol{R}) \, \mathrm{d}V_{\mathrm{n}} \; . \tag{15}
$$

For elastic scattering, the state of the nucleus is not changed. Let the nucleus be in its ground state ϕ. Then we have $\phi_{\mathrm{f}} = \phi_{\mathrm{i}} = \phi$, and the product $Z\phi^*\phi(\boldsymbol{R})$ is the density distribution of the protons in the nucleus. Instead of the wave functions, we can introduce the charge density $\varrho_{\mathrm{p}}(\boldsymbol{R})$ of the atomic nucleus (more accurately: $\varrho_{\mathrm{p}}(\boldsymbol{R})$ is the charge density without the factor e, which we have explicitly taken out):

$$
Z\phi^*\phi = \varrho_{\mathrm{p}}(\boldsymbol{R}) \quad \text{with} \quad \int \varrho_{\mathrm{p}}(\boldsymbol{R}) \, \mathrm{d}V_{\mathrm{n}} = Z \; . \tag{16}
$$

For further simplification we assume *spherical symmetry of the charge distribution*:

$$
\varrho_{\mathrm{p}}(\boldsymbol{R}) = \varrho_{\mathrm{p}}(R) \; . \tag{17}
$$

Example 11.8

This assumption is only valid for atomic nuclei in the vicinity of the magic numbers. The others are prolately (cigar-like) deformed. Therefore we obtain for the matrix element

$$\begin{aligned}\langle\psi_{\mathrm{f}}|\hat{V}|\psi_{\mathrm{i}}\rangle &= \frac{4\pi e^2}{s^2}\int \varrho_{\mathrm{p}}(R)\,\mathrm{e}^{\mathrm{i}\boldsymbol{s}\cdot\boldsymbol{R}}\,\mathrm{d}V_{\mathrm{k}} \\ &= \frac{4\pi e^2}{s^2}F(\boldsymbol{s}) \ . \end{aligned} \tag{18}$$

The quantity $F(\boldsymbol{s})$ is called a *form factor*. It is the *Fourier-transformed* charge distribution and reflects the deviation of the nuclear charge distribution from point structure. Indeed, if the nucleus is assumed to be pointlike (i.e. $\varrho_{\mathrm{p}}(\boldsymbol{R}) = \delta^3(\boldsymbol{R})$), we get $F = 1$.

The form factor $F(\boldsymbol{s}) = \int \varrho_{\mathrm{p}}(R)\,\mathrm{e}^{\mathrm{i}\boldsymbol{s}\cdot\boldsymbol{R}}\,\mathrm{d}V_{\mathrm{k}}$ can be further evaluated by again introducing spherical coordinates and using the axis defined by $\boldsymbol{s}$ as the polar axis. Hence, we get

$$\mathrm{d}V_{\mathrm{k}} = R^2\sin\vartheta\,\mathrm{d}R\,\mathrm{d}\vartheta\,\mathrm{d}\varphi \quad \text{and} \quad \boldsymbol{s}\cdot\boldsymbol{R} = Rs\cos\vartheta \ , \tag{19}$$

and therefore

$$\begin{aligned}F(\boldsymbol{s}) &= 2\pi\int_0^\infty\int_0^\pi \varrho_{\mathrm{p}}(R)\,\mathrm{e}^{\mathrm{i}sR\cos\vartheta}R^2\sin\vartheta\,\mathrm{d}R\,\mathrm{d}\vartheta \\ &= 2\pi\int_0^\infty \varrho_{\mathrm{p}}(R)\left(-\int_0^\pi \mathrm{e}^{\mathrm{i}sR\cos\vartheta}\,\mathrm{d}(\cos\vartheta)\right)R^2\,\mathrm{d}R \\ &= 2\pi\int_0^\infty \varrho_{\mathrm{p}}(R)\frac{1}{\mathrm{i}sR}(\mathrm{e}^{\mathrm{i}sR}-\mathrm{e}^{-\mathrm{i}sR})R^2\,\mathrm{d}R \\ &= \frac{4\pi}{s}\int_0^\infty \varrho_{\mathrm{p}}(R)\sin(sR)R\,\mathrm{d}R \ . \end{aligned} \tag{20}$$

The last integral can be calculated only if the charge distribution $\varrho_{\mathrm{p}}(R)$ is known. We will consider this point later (see Exercise 11.9). Our current result can be summarized as

$$\langle\psi_{\mathrm{f}}\,|V|\,\psi_{\mathrm{i}}\rangle = \frac{4\pi e^2}{s^2}F(s) \ , \tag{21}$$

and we can turn to the *calculation of the density of states*. The orthogonality and closure relation (7) implies that $n(\alpha)$ equals $(2\pi)^{-3}$ if α is identified with $\boldsymbol{k}$. In other words, in the space of the vectors $\boldsymbol{k}$, the density of states is constant and equal to $(2\pi)^{-3}$, i.e. the number of states in the interval $[\boldsymbol{k}, \boldsymbol{k}+\mathrm{d}\boldsymbol{k}]$ is equal to $(2\pi)^{-3}\mathrm{d}^3k$. If we had used plane waves normalized on δ functions from the beginning, the density would have been $n(k) = 1$.

Example 11.8

Now we are interested in states with momenta pointing in a certain direction Ω. These momenta differ from each other in their energy only. Therefore the density of these states is $\varrho(\Omega, E)$, i.e. $\varrho(\Omega, E)\,\mathrm{d}\Omega\,\mathrm{d}E$ is equal to the number of states which have a momentum pointing into the solid angle $[\Omega, \Omega + \mathrm{d}\Omega]$ and whose energy lies in the interval $[E, E + \mathrm{d}E]$. Thus we get

$$\varrho(\Omega, E)\,\mathrm{d}\Omega\,\mathrm{d}E = \frac{\mathrm{d}^3k}{(2\pi)^3} \ . \tag{22}$$

By introducing spherical coordinates in the $\boldsymbol{k}$ space

$$\mathrm{d}^3k = k^2\,\mathrm{d}k\,\mathrm{d}\Omega \ , \quad \text{we get} \tag{23}$$

$$\varrho(\Omega, E)\,\mathrm{d}\Omega\,\mathrm{d}E = \frac{k^2\,\mathrm{d}k}{(2\pi)^3}\,\mathrm{d}\Omega \ . \tag{24}$$

We note that the density of states is independent of Ω, which allows us finally to write $\varrho(\Omega, E) = \varrho(E)$. Substituting the momentum p for k, the density of states $\varrho(E)$ becomes

$$\varrho(E) = \frac{p^2\,\mathrm{d}p}{(2\pi\hbar)^3\,\mathrm{d}E} = \frac{p^2}{(2\pi\hbar)^3}\frac{1}{\mathrm{d}E/\mathrm{d}p} \ . \tag{25}$$

To calculate the derivative, the energy conditions of the collision have to be examined. We assume the electron is very fast and therefore proceed from the relativistic energy–momentum relation

$$\sqrt{p_0^2c^2 + m^2c^4} + Mc^2 = \sqrt{p^2c^2 + m^2c^4} + \sqrt{\hbar^2s^2c^2 + M^2c^4} = E \ . \tag{26}$$

If the kinetic energy of the electron is large enough compared to the rest energy, the term mc^2 can be neglected and we obtain

$$\frac{E}{c} = p_0 + Mc = p + \sqrt{\hbar^2s^2 + M^2c^2} \ . \tag{27}$$

According to the definition of $\boldsymbol{s}$, we get

$$\hbar^2s^2 = p_0^2 + p^2 - 2p_0p\cos\Theta \ . \tag{28}$$

Because of the great mass difference between nucleus and electron, the energy transfer can be considered small compared to Mc^2, and therefore $p \approx p_0$ and $(p - p_0)^2 \approx 0$, so that $p^2 + p_0^2 = 2pp_0$ results. Thus we get

$$\hbar^2s^2 = 2p_0p(1 - \cos\Theta) = 4p_0p\sin^2\frac{\Theta}{2} \ . \tag{29}$$

Example 11.8

From relation (27) we find for E/c that

$$(p_0 - p + Mc)^2 = \hbar^2 s^2 + mc^2$$
$$= 4p_0 p \sin^2\frac{\Theta}{2} + M^2c^2 \ . \tag{30}$$

On the other hand, we have

$$(p_0 - p + Mc)^2 = (p_0 - p)^2 + 2(p_0 - p)Mc + M^2c^2$$
$$\approx 2(p_0 - p)Mc + M^2c^2 \ , \tag{31}$$

because $p_0 - p \approx 0$ and thus the square $(p - p_0)^2$ is vanishingly small. If we compare the two last relations, we obtain

$$(p_0 - p)Mc \approx 2p_0 p \sin^2\frac{\Theta}{2} \ , \tag{32}$$

and with this, finally,

$$p = \frac{p_0}{1 + (2p_0/Mc)\sin^2\Theta/2} \ . \tag{33}$$

Now the expression $\mathrm{d}E/\mathrm{d}p$ has to be calculated. For this purpose we start with

$$E = pc + \sqrt{\hbar^2 s^2 c^2 + M^2 c^4} \ , \tag{34}$$

from which we get

$$\frac{\mathrm{d}E}{\mathrm{d}p} = c + \frac{\hbar^2 c^2 (\mathrm{d}s^2/\mathrm{d}p)}{2\sqrt{\hbar^2 s^2 c^2 + M^2 c^4}}$$
$$\approx c + \frac{\hbar^2 c^2}{2Mc^2}\frac{\mathrm{d}s^2}{\mathrm{d}p} \tag{35}$$

and, if we make use of (29) and (33),

$$\frac{\mathrm{d}E}{\mathrm{d}p} \approx c\left(1 + \frac{2p_0}{Mc}\sin^2\frac{\Theta}{2}\right) = c\frac{p_0}{p} \ . \tag{36}$$

Finally, we get the following expression for the density of states (25):

$$\varrho(E) = \frac{p^2}{(2\pi\hbar)^3}\frac{p}{cp_0} = \frac{p^3}{(2\pi\hbar)^3}\frac{1}{cp_0} \ . \tag{37}$$

With the help of Fermi's golden rule expressed in (12) of Example 11.7, and in (21) and (37), the *transition probability per unit time* is given by

$$P_{\mathrm{i}\to\mathrm{f}} = \frac{2\pi}{\hbar}\left|\langle\psi_\mathrm{f}|\hat{V}|\psi_\mathrm{i}\rangle\right|^2\varrho(E) = \frac{2\pi}{\hbar}\left(\frac{4\pi e^2}{s^2}\right)^2 |F(s)|^2 \frac{p^3}{(2\pi\hbar)^3}\frac{1}{cp_0}$$
$$= \frac{4(2\pi)^3 e^4}{\hbar}|F(s)|^2 \frac{1}{(4p_0 p/\hbar^2)^2 \sin^4\Theta/2}\frac{p^3}{(2\pi\hbar)^3}\frac{1}{cp_0}$$
$$= \frac{4e^4|F(s)|^2}{16p_0^2 p^2 \sin^4\Theta/2}\frac{p^3}{cp_0} = \left(\frac{e^2}{2p_0}\right)^2 \frac{1}{c}\frac{p}{p_0}\frac{1}{\sin^4\Theta/2}|F(s)|^2 \ . \tag{38}$$

The transition probability itself cannot be measured directly, but a quantity can be observed, which is called a cross-section or, more accurately, a *scattering cross-section*, and is denoted by σ or $d\sigma_i$ respectively (see figure). $d\sigma_{i\to f}$ is the *number of particles scattered per unit time and per unit of the incoming particle current into the section of the solid angle* $(\Omega, \Omega + d\Omega)$. Since the states $|\boldsymbol{k}\rangle$ represent particles, the current of which is v, we obtain for the *differential cross-section*

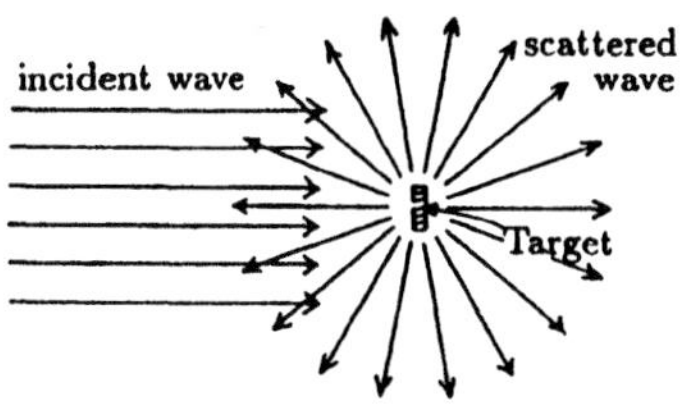

The incident particle wave and the scattered wave

$$d\sigma_{i\to f} = \frac{p_{i\to f}\, d\Omega}{v_i} \,. \tag{39}$$

The electron velocity v_i can be set approximately to the velocity of light c; thus we write

$$\frac{d\sigma_{i\to f}}{d\Omega} = \frac{1}{c} p_{i\to f} = \left(\frac{e^2}{2p_0 c}\right)^2 \frac{p}{p_0} \frac{1}{\sin^4 \Theta/2} |F(s)|^2 \,. \tag{40}$$

This is the extended *Rutherford scattering formula for the cross-section.* Substituting p/p_0 from (33) yields finally

$$\frac{d\sigma_{i\to f}}{d\Omega} = \left(\frac{e^2}{2p_0 c}\right)^2 \frac{1}{\sin^4 \Theta/2} |F(s)|^2 \times \frac{1}{1 + (2p_0/Mc)\sin^2 \Theta/2} \,. \tag{41}$$

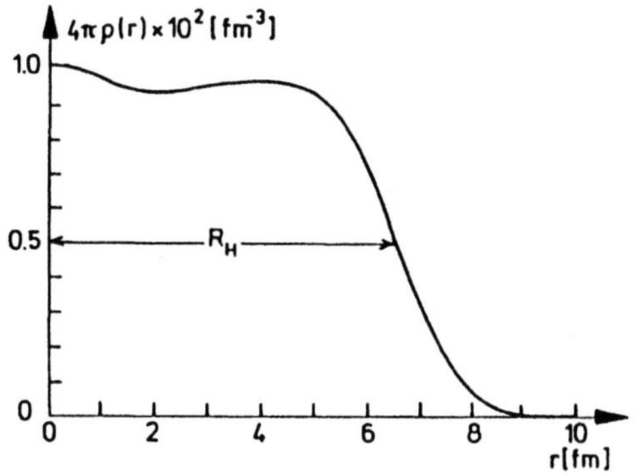

Fig. 11.8. The charge distribution $4\pi\varrho(r)$ of the lead nucleus, determined by elastic electron scattering. $\varrho(r)$ is approximately constant inside the nucleus as far as $r \approx 5$ fm and decreases in the surface region with a thickness of ≈ 2 fm. Other nuclei show similar behaviour: whereas the surface thickness is nearly equal for all nuclei, the radius R_H, at which $4\pi\varrho(r)$ has its half-maximum value, changes according to $R_H = r_0 A^{1/3}$, where $r_0 \approx 1.2$ fm, and A is the number of nucleons

Example 11.8

The effect of the recoiling nucleus is taken into account by the last factor. If the atomic nucleus is very heavy ($M \to \infty$), this factor is nearly 1, corresponding to a scattering without momentum transfer. The form factor $|F(s)|^2$ takes into consideration properties (extension) of the nuclear charge density. It can be experimentally deduced using (41) and comparing it with the experimentally determined cross-section (by measuring the differential cross-section). From the thus determined form factor, the charge distribution can be calculated according to (20). Robert ***Hofstadter***, who made systematic measurements of this kind, was awarded the Nobel Prize for this work in 1961. Some of the most accurate charge distributions of atomic nuclei were measured in the same way by Peter ***Brix*** at the electron accelerator in Darmstadt.

By way of example, we show the charge distribution of the lead nucleus (see previous figure). It is nearly constant inside the nucleus and has a surface thickness of $\approx 2\,\mathrm{fm}$.

EXERCISE

11.9 Limit of Small Momentum Transfer

Problem. Show that electron scattering with small momentum transfer permits the determination of the total charge and mean quadratic radius of atomic nuclei only.

Solution. To solve this problem, we start with the structure function (20) in Example 11.8:

$$F(s) = \frac{4\pi}{s} \int \varrho_{\mathrm{p}}(R) \sin(sR) R \,\mathrm{d}R \ . \tag{1}$$

Assuming small momentum transfer s, or, more accurately, $sR \ll 1$, we can expand $\sin(sR)$ to obtain

$$\sin(sR) \approx sR - \frac{(sR)^3}{6} \ . \tag{2}$$

Thus $F(s)$ becomes

$$F(s) = 4\pi \int_0^\infty \varrho_{\mathrm{p}}(R) R^2 \,\mathrm{d}R - \frac{2\pi}{3} s^2 \int_0^\infty \varrho_{\mathrm{p}}(R) R^4 \,\mathrm{d}R \equiv Z - \frac{2\pi}{3} s^2 \left\langle R^2 \right\rangle \ . \tag{3}$$

The first term is just the total charge Z of the nucleus, while the second one contains the mean quadratic radius. To measure more details of the charge distribution $\varrho_{\mathrm{p}}(R)$, the momentum transfer has to be increased [compare with (29) in Example 11.8]:

$$(\hbar s)^2 = 4 p_0 p \sin^2 \frac{\Theta}{2} \ . \tag{4}$$

Exercise 11.9

This can be done by increasing the energy E of the electron and simultaneously increasing the momentum $p_0 \approx p$. Of course, we should take the best scattering angle possible ($\theta = 180°$), i.e. we must detect the backward-scattered electrons at higher energy. Then the next term of the sine expansion becomes important, so that one gets

$$\sin(sR) \approx sR - \frac{(sR)^3}{3!} + \frac{(sR)^5}{5!} , \tag{5}$$

with the following result:

$$F(s) = 4\pi \langle R \rangle - \frac{2\pi}{3} s^2 \left\langle R^2 \right\rangle + \frac{4\pi}{5!} s^4 \left\langle R^4 \right\rangle . \tag{6}$$

We recognize that the different factors in front of the powers s^{2n} of the form factor reflect the *higher moments of the charge distribution.*

EXERCISE

11.10 Properties of the Function $f(t, \omega)$

Problem. Show that the function

$$f(t, \omega) = 1/\omega^2 |e^{i\omega t} - 1|^2$$

fulfils the following relations:

$$\int_{-\infty}^{\infty} f(t, \omega)\, d\omega = 2\pi t \quad \text{and}$$

$$\lim_{t\to\infty} \frac{f(t, \omega)}{t} = 2\pi\delta(\omega) .$$

Solution. According to (11.69) the function $f(t, \omega)$ is defined as

$$f(t, \omega) = \frac{2}{\omega^2}(1 - \cos \omega t) ; \tag{1}$$

therefore

$$\int_{-\infty}^{\infty} f(t, \omega)\, d\omega = 2 \int_{-\infty}^{\infty} \frac{1 - \cos \omega t}{\omega^2}\, d\omega . \tag{2}$$

By substituting $\omega t = x$, the integral becomes

$$\int_{-\infty}^{\infty} f(t, \omega)\, d\omega = 2t \int_{-\infty}^{\infty} \frac{1 - \cos x}{x^2}\, dx , \tag{3}$$

Exercise 11.10

which can be integrated by parts:

$$\int_{-\infty}^{\infty} \frac{1-\cos x}{x^2}\,\mathrm{d}x = 2\int_{0}^{\infty} \frac{\sin x}{x}\,\mathrm{d}x \ . \tag{4}$$

With the help of the relation

$$\frac{1}{x} = \int_{0}^{\infty} \mathrm{e}^{-ux}\,\mathrm{d}u \quad \text{for} \quad x > 0 \ , \tag{5}$$

and interchanging the integration, the integral (4) can be evaluated:

$$\begin{aligned}\int_{0}^{\infty} \frac{\sin x}{x}\,\mathrm{d}x &= \int_{0}^{\infty} \sin x \int_{0}^{\infty} \mathrm{e}^{-ux}\,\mathrm{d}u\,\mathrm{d}x \\ &= \int_{0}^{\infty} \mathrm{d}u \int_{0}^{\infty} \mathrm{e}^{-ux} \sin x\,\mathrm{d}x \\ &= \int_{0}^{\infty} \frac{\mathrm{d}u}{1+u^2} = \arctan u\Big|_0^{\infty} = \frac{\pi}{2} \ .\end{aligned} \tag{6}$$

Thus we get

$$\int_{-\infty}^{\infty} f(t,\omega)\,\mathrm{d}\omega = 2t \times 2 \times \frac{\pi}{2} = 2\pi t \ . \tag{7}$$

To solve the second part of the problem, we use the representation of the δ function (see Example 5.2):

$$\delta(\omega) = \frac{1}{\pi} \lim_{t\to\infty} \frac{1-\cos \omega t}{\omega^2 t} \ . \tag{8}$$

With the help of (1), we obtain

$$\begin{aligned}\lim_{t\to\infty} \frac{f(t,\omega)}{t} &= \lim_{t\to\infty} \frac{2}{\omega^2 t}(1-\cos \omega t) \\ &= 2\pi \lim_{t\to\infty} \frac{1}{\pi} \frac{(1-\cos \omega t)}{\omega^2 t} \Rightarrow 2\pi\delta(\omega) \ .\end{aligned} \tag{9}$$

EXERCISE

11.11 Elementary Theory of the Dielectric Constant

Let $\hat{H}_0$ be the Hamiltonian of an electron with charge $-e$, e.g. in a molecule (to simplify the problem, the spectrum is assumed to be discrete). An incoming plane monochromatic linearly polarized electromagnetic wave shall not be influenced by the polarization of the molecule, i.e. its frequency ω shall be clearly different from all absorption lines. It can also be proved that the contribution of the magnetic part of the wave produces negligibly small effects.

Problem. (a) Under these circumstances the wave can be described by a homogeneous external potential, which is periodic in time and has the amplitude F_0 and the frequency ω. Find the related Schrödinger equation, if the z axis points in the direction of the oscillation.

(b) Let ψ_0 be the ground state of $\hat{H}_0$ with the energy E_0. Take

$$\psi(\boldsymbol{x},t) = \psi_0(\boldsymbol{x})\exp\left(-\frac{\mathrm{i}}{\hbar}E_0 t\right) + F_0\psi^{(1)}(\boldsymbol{x},t) \tag{1}$$

and find the first correction $\psi^{(1)}$ to the "stationary ground-state" in the time-periodic potential.

Hint: Set

$$\begin{aligned}\psi^{(1)}(\boldsymbol{x},t) =& \omega_+(\boldsymbol{x})\exp\left[-\frac{\mathrm{i}}{\hbar}(E_0+\hbar\omega)t\right]\\ &+\omega_-(\boldsymbol{x})\exp\left[-\frac{\mathrm{i}}{\hbar}(E_0-\hbar\omega)t\right] .\end{aligned} \tag{2}$$

(c) In the absence of an external field, let the molecule be in the ground state ψ_0, without a dipole moment, i.e.

$$\langle \boldsymbol{p}_0\rangle = -e\int \psi_0^* \boldsymbol{r}\psi_0\,\mathrm{d}V = 0 . \tag{3}$$

To calculate the dielectric constant, start with the definition for the dielectric constant ε via the relation

$$\varepsilon \boldsymbol{E} = \boldsymbol{E} + 4\pi \boldsymbol{P} , \tag{4}$$

and insert the polarization $\boldsymbol{P}$ in a nonconducting solid given by

$$\boldsymbol{P} = P\boldsymbol{e}_z = N\bar{p}\boldsymbol{e}_z , \tag{5}$$

where N is the density of the molecules (i.e. numbers of molecules per cm^3) and $\bar{p}$, the part of the mean dipole moment $-e\int \psi^* z\psi\,\mathrm{d}V$, linear in F_0, of a single molecule oriented in the z direction. Calculate the dielectric constant ε.

Solution. (a) For the electric field, we write

$$\boldsymbol{E} = F_0 \sin\omega t\,\boldsymbol{e}_z .$$

Exercise 11.11

The related potential $\phi(\boldsymbol{r},t)$ has to satisfy

$$\boldsymbol{E} = -\operatorname{grad}\phi \ , \tag{6}$$

so that $\phi(\boldsymbol{r},t)$ can be written

$$\phi(\boldsymbol{r},t) = \phi(z,t) = -zF_0 \sin\omega t \ . \tag{7}$$

Thus the electron feels the potential energy

$$V(\boldsymbol{r},t) = -e\phi(\boldsymbol{r},t) = ezF_0 \sin\omega t \tag{8}$$

and the Schrödinger equation for the molecular electrons can be written

$$(\hat{H}_0 + ezF_0 \sin\omega t)\psi(\boldsymbol{r},t) = \mathrm{i}\hbar\frac{\partial}{\partial t}\psi(\boldsymbol{r},t) \ . \tag{9}$$

(b) With the given formulation (1) for ψ, the Schrödinger equation reads

$$\begin{aligned}(\hat{H}_0 + ezF_0 \sin\omega t)&\left[\psi_0(\boldsymbol{r})\exp\left(-\frac{\mathrm{i}}{\hbar}E_0t\right) + F_0\psi^{(1)}(\boldsymbol{r},t)\right]\\ &= E_0\psi_0(\boldsymbol{r})\exp\left(-\frac{\mathrm{i}}{\hbar}E_0t\right) + \mathrm{i}\hbar F_0\frac{\partial}{\partial t}\psi^{(1)}(\boldsymbol{r},t) \ .\end{aligned} \tag{10}$$

Comparing the terms linear in F_0 on both sides, we obtain

$$\begin{aligned}&ezF_0 \sin\omega t\psi_0 \exp\left(-\frac{\mathrm{i}}{\hbar}E_0t\right) + F_0\hat{H}_0\psi^{(1)}(\boldsymbol{r},t) = \mathrm{i}\hbar F_0\frac{\partial}{\partial t}\psi^{(1)}(\boldsymbol{r},t)\\ &\Rightarrow \left[\hat{H}_0 - \mathrm{i}\hbar\frac{\partial}{\partial t}\right]\psi^{(1)} = -ez\sin\omega t\psi_0\exp\left(-\frac{\mathrm{i}}{\hbar}E_0t\right) \ .\end{aligned} \tag{11}$$

If we use the hint for $\psi^{(1)}$, and note that

$$\sin\omega t = \frac{1}{2\mathrm{i}}(\mathrm{e}^{\mathrm{i}\omega t} - \mathrm{e}^{-\mathrm{i}\omega t}) \ ,$$

we get

$$\begin{aligned}&\hat{H}_0w_+\exp\left[-\frac{\mathrm{i}}{\hbar}(E_0+\hbar\omega)t\right] + \hat{H}_0w_-\exp\left[-\frac{\mathrm{i}}{\hbar}(E_0-\hbar\omega)t\right]\\ &\quad -(E_0+\hbar\omega)w_+\exp\left[-\frac{\mathrm{i}}{\hbar}(E_0+\hbar\omega)t\right]\\ &\quad -(E_0-\hbar\omega)w_-\exp\left[-\frac{\mathrm{i}}{\hbar}(E_0-\hbar\omega)t\right]\\ &= -\frac{ez}{2\mathrm{i}}\psi_0\exp\left(-\frac{\mathrm{i}}{\hbar}E_0t\right)(\mathrm{e}^{\mathrm{i}\omega t} - \mathrm{e}^{-\mathrm{i}\omega t}) \ .\end{aligned} \tag{12}$$

The common factor $\exp[-(\mathrm{i}/\hbar)E_0t]$ can be dropped; the functions $\mathrm{e}^{\mathrm{i}\omega t}$ and $\mathrm{e}^{-\mathrm{i}\omega t}$ are linearly independent, i.e. their coefficients have to vanish. This yields

$$\begin{aligned}\hat{H}_0w_+ - (E_0+\hbar\omega)w_+ &= \frac{ez}{2\mathrm{i}}\psi_0 \ ,\\ \hat{H}_0w_- - (E_0-\hbar\omega)w_- &= -\frac{ez}{2\mathrm{i}}\psi_0 \ .\end{aligned} \tag{13}$$

Now $E_0 + \hbar\omega$ is *not* an eigenvalue of $\hat{H}_0$: otherwise, ω would be an absorption frequency; nor is $E_0 - \hbar\omega$, because E_0 is the lowest eigenvalue of $\hat{H}_0$ (the ground state). Therefore the equations have no homogeneous solution and can be solved without ambiguity. Let $E_j (j = 0, 1, 2, \dots)$ be the eigenvalues of $\hat{H}_0$ (which we have assumed to be discrete) and $\varphi_j(\boldsymbol{r})$, the related eigenfunctions. Then w_+ can be expanded:

Exercise 11.11

$$w_+ = \sum_{j=0}^{\infty} C_j^{(+)} \varphi_j \ . \tag{14}$$

Because of $\hat{H}_0 \varphi_j = E_j \varphi_j$, we obtain

$$\sum_{j=0}^{\infty} C_j^{(+)} E_j \varphi_j - (E_0 + \hbar\omega) \sum_{j=0}^{\infty} C_j^{(+)} \varphi_j = +\frac{ez}{2\mathrm{i}} \psi_0 \ . \tag{15}$$

Multiplication by φ_k^* and integration yield

$$C_k^{(+)} (E_k - E_0 - \hbar\omega) = +\frac{1}{2\mathrm{i}} \langle \varphi_k | ez | \psi_0 \rangle \quad \text{and} \tag{16}$$

$$w_+(\boldsymbol{r}) = +\frac{e}{2\mathrm{i}} \sum_{j=1}^{\infty} \frac{\langle \varphi_j | z | \psi_0 \rangle}{E_j - E_0 - \hbar\omega} \varphi_j(\boldsymbol{r}) \tag{17}$$

and, analogously,

$$w_-(\boldsymbol{r}) = -\frac{e}{2\mathrm{i}} \sum_{j=1}^{\infty} \frac{\langle \varphi_j | z | \psi_0 \rangle}{E_j - E_0 + \hbar\omega} \varphi_j(\boldsymbol{r}) \ . \tag{18}$$

Therefore we can write the wave function as

$$\psi(\boldsymbol{r}, t) = \left[\psi_0(\boldsymbol{r}) + \frac{eF_0}{2\mathrm{i}} \sum_{j=1}^{\infty} \varphi_j(\boldsymbol{r}) \langle \varphi_j | z | \psi_0 \rangle \right.$$
$$\left. \times \left(\frac{\mathrm{e}^{-\mathrm{i}\omega t}}{E_j - E_0 - \hbar\omega} - \frac{\mathrm{e}^{\mathrm{i}\omega t}}{E_j - E_0 + \hbar\omega} \right) \right] \exp\left(-\frac{\mathrm{i}}{\hbar} E_0 t \right) \ . \tag{19}$$

(c) The mean dipole moment in the field direction is

$$-e \langle \psi | z | \psi \rangle = -\frac{e^2 F_0}{2\mathrm{i}} \sum_{j=1}^{\infty} |\langle \varphi_j | z | \psi_0 \rangle|^2$$
$$\times \left(\frac{\mathrm{e}^{-\mathrm{i}\omega t}}{E_j - E_0 - \hbar\omega} - \frac{\mathrm{e}^{\mathrm{i}\omega t}}{E_j - E_0 + \hbar\omega} \right) + \frac{e^2 F_0}{2\mathrm{i}} \sum_{j=1}^{\infty} |\langle \varphi_j | z | \psi_0 \rangle|^2$$
$$\times \left(\frac{\mathrm{e}^{\mathrm{i}\omega t}}{E_j - E_0 - \hbar\omega} - \frac{\mathrm{e}^{-\mathrm{i}\omega t}}{E_j - E_0 + \hbar\omega} \right) + O(F_0^2) \ , \tag{20}$$

Exercise 11.11

and hence

$$\bar{p} = \frac{e^2 F_0}{2\mathrm{i}} \sum_{j=1}^{\infty} \left|\langle\varphi_j|z|\psi_0\rangle\right|^2 2\mathrm{i} \sin\omega t \frac{2(E_j - E_0)}{(E_j - E_0)^2 - \hbar^2\omega^2} \; . \tag{21}$$

In the absence of the field ($F_0 = 0$), only the ground state $\psi_0 = \varphi_{j=0}$ is populated and whose dipole moment vanishes:

$$\langle\psi_0 \, |z| \, \psi_0\rangle = 0 \; .$$

Hence, it is the electric field of the light wave which partly polarizes the atoms or the molecules. This is quite reasonable and may be intuitively expected.

With $\hbar\omega_j = E_j - E_0$ and the so-called *dipole strength*

$$f_j = \frac{2m_\mathrm{e}}{\hbar} \left|\langle\varphi_j|z|\psi_0\rangle\right|^2 \omega_j$$

(where m_e = electron mass), we obtain for the dielectric constant

$$\begin{aligned} \varepsilon F_0 \boldsymbol{e}_z \sin\omega t &= F_0 \boldsymbol{e}_z \sin\omega t + \boldsymbol{e}_z \frac{8\pi N e^2}{\hbar} \\ &\quad \times \sum_{j=1}^{\infty} \left|\langle\phi_j|z|\psi_0\rangle\right|^2 \frac{\omega_j}{\omega_j^2 - \omega^2} F_0 \sin\omega t \\ &= \left(1 + \frac{4\pi N e^2}{m_\mathrm{e}} \sum_{j=1}^{\infty} \frac{f_j}{\omega_j^2 - \omega^2}\right) F_0 \sin\omega t \boldsymbol{e}_z \; . \end{aligned} \tag{22}$$

Hence,

$$\varepsilon(\omega) = 1 + \frac{4\pi N e^2}{m_\mathrm{e}} \sum_{j=1}^{\infty} \frac{f_j}{\omega_j^2 - \omega^2} \; . \tag{23}$$

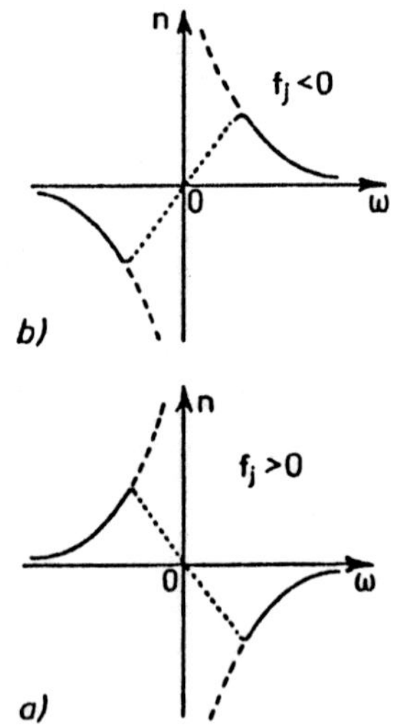

The refractive index n in the range of anomalous dispersion in the case of (**a**) positive and (**b**) negative refraction

This corresponds to the classical expression of the dielectric constant.

Since, to a good approximation, the refraction index n is related to the dielectric constant by $n^2 = \varepsilon$, the above formula (23) represents the quantum-mechanical calculation of the *refraction index* as well. Contrary to the classical expression, the quantum-mechanical oscillator strength

$$f_j = \frac{2m_e}{\hbar^2} \left|\langle\varphi_j \, |z| \, \varphi_0\rangle\right|^2 (E_j - E_0) \tag{24}$$

can also have negative values if the atom or molecule is in an excited state in the beginning. This leads to the phenomenon of *negative refraction* (see figure).

11.7 Biographical Notes

STARK, Johannes, German physicist, *15.4.1874 Schickenhof, in Thansüß, district of Amberg, † 21.6.1957 Traunstein. S. became a professor in Hannover; in 1909 he went to Aachen, in 1917, to Greifswald and in 1920, to Würzburg. He founded the "Jahrbuch der Radioaktivität und Elektronik" in 1904 and discovered in 1905 the (optical) *Doppler effect* in so-called channel rays and in 1913 the *Stark effect.* He was awarded the Nobel Prize in 1919. In 1933 he became the president of the "Notgemeinschaft der Deutschen Wissenschaft". He was a friend of P. Lenard and a supporter of "German physics"; thus he dismissed quantum theory and the theory of relativity as "the product of Jewish thinking".

HOFSTADTER, Robert, American physicist, *5.2.1915 New York. H. is a professor at Stanford University, California. He has considered problems of molecular structure and has contributed to the development of scintillation and crystal counters. H. proved that the proton and the neutron have a finite extension and structure. By examining electron scattering by atomic nuclei he succeeded in finding the charge distribution not only for the proton and neutron, but also systematically for many other nuclei of the Periodic System. For this research, H. was awarded the Nobel Prize in Physics in 1961, together with R. Mössbauer.

BRIX, Peter, German physicist, *20.10.1918 Kappeln/Schlei. B. was a professor from 1957 to 1973 in Darmstadt, then director at the Max-Planck-Institute für Kernphysik in Heidelberg. Together with Kopfermann, he examined the isotope shift and built the first German electron linear accelerator in Darmstadt with which he and his colleagues measured accurately the charge distribution of atomic nuclei.

12. Spin

We have often mentioned the spin of the electron in our previous considerations. In this chapter we want to discuss the experimental evidence for the existence of spin. Furthermore, we shall develop its mathematical description.

Like the Pauli principle, spin is a phenomenon which first occurred in quantum mechanics and has no analogy in classical physics. The electron was the first elementary particle whose spin was detected. Several experiments, which could not be classically interpreted, motivated ***Goudsmit*** and ***Uhlenbeck*** in 1925 to hypothesize:

> *Every electron has an intrinsic angular momentum (spin) of* $\frac{1}{2}\hbar$*, which corresponds to a magnetic moment of one Bohr magneton,* $\mu_B = |e|\hbar/2mc$.

In the following, we briefly discuss three special experiments.

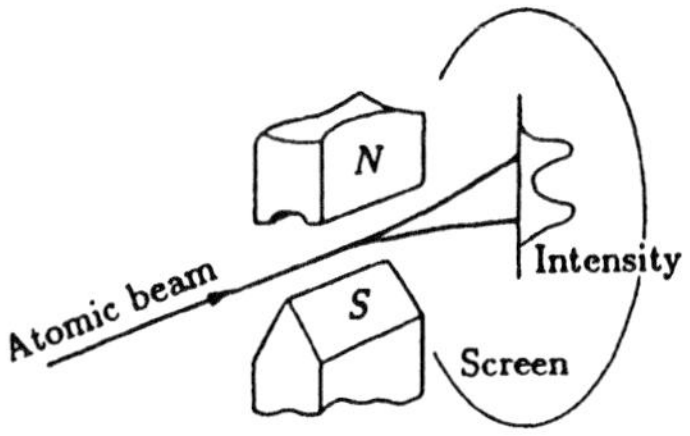

Fig. 12.1. The Stern–Gerlach experiment

In the first section we mentioned the Stern–Gerlach experiment (1922) as an example for the quantization of the angular momentum.[1] Figure 12.1 shows the principle of this experiment. A beam of hydrogen atoms (in the original experiment, silver atoms) is sent through an inhomogeneous magnetic field. The atoms are in the ground state, which implies that the electrons are in the 1s state; thus they have no orbital angular momentum. Therefore the atoms should not have any magnetic moment. However, *a splitting of the beam into two components is observed*, as the distribution of the intensity given in Sect. 1.6 shows. This splitting has its origin in a force[2]

$$\boldsymbol{F} = -\nabla(-\boldsymbol{M}\cdot\boldsymbol{B}) = \nabla(\boldsymbol{M}\cdot\boldsymbol{B}) = (\boldsymbol{M}\cdot\nabla)\boldsymbol{B} \ , \qquad (12.1)$$

which acts on the magnetic moment $\boldsymbol{M}$ in the inhomogeneous magnetic field $\boldsymbol{B}$. This splitting gives rise to the assumption that the electron has an intrinsic mag-

---

[1] This experiment was performed at the Institute of Physics at the University of Frankfurt a. M. At that time, O. Stern was a "Privatdozent" working with Prof. E. Madelung at the Institute of Theoretical Physics, and W. Gerlach a university lecturer at the Institute of Experimental Physics.

[2] See J.D. Jackson: *Classical Electrodynamics*, 2nd ed. (Wiley, New York 1975) and W. Greiner: *Classical Electrodynamics* (Springer, New York 1998).

netic moment. Because the beam is split into components of equal intensity, it follows that all electrons have a magnetic moment with the same absolute value. They also have two possible orientations, i.e. parallel or antiparallel to the magnetic field.

In principle it is also possible that the magnetic moment originates in the nucleus. But we shall later see that the ratio of the magnetic moments of the nuclei to the Bohr magneton is approximately equal to that of the inverse of the corresponding masses ($m_{\text{electron}}/m_{\text{proton}}$). Indeed, a more careful analysis of the Stern–Gerlach experiment reveals a fine structure of the lines caused by the magnetic moments of the nuclei.

12.1 Doublet Splitting

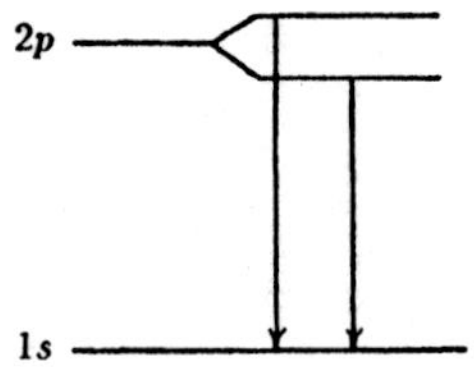

Fig. 12.2. Doublet splitting

A further proof of the existence of electron spin is given by the multiplet structure of atomic spectra. Let us take, for instance, the doublet splitting of sodium. Sodium has one valence electron. The transition of this electron from the first excited state to the ground state (2p → 1s; see Fig. 12.2) leads to two adjacent spectral lines of 5890 Å and 5896 Å.

Although the 2p level is three-fold degenerate ($m = 0, \pm 1$), this degeneracy can be removed by an external magnetic field. But doublet splitting can already be observed without an external magnetic field. This can be explained by assuming the existence of an electron spin in the following way.

Owing to the magnetic field originating from the orbital motion of the electrons, the intrinsic magnetic moment – stemming from the spin – orients itself in two energetically different positions. This is analogous to the splitting of the beam in the Stern–Gerlach experiment. Doublet splitting follows from the two orientation possibilities. The splitting of the spectral lines caused by the spin is observed in all atoms and is called *multiplet structure*.

The magnitude of the magnetic moment of the electron caused by the orbital motion can be determined experimentally. It is a multiple of the *Bohr magneton*:

$$|\boldsymbol{M}| = \mu_{\text{B}} = |e|\,\hbar/2mc \; . \tag{12.2}$$

Indeed, classically, the magnetic moment caused by the orbital motion is given by the formula[3]

$$\boldsymbol{M} = \frac{1}{2c}\int \boldsymbol{r}' \times \boldsymbol{j}(\boldsymbol{r}')\,\mathrm{d}V' = \frac{1}{2c}\boldsymbol{r} \times \boldsymbol{I} = \frac{q}{2c}\boldsymbol{r} \times \boldsymbol{v} = \frac{q}{2mc}\boldsymbol{L} \; , \tag{12.3}$$

[3] See J.D. Jackson: *Classical Electrodynamics*, 2nd ed. (Wiley, New York 1975) and W. Greiner: *Classical Electrodynamics* (Springer, New York 1998).

where q is the charge, $\boldsymbol{v}$ the velocity, m the mass, $\boldsymbol{j}(\boldsymbol{r}') = q\boldsymbol{v}(\boldsymbol{r}')\delta(\boldsymbol{r}' - \boldsymbol{r})$ is the current density and $\boldsymbol{L}$ the orbital angular momentum of the particle.

As we have seen in Chap. 4, the z component of the orbital angular momentum is quantized according to

$$m_l = 0, \pm 1, \dots, \pm l \; . \tag{12.4}$$

Hence, we expect from (12.3) that $M_z = \mu_B m_l$. For each angular momentum $l\hbar$ there are $2l+1$ possibilities for adjusting the magnetic moment [i.e. $(2l+1)$ values for m_l – see (12.4)]. In the case of spin momentum, only two such different orientations occur. Therefore we conclude by analogy that the component of the spin parallel to the field is only half of Planck's constant:

$$S_z = \tfrac{1}{2}\hbar \quad \text{and} \quad S_z = -\tfrac{1}{2}\hbar \; . \tag{12.5}$$

Since from $(2l_s + 1) = 2$ it follows that $l_s = 1/2$, it is clear that the spin of the electron, i.e. its intrinsic (rotational) angular momentum, is $\frac{1}{2}\hbar$. This would indeed explain the *two* orientations observed both in the Stern–Gerlach experiment and also in doublet splitting.

Thus the spin of the electron is said to be half-integer. This property of the spin marks a further difference between spin and orbital angular momentum. An orbital angular momentum of $l\hbar$ has a maximal magnetic moment of $l\mu_B$ [see (12.3)]. According to the measurements, the spin of an electron with a magnitude $\frac{1}{2}\hbar$ also implies a magnetic moment of μ_B. This is most surprising because we would have expected the spin magnetic moment to be only $\frac{1}{2}\mu_B$.

To solve this dilemma, we introduce a new factor g. The connection between angular momentum and magnetic moment is generally written as

$$\boldsymbol{M} = g(q/2mc)\boldsymbol{J} \; , \tag{12.6}$$

where $\boldsymbol{J}$ denotes the orbital angular momentum or the spin and q is the charge of the particle. The quantity g is called the *gyromagnetic factor* or g factor. For the orbital angular momentum we have $g = 1$ [see (12.3)]. For the spin, however, $g = 2$. Since the electron is negatively charged, $q = -e$, its magnetic moment is always antiparallel to the angular momentum. If we combine the angular momentum $\boldsymbol{L}$ and the spin $\boldsymbol{S}$ into a total angular momentum $\boldsymbol{J}$, the resulting total magnetic moment $\boldsymbol{M}$ is not parallel to the angular momentum, because of the differing gyromagnetic factors (see Fig. 12.3). The total magnetic moment $\boldsymbol{M}$ precesses around the total angular momentum $\boldsymbol{J}$. Therefore, after averaging over time, only the component in the $\boldsymbol{J}$ direction remains. (In Example 12.3 we shall discuss this fact in more detail.)

In the case of atomic nuclei, the "*nuclear magneton*" is commonly used as a unit for the magnetic moment. It differs from the Bohr magneton in that it replaces the electron mass m in the denominator of (12.2) by the proton mass m_p.

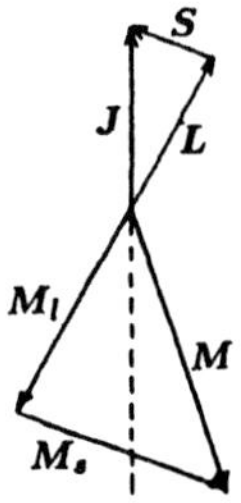

Fig. 12.3. Vector addition of the orbital ($\boldsymbol{L}$) and spin ($\boldsymbol{S}$) angular momentum and the corresponding magnetic moments $\boldsymbol{M}_l$ and $\boldsymbol{M}_s$, respectively. The resulting total magnetic moment $\boldsymbol{M}$ is – because of the spin g factor $g = 2$ – not collinear with the total angular momentum $\boldsymbol{J}$. This leads to a precession of $\boldsymbol{M}$ around $\boldsymbol{J}$

Therefore the nuclear magnetic moments are roughly three orders of magnitude smaller than those of the electron. The interaction between the nuclear moment and the moment of the electron leads to the *hyperfine structure of the spectra.*

12.2 The Einstein–de Haas Experiment

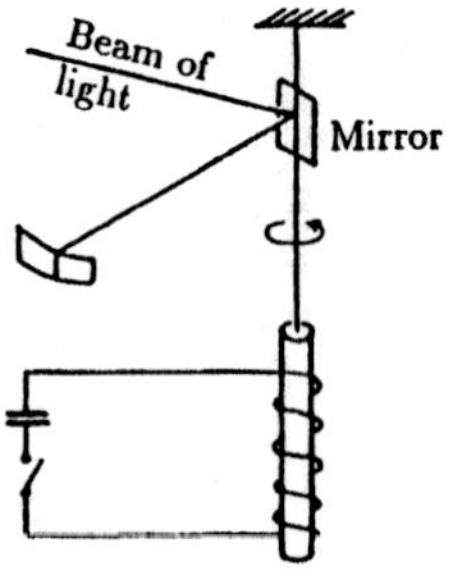

Fig. 12.4. Setup of the Einstein–de Haas experiment

If an iron bar is magnetized, not only the elementary magnetic moments, but also the elementary angular moment which cause them, change and become oriented. Because of the conservation of the angular momentum, the iron bar as a whole must change its macroscopic angular momentum, too. From the magnetization and the angular momentum of the bar, the gyromagnetic ratio can be determined.

This basic idea led Einstein and ***de Haas*** in 1915 to an experiment that makes it possible to measure the gyromagnetic ratio (see Fig. 12.4). An iron bar is hung from a string in such a way that it can rotate about its axis. From the torsion vibrations of the string, the angular momentum $\boldsymbol{L}$, which the bar gets as a whole when it is magnetized, can be measured. Suppose that N electrons with their elementary angular momentum $\boldsymbol{j}$ contribute to the magnetization. Then

$$N\boldsymbol{j}+\boldsymbol{L}=0 \rightarrow \boldsymbol{L}=-N\boldsymbol{j} \tag{12.7}$$

holds and correspondingly we get for the magnetization

$$\begin{aligned} \boldsymbol{M}_{\mathrm{bar}} &= N\boldsymbol{M}_{\mathrm{electron}} = Ng(q/2mc)\boldsymbol{j} = -g(q/2mc)\boldsymbol{L} \\ &= +g(e/2mc)\boldsymbol{L} = +g(\mu_{\mathrm{B}}/\hbar)\boldsymbol{L} \ . \end{aligned} \tag{12.8}$$

By measuring the macroscopic quantities $\boldsymbol{L}$ and $\boldsymbol{M}_{\mathrm{bar}}$, we can determine the gyromagnetic ratio g of the elementary magnetic moments and the angular momenta which cause them. $\boldsymbol{L}$ is measured by the deflection of the light ray (see Fig. 12.4); $\boldsymbol{M}_{\mathrm{bar}}$ can be determined by the residual magnetism (after previous gauging).

As a result of the Einstein–de Haas experiment, a magnetic moment of $+2\mu_{\mathrm{B}}$ was found; thus $g=2$. The negative sign ($q=-e$) results from the negative charge of the electrons. The magnitude of two Bohr magnetons excludes the orbital angular momentum as the source of the ferromagnetism and can only be explained by the existence of the electron spin.

A further conclusion to be drawn from this experiment is that the magnetization of the iron bar does not originate from elementary magnetic monopoles, but stems from electric currents caused by angular momenta.

We also note in passing that a precise determination of the g factor of the electron follows from measuring the doublet splitting and also from the Rabi experiment, which we shall discuss later in Example 12.2.

12.3 The Mathematical Description of Spin

Spin is an angular momentum; therefore its mathematical description is analogous to the formalism of the orbital angular momentum, which we became acquainted with earlier.[4] In this section we will deal with some peculiar features resulting from the facts that spin is a half-integer and that it can orient itself in one of two ways.

Experiments suggest the existence of a spin vector $\hat{\boldsymbol{S}} = \{\hat{S}_x, \hat{S}_y, \hat{S}_z\}$, which has three components: $\hat{S}_x$, $\hat{S}_y$, and $\hat{S}_z$; it should be an angular-momentum vector operator. The characteristic feature of angular-momentum operators is their commutation relations. Therefore we require that the $\hat{S}_x, \hat{S}_y, \hat{S}_z$ *obey the same commutation relations as the operators* $\hat{L}_x, \hat{L}_y, \hat{L}_z$ of the orbital angular momentum. This is the manifestation of the spin as an angular momentum. Hence,

$$\begin{aligned} \hat{S}_x\hat{S}_y - \hat{S}_y\hat{S}_x &= \mathrm{i}\hbar\hat{S}_z \ , \\ \hat{S}_y\hat{S}_z - \hat{S}_z\hat{S}_y &= \mathrm{i}\hbar\hat{S}_x \ , \\ \hat{S}_z\hat{S}_x - \hat{S}_x\hat{S}_z &= \mathrm{i}\hbar\hat{S}_y \ , \end{aligned} \tag{12.9}$$

or, using the abbreviations $\{\hat{S}_1, \hat{S}_2, \hat{S}_3\} \equiv \{\hat{S}_x, \hat{S}_y, \hat{S}_z\}$,

$$\hat{S}_i\hat{S}_j - \hat{S}_j\hat{S}_i = \mathrm{i}\hbar\varepsilon_{ijk}\hat{S}_k \ . \tag{12.10}$$

Here, ε_{ijk} is the complete antisymmetric tensor

$$\varepsilon_{ijk} = \begin{cases} 1 & \text{for even permutations of } 1, 2, 3 \\ 0 & \text{for 2 or more equal indices} \\ -1 & \text{for odd permutations of } 1, 2, 3 \ . \end{cases} \tag{12.11}$$

With this ε_{ijk} tensor, e.g. the cross product of two vectors $\boldsymbol{A} = \{A_i\}$ and $\boldsymbol{B} = \{B_i\}$ can be written as

$$(\boldsymbol{A} \times \boldsymbol{B})_k = \sum_{i,j} \varepsilon_{ijk} A_i B_j \ . \tag{12.12}$$

Furthermore, the operators $\hat{S}_i$ should be Hermitian, i.e. $\hat{S}_i = \hat{S}_i^+$, to guarantee that their expectation values are real.

For the representation of the operators, it is customary to use the ***Pauli** matrices* $\hat{\sigma}_i$. To exclude the factor $\frac{1}{2}\hbar$ from the equations, we define them in the following way:

$$\hat{S}_x = \tfrac{1}{2}\hbar\hat{\sigma}_x \ , \quad \hat{S}_y = \tfrac{1}{2}\hbar\hat{\sigma}_y \ , \quad \hat{S}_z = \tfrac{1}{2}\hbar\hat{\sigma}_z \ . \tag{12.13}$$

[4] We treat the algebra of angular momentum in great detail when discussing symmetries in W. Greiner, B. Müller: *Quantum Mechanics – Symmetries*, 2nd ed. (Springer, Berlin, Heidelberg 1994).

Thus, the commutation relations (12.9) take the following form:

$$\hat{\sigma}_x\hat{\sigma}_y - \hat{\sigma}_y\hat{\sigma}_x = 2\mathrm{i}\hat{\sigma}_z \ ,$$
$$\hat{\sigma}_y\hat{\sigma}_z - \hat{\sigma}_z\hat{\sigma}_y = 2\mathrm{i}\hat{\sigma}_x \ ,$$
$$\hat{\sigma}_z\hat{\sigma}_x - \hat{\sigma}_x\hat{\sigma}_z = 2\mathrm{i}\hat{\sigma}_y \ ; \tag{12.14}$$

or, in a more compact form,

$$\hat{\sigma}_i\hat{\sigma}_j - \hat{\sigma}_j\hat{\sigma}_i = 2\mathrm{i}\varepsilon_{ijk}\hat{\sigma}_k \ . \tag{12.15}$$

Because the spin components $\hat{S}_i$ have, according to their two possible orientations, only the two eigenvalues, $\pm\frac{1}{2}\hbar$, the spin matrices must be 2×2 matrices which have – as we know – exactly two eigenvalues. In the following, we take the z direction as the "direction of quantization". Then the z axis is the axis which the orientation of the spin is related to. Mathematically, this means that the spin functions are given as the eigenfunctions of the matrix $\hat{\sigma}_z$.

The matrix $\hat{\sigma}_z$ is diagonal in its eigenrepresentation and has the eigenvalues ± 1 as diagonal elements:

$$\hat{\sigma}_z = \begin{pmatrix} 1 & 0 \\ 0 & -1 \end{pmatrix} \quad \text{and} \quad \hat{\sigma}_z^2 = \begin{pmatrix} 1 & 0 \\ 0 & 1 \end{pmatrix} = \mathbb{1} \ . \tag{12.16}$$

For the matrices $\hat{\sigma}_x$ and $\hat{\sigma}_y$, the analogous relations hold in their eigenrepresentations. Since the unit matrix remains unchanged when we change the representation, the identity

$$\hat{\sigma}_x^2 = \hat{\sigma}_y^2 = \hat{\sigma}_z^2 = 1 \tag{12.17}$$

holds generally. To obtain the matrices $\hat{\sigma}_x$ and $\hat{\sigma}_y$ in the eigenrepresentations of $\hat{\sigma}_z$, we start from the commutation relations (12.14). Multiplication of the second equation in (12.14) from the left and from the right by $\hat{\sigma}_y$, and addition of both equations yields

$$\begin{aligned} 2\mathrm{i}(\hat{\sigma}_x\hat{\sigma}_y + \hat{\sigma}_y\hat{\sigma}_x) &= (\hat{\sigma}_y\hat{\sigma}_z - \hat{\sigma}_z\hat{\sigma}_y)\hat{\sigma}_y + \hat{\sigma}_y(\hat{\sigma}_y\hat{\sigma}_z - \hat{\sigma}_z\hat{\sigma}_y) \\ &= \hat{\sigma}_y^2\hat{\sigma}_z - \hat{\sigma}_z\hat{\sigma}_y^2 = 0 \end{aligned} \tag{12.18}$$

if we also take (12.17) into account. This means that, independently of the representation,

$$\hat{\sigma}_x\hat{\sigma}_y + \hat{\sigma}_y\hat{\sigma}_x = 0 \tag{12.19}$$

holds, and likewise for the other components.

The Pauli matrices are *anticommuting*. These relations can also be written in the following form:

$$[\hat{\sigma}_x, \hat{\sigma}_y]_+ = [\hat{\sigma}_y, \hat{\sigma}_z]_+ = [\hat{\sigma}_z, \hat{\sigma}_x]_+ = 0 \ . \tag{12.20}$$

Relations (12.17) and (12.20) can be combined in a more compact form:

$$\hat{\sigma}_i\hat{\sigma}_j+\hat{\sigma}_j\hat{\sigma}_i=2\delta_{ij} \ . \qquad (12.21)$$

To calculate the matrices $\hat{\sigma}_x$ and $\hat{\sigma}_y$ explicitly, we write the following:

$$\hat{\sigma}_x=\begin{pmatrix}a_{11} & a_{12}\\ a_{21} & a_{22}\end{pmatrix} \ , \quad \hat{\sigma}_y=\begin{pmatrix}b_{11} & b_{12}\\ b_{21} & b_{22}\end{pmatrix} \ . \qquad (12.22)$$

From the anticommutation relations (12.20) of $\hat{\sigma}_x$ and $\hat{\sigma}_y$ with $\hat{\sigma}_z$, we get

$$\begin{pmatrix}a_{11} & a_{12}\\ -a_{21} & -a_{22}\end{pmatrix}=\begin{pmatrix}-a_{11} & a_{12}\\ -a_{21} & a_{22}\end{pmatrix} \ , \qquad (12.23)$$

and thus $a_{11}=a_{22}=0$. Therefore

$$\hat{\sigma}_x=\begin{pmatrix}0 & a_{12}\\ a_{21} & 0\end{pmatrix} \ . \qquad (12.24)$$

Since the matrices should be Hermitian, $\hat{\sigma}_x=\hat{\sigma}_x^+=\hat{\tilde{\sigma}}_x^*$ is valid and hence

$$a_{21}=a_{12}^* \ , \qquad (12.25)$$

so that

$$\hat{\sigma}_x=\begin{pmatrix}0 & a_{12}\\ a_{12}^* & 0\end{pmatrix} \quad \text{and} \quad \hat{\sigma}_x^2\begin{pmatrix}|a_{12}|^2 & 0\\ 0 & |a_{12}|^2\end{pmatrix} \ . \qquad (12.26)$$

Because of (12.17), $\hat{\sigma}_x^2$ has to be $\hat{\mathbf{1}}$; it follows that

$$|a_{12}|^2=1 \ . \qquad (12.27)$$

The appropriate way of writing the matrix element is $\mathrm{e}^{\mathrm{i}\alpha}$, α being real.

For the matrix $\hat{\sigma}_y$ we can proceed in an analogous way; i.e. we write the matrices in the form

$$\hat{\sigma}_x=\begin{pmatrix}0 & \mathrm{e}^{\mathrm{i}\alpha}\\ \mathrm{e}^{-\mathrm{i}\alpha} & \end{pmatrix} \ , \quad \hat{\sigma}_y=\begin{pmatrix}0 & \mathrm{e}^{\mathrm{i}\beta}\\ \mathrm{e}^{-\mathrm{i}\beta} & 0\end{pmatrix} \ . \qquad (12.28)$$

Applying the anticommutation relation (12.20) to $\hat{\sigma}_x$ and $\hat{\sigma}_y$ leads to

$$\begin{pmatrix}\mathrm{e}^{\mathrm{i}(\alpha-\beta)} & 0\\ 0 & \mathrm{e}^{-\mathrm{i}(\alpha-\beta)}\end{pmatrix}=-\begin{pmatrix}\mathrm{e}^{-\mathrm{i}(\alpha-\beta)} & 0\\ 0 & \mathrm{e}^{\mathrm{i}(\alpha-\beta)}\end{pmatrix} \ , \qquad (12.29)$$

or

$$\mathrm{e}^{\mathrm{i}(\alpha-\beta)}=-\mathrm{e}^{-\mathrm{i}(\alpha-\beta)} \rightarrow \mathrm{e}^{2\mathrm{i}(\alpha-\beta)}=-1 \ . \qquad (12.30)$$

Thus $\alpha - \beta = \frac{1}{2}\pi$. All relations can be satisfied if only the last one is fulfilled. Therefore we set

$$\alpha = 0\ , \quad \beta = -\tfrac{1}{2}\pi\ , \tag{12.31}$$

and get the Pauli matrices in the $\hat{\sigma}_z$ representation:

$$\hat{\sigma}_x = \begin{pmatrix} 0 & 1 \\ 1 & 0 \end{pmatrix}\ , \quad \hat{\sigma}_y = \begin{pmatrix} 0 & -\mathrm{i} \\ \mathrm{i} & 0 \end{pmatrix}\ , \quad \hat{\sigma}_z = \begin{pmatrix} 1 & 0 \\ 0 & -1 \end{pmatrix}\ . \tag{12.32}$$

The unit matrix together with the Pauli matrices are four linearly independent matrices, which can be taken as a basis in the space of the two-dimensional matrices (compare with Exercise 13.1). They are also suitable for the description of other physical quantities which appear only in two states. Precisely for this reason, we again find the Pauli matrices in the formulation of the *isotopic spin*, which describes the states "proton" and "neutron" of a nucleon.

The total spin is

$$\begin{aligned} S^2 &= \hat{S}_x^2 + \hat{S}_y^2 + \hat{S}_z^2 = \frac{\hbar^2}{4}(\hat{\sigma}_x^2 + \hat{\sigma}_y^2 + \hat{\sigma}_z^2) = \frac{3}{4}\hbar^2\hat{\mathbf{1}} \\ &= \frac{1}{2}\left(\frac{1}{2}+1\right)\hbar^2\hat{\mathbf{1}}\ , \end{aligned} \tag{12.33}$$

in complete analogy to the formalism of the orbital angular momentum. From the commutation relations (12.9), it also follows that $[S^2, \hat{S}_i]_- = 0$, i.e. each spin component commutes with the square of the total spin.[5] Of course, this also follows from (12.32) and (12.33), because the unit matrix commutes with every other matrix. Since $\hat{S}^2$ is proportional to the unit matrix, it is immediately obvious that

$$[\hat{S}^2, \hat{S}_i]_- = 0$$

holds for all $\hat{S}_i$.

12.4 Wave Functions with Spin

By taking the spin into account, we assign a further degree of freedom to a particle. To describe this degree of freedom, we additionally introduce the component of the spin in the z direction, S_z, as an argument of the wave function. The component S_z can take only two values, namely $\pm\frac{1}{2}\hbar$. Therefore the wave function has the following coordinate representation:

$$\psi = \psi(\boldsymbol{r}, S_z, t)\ . \tag{12.34}$$

[5] To this end the conclusions reached in Chap. 4 can be repeated step by step.

Since S_z takes on only two values, it is useful to denote the wave functions with spin as column vectors with two components (*spinors*). This concurs with the fact that the spin operators $\hat{S}_i$ are represented by 2×2 matrices. The two components of the spinor are

$$\psi_1(\boldsymbol{r}, t) = \psi(\boldsymbol{r}, +\tfrac{1}{2}\hbar, t) \ , \quad \psi_2(\boldsymbol{r}, t) = \psi(\boldsymbol{r}, -\tfrac{1}{2}\hbar, t) \ , \tag{12.35}$$

while the complete wave function is

$$\begin{aligned} \Psi &= \begin{pmatrix} \psi_1(\boldsymbol{r}, t) \\ \psi_2(\boldsymbol{r}, t) \end{pmatrix} = \psi_1(\boldsymbol{r}, t)\chi_+ + \psi_2(\boldsymbol{r}, t)\chi_- \\ &= \psi_1(\boldsymbol{r}, t)\begin{pmatrix} 1 \\ 0 \end{pmatrix} + \psi_2(\boldsymbol{r}, t)\begin{pmatrix} 0 \\ 1 \end{pmatrix} \ . \end{aligned} \tag{12.36}$$

The introduction of product functions for both components, i.e.

$$\psi_1(\boldsymbol{r}, t)\chi_+ = \psi_1(\boldsymbol{r}, t)\begin{pmatrix} 1 \\ 0 \end{pmatrix} \quad \text{and} \quad \psi_2(\boldsymbol{r}, t)\chi_- = \psi_2(\boldsymbol{r}, t)\begin{pmatrix} 0 \\ 1 \end{pmatrix} \ , \tag{12.37}$$

is particularly convenient. The functions $\chi_\pm$ indicate only the state of the spin, i.e. "spin up" or "spin down". $|\psi_1|^2$ is obviously the probability of finding an electron with spin up at the location $\boldsymbol{r}$ and the time t. Correspondingly, $|\psi_2|^2$ is the probability of finding an electron with spin down at the location $\boldsymbol{r}$ and time t. This interpretation suggests that the total probability of finding the electron independently of its spin direction must be 1; thus

$$\begin{aligned} \int (|\psi_1(\boldsymbol{r}, t)|^2 + |\psi_2(\boldsymbol{r}, t)|^2)\,\mathrm{d}V &= \int (\psi_1^* \psi_2^*)\begin{pmatrix} \psi_1 \\ \psi_2 \end{pmatrix}\mathrm{d}V \\ &= \int \Psi^+ \Psi(\boldsymbol{r}, t)\,\mathrm{d}V = 1 \ . \end{aligned} \tag{12.38}$$

Spinor notation offers a clear formulation of how the spin operators, written as Pauli matrices, act on spinors. The eigenstates of the operator $\hat{\sigma}_z$ are

$$\hat{\sigma}_z \begin{pmatrix} \psi_1 \\ 0 \end{pmatrix} = \begin{pmatrix} 1 & 0 \\ 0 & -1 \end{pmatrix}\begin{pmatrix} \psi_1 \\ 0 \end{pmatrix} = (+1)\begin{pmatrix} \psi_1 \\ 0 \end{pmatrix} \quad \text{and} \tag{12.39}$$

$$\hat{\sigma}_z \begin{pmatrix} 0 \\ \psi_2 \end{pmatrix} = \begin{pmatrix} 1 & 0 \\ 0 & -1 \end{pmatrix}\begin{pmatrix} 0 \\ \psi_2 \end{pmatrix} = (-1)\begin{pmatrix} 0 \\ \psi_2 \end{pmatrix} \ . \tag{12.40}$$

The spin functions $\chi_\pm$ are *unit spinors*:

$$\chi_+ = \begin{pmatrix} 1 \\ 0 \end{pmatrix} \quad \text{and} \quad \chi_- = \begin{pmatrix} 0 \\ 1 \end{pmatrix} \ . \tag{12.41}$$

Obviously,

$$\begin{pmatrix}\psi_1\\0\end{pmatrix}=\psi_1\chi_+\ , \quad \text{or} \quad \begin{pmatrix}0\\\psi_2\end{pmatrix}=\psi_2\chi_-\ . \tag{12.42}$$

The unit spinors are, as can easily be seen, eigenfunctions of the spin operator $\hat{\sigma}_z$ with the eigenvalues $+1$ and -1, respectively:

$$\hat{\sigma}_z\chi_+=(+1)\chi_+ \quad \text{and} \quad \hat{\sigma}_z\chi_-=(-1)\chi_+\ . \tag{12.43}$$

Let us write down an arbitrary spin operator in the form

$$\hat{S}=\begin{pmatrix}S_{11} & S_{12}\\ S_{21} & S_{22}\end{pmatrix}\ . \tag{12.44}$$

If we use the matrix representation, an operator acts on a spin function by matrix multiplication (see Chap. 10):

$$\Phi=\hat{S}\Psi\ , \quad \text{where} \quad \Phi=\begin{pmatrix}\varphi_1\\\varphi_2\end{pmatrix} \quad \text{and} \quad \Psi=\begin{pmatrix}\psi_1\\\psi_2\end{pmatrix}\ . \tag{12.45}$$

In a more detailed form, this relation reads

$$\begin{pmatrix}\varphi_1\\\varphi_2\end{pmatrix}=\begin{pmatrix}S_{11} & S_{12}\\ S_{21} & S_{22}\end{pmatrix}\begin{pmatrix}\psi_1\\\psi_2\end{pmatrix}=\begin{pmatrix}S_{11}\psi_1+S_{12}\psi_2\\ S_{21}\psi_1+S_{22}\psi_2\end{pmatrix}\ , \tag{12.46}$$

or, for each component,

$$\begin{aligned}\varphi_1&=S_{11}\psi_1+S_{12}\psi_2\ ,\\ \varphi_2&=S_{21}\psi_1+S_{22}\psi_2\ .\end{aligned} \tag{12.47}$$

The average value of an operator is defined by [compare with (10.96)]

$$\langle\hat{S}\rangle=\int\Psi^+\hat{S}\Psi\,\mathrm{d}V\ . \tag{12.48}$$

If the wave functions are spinors, we have to use the Hermitian-conjugated wave functions, instead of the complex-conjugated [see (10.27)], i.e.

$$\begin{pmatrix}\psi_1\\\psi_2\end{pmatrix}^+=(\psi_1^*\psi_2^*)\ . \tag{12.49}$$

The average value is then easily calculated as

$$\langle\hat{S}(t)\rangle=\int\langle\hat{S}(\boldsymbol{r},t)\rangle\,\mathrm{d}V=\int\Psi^+\hat{S}(\boldsymbol{r},t)\Psi\,\mathrm{d}V\ , \tag{12.50}$$

with

$$\begin{aligned}\langle\hat{S}(\boldsymbol{r},t)\rangle&=\Psi^+S\Psi\ ,\\ \langle\hat{S}(\boldsymbol{r},t)\rangle&=(\psi_1^*\psi_2^*)\begin{pmatrix}S_{11} & S_{12}\\ S_{21} & S_{22}\end{pmatrix}\begin{pmatrix}\psi_1\\\psi_2\end{pmatrix}\ ,\\ \langle\hat{S}(\boldsymbol{r},t)\rangle&=\psi_1^*S_{11}\psi_1+\psi_1^*S_{12}\psi_2+\psi_2^*S_{21}\psi_1+\psi_2^*S_{22}\psi_2\ .\end{aligned} \tag{12.51}$$

Here, $\langle\hat{S}(\boldsymbol{r},t)\rangle$ is the average value of the spin operator (averaged over the spin directions) at the location $\boldsymbol{r}$ and time t. On the other hand, $\langle\hat{S}(t)\rangle$ is the average over the spin directions *and* each location at time t.

Let us apply the above to calculate the average value over both possible spin states of the Pauli matrices. For the x component we have

$$\langle\sigma_x(\boldsymbol{r},t)\rangle = \Psi^+\hat{\sigma}_x\Psi = (\psi_1^*\psi_2^*)\begin{pmatrix}0 & 1\\ 1 & 0\end{pmatrix}\begin{pmatrix}\psi_1\\ \psi_2\end{pmatrix} = \psi_1^*\psi_2+\psi_2^*\psi_1 \ . \quad (12.52)$$

In an analogous way we get for the y component

$$\langle\hat{\sigma}_y\rangle = (\psi_1^*\psi_2^*)\begin{pmatrix}0 & -\mathrm{i}\\ \mathrm{i} & 0\end{pmatrix}\begin{pmatrix}\psi_1\\ \psi_2\end{pmatrix} = -\mathrm{i}\psi_1^*\psi_2+\mathrm{i}\psi_2^*\psi_1 \quad \text{and} \quad (12.53)$$

$$\langle\hat{\sigma}_z\rangle x = (\psi_1^*\psi_2^*)\begin{pmatrix}1 & 0\\ 0 & -1\end{pmatrix}\begin{pmatrix}\psi_1\\ \psi_2\end{pmatrix} = \psi_1^*\psi_1-\psi_2^*\psi_2 \ . \quad (12.54)$$

12.5 The Pauli Equation

In Chap. 9 we developed the Hamiltonian for the motion of an electron (charge e) in an electromagnetic field in the absence of spin. This Hamiltonian reads

$$\hat{H}_0 = \frac{1}{2m}\left(\hat{\boldsymbol{p}}-\frac{e}{c}\boldsymbol{A}\right)^2 + e\phi \ , \quad (12.55)$$

where $\boldsymbol{A}$ is the vector and ϕ the Coulomb potential. Since the spin interacts with the magnetic field, the electron gains additional potential energy. The magnetic moment reads

$$\hat{\boldsymbol{M}} = g\left(\frac{-|e|}{2mc}\right)\hat{\boldsymbol{S}} = 2\left(\frac{-|e|}{2mc}\right)\hat{\boldsymbol{S}} = -\mu_{\mathrm{B}}\hat{\boldsymbol{\sigma}} \ , \quad (12.56)$$

$\mu_{\mathrm{B}} = (|e|\hbar/2mc)$ and the potential energy in the magnetic field[6] is

$$U = -\boldsymbol{M}\cdot\boldsymbol{B} \ . \quad (12.57)$$

The Hamiltonian of an electron with spin takes the following form:

$$\hat{H} = \hat{H}_0 + \mu_{\mathrm{B}}\hat{\boldsymbol{\sigma}}\cdot\boldsymbol{B} \ . \quad (12.58)$$

We can use the information about the g factor of the electron ($g=2$), which was discussed earlier, precisely at this point. With this Hamiltonian [(12.55) and

[6] See J.D. Jackson: *Classical Electrodynamics*, 2nd ed. (Wiley, New York 1975) and W. Greiner: *Classical Electrodynamics* (Springer, New York 1998).

(12.58)], we get the Schrödinger equation of a particle with spin, known as the *Pauli equation*,

$$\left[\frac{1}{2m}\left(\hat{p}-\frac{e}{c}A\right)^2+e\phi+\mu_B\hat{\sigma}\cdot B\right]\Psi=\mathrm{i}\hbar\frac{\partial\Psi}{\partial t}\ , \tag{12.59}$$

where

$$\Psi=\begin{pmatrix}\psi_1\\ \psi_2\end{pmatrix} \tag{12.60}$$

are the *spinor wave functions*. We call such two-component wave functions simply *spinors*; sometimes they are called *two-spinors* to distinguish them from the *four-spinors* occurring in relativistic quantum theory.[7]

Thus the Pauli equation is a system of two coupled differential equations for ψ_1 and ψ_2, describing electrons with the z component of their spin up or down, respectively. Because of the form of the Pauli spin matrices, we can easily see that the system (12.59) is decoupled for $\hat{\sigma}_z$ and only coupled by $\hat{\sigma}_x$ and $\hat{\sigma}_y$.

In the following we shall calculate the current density which results from the spinor equation (12.59). To do this, we write it in the form

$$\mathrm{i}\hbar\frac{\partial\Psi}{\partial t}=\hat{H}_0\Psi+\mu_B\hat{\sigma}\cdot B\Psi\ . \tag{12.61}$$

The adjoint equation of (12.61) reads

$$-\mathrm{i}\hbar\frac{\partial\Psi^+}{\partial t}=\hat{H}_0^*\Psi^++\mu_B(\hat{\sigma}\cdot B\Psi)^+=\hat{H}_0^*\Psi^++\mu_B\Psi^+\hat{\sigma}\cdot B\ , \tag{12.62}$$

because $\hat{\sigma}$ is Hermitian and the magnetic field B is real. Now we multiply (12.61) from the left by Ψ^+ and (12.62) from the right by Ψ. Subtraction of the equations yields

$$\mathrm{i}\hbar\frac{\partial}{\partial t}\Psi^+\Psi=\Psi^+(\hat{H}_0\Psi)-(\hat{H}_0^*\Psi^+)\Psi\ . \tag{12.63}$$

If we insert $\hat{H}_0$, all parts which contain no operators drop out, i.e.we are left with

$$\begin{aligned}\mathrm{i}\hbar\frac{\partial}{\partial t}\Psi^+\Psi=&-\frac{\hbar^2}{2m}[\Psi^+\nabla^2\Psi-(\nabla^2\Psi^+)\Psi]+\frac{\mathrm{i}\hbar e}{2mc}\{\Psi^+(\nabla\cdot A+A\cdot\nabla)\Psi\\&+[(\nabla\cdot A+A\cdot\nabla)\Psi^+]\Psi\}\ .\end{aligned} \tag{12.64}$$

The first term on the right-hand side can be transformed into

$$\Psi^+\nabla^2\Psi-\Psi\nabla^2\Psi^+=\mathrm{div}(\Psi^+\nabla\Psi-\Psi\nabla\Psi^+)\ . \tag{12.65}$$

[7] See Chap. 13 and W. Greiner: *Relativistic Quantum Mechanics – Wave Equations* 3rd ed. (Springer, Berlin, Heidelberg 1994).

For the second term (we must pay attention to the order. Ψ^+ is the first and Ψ the second factor!),

$$\begin{aligned}
&\Psi^+(\nabla\cdot A + A\cdot\nabla)\Psi + [(\nabla\cdot A + A\cdot\nabla)\Psi^+]\Psi \\
&\quad = 2\Psi^+\Psi \operatorname{div} A + 2A\cdot(\Psi^+\nabla\Psi + (\nabla\Psi^+)\Psi) \\
&\quad = 2\Psi^+\Psi \operatorname{div} A + 2A\cdot\nabla(\Psi^+\Psi) = 2\operatorname{div}(A\Psi^+\Psi) \ , \qquad (12.66)
\end{aligned}$$

is valid. Therefore from (12.63) it follows that

$$\mathrm{i}\hbar\frac{\partial}{\partial t}\Psi^+\Psi = -\frac{\hbar^2}{2m}\operatorname{div}[\Psi^+\nabla\Psi - (\nabla\Psi^+)\Psi] + \frac{\mathrm{i}\hbar e}{mc}\operatorname{div}(A\Psi^+\Psi) \ . \qquad (12.67)$$

This is the continuity equation in the form

$$\frac{\partial w}{\partial t} + \operatorname{div} j = 0 \ , \quad \text{where} \qquad (12.68)$$

$$w = \Psi^+\Psi \qquad (12.69)$$

is the *probability density* and

$$j = -\frac{\mathrm{i}\hbar}{2m}[\Psi^+\nabla\Psi - (\nabla\Psi^+)\Psi] - \frac{e}{mc}A\Psi^+\Psi \qquad (12.70)$$

is the *current density* of the electrons.

Now we insert the two-component wave function

$$\Psi = \begin{pmatrix}\psi_1\\ \psi_2\end{pmatrix} \quad \text{and} \quad \Psi^+ = (\psi_1^*, \psi_2^*) \qquad (12.71)$$

and arrive at

$$w = (\psi_1^*\psi_1 + \psi_2^*\psi_2) \qquad (12.72)$$

and

$$\begin{aligned}
j &= \frac{\mathrm{i}\hbar}{2m}(\psi_1\nabla\psi_1^* + \psi_2\nabla\psi_2^* - \psi_1^*\nabla\psi_1 - \psi_2^*\nabla\psi_2) \\
&\quad - \frac{e}{mc}A(\psi_1^*\psi_1 + \psi_2^*\psi_2) \qquad (12.73)
\end{aligned}$$

or, rearranging,

$$\begin{aligned}
j &= \frac{\mathrm{i}\hbar}{2m}(\psi_1\nabla\psi_1^* - \psi_1^*\nabla\psi_1) - \frac{e}{mc}A\psi_1^*\psi_1 \\
&\quad + \frac{\mathrm{i}\hbar}{2m}(\psi_2\nabla\psi_2^* - \psi_2^*\nabla\psi_2) - \frac{e}{mc}A\psi_2^*\psi_2 \ . \qquad (12.74)
\end{aligned}$$

It turns out that both the probability density and the current density are composed additively of the parts of the two different spin directions; this is reasonable. Multiplication of the *particle current density* j by the charge e yields the *electrical current density* j_e.

The current density j_e does not contain the spin; rather it is the current density caused by the *orbital motion of the electrons* (with a different spin). However, the spin of an electron also causes a magnetic moment, which can be expressed by a corresponding current. We shall call this part of the current density j_s the *spin current density*. This current density cannot occur in a continuity equation in which the charge conservation is expressed by convection currents.

To calculate the spin current density j_s, we start with Maxwell's equations. For the curl of the $\boldsymbol{B}$ field the following well-known relation[8] exists:

$$\operatorname{curl} \boldsymbol{B} = (4\pi/c)(\boldsymbol{j}_e + c \operatorname{curl} \langle \boldsymbol{M} \rangle) . \tag{12.75}$$

We have replaced the *magnetization* $\langle \boldsymbol{M} \rangle$ in this case by the averaged density of the magnetic moment $\langle \boldsymbol{M} \rangle$, where averaging over the spin states is meant. The magnetic dipole density is given by

$$\langle \boldsymbol{M} \rangle = -\mu_B \Psi^+ \hat{\boldsymbol{\sigma}} \Psi , \tag{12.76}$$

and thus

$$\begin{aligned} \operatorname{curl} \boldsymbol{B} &= 4\pi/c \boldsymbol{j} = (4\pi/c)(\boldsymbol{j}_e - c\mu_B \operatorname{curl} \Psi^+ \hat{\boldsymbol{\sigma}} \Psi \\ &= (4\pi/c)(\boldsymbol{j}_e + \boldsymbol{j}_s) . \end{aligned} \tag{12.77}$$

The contribution

$$\boldsymbol{j}_s = -c\mu_B \operatorname{curl} \Psi^+ \hat{\boldsymbol{\sigma}} \Psi \tag{12.78}$$

is the current causing the magnetic moments of the electrons, and is equivalent to them.

EXERCISE

12.1 Spin Precession in a Homogeneous Magnetic Field

Problem. Determine the precession of the spin in a homogeneous magnetic field (see figure).

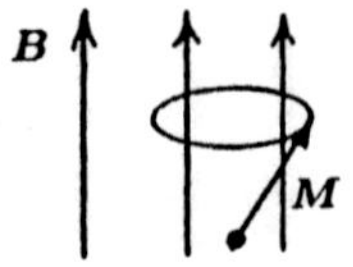

Particle with spin and magnetic moment $\boldsymbol{M}$ precessing in a magnetic field $\boldsymbol{B}$

Solution. If a charged body moves in a homogeneous magnetic field, it circulates around the direction of the field with the frequency $\omega = 2\omega_L = -eB/mc$. Here the charge of the electron is $(-e)$. This follows from the fact that the Lorentz

[8] See J.D. Jackson: *Classical Electrodynamics*, 2nd ed. (Wiley, New York 1975) and W. Greiner: *Classical Electrodynamics* (Springer, New York 1998).

Exercise 12.1

force balances the centrifugal force:

$$-eBv/c = mr\omega^2 \ . \tag{1}$$

Thus

$$\omega = -eB/mc \ , \tag{2}$$

whereby $\omega_L = -eB/2mc$ is the so-called ***Larmor** frequency*.

The spin function at time $t = 0$ reads

$$\chi = a_0\chi_+ + b_0\chi_- \ . \tag{3}$$

If we take for the constants $a_0 = e^{i\gamma}\cos(\Theta/2)$ and $b_0 = e^{i\delta}\sin(\Theta/2)$, the normalization condition

$$|a_0|^2 + |b_0|^2 = 1 \tag{4}$$

is obviously fulfilled. Now we shall calculate the frequency of precession of the spin in the magnetic field $\boldsymbol{B} = \{0, 0, B_z\}$.

Let us assume that the electron is fixed at a certain location and its spin is the only degree of freedom. That part of the Pauli equation (12.59) which contains the spin yields

$$\begin{aligned} i\hbar\frac{\partial\chi}{\partial t} &= \mu_B\hat{\boldsymbol{\sigma}}\cdot\boldsymbol{B}\chi = -\frac{e\hbar}{2mc}\hat{\sigma}_z B_z\chi \\ i\hbar\frac{\partial\chi}{\partial t} &= \hbar\omega_L\hat{\sigma}_z\chi \ . \end{aligned} \tag{5}$$

The spin function written as a column vector reads

$$\chi = a\chi_+ + b\chi_- = a\begin{pmatrix}1\\0\end{pmatrix} + b\begin{pmatrix}0\\1\end{pmatrix} = \begin{pmatrix}a\\b\end{pmatrix} \ . \tag{6}$$

Inserting this into (5) results in

$$i\begin{pmatrix}\dot{a}\\\dot{b}\end{pmatrix} = \omega_L\begin{pmatrix}1 & 0\\0 & -1\end{pmatrix}\begin{pmatrix}a\\b\end{pmatrix} = \omega_L\begin{pmatrix}a\\-b\end{pmatrix} \ , \tag{7}$$

and thus

$$\dot{a} = -i\omega_L a \ , \quad \dot{b} = i\omega_L b \ . \tag{8}$$

After integration we obtain

$$a = a_0\,e^{-i\omega_L t} \ , \quad b = b_0\,e^{i\omega_L t} \ . \tag{9}$$

The time-dependent spin function thus reads

$$\chi = \begin{pmatrix} e^{-i\omega_L t}e^{i\gamma}\cos(\Theta/2) \\ e^{i\omega_L t}e^{i\delta}\sin(\Theta/2) \end{pmatrix} \ . \tag{10}$$

Exercise 12.1

The expectation value of the spin is obtained from

$$\langle\hat{S}\rangle = \frac{\hbar}{2}\langle\hat{\sigma}\rangle = \frac{\hbar}{2}\chi^+\hat{\sigma}\chi$$
$$= \frac{\hbar}{2}(\chi^+\hat{\sigma}_x\chi, \chi^+\hat{\sigma}_y\chi, \chi^+\hat{\sigma}_z\chi) \ . \quad (11)$$

Inserting the Pauli spin matrices from (12.32) and χ of (10) into (11) leads to

$$\langle\hat{S}\rangle = \frac{\hbar}{2}[\cos(2\omega_L t + \delta - \gamma)\sin\Theta, \sin(2\omega_L t + \delta - \gamma)\sin\Theta, \cos\Theta] \ . \quad (12)$$

Obviously, the spin component in the field direction S_z is conserved while the spin precesses around the z axis with twice the Larmor frequency $2\omega_L$. This is due to the gyromagnetic factor 2 of the spin. In contrast to this, the average value of the orbital angular momentum precesses with only the frequency ω_L around the z axis [see Example 12.3, (15)ff.].

EXAMPLE

12.2 The Rabi Experiment (Spin Resonance)

To measure the nuclear magnetic moment, ***Rabi*** developed the method of *spin resonance*.The scheme of this experiment is sketched in the figure below.

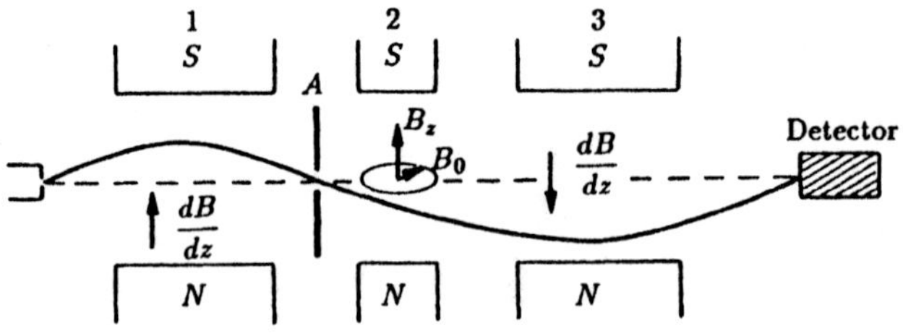

Setup of the Rabi experiment. The magnetic fields of magnets 1 and 3 have strong gradients in opposite directions, while the field of magnet 2 is homogeneous

If the particles reach the inhomogeneous magnetic field 1, they will be deflected, depending on their spin orientation, in such a way that the slit A can be passed only by particles with a certain spin direction. The homogeneous field 2 has no influence on the deflection of the particles. Thereafter, the particles enter field 3, which has a field gradient opposite to that of field 1. Field 3 cancels the original deflection, so that the particles reach the detector.

If the homogeneous field 2 is superposed by an oscillating field, which leads to a *spin flip*, then the particles will be deflected in field 3 in the wrong direction and will not reach the detector. From the frequency of the oscillating field (the

Example 12.2

resonance frequency), leading to a minimum in the beam intensity at the detector, the magnetic moment of the particle can be calculated.

Now we want to investigate mathematically the behaviour of a particle with spin $\pm\frac{1}{2}\hbar$ in an inhomogeneous magnetic field B_z on which a weak oscillating field is superposed. The magnetic field should have the following form:

$$\boldsymbol{B} = (B_0 \cos\omega_0 t,\, B_0 \sin\omega_0 t,\, B_z) \ . \qquad (1)$$

We write the following for the spin function:

$$\chi(t) = a(t)\,\mathrm{e}^{-\mathrm{i}\tilde{\omega}t}\chi_+ + b(t)\,\mathrm{e}^{\mathrm{i}\tilde{\omega}t}\chi_- \ , \qquad (2)$$

where $\tilde{\omega} = -\mu B_z/\hbar$ is the Larmor frequency for the precession of the spins around the homogeneous field. The factor μ denotes the connection between the magnetic moment and the spin of the particle:

$$\boldsymbol{M} = \boldsymbol{\sigma}\mu \ . \qquad (3)$$

We start from the spin-dependent part of the Pauli equation:

$$\mathrm{i}\hbar\frac{\partial\chi}{\partial t} = -\boldsymbol{B}\cdot\boldsymbol{M}\chi = -\mu\boldsymbol{B}\cdot\hat{\boldsymbol{\sigma}}\chi \ . \qquad (4)$$

Inserting (1) and (2) into (4) yields

$$\begin{aligned}\mathrm{i}\hbar\frac{\partial}{\partial t}\begin{pmatrix} a(t)\,\mathrm{e}^{-\mathrm{i}\tilde{\omega}t} \\ b(t)\,\mathrm{e}^{\mathrm{i}\tilde{\omega}t}\end{pmatrix} &= -\mu(B_0\cos\omega_0 t\hat{\sigma}_x \\ &\quad + B_0\sin\omega_0 t\hat{\sigma}_y + B_z\hat{\sigma}_z)\begin{pmatrix} a(t)\,\mathrm{e}^{-\mathrm{i}\tilde{\omega}t} \\ b(t)\,\mathrm{e}^{\mathrm{i}\tilde{\omega}t}\end{pmatrix} \ . \end{aligned} \qquad (5)$$

Now explicitly inserting the Pauli matrices and calculating the derivatives yields

$$\begin{aligned}\mathrm{i}\hbar\begin{pmatrix}\dot{a}\,\mathrm{e}^{-\mathrm{i}\tilde{\omega}t} \\ \dot{b}\,\mathrm{e}^{\mathrm{i}\tilde{\omega}t}\end{pmatrix} + \hbar\tilde{\omega}\begin{pmatrix} a\,\mathrm{e}^{-\mathrm{i}\tilde{\omega}t} \\ -b\,\mathrm{e}^{\mathrm{i}\tilde{\omega}t}\end{pmatrix} &= -\mu B_0\cos\omega_0 t\begin{pmatrix} b\,\mathrm{e}^{\mathrm{i}\tilde{\omega}t} \\ a\,\mathrm{e}^{-\mathrm{i}\tilde{\omega}t}\end{pmatrix} \\ &\quad -\mathrm{i}\mu B_0\sin\omega_0 t\begin{pmatrix} -b\,\mathrm{e}^{\mathrm{i}\tilde{\omega}t} \\ a\,\mathrm{e}^{-\mathrm{i}\tilde{\omega}t}\end{pmatrix} - \mu B_z\begin{pmatrix} a\,\mathrm{e}^{-\mathrm{i}\tilde{\omega}t} \\ -b\,\mathrm{e}^{\mathrm{i}\tilde{\omega}t}\end{pmatrix} \ . \end{aligned} \qquad (6)$$

The last terms on both sides cancel each other because $\hbar\tilde{\omega} = -\mu B_z$. Together with $\hbar\tilde{\omega}' = -\mu B_0$, we can write for both components of the spinor χ

$$\dot{a} = -\mathrm{i}\tilde{\omega}' b\,\mathrm{e}^{\mathrm{i}(2\tilde{\omega}-\omega_0)t} \ , \qquad (7)$$

$$\dot{b} = -\mathrm{i}\tilde{\omega}' a\,\mathrm{e}^{-\mathrm{i}(2\tilde{\omega}-\omega_0)t} \ . \qquad (8)$$

These equations can be decoupled by taking the derivative of the first one and eliminating b and $\dot{b}$:

$$\ddot{a} - \mathrm{i}(2\tilde{\omega}-\omega_0)\dot{a} + \tilde{\omega}'^2 a = 0 \ . \qquad (9)$$

Example 12.2

To solve this homogeneous differential equation, we try writing $a \propto e^{i\omega t}$ and thus get the characteristic equation for ω having the solutions

$$\omega_{1,2} = \tilde{\omega} - \frac{\omega_0}{2} \pm \sqrt{(\tilde{\omega} - \omega_0/2)^2 + \tilde{\omega}'^2} \tag{10}$$

or

$$\omega_{1,2} = \Omega \pm \delta \ , \quad \Omega = \tilde{\omega} - \frac{\omega_0}{2} \ ,$$
$$\delta = \sqrt{(\tilde{\omega} - \omega_0/2)^2 + \tilde{\omega}'^2} \ . \tag{11}$$

The common solution for the coefficient a is therefore given by

$$a(t) = a_1 \, e^{i(\Omega+\delta)t} + a_2 \, e^{i(\Omega-\delta)t} \ . \tag{12}$$

We choose the initial conditions in such a way that the particle at $t = 0$ is in the spin state χ_+, i.e. $|a(t=0)|^2 = 1$ and $b(t=0) = 0$. Thus from (12) it follows that

$$a_1 + a_2 = 1 \ . \tag{13}$$

Together with (12) and (7) we get for the coefficient b

$$b(t) = \frac{-e^{-2i\Omega t}}{\tilde{\omega}'} [a_1(\Omega+\delta)\, e^{i(\Omega+\delta)t} + a_2(\Omega-\delta)\, e^{i(\Omega-\delta)t}] \ . \tag{14}$$

Starting from the initial conditions, we can now calculate the coefficients a_1 and a_2. From $b(t=0) = 0$ it follows that

$$a_1(\Omega+\delta) + a_2(\Omega-\delta) = 0 \ . \tag{15}$$

Together with (13) we get

$$a_1 = \frac{1}{2}\left(1 - \frac{\Omega}{\delta}\right) \ , \quad a_2 = \frac{1}{2}\left(1 + \frac{\Omega}{\delta}\right) \ . \tag{16}$$

Both amplitudes $a(t)$ and $b(t)$ are given by

$$a(t) = \left(\cos \delta t - i\frac{\Omega}{\delta} \sin \delta t\right) e^{i\Omega t} \ ,$$
$$b(t) = -i\frac{\tilde{\omega}'}{\delta} \sin \delta t \, e^{-i\Omega t} \ . \tag{17}$$

From the terms in (2), returning to the spin, we realize that the quantity $|b(t)|^2$ is the probability of finding the particle in the state χ_- at time t:

$$|b(t)|^2 = (\tilde{\omega}'^2/\delta^2) \sin^2 \delta t \ . \tag{18}$$

Let t_0 be the time the particle needs to pass through the oscillating field. The experimental data are to be adjusted in such a way that after this time, the largest

Example 12.2

possible number of particles are in the state χ_-. At this time, the maximum of the spin-flip probability $|b(t)|^2$ will be reached. From $\mathrm{d}|b|^2/\mathrm{d}t = 0$ it follows that

$$\sin\delta t\cos\delta t = 0\ . \tag{19}$$

The $\sin^2\delta t$ curve of (18) has its maxima at the same locations at which the sine function has extrema. Hence, the maximum, which we are looking for, is given by the zero of the cosine function.

Therefore, at time t_0, it holds that

$$\delta t_0 = \frac{\pi}{2} \quad \text{or} \quad t_0 = \frac{\pi}{2\sqrt{(\tilde{\omega}-\omega_0/2)^2+\tilde{\omega}'^2}}\ . \tag{20}$$

The time t_0 is fixed by the velocity of the particle and the size of the area which the oscillating field occupies. Equation (20) contains, in addition, the data on the magnetic field and the unknown magnetic moment μ, which can thus be determined.

EXAMPLE

12.3 The Simple Zeeman Effect (Weak Magnetic Fields)

As a further example of the application of the Pauli equation, we consider the splitting of the spectral lines in a weak magnetic field. Here we shall treat the simple ***Zeeman*** effect, i.e. we neglect the spin–orbit interaction.

The spin–orbit interaction leads to the fine structure of the spectra, a further splitting, which we shall not take into account here.[9]

The magnetic field should be homogeneous and possess only a z component:

$$\boldsymbol{B} = \{0, 0, B\}\ . \tag{1}$$

In this case, we may express it by a vector potential

$$\boldsymbol{A} = \{-\tfrac{1}{2}By, \tfrac{1}{2}Bx, 0\}\ , \tag{2}$$

as can easily be shown by the relation

$$\boldsymbol{B} = \operatorname{curl}\boldsymbol{A}\ . \tag{3}$$

We denote the Coulomb potential by ϕ.

[9] We discuss this topic in W. Greiner: *Relativistic Quantum Mechanics — Wave Equations*, 3rd ed. (Springer, Berlin, Heidelberg 2000), where it follows naturally from the Dirac equation.

Example 12.3

Again we start from the Pauli equation for a particle with the charge e. The Hamiltonian reads

$$\hat{H} = \frac{1}{2m}\left(\hat{\boldsymbol{p}} - \frac{e}{c}\boldsymbol{A}\right)^2 + e\phi - \frac{e\hbar}{2mc}\hat{\boldsymbol{\sigma}} \cdot \boldsymbol{B} \ . \tag{4}$$

Since the magnetic field is weak, we neglect the term with $\boldsymbol{A}^2$ and get, using $\operatorname{div} \boldsymbol{A} = 0$,

$$\left(\frac{\hat{\boldsymbol{p}}^2}{2m} - \frac{e}{mc}\boldsymbol{A} \cdot \hat{\boldsymbol{p}} + e\phi - \frac{e\hbar}{2mc}\boldsymbol{B} \cdot \hat{\boldsymbol{\sigma}}\right)\Psi = \mathrm{i}\hbar\frac{\partial}{\partial t}\Psi \ . \tag{5}$$

Instead of the term $\boldsymbol{A} \cdot \hat{\boldsymbol{p}}$, we introduce the angular-momentum operator. According to (2) we get

$$\boldsymbol{A} \cdot \hat{\boldsymbol{p}} = -\frac{B}{2}(y\hat{p}_x - x\hat{p}_y) = \mathrm{i}\hbar\frac{B}{2}\left(y\frac{\partial}{\partial x} - x\frac{\partial}{\partial y}\right) = \frac{B}{2}\hat{L}_z \ .$$

Together with $\hat{H}_0 = \hat{\boldsymbol{p}}^2/2m + e\phi$, (5) leads to

$$\mathrm{i}\hbar\frac{\partial}{\partial t}\Psi = \hat{H}_0\Psi - \frac{eB}{2mc}(\hat{L}_z + \hbar\hat{\sigma}_z)\Psi \ . \tag{6}$$

Since we are interested only in the energies of the stationary states, we write the following for the wave function:

$$\psi(\boldsymbol{r}, t) = \psi(\boldsymbol{r})\exp\left(-\frac{\mathrm{i}}{\hbar}Et\right) \ . \tag{7}$$

Thus (6) can be transformed into the eigenvalue equation

$$\hat{H}_0\Psi - \frac{eB}{2mc}(\hat{L}_z + \hbar\hat{\sigma}_z)\Psi = E\Psi \ . \tag{8}$$

Taking the Larmor frequency $\omega_{\mathrm{L}} = -eB/2mc$, and applying the spinor notation, we get

$$\hat{H}_0\begin{pmatrix}\psi_1\\ \psi_2\end{pmatrix} + \omega_{\mathrm{L}}\left[\hat{L}_z + \hbar\begin{pmatrix}1 & 0\\ 0 & -1\end{pmatrix}\right]\begin{pmatrix}\psi_1\\ \psi_2\end{pmatrix} = E\begin{pmatrix}\psi_1\\ \psi_2\end{pmatrix}$$

Both spinor components are decoupled (since $\hat{\sigma}_z$ is diagonal) and yield the equations

$$\begin{aligned}\hat{H}_0\psi_1 + \omega_{\mathrm{L}}(\hat{L}_z + \hbar)\psi_1 &= E\psi_1 \ ,\\ \hat{H}_0\psi_2 + \omega_{\mathrm{L}}(\hat{L}_z - \hbar)\psi_2 &= E\psi_2 \ .\end{aligned} \tag{9}$$

If the magnetic field were absent, we would get the eigenstates of $\hat{H}_0$ as solutions – identical solutions, in fact – for both spinor components, as a look at (9) tells us:

$$\psi_1 = \psi_2 = \psi_{nlm} = R_{nl}(r)Y_{lm}(\theta, \phi) \ . \tag{10}$$

Example 12.3

Since the wave function ψ_{nlm} is an eigenfunction of $\hat{L}_z$,

$$\hat{L}_z\psi_{nlm} = \hbar m\psi_{nlm} \; ; \tag{11}$$

ψ_{nlm} is also an eigenfunction of the complete equations (9). Thus the wave functions are not altered in the approximation ($A^2 = 0$) we have used.

Together with the eigenvalue equation of the operator $\hat{H}_0$

$$\hat{H}_0\psi_{nlm} = E^0_{nl}\psi_{nlm}$$

and the relations (10) and (11), we get from (9) the energy eigenvalues:

$$E'_{nlm} = E^0_{nl} + \omega_L\hbar(m+1) \quad \text{for} \quad \Psi = \begin{pmatrix}\psi_{nlm}\\ 0\end{pmatrix}$$

and

$$E''_{nlm} = E^0_{nl} + \omega_L\hbar(m-1) \quad \text{for} \quad \Psi = \begin{pmatrix}0\\ \psi_{nlm}\end{pmatrix} \; . \tag{12}$$

Because of the magnetic field, the energy depends on the orientation of the magnetic moment with respect to the field direction. Levels which are degenerate when the magnetic field is missing then split up. The two-fold splitting of the s states, which have no orbital magnetic moment, is proof of the existence of spin (the Stern–Gerlach experiment).

The following figure shows the splitting of a ψ_{100} and a ψ_{21m} state.

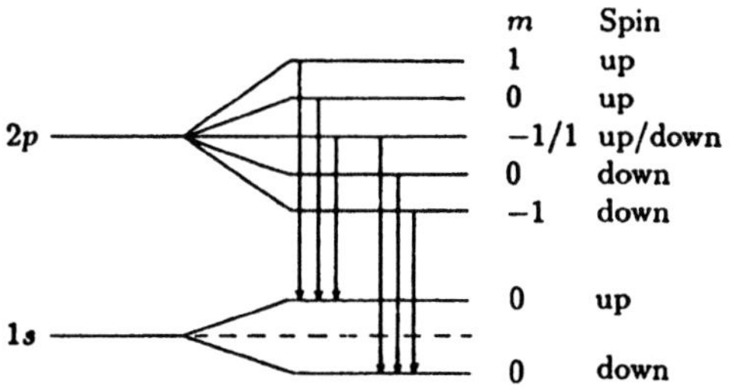

Splitting of the 1s and 2p levels in a magnetic field (the Zeeman effect)

The 2p state thus splits into five levels; one of them is twofold degenerate.

Since the interaction of the spin with the light wave emitted when a transition occurs is small, the spin is not altered. Therefore only transitions between states of *equal spin direction* occur; these transitions are indicated in the figure. (Commonly, the dipole selection rule $\Delta m = \pm 1, 0$ holds.)

We get the transition frequencies by using the differences of the energies (12). Since the spin direction does not change, we have

$$\begin{aligned}\hbar\omega &= E_{n'l'm'} - E_{n''l''m''}\\ &= E^0_{n'l'} - E^0_{n''l''} + \omega_L\hbar(m' - m'')\\ &= \omega_0 + \omega_L(m' - m'') \; ,\end{aligned}$$

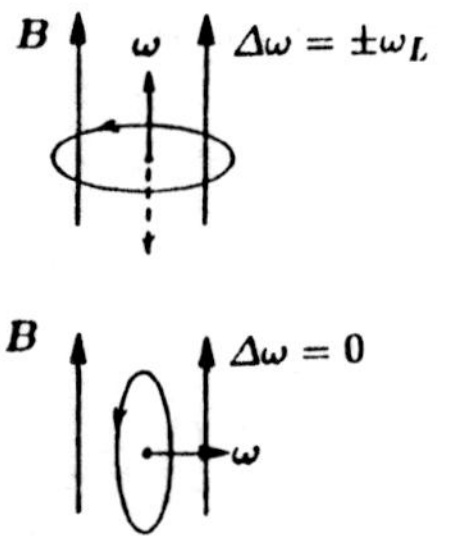

The classical understanding of the Zeeman effect

where ω_0 is the transition frequency if the magnetic field is missing. Since the difference is $m' - m''$, $= \pm 1, 0$, we get two spectral lines in the magnetic field, shifted from the original one by $\pm\omega_{\rm L}$.

This result coincides precisely with the classical theory of the Zeeman effect (see figure). Here, circular motion of an electron in a magnetic field is investigated. The centrifugal force $mr\omega^2$ and the Lorentz force $\pm er\omega B/c$ act on the electron, depending on the direction of motion.

Thus

$$mr\omega^2 \pm er\omega B/c = mr(\omega \pm \Delta\omega)^2$$

holds. If we neglect the second-order term $\propto (\Delta\omega)^2$ we get

$$\Delta\omega = eB/2mc = \omega_{\rm L} \ .$$

The decomposition of the circular motion is illustrated in the figure. The motion leading to a shift in the frequency proceeds in the plane perpendicular to the field. $\Delta\omega = 0$ corresponds to a motion parallel to the field.

We noted earlier that an angular momentum of a classical charged particle in a magnetic field precesses with the Larmor frequency around the direction of the magnetic field (see Exercise 12.1). It is possible also to identify a precession by treating the Zeeman effect quantum-mechanically.

The Hamiltonian in (6) can be written in the form

$$\hat{H} = \hat{H}_0 + \omega_{\rm L}\hat{L}_z + 2\omega_{\rm L}\hat{S}_z \ . \tag{13}$$

The rate of change of the angular momenta results from the commutation relations

$$\frac{{\rm d}\hat{L}_x}{{\rm d}t} = \frac{\rm i}{\hbar}[\hat{H}, L_x] \ ,$$

and analogously for the other components as well as for the spin. The components of the orbital angular momentum commute with $\hat{H}_0$ and $\hat{S}_z$; therefore, only the commutation relation containing $\hat{L}_z$ remains. Together with the commutation relations of (4.65), it follows that

$$\frac{{\rm d}\hat{L}_z}{{\rm d}t} = -\omega_{\rm L}\hat{L}_y \ , \quad \frac{{\rm d}\hat{L}_y}{{\rm d}t} = +\omega_{\rm L}\hat{L}_x \ , \quad \frac{{\rm d}\hat{L}_z}{{\rm d}t} = 0 \ . \tag{14}$$

The second derivatives with respect to time follow immediately from these equations:

$$\frac{{\rm d}^2\hat{L}_x}{{\rm d}t^2} = -\omega_L^2\hat{L}_x \ , \quad \frac{{\rm d}^2\hat{L}_y}{{\rm d}t^2} = -\omega_L^2\hat{L}_y \ . \tag{15}$$

Example 12.3

We know that the expectation values fulfil the same relations as the operators. As can easily be calculated, (14) and (15) have the same solutions:

$$\langle L_x \rangle = A \sin(\omega_{\rm L} t + \phi) \ ,$$
$$\langle L_y \rangle = -A \cos(\omega_{\rm L} t + \phi) \ ,$$
$$\langle L_z \rangle = \text{const} \ . \qquad (16)$$

The same commutation relations also hold for the spin. The spin operator commutes with $\hat{H}_0$ and $\hat{L}_z$, but not with the term containing $\hat{S}_z$ in (13). We get relations equivalent to (14) and (15), whereby, corresponding to (13), $\omega_{\rm L}$ is replaced by $2\omega_{\rm L}$ (cf. Exercise 12.1):

$$\langle S_x \rangle = A \sin(2\omega_{\rm L} t + \phi) \ ,$$
$$\langle S_y \rangle = -A \cos(2\omega_{\rm L} t + \phi) \ ,$$
$$\langle S_z \rangle = \text{const} \ . \qquad (17)$$

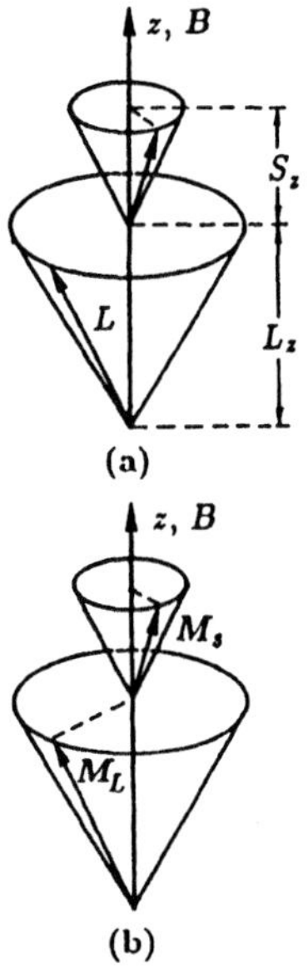

Precession of the orbital angular momentum and the spin (a), as well as their corresponding magnetic momenta (b), around the magnetic field (z axis)

As indicated in the figure, these equations imply that the components of orbital and spin angular momentum parallel to the magnetic field (L_z and S_z, respectively) are constants of motion. On the other hand, the components orthogonal to the magnetic field, $\boldsymbol{L}_\perp = (L_x, L_y)$ and $\boldsymbol{S}_\perp = (S_x, S_y)$, rotate with the Larmor frequency, $\omega_{\rm L}$ and $2\omega_{\rm L}$, respectively.

Since we have neglected the coupling between spin and orbital angular momentum here, both vectors precess independently around the magnetic field. The z component of the orbital angular momentum L_z, and that of the spin S_z, remain, as mentioned above, constant. We should note that the spin rotates twice as fast as the orbital angular momentum. Taking into account the corresponding gyromagnetic factors (see Sect. 12.1), the magnetic moment $\hat{\boldsymbol{M}}$, given by

$$\hat{\boldsymbol{M}} = \hat{\boldsymbol{M}}_L + \hat{\boldsymbol{M}}_S = (\mu_{\rm B}/\hbar)(\hat{\boldsymbol{L}} + 2\hat{\boldsymbol{S}}) \ , \qquad (18)$$

behaves analogously. Owing to the absence of $\boldsymbol{LS}$ coupling, we directly get for the z component of $\hat{\boldsymbol{M}}$

$$M_z = (\mu_{\rm B}/\hbar)(L_z + 2S_z) \ . \qquad (19)$$

12.6 Biographical Notes

GOUDSMIT, Samuel Abraham, American physicist of Dutch origin, *11.7.1902, †4.12.1978. G. taught from 1928 to 1941 at the University of Michigan in Ann Arbor, and was a member of the Massachusetts Institute of Technology in Cambridge, Mass. from 1941 to 1946. From 1948 he worked at the Brookhaven National Laboratory in Upton, N.Y., in particular on the structure of atomic spectra. To interpret the latter, in 1925, together with G. Uhlenbeck, he introduced the spin of the electron into quantum theory. This concept proved to be of much greater importance than the discoverer expected. In 1944/45, G. was the leader of a secret mission ("Alsos") to investigate the German project on atomic energy. He received the Max Planck medal of the German Physical Society in 1964.

UHLENBECK, Georg Eugen, Dutch-American physicist, *Batavia 6.12.1900. U., a university professor in Ultrecht and Ann Arbor, introduced in 1925, together with S. A. Goudsmit the hypothesis of "spin" as an intrinsic rotation of the electron. U. published, among other works, "Over statistische methoden in de theorie der quanta" in 1927 and received, together with S. A. Goudsmit, the Max Planck medal of the German Physical Society in 1964 [BR].

de HAAS, Wonder Johannes, Dutch physicist, *Lisse at Leiden 2.5.1878, †Bilthoven 26.4.1960. H. was a co-worker at the "Physikalisch Technische Reichsanstalt" in Berlin from 1913 until 1915. Together with Einstein, he demonstrated the *Einstein de Haas effect* in 1915, i.e. the occurrence of a torque when an iron bar is magnetized in different directions. The verification of this effect was considered a confirmation of the existence of the Ampére molecular currents. After he had been a teacher at a secondary school in Deventer and the "Konservator" of the Texler Foundation in Haarlem, H. was a university professor at the Technical University in Delft and of the University of Groningen. From 1924 to 1948, he was the successor of H. Kamerlingh Onnes and together with W. H. Keesom, was joint director of the low-temperature laboratory in Leiden. There, together with his students, he performed basic investigations of paramagnetism at very low temperatures, superfluidity of helium, and superconductivity. In 1927, simultaneously, but independently of W. F. Giauque, H. applied the procedure of the adiabatic dimagnetization of paramagnetic salts to produce temperatures far below 1 K. This procedure had been suggested by P. Debye in 1926. Furthermore, in 1930, he discovered, together with his assistant J. van Alphen, the effect named after both discoverers. This effect is of importance for the investigation of the behaviour of electrons in metals [BR].

PAULI, Wolfgang, Austrian-German-Swiss physicist, *Vienna 4.12.1900, †Zürich 15.12.1958. As a fifth-semester student of A. Sommerfeld in Munich, P. wrote a summary on the theory of relativity for the Mathemat. Enzyklopaedia. In 1921 he proved in his Ph. D. thesis that quantum theory at that time was still incorrect. In his discussions with W. Heisenberg, M. Born and N. Bohr, P. contributed substantially to the development of *matrix mechanics*. At the beginning of 1926 he applied the new theory successfully to the hydrogen atom. In 1924, P. discovered the *exclusion principle (Pauli principle)*, for which he got the Nobel Prize in 1945. In the same year he postulated the existence of nuclear spin to explain hyperfine structure. In 1927 he set up the field equations for the electron, which included spin in nonrelativistic form; in the following years, together with Heisenberg, he made initial contributions to quantum field theory. After periods in which he worked in Göttingen, Copenhagen, and Hamburg, P. returned in 1928 as a professor to Zürich at the ETH. In 1930 he put forward the neutrino hypothesis. From 1940 to 1945, while working in the United States, he was concerned especially with meson theory. In 1946 he returned to Zürich, where he devoted himself primarily to quantum field theory and particle physics. In 1953, he began discussions with Heisenberg on the *unified theory of matter* ("Weltformel"), which the latter had developed. P. greatly influenced the physics of his time. With his profound analysis of the epistemological suppositions of science and his criticism of obscurity, he was considered the "conscience of physics". [BR]

LARMOR, Sir Joseph, English physicist and mathematician, *Magheragall, Co. Antrim, Ireland 11.7.1857, † 1942. From 1903 L. was a professor of mathematics at Cambridge University. He worked on problems in theoretical physics, especially on the theory of the electron, in the course of which he discovered the so-called Larmor precession. He made important contributions to relativity theory, and wrote *Aether and Matter* (1900).

RABI, Isaac Isidor, American physicist, *Rymanov (Galicia) 29.7.1898, † 1988. R. was a professor at Columbia University in New York from 1929. By suitably changing the molecular beam method discovered by O. Stern, R. could detect in 1933/34 the nuclear spin of sodium and determine the nuclear magnetic moments and the hyperfine structure of the spectral lines. R. developed the resonance method to determine the magnetic properties of atomic nuclei. In 1944, he was awarded the Nobel Prize in physics. During the Second World War, R. participated in the development of radar. [BR]

ZEEMAN, Pieter, Dutch physicist, *Zonnemaire (at Zierikezee) 25.5.1865, † Amsterdam 9.10.1943. Z. was a university professor in Amsterdam. In 1895, he discovered and studied the *Zeeman effect*, which had already been observed ten years earlier by Charles Jean Baptiste Fievez. In 1902, together with H. A. Lorentz, who gave an explanation of the Z. effect on the basis of his so-called *electron theory* – meanwhile outdated – Z. received the Nobel Prize in physics. [BR]

13. A Nonrelativistic Wave Equation with Spin

In this chapter we introduce a new method of deducing – in a systematic, theoretical manner – the Pauli equation for the electron *with the correct g factor*. In contrast to earlier derivations, we do not refer to empirical facts, but develop the new theoretical concept of the *linearization of the wave equation*.

What is meant by this will become clear in the next few sections. Conceptually we are dealing with the same method which will be used later on in relativistic quantum theory to derive the Dirac equation from the Klein–Gordon equation. *Levy-Leblond*,[1] for example, performed such a linearization for the Schrödinger equation. Here, we partially follow his argumentation, nevertheless abandoning it at some points, in order to demonstrate the ideas more easily and clearly.

13.1 The Linearization of the Schrödinger Equation

First we abbreviate the Schrödinger operator by

$$\hat{S} \equiv \mathrm{i}\hbar\frac{\partial}{\partial t} + \frac{\hbar^2}{2m}\Delta = \hat{E} - \frac{\hat{\boldsymbol{p}}^2}{2m} \ . \tag{13.1}$$

The free Schrödinger equation then reads

$$\hat{S}\psi = 0 \ . \tag{13.2}$$

It is asymmetric with respect to time ($\partial/\partial t$) and space derivatives ($\partial/\partial x$). This is because the former appears linearly in $\hat{S}$, while $\hat{S}$ is quadratic in $\hat{\boldsymbol{p}}$. To remove this asymmetry, we try to construct a wave equation of the general form

$$\hat{\Theta}\psi = (\hat{A}\hat{E} + \hat{\boldsymbol{B}}\cdot\hat{\boldsymbol{p}} + \hat{C})\psi = 0 \ . \tag{13.3}$$

Here $\hat{A}$, $\hat{\boldsymbol{B}}$ and $\hat{C}$ are to be linear operators (matrices) which still have to be determined, but which no longer depend on $\hat{E}$ or $\hat{\boldsymbol{p}}$. According to (13.2), we

[1] J. M. Levy-Leblond: Comm. Math. Phys. **6**, 286 (1967).

further require that the solutions ψ of (13.3) simultaneously be solutions of the Schrödinger equation. This means that the equations

$$\hat{\Theta}\psi = 0 \tag{13.4a}$$

and

$$\hat{S}\psi = 0 \tag{13.4b}$$

must be simultaneously valid. Then an operator

$$\hat{\Theta}' = \hat{A}'\hat{E} + \hat{\boldsymbol{B}}' \cdot \hat{\boldsymbol{p}} + \hat{C}' \tag{13.5}$$

must exist so that the multiplication of (13.4a) by $\hat{\theta}'$ again yields the Schrödinger equation (13.4b), i.e.

$$\hat{\Theta}'\hat{\Theta} = 2m\hat{S} \ . \tag{13.6}$$

The factor $2m$ is actually arbitrary, but will prove useful later. The operators $\hat{A}'$, $\hat{\boldsymbol{B}}'$ and $\hat{C}'$, introduced in (13.5), again shall not contain $\hat{E}$ and $\hat{\boldsymbol{p}}$. They still have to be determined, as do $\hat{A}$, $\hat{\boldsymbol{B}}$ and $\hat{C}$. If our procedure proves unsuccessful or impossible, we would be unable to find operators $\hat{A}$, $\hat{\boldsymbol{B}}$ and $\hat{C}$. If we are successful (and indeed, this will become apparent), then the equation $\Theta\psi = 0$ represents a more or less equivalent wave equation to the Schrödinger equation, but linear in both $\hat{E}$ and $\hat{\boldsymbol{p}}$. Then we speak of (13.3) as the *linearized Schrödinger equation*.

To construct $\hat{A}$, $\hat{A}'$, $\hat{\boldsymbol{B}}$, $\hat{\boldsymbol{B}}'$, $\hat{C}$ and $\hat{C}'$, we multiply the expressions (13.3) and (13.5) for $\hat{\Theta}$ on $\hat{\Theta}'$ and carry through, according to (13.6), a comparison of coefficients with $2m\hat{S}$. We obtain

$$\left(\hat{A}'\hat{E} + \sum_{i=1}^{3} \hat{B}_i'\hat{p}_i + \hat{C}'\right)\left(\hat{A}\hat{E} + \sum_{j=1}^{3} \hat{B}_j\hat{p}_j + \hat{C}\right) \stackrel{!}{=} 2m\hat{E} - \sum_{k=1}^{3} \hat{p}_k^2 \ , \tag{13.7}$$

and thus

$$\begin{aligned}
&\hat{A}'\hat{A} = 0 \ , && \hat{A}'\hat{B}_i + \hat{B}_i'\hat{A} = 0 \ , \\
&\hat{A}'\hat{C} + \hat{C}'\hat{A} = 2m \ , && \hat{B}_i'\hat{B}_j + \hat{B}_j'\hat{B}_i = -2\delta_{ij} \ , \\
&\hat{C}'\hat{C} = 0 \ , && \hat{C}'\hat{B}_i + \hat{B}_i'\hat{C} = 0 \quad (i, j = 1, 2, 3) \ .
\end{aligned} \tag{13.8}$$

To simplify these conditions, we define the new operators

$$\begin{aligned}
&\hat{B}_4 = \mathrm{i}\left(\hat{A} + \frac{1}{2m}\hat{C}\right) \ , && \hat{B}_4' = \mathrm{i}\left(\hat{A}' + \frac{1}{2m}\hat{C}'\right) \ , \\
&\hat{B}_5 = \hat{A} - \frac{1}{2m}\hat{C} \ , && \hat{B}_5' = \hat{A}' - \frac{1}{2m}\hat{C}' \ .
\end{aligned} \tag{13.9}$$

So (13.8) becomes

$$\hat{B}_\mu'\hat{B}_\nu + \hat{B}_\nu'\hat{B}_\mu = -2\delta_{\mu\nu} \quad (\mu, \nu) = 1 \text{ to } 5 \ . \tag{13.10}$$

These relations can still be changed into another form which is more customary in relativistic quantum mechanics. Let $\hat{M}$ be a nonsingular, arbitrary operator (with $\hat{M}\hat{M}^{-1} = 1$). Then we choose

$$\begin{aligned} \hat{B}_\alpha &= \hat{M}\hat{\gamma}_\alpha \ , \quad \hat{B}'_\alpha = -\hat{\gamma}_\alpha \hat{M}^{-1} \quad (\alpha = 1, \dots, 4) \ , \\ \hat{B}_5 &= -\mathrm{i}\hat{M} \ , \quad \hat{B}'_5 = -\mathrm{i}\hat{M}^{-1} \ . \end{aligned} \tag{13.11}$$

The following *anticommutation relations* follow by insertion of (13.11) into (13.10):

$$\hat{\gamma}_\alpha\hat{\gamma}_\beta + \hat{\gamma}_\beta\hat{\gamma}_\alpha = 2\delta_{\alpha\beta} \quad (\alpha, \beta = 1, \dots, 4) \ . \tag{13.12}$$

We should note that these relations are valid only for four operators $\hat{\gamma}_\alpha$, while in (13.10) there are *five* operators $\hat{B}_\mu$ present. It can easily be seen that the definitions (13.11) automatically fulfil the anticommutation relations (13.10) for the case that one or both indices are equal to 5 (i.e. $\mu = 5$ or $\nu = 5$, or $\mu = \nu = 5$).

For example, we calculate

$$\begin{aligned} \hat{B}'_5\hat{B}_\nu + \hat{B}'_\nu\hat{B}_5 &= -\mathrm{i}\hat{M}^{-1}\hat{M}\hat{\gamma}_\nu - \hat{\gamma}_\nu\hat{M}^{-1}(-\mathrm{i})\hat{M} \\ &= -\mathrm{i}(\hat{\gamma}_\nu - \hat{\gamma}_\nu) = 0 \quad \text{for} \quad \nu = 1, 2, 3, 4 \ , \end{aligned} \tag{13.13}$$

and

$$\hat{B}'_5\hat{B}_5 + \hat{B}_5\hat{B}'_5 = -\mathrm{i}\hat{M}^{-1}(-\mathrm{i})\hat{M} + (-\mathrm{i})\hat{M}(-\mathrm{i})\hat{M}^{-1} = 2\mathrm{i}^2 = -2 \ . \tag{13.14}$$

So the five operators $\hat{B}_\nu$ will be replaced by the four $\hat{\gamma}_\alpha$ and the arbitrarily chosen operator $\hat{M}$ (which indeed must not be singular because it must have an inverse operator $\hat{M}^{-1}$).

The anticommutation relations (13.12) define an algebra, which is known in the literature as *Clifford algebra*. It can be represented by matrices and leads to the algebra of the complex 4×4 matrices (of particular importance in relativistic quantum theory) as a special representation.

In order to obtain an explicit representation for the $\hat{\gamma}_\alpha$, and thus for the $\hat{B}_\nu$, we observe (13.12) more carefully, and immediately verify that[2]

$$\gamma_1^2 = \gamma_2^2 = \gamma_3^2 = \gamma_4^2 = 1 \quad \text{and} \tag{13.15}$$

$$\gamma_\alpha\gamma_\beta = -\gamma_\beta\gamma_\alpha \quad \text{for} \quad \alpha \neq \beta \tag{13.16}$$

must hold. In other words, the squares of the γ_α are 1 and the different γ operators anticommute.

It follows from the former that the eigenvalues of γ_α have to be ± 1. Matrices representing the γ's have to be quadratic according to (13.15). And from (13.16),

[2] From now on, we shall omit the operator sign on the γ_α operators, keeping in mind, however, their operator character.

it follows that the traces of these matrices have to vanish, because for $\alpha \neq \beta$, we have

$$\gamma_\beta\gamma_\alpha = -\gamma_\alpha\gamma_\beta \Rightarrow \gamma_\alpha = -\gamma_\beta\gamma_\alpha\gamma_\beta \Rightarrow$$
$$\mathrm{tr}\gamma_\alpha = -\mathrm{tr}\gamma_\beta\gamma_\alpha\gamma_\beta = -\mathrm{tr}\gamma_\alpha\gamma_\beta^2 = -\mathrm{tr}\gamma_\alpha \ , \tag{13.17}$$

and thus $\mathrm{tr}\gamma_\alpha = 0$. In the last step we have made use of trace $\hat{A}\hat{B} = \sum_{i,k} A_{ik}B_{ki} = \mathrm{tr}\hat{B}\hat{A}$ and the fact that $\gamma_\beta^2 = 1$, according to (13.15). As the trace is just the sum over the eigenvalues, the numbers of positive and negative eigenvalues have to be equal. Therefore the γ matrices have to be of even dimension. The smallest even dimension, $N = 2$, has to be excluded, because in 2×2 matrix space, there is only room for 3 anticommuting matrices $\hat{\sigma}_i$ and the unit matrix. The $\hat{\sigma}_i$, $i = 1, 2, 3$, are the well-known Pauli matrices, which anticommute according to (12.21). (In Exercise 13.1 we will show that the three $\hat{\sigma}_i$ and the 2×2 unit matrix $\mathbb{1}$ completely span the space of 2×2 matrices.)

Thus we conclude that the smallest dimension under which the conditions listed above on the 4 anticommuting matrices γ_α can be fulfilled is $N = 4$. Because of the properties of the Pauli matrices, described in Chap. 12, it is not difficult to give the following representation for the γ_α:

$$\gamma_i = \begin{pmatrix} 0 & \hat{\sigma}_i \\ \hat{\sigma}_i & 0 \end{pmatrix} \ , \quad (i = 1, 2, 3) \ , \quad \gamma_4 = \begin{pmatrix} \mathbb{1} & 0 \\ 0 & -\mathbb{1} \end{pmatrix} \ . \tag{13.18}$$

Here, 0, $\mathbb{1}$ and the $\hat{\sigma}_i$ indicate 2×2 submatrices; thus (13.18) is an abbreviation. Explicitly, (13.18) reads

$$\gamma_1 = \begin{pmatrix} 0 & 0 & 0 & 1 \\ 0 & 0 & 1 & 0 \\ 0 & 1 & 0 & 0 \\ 1 & 0 & 0 & 0 \end{pmatrix} \ , \quad \gamma_2 = \begin{pmatrix} 0 & 0 & 0 & -\mathrm{i} \\ 0 & 0 & \mathrm{i} & 0 \\ 0 & -\mathrm{i} & 0 & 0 \\ \mathrm{i} & 0 & 0 & 0 \end{pmatrix} \ ,$$

$$\gamma_3 = \begin{pmatrix} 0 & 0 & 1 & 0 \\ 0 & 0 & 0 & -1 \\ 1 & 0 & 0 & 0 \\ 0 & -1 & 0 & 0 \end{pmatrix} \ , \quad \gamma_4 = \begin{pmatrix} 1 & 0 & 0 & 0 \\ 0 & 1 & 0 & 0 \\ 0 & 0 & -1 & 0 \\ 0 & 0 & 0 & -1 \end{pmatrix} \ . \tag{13.19}$$

The validity of relations (13.15) and (13.16) can easily be checked. For example

$$\gamma_i^2 = \begin{pmatrix} 0 & \hat{\sigma}_i \\ \hat{\sigma}_i & 0 \end{pmatrix}\begin{pmatrix} 0 & \hat{\sigma}_i \\ \hat{\sigma}_i & 0 \end{pmatrix} = \begin{pmatrix} \hat{\sigma}_i^2 & 0 \\ 0 & \hat{\sigma}_i^2 \end{pmatrix} = \begin{pmatrix} \mathbb{1} & 0 \\ 0 & \mathbb{1} \end{pmatrix} = \mathbb{1}_4 \ ,$$

$$\gamma_4^2 = \begin{pmatrix} \mathbb{1} & 0 \\ 0 & -\mathbb{1} \end{pmatrix}\begin{pmatrix} \mathbb{1} & 0 \\ 0 & -\mathbb{1} \end{pmatrix} = \begin{pmatrix} \mathbb{1} & 0 \\ 0 & \mathbb{1} \end{pmatrix} = \mathbb{1}_4 \ ,$$

$$
\begin{aligned}
\gamma_i\gamma_j+\gamma_j\gamma_i &= \begin{pmatrix} 0 & \hat{\sigma}_i \\ \hat{\sigma}_i & 0 \end{pmatrix}\begin{pmatrix} 0 & \hat{\sigma}_j \\ \hat{\sigma}_j & 0 \end{pmatrix}+\begin{pmatrix} 0 & \hat{\sigma}_j \\ \hat{\sigma}_j & 0 \end{pmatrix}\begin{pmatrix} 0 & \hat{\sigma}_i \\ \hat{\sigma}_i & 0 \end{pmatrix} \\
&= \begin{pmatrix} \hat{\sigma}_i\hat{\sigma}_j+\hat{\sigma}_j\hat{\sigma}_i & 0 \\ 0 & \hat{\sigma}_i\hat{\sigma}_j+\hat{\sigma}_j\hat{\sigma}_i \end{pmatrix}\begin{pmatrix} 2\delta_{ij} & 0 \\ 0 & 2\delta_{ij} \end{pmatrix} \\
&= 2\begin{pmatrix} \mathbb{1} & 0 \\ 0 & \mathbb{1} \end{pmatrix}\delta_{ij}\ , \quad \text{for} \quad i,j=1,2,3\ ,
\end{aligned} \tag{13.20}
$$

$$
\begin{aligned}
\gamma_i\gamma_4+\gamma_4\gamma_i &= \begin{pmatrix} 0 & \hat{\sigma}_i \\ \hat{\sigma}_i & 0 \end{pmatrix}\begin{pmatrix} \mathbb{1} & 0 \\ 0 & -\mathbb{1} \end{pmatrix} \\
&\quad +\begin{pmatrix} \mathbb{1} & 0 \\ 0 & -\mathbb{1} \end{pmatrix}\begin{pmatrix} 0 & \hat{\sigma}_i \\ \hat{\sigma}_i & 0 \end{pmatrix}=\begin{pmatrix} 0 & 0 \\ 0 & 0 \end{pmatrix}=0\ .
\end{aligned} \tag{13.21}
$$

To obtain a matrix representation for the $\hat{B}_\nu$ in accordance with (13.11), we choose

$$
\hat{M}=\begin{pmatrix} 0 & \mathbb{1} \\ \mathbb{1} & 0 \end{pmatrix}=\begin{pmatrix} 0 & 0 & 1 & 0 \\ 0 & 0 & 0 & 1 \\ 1 & 0 & 0 & 0 \\ 0 & 1 & 0 & 0 \end{pmatrix}=\hat{M}^{-1}\ . \tag{13.22}
$$

The relation

$$
\hat{M}\hat{M}^{-1}=\begin{pmatrix} \mathbb{1} & 0 \\ 0 & \mathbb{1} \end{pmatrix}
$$

is obvious. We continue calculating:

$$
\begin{aligned}
\hat{B}_i &= \hat{M}\gamma_i=\begin{pmatrix} 0 & \mathbb{1} \\ \mathbb{1} & 0 \end{pmatrix}\begin{pmatrix} 0 & \hat{\sigma}_i \\ \hat{\sigma}_i & 0 \end{pmatrix}=\begin{pmatrix} \hat{\sigma}_i & 0 \\ 0 & \hat{\sigma}_i \end{pmatrix} \quad \text{for} \quad i=1,2,3\ , \\
\hat{B}_4 &= \hat{M}\gamma_4=\begin{pmatrix} 0 & \mathbb{1} \\ \mathbb{1} & 0 \end{pmatrix}\begin{pmatrix} \mathbb{1} & 0 \\ 0 & -\mathbb{1} \end{pmatrix}\begin{pmatrix} 0 & -\mathbb{1} \\ \mathbb{1} & 0 \end{pmatrix}\ , \\
\hat{B}_5 &= -\mathrm{i}\hat{M}=-\mathrm{i}\begin{pmatrix} 0 & \mathbb{1} \\ \mathbb{1} & 0 \end{pmatrix}\ .
\end{aligned} \tag{13.23}
$$

As was just mentioned, the Pauli matrices $\hat{\sigma}_i$ and the 2×2 unit matrix $\mathbb{1}$ completely span the space of the 2×2 matrices. This means that any arbitrary 2×2 matrix can be expressed by the $\hat{\sigma}_i$ and $\mathbb{1}$. We will show this in the following exercise.

EXERCISE

13.1 Completeness of the Pauli Matrices

Problem. Show that every 2×2 matrix $\begin{pmatrix} u_{11} & u_{12} \\ u_{21} & u_{22}\end{pmatrix}$ can be expressed by $\mathbb{1}$ and $\hat{\sigma}_i$.

Solution. First we write down the proposition

$$\begin{aligned}\begin{pmatrix} u_{11} & u_{12} \\ u_{21} & u_{22}\end{pmatrix} &= \sum_{i=1}^{3} a_i\hat{\sigma}_i + a_4\mathbb{1} \\ &= a_1\begin{pmatrix} 0 & 1 \\ 1 & 0\end{pmatrix} + a_2\begin{pmatrix} 0 & -\mathrm{i} \\ \mathrm{i} & 0\end{pmatrix} + a_3\begin{pmatrix} 1 & 0 \\ 0 & -1\end{pmatrix} + a_4\begin{pmatrix} 1 & 0 \\ 0 & 1\end{pmatrix} \\ &= \begin{pmatrix} (a_3+a_4) & (a_1-\mathrm{i}a_2) \\ (a_1+\mathrm{i}a_2) & (-a_3+a_4)\end{pmatrix} \; . \end{aligned} \tag{1}$$

Both matrices have to be equal in each element. Then we get the following system of equations:

$$\begin{aligned} u_{11} &= 0a_1 + 0a_2 + a_3 + a_4 \; , \\ u_{12} &= a_1 - \mathrm{i}a_2 + 0a_3 + 0a_4 \; , \\ u_{21} &= a_1 + \mathrm{i}a_2 + 0a_3 + 0a_4 \; , \\ u_{22} &= 0a_1 + 0a_2 - a_3 + a_4 \; , \end{aligned} \tag{2}$$

with the determinant of coefficients

$$\begin{vmatrix} 0 & 0 & 1 & 1 \\ 1 & -\mathrm{i} & 0 & 0 \\ 1 & \mathrm{i} & 0 & 0 \\ 0 & 0 & -1 & 1 \end{vmatrix} = 4\mathrm{i} \neq 0 \; , \tag{3}$$

which is always nonzero. Hence, a nontrivial solution always exists; i.e. not all coefficients a_i vanish, proving the proposition; $\hat{\sigma}_i$ and $\mathbb{1}$ span the whole space of 2×2 matrices!

EXERCISE

13.2 A Computation Rule for Pauli Matrices

Problem. Let $\boldsymbol{A}$ and $\boldsymbol{B}$ be arbitrary vectors. Prove the relation

$$(\hat{\boldsymbol{\sigma}}\cdot\boldsymbol{A})(\hat{\boldsymbol{\sigma}}\cdot\boldsymbol{B}) = \boldsymbol{A}\cdot\boldsymbol{B} + \mathrm{i}\hat{\boldsymbol{\sigma}}\cdot(\boldsymbol{A}\times\boldsymbol{B}) \; . \tag{1}$$

Solution. The commutation relations for the $\hat{\sigma}_i$ are *Exercise 13.2*

$$\hat{\sigma}_i\hat{\sigma}_j = \mathrm{i}\varepsilon_{ijk}\hat{\sigma}_k + \delta_{ij} \ , \quad \text{where}$$

$$\varepsilon_{ijk} = \begin{cases} 1 & \text{even permutation of } 1,2,3 \\ -1 & \text{odd permutation of } 1,2,3 \\ 0 & \text{otherwise} \ . \end{cases} \tag{2}$$

Addition (or subtraction) then gives

$$\begin{aligned} \hat{\sigma}_i\hat{\sigma}_j - \hat{\sigma}_j\hat{\sigma}_i &= 2\mathrm{i}\varepsilon_{ijk}\hat{\sigma}_k \ , \\ \hat{\sigma}_i\hat{\sigma}_j + \hat{\sigma}_j\hat{\sigma}_i &= 2\delta_{ij} \ . \end{aligned} \tag{3}$$

We write out the scalar product

$$(\hat{\boldsymbol{\sigma}}\cdot\boldsymbol{A})(\hat{\boldsymbol{\sigma}}\cdot\boldsymbol{B}) = \left(\sum_{i=1}^{3}\hat{\sigma}_i A_i\right)\left(\sum_{j=1}^{3}\hat{\sigma}_j B_j\right) \ . \tag{4}$$

For the individual components we can write

$$\hat{\sigma}_i A_i \hat{\sigma}_j B_j = A_i B_j(\mathrm{i}\varepsilon_{ijk}\hat{\sigma}_k + \delta_{ij}) \ , \tag{5}$$

and

$$\sum_{ij} A_i B_j \delta_{ij} = \sum_i A_i B_i$$

is just the scalar product $\boldsymbol{A}\cdot\boldsymbol{B}$. In the first term, the sum can be expanded over k without making any changes:

$$\sum_{i,j}\varepsilon_{ijk}A_i B_j\hat{\sigma}_k = \sum_{i,j,k}\varepsilon_{ijk}A_i B_j\hat{\sigma}_k \ , \tag{6}$$

because, for example, for $i, j = 1, 2$, k has to be equal to 3 and the additional terms of the supplementary summation over k with $k = 1, 2$ vanish identically. Now $\varepsilon_{ijk}A_i B_j$ are just the components of the vector product $\boldsymbol{A}\times\boldsymbol{B}$. Therefore we have

$$\begin{aligned} \sum_{i,j}\varepsilon_{ijk}A_i B_j\hat{\sigma}_k &= \sum_{i,j,k}\varepsilon_{ijk}A_i B_j\hat{\sigma}_k \\ &= \sum_k(\boldsymbol{A}\times\boldsymbol{B})_k\hat{\sigma}_k = \hat{\boldsymbol{\sigma}}\cdot(\boldsymbol{A}\times\boldsymbol{B}) \ . \end{aligned} \tag{7}$$

Altogether we get

$$(\hat{\boldsymbol{\sigma}}\cdot\boldsymbol{A})(\hat{\boldsymbol{\sigma}}\cdot\boldsymbol{B}) = \boldsymbol{A}\cdot\boldsymbol{B} + \mathrm{i}\hat{\boldsymbol{\sigma}}\cdot(\boldsymbol{A}\times\boldsymbol{B}) \ . \tag{8}$$

It now follows from (13.3) that the wave function

$$\psi = \begin{pmatrix} \varphi \\ \chi \end{pmatrix} \equiv \begin{pmatrix} \varphi_1 \\ \varphi_2 \\ \chi_1 \\ \chi_2 \end{pmatrix} \tag{13.24}$$

must have four components, because $\hat{A}$, $\hat{\boldsymbol{B}}$ and $\hat{C}$ are 4×4 matrices. Here, $\varphi = \begin{pmatrix} \varphi_1 \\ \varphi_2 \end{pmatrix}$ and $\chi = \begin{pmatrix} \chi_1 \\ \chi_2 \end{pmatrix}$ are two-component spinors, which together form the fourcomponent spinor ψ.

We now solve (13.9) for $\hat{A}$ and $\hat{C}$:

$$\hat{A} = \tfrac{1}{2}(B_5 - \mathrm{i}B_4) \ , \quad \hat{C} = -m(B_5 + \mathrm{i}B_4) \ , \tag{13.25}$$

and thus

$$\hat{A} = -\mathrm{i}\begin{pmatrix} 0 & 0 \\ \mathbf{1} & 0 \end{pmatrix} \quad \text{and} \quad \hat{C} = 2m\mathrm{i}\begin{pmatrix} 0 & \mathbf{1} \\ 0 & 0 \end{pmatrix} \ . \tag{13.26}$$

In the next step, the matrices $\hat{A}$, $\hat{\boldsymbol{B}}$ and $\hat{C}$ are inserted into the equation of motion (13.3), giving

$$\left[-\mathrm{i}\begin{pmatrix} 0 & 0 \\ \mathbf{1} & 0 \end{pmatrix}\hat{E} + \begin{pmatrix} \hat{\boldsymbol{\sigma}} & 0 \\ 0 & \hat{\boldsymbol{\sigma}} \end{pmatrix}\cdot\hat{\boldsymbol{p}} + 2m\mathrm{i}\begin{pmatrix} 0 & \mathbf{1} \\ 0 & 0 \end{pmatrix}\right]\begin{pmatrix} \varphi \\ \chi \end{pmatrix} = 0 \ . \tag{13.27}$$

Writing this matrix equation by components, we obtain the coupled system of equations for the two-component spinors χ and φ,

$$\hat{\boldsymbol{\sigma}}\cdot\hat{\boldsymbol{p}}\varphi + 2m\mathrm{i}\chi = 0 \ , \quad \hat{\boldsymbol{\sigma}}\cdot\hat{\boldsymbol{p}}\chi - \mathrm{i}\hat{E}\varphi = 0 \ , \tag{13.28}$$

where $\hat{\boldsymbol{\sigma}}$ is the vector with the components $\hat{\sigma}_i$: $\hat{\boldsymbol{\sigma}} = \{\hat{\sigma}_1, \hat{\sigma}_2, \hat{\sigma}_3\}$.

EXERCISE

13.3 Spinors Satisfying the Schrödinger Equation

Problem. Show that the two-spinors φ and χ satisfy the ordinary Schrödinger equation.

Solution. We first eliminate $\chi = -(\hat{\boldsymbol{\sigma}}\cdot\hat{\boldsymbol{p}}/2m\mathrm{i})\varphi$ and get from (13.28) that

$$\left[-\mathrm{i}E - \frac{(\hat{\boldsymbol{\sigma}}\cdot\hat{\boldsymbol{p}})(\hat{\boldsymbol{\sigma}}\cdot\hat{\boldsymbol{p}})}{2m\mathrm{i}}\right]\varphi = 0 \ . \tag{1}$$

Exercise 13.3

Since $(\hat{\boldsymbol{\sigma}}\cdot\hat{\boldsymbol{p}})(\hat{\boldsymbol{\sigma}}\cdot\hat{\boldsymbol{p}})=\hat{\boldsymbol{p}}^2$, we obtain

$$\left(\hat{E}-\frac{\hat{\boldsymbol{p}}^2}{2m}\right)\varphi=0\ . \qquad (2)$$

This is the Schrödinger equation for φ.

Now we eliminate $\hat{E}\varphi=(\hat{\boldsymbol{\sigma}}\cdot\hat{\boldsymbol{p}})\chi/\mathrm{i}$ from the second equation in (13.28), and insert the result into the first equation in (13.28). Multiplying that result by $\hat{E}$ yields

$$[(1/\mathrm{i})(\hat{\boldsymbol{\sigma}}\cdot\hat{\boldsymbol{p}})(\hat{\boldsymbol{\sigma}}\cdot\hat{\boldsymbol{p}})+2m\mathrm{i}\hat{E}]\chi=0 \quad \text{or}$$

$$\left(\hat{E}-\frac{\hat{\boldsymbol{p}}^2}{2m}\right)\chi=0\ . \qquad (3)$$

Therefore χ also satisfies the Schrödinger equation, as was to be shown.

In Exercise 13.3 we show that the four-spinor $\psi=\begin{pmatrix}\varphi\\ \chi\end{pmatrix}$ of the linearized Schrödinger equation [(13.3), (13.28)] indeed satisfies the ordinary Schrödinger equation, as we required. Therefore the energy eigenvalues are in both cases also $E=p^2/2m$. After eliminating χ, the corresponding eigenvectors take the form

$$\psi=\begin{bmatrix}\varphi\\ (-\hat{\boldsymbol{\sigma}}\cdot\hat{\boldsymbol{p}}/2m\mathrm{i})\varphi\end{bmatrix}\ . \qquad (13.29)$$

Here, it would seem that the wave function ψ with the lower component χ contains redundant information. That this is not valid in general will now be demonstrated by considering the coupling with an external electromagnetic field.

13.2 Particles in an External Field and the Magnetic Moment

The gauge invariance of the Schrödinger equation requires the substitution (minimal coupling – see Chap. 9)

$$\mathrm{i}\hbar\frac{\partial}{\partial t}\to\mathrm{i}\hbar\frac{\partial}{\partial t}-eV(\boldsymbol{x},t) \quad\text{and}\quad -\mathrm{i}\hbar\nabla\to\mathrm{i}\hbar\nabla-\frac{e}{c}\boldsymbol{A}(\boldsymbol{x},t)\ , \qquad (13.30)$$

or, in Lorentz covariant notation,

$$-\mathrm{i}\hbar\frac{\partial}{\partial x_\mu}\to\left(-\mathrm{i}\hbar\frac{\partial}{\partial x_\mu}-\frac{e}{c}A_\mu\right)\ ,\quad \mu=1,2,3,4\ , \qquad (13.31)$$

where the four-potential is given by

$$\hat{A} = \{A_\mu\} = \{\hat{A}, \mathrm{i}V\} \ . \tag{13.32}$$

Here, e is the electric charge of the particle, $V(\boldsymbol{x}, t)$ is the Coulomb potential and $\boldsymbol{A}(\boldsymbol{x}, t)$ is the vector potential. Let us remember the essential argument. A gauge transformation is described by

$$A'_\mu = A_\mu + \frac{\partial f}{\partial x_\mu} \tag{13.33}$$

with an arbitrary function $f(x_\mu)$. The minimal coupling (13.30), together with the phase transformation for the wave equation

$$\psi' = \psi \exp\left[-\frac{e}{\mathrm{i}\hbar c} f(x_\mu)\right] \ , \tag{13.34}$$

then lead to

$$\begin{aligned}
&\left(-\mathrm{i}\hbar\frac{\partial}{\partial x_\mu} - \frac{e}{c}A'\mu\right)\psi' \\
&\quad = \left(-\mathrm{i}\hbar\frac{\partial}{\partial x_\mu} - \frac{e}{c}A_\mu - \frac{e}{c}\frac{\partial f}{\partial x_\mu}\right)\psi \exp[-(e/\mathrm{i}\hbar c)f] \\
&\quad = \left[\left(-\mathrm{i}\hbar\frac{\partial}{\partial x_\mu} - \frac{e}{c}A_\mu\right)\psi\right]\exp[-(e/\mathrm{i}\hbar c)f] \\
&\qquad + \left(\frac{e}{c}\frac{\partial f}{\partial x_\mu} - \frac{e}{c}\frac{\partial f}{\partial x_\mu}\right)\psi \exp[-(e/\mathrm{i}\hbar c)f] \\
&\quad = \left[\left(-\mathrm{i}\hbar\frac{\partial}{\partial x_\mu} - \frac{e}{c}A\mu\right)\psi\right]\exp[-(e/\mathrm{i}\hbar c)f] \ .
\end{aligned} \tag{13.35}$$

This means that a gauge transformation can be absorbed with the state-independent phase $\exp[-(e/\mathrm{i}\hbar c)f(\boldsymbol{r}, t)]$ and therefore does not change the physics (matrix elements, expectation values etc.). So the minimal coupling (13.30) leads to gauge-invariant quantum theories.

With (13.30), the free equations of motion (13.28) become

$$\begin{aligned}
&\hat{\boldsymbol{\sigma}}\cdot\left(\hat{\boldsymbol{p}} - \frac{e}{c}\boldsymbol{A}\right)\chi - \mathrm{i}(\hat{E} - eV)\varphi = 0 \ , \\
&\hat{\boldsymbol{\sigma}}\cdot\left(\hat{\boldsymbol{p}} - \frac{e}{c}\boldsymbol{A}\right)\varphi + \mathrm{i}2m\chi = 0 \ .
\end{aligned} \tag{13.36}$$

Again we eliminate $\chi = -[\hat{\boldsymbol{\sigma}}\cdot(\boldsymbol{p} - e\boldsymbol{A}/c)/2m\mathrm{i}]\varphi$ and get

$$\left[-\mathrm{i}(\hat{E} - eV) - \frac{1}{2m\mathrm{i}}\hat{\boldsymbol{\sigma}}\cdot\left(\hat{\boldsymbol{p}} - \frac{e}{c}\boldsymbol{A}\right)\hat{\boldsymbol{\sigma}}\cdot\left(\hat{\boldsymbol{p}} - \frac{e}{c}\boldsymbol{A}\right)\right]\varphi = 0 \ , \tag{13.37}$$

$$\left[\hat{E} - eV - \frac{1}{2m}\hat{\boldsymbol{\sigma}}\cdot\left(\hat{\boldsymbol{p}} - \frac{e}{c}\boldsymbol{A}\right)\hat{\boldsymbol{\sigma}}\cdot\left(\hat{\boldsymbol{p}} - \frac{e}{c}\boldsymbol{A}\right)\right]\varphi = 0 \ . \tag{13.38}$$

Using once more the identity

$$(\hat{\sigma}\cdot\boldsymbol{\pi})(\hat{\sigma}\cdot\boldsymbol{\pi}) = \boldsymbol{\pi}^2 + \mathrm{i}\hat{\sigma}\cdot(\boldsymbol{\pi}\times\boldsymbol{\pi}) \ , \tag{13.39}$$

we thus obtain

$$\begin{aligned}&\hat{\sigma}\cdot\left(\hat{p}-\frac{e}{c}A\right)\hat{\sigma}\cdot\left(\hat{p}-\frac{e}{c}A\right)\\&\quad=\left(\hat{p}-\frac{e}{c}A\right)^2+\mathrm{i}\hat{\sigma}\cdot\left[\left(\hat{p}-\frac{e}{c}A\right)\times\left(\hat{p}-\frac{e}{c}A\right)\right] \ .\end{aligned} \tag{13.40}$$

The last term reduces to

$$\begin{aligned}\left(\hat{p}-\frac{e}{c}A\right)\times\left(\hat{p}-\frac{e}{c}A\right) &= -\frac{e}{c}(\hat{p}\times A + A\times\hat{p})\\&= -\frac{e}{c}[(\hat{p}\times A) - A\times\hat{p} + A\times\hat{p}]\\&= -\frac{\mathrm{e}}{c}(\hat{p}\times A) \ ,\end{aligned} \tag{13.41}$$

so that at last (13.38) can be written as

$$\left[E - eV - \frac{1}{2m}\left(\hat{p}-\frac{e}{c}A\right)^2 + \frac{\mathrm{i}e}{2mc}\hat{\sigma}\cdot(\hat{p}\times A)\right]\varphi = 0 \ . \tag{13.42}$$

Now $\hat{p} = -\mathrm{i}\hbar\nabla$ and $B = \nabla\times A$. Hence, (13.42) becomes

$$\left[\hat{E} - eV - \frac{1}{2m}\left(\hat{p}-\frac{e}{c}A\right)^2 \frac{e\hbar}{2mc}\hat{\sigma}\cdot B\right]\varphi = 0 \ . \tag{13.43}$$

This is just the well-known Pauli equation! See Chap. 12.

The last term in the equation of motion (13.43) is the interaction energy of the magnetic field with the intrinsic magnetic moment of the particle

$$\hat{\mu} = \frac{e\hbar}{2mc}\hat{\sigma} \ , \tag{13.44}$$

or, because the spin operator of the particle is $\hat{S} = (1/2)\hat{\sigma}$,

$$\hat{\mu} = \frac{e\hbar}{mc}\hat{S} = g_{\mathrm{spin}}\ \mu_{\mathrm{B}}\hat{S} = 2\mu_{\mathrm{B}}\hat{S} \ . \tag{13.45}$$

The factor g_{spin} is called the *gyromagnetic ratio* or *gyromagnetic factor* and turns out to be twice as large as that coming from the *orbital motion*. The ratio $g_{\mathrm{spin}}/g_{\mathrm{orbit}}$ is called the *spin-Landé factor* g_{s}. For the particle in question, g_{s} is therefore 2.

Thus a completely nonrelativistic linearized theory predicts the correct intrinsic magnetic moment of a spin-$\frac{1}{2}$ particle.

In contrast to this, almost all textbooks falsely claim that the anomalous magnetic moment is due to *relativistic* properties. The existence of spin is therefore *not* a relativistic effect, as is often asserted, but is a *consequence of the linearization of the wave equations*. This can be philosophically expressed as follows: obviously, the good Lord wrote the field equations in linearized form, i.e. in the nonrelativistic case, as a system of two coupled differential equations of first order, and then coupled the electromagnetic field minimally. He did *not* write them as a differential equation of second order (the Schrödinger equation).

We have successfully derived the Pauli equation from the Schrödinger equation. Whereas in the heuristic derivation of the Pauli equation presented in Chap. 12 the spin degree of freedom was introduced "ad hoc", in the derivation presented here we have only postulated the linearization of the equations of motion. Everything else that followed was just a consequence of this postulate.

14. Elementary Aspects of the Quantum-Mechanical Many-Body Problem

If we consider a system of more than one particle, we derive its Hamiltonian, describing it quantum-mechanically in the usual manner from the Hamiltonian function of the system in classical mechanics. The Hamiltonian function

$$H=\sum_{i=1}^{N}\left(\frac{p_i^2}{2m_i}+V_i(\boldsymbol{r}_i,t)\right)+\sum_{i\neq k}V_{ik}(\boldsymbol{r}_i,\boldsymbol{r}_k) \tag{14.1}$$

describes a system of N particles with mass m_i. Here, $V_i(\boldsymbol{r}_i,t)$ is the externally given potential (the so-called *one-particle potential*), in which the ith particle moves; it can, for example, mean the external electric potential. $V_{ik}(\boldsymbol{r}_i,\boldsymbol{r}_k)$ stands for the interaction potentials between two particles i and k; it can, for example, be their mutual Coulomb interaction. To get the Hamiltonian, we replace the momenta by the corresponding differential operators

$$\boldsymbol{p}_i\to\hat{\boldsymbol{p}}_i=\frac{\hbar}{\mathrm{i}}\boldsymbol{\nabla}_i\ , \tag{14.2}$$

where the index i of the nabla operator specifies that the gradient has to be determined at the location of the particle i, i.e. $\boldsymbol{\nabla}_i$ only acts on the coordinates of the ith particle. Consequently the momentum operators of different particles commute, i.e. $[\hat{\boldsymbol{p}}_i,\hat{\boldsymbol{p}}_j]_-=0$ for all i,j. Thus the many-particle Hamiltonian reads

$$\hat{H}=\sum_{i=1}^{N}\left(-\frac{\hbar^2}{2m_i}\Delta_i+V_i(\boldsymbol{r}_i,t)\right)+\sum_{i\neq k}V_{ik}(\boldsymbol{r}_i,\boldsymbol{r}_k)\ . \tag{14.3}$$

This is obviously a generalization of the Hamiltonian for one particle. We can now formulate a many-particle Schrödinger equation

$$\hat{H}\psi=\mathrm{i}\hbar\frac{\partial}{\partial t}\psi\ ,$$

where the wave function now depends on the $3N$ coordinates of all particles and on time:

$$\begin{aligned}\psi&=\psi(\boldsymbol{r}_1,\dots,\boldsymbol{r}_N,t)\\&=\psi(x_1,y_1,z_1,\dots,x_k,y_k,z_k,\dots,x_N,y_N,z_N,t)\ .\end{aligned} \tag{14.4}$$

The treatment of this many-body problem confronts the same difficulties in quantum mechanics as in classical physics because of the complexity compared to the one-particle problem.

The wave equation is defined in a space with $3N$ dimensions, in the so-called *configuration space* of the system. The name of this fictitious space originates from the fact that the specification of the coordinates of a special point in this space means the specification of the three-dimensional coordinates of the position (x_k, y_k, z_k) for all particles of the system $(k = 1, 2, \dots, N)$, and thus determines the state (configuration) of all particles in three-dimensional space. Therefore a point in configuration space with $3N$ coordinates $(x_1, y_1, z_1, \dots, x_N, y_N, z_N)$ is also called the *configuration point* of the system.

We denote an infinitesimally small volume element in the configuration space by $\mathrm{d}V$:

$$\begin{aligned} \mathrm{d}V &= \mathrm{d}V_1\,\mathrm{d}V_2 \cdots \mathrm{d}V_k \cdots \mathrm{d}V_N \\ &= \mathrm{d}x_1\,\mathrm{d}y_1\,\mathrm{d}z_1 \cdots \mathrm{d}x_k\,\mathrm{d}y_k\,\mathrm{d}z_k \cdots \mathrm{d}x_N\,\mathrm{d}y_N\,\mathrm{d}z_N \ . \end{aligned} \tag{14.5}$$

Then the quantity

$$w(x_1, y_1, z_1, \dots, x_k, y_k, z_k, \dots, x_N, y_N, z_N, t)\,\mathrm{d}V = \psi^*\psi\,\mathrm{d}V \tag{14.6}$$

is the probability that the system can be found at time t in the volume element $\mathrm{d}V$ of configuration space. This means w is the probability density of the configuration of the system, in which at time t the coordinates of the first particle lie between $x_1, x_1 + \mathrm{d}x_1$; $y_1, y_1 + \mathrm{d}y_1$; $z_1, z_1 + \mathrm{d}z_1$; and of the kth particle between $x_k, x_k + \mathrm{d}x_k$; $y_k, y_k + \mathrm{d}y_k$; $z_k, z_k + \mathrm{d}z_k$; etc. Besides the volume element, we also examine the volume elements in the subspaces of the kind $\mathrm{d}\Omega_k$, $\mathrm{d}\Omega_{kj}, \dots$ etc., which are defined by

$$\begin{aligned} \mathrm{d}V &= \mathrm{d}x_k\,\mathrm{d}y_k\,\mathrm{d}z_k\,\mathrm{d}\Omega_k = \mathrm{d}V_k\,\mathrm{d}\Omega_k \ , \\ \mathrm{d}V &= \mathrm{d}x_k\,\mathrm{d}y_k\,\mathrm{d}z_k\,\mathrm{d}x_j\,\mathrm{d}y_j\,\mathrm{d}z_j\,\mathrm{d}\Omega_{kj} = \mathrm{d}V_k\,\mathrm{d}V_j\,\mathrm{d}\Omega_{kj} \ , \quad \text{etc} \ . \end{aligned} \tag{14.7}$$

Integrating (14.6) with respect to the coordinates of all particles, excluding the particle k, i.e. over $\mathrm{d}\Omega_k$, we thus find that the probability density of the kth particle lies between $x_k, x_k + \mathrm{d}x_k$; $y_k, y_k + \mathrm{d}y_k$; $z_k, z_k + \mathrm{d}z_k$; and all other particles are in arbitrary positions. In other words, we find the probability in such a way that the kth particle is near a given position in space. Denoting this probability by $w(x_k, y_k, z_k, t)$, we obtain

$$w(x_k, y_k, z_k, t)\,\mathrm{d}x_k\,\mathrm{d}y_k\,\mathrm{d}z_k = \mathrm{d}x_k\,\mathrm{d}y_k\,\mathrm{d}z_k \int \psi^*\psi\,\mathrm{d}\Omega_k \ . \tag{14.8}$$

In a similar way, the quantity

$$\begin{aligned} &w(x_k, y_k, z_k, x_j, y_j, z_j, t)\,\mathrm{d}x_k\,\mathrm{d}y_k\,\mathrm{d}z_k\,\mathrm{d}x_j\,\mathrm{d}y_j\,\mathrm{d}z_j \\ &= \mathrm{d}x_k\,\mathrm{d}y_k\,\mathrm{d}z_k\,\mathrm{d}x_j\,\mathrm{d}y_j\,\mathrm{d}z_j \int \psi^*\psi\,\mathrm{d}\Omega_{kj} \end{aligned} \tag{14.9}$$

is the probability that the kth particle lies at the point x_k, y_k, z_k and the jth one, simultaneously, at the point x_j, y_j, z_j. If we know the wave equation ψ expressed in configuration space, we thus can determine the probability of a given configuration (14.6) of the system, the probability of the position of any given particle (14.8) and, finally, the probability of the position of any given pair of particles (14.9) etc. In the same manner, the probabilities for the value of an arbitrary quantity can be calculated according to the general formulae of quantum mechanics by expanding ψ in terms of eigenfunctions of any operator of interest to us.

We assume that the wave function $\psi(x_1, \dots, z_N, t)$, like the wave function for one particle, satisfies the Schrödinger equation

$$\mathrm{i}\hbar\frac{\partial\psi}{\partial t} = \hat{H}\psi \ , \tag{14.10}$$

where $\hat{H}$ is the Hamiltonian (14.3) of the particle system. As stated earlier in (14.1), in analogy to the classical Hamiltonian for a system of N particles with masses $m_i, \dots, m_k, \dots, m_N$,

$$\hat{H} = \sum_{i=1}^{N}\left(\frac{\hat{p}_i}{2m_i} + V_i(x_i, y_i, z_i, t)\right) + \sum_{i\neq k=1}^{N} V_{ik}(x_i, y_i, z_i, x_k, y_k, z_k) \ ,$$

where $V_i(x_i, y_i, z_i, t)$ – as just mentioned – is the potential energy of particle i in the external field and $V_{ik}(x_i, \dots, z_k)$ is the interaction energy between the particles i and k, the Hamiltonian takes the following form:

$$\begin{aligned}\hat{H} = &\sum_{i=1}^{N}\left(-\frac{\hbar^2}{2m_i}\nabla_i^2 + V_i(x_i, y_i, z_i, t)\right)\\ &+ \sum_{i\neq k=1}^{N} V_{ik}(x_i, y_i, z_i, x_k, y_k, z_k) \ ,\end{aligned} \tag{14.11}$$

whereby

$$\nabla_i^2 = \frac{\partial^2}{\partial x_i^2} + \frac{\partial^2}{\partial y_i^2} + \frac{\partial^2}{\partial z_i^2}$$

is the Laplace operator acting on the ith particle. The Hamiltonian operator can also be written down in the presence of a magnetic field and spin. It is equal to the sum of the Hamiltonians of the single particles plus the terms which determine the mutual interaction.

From (14.10) we can get the equation of continuity for the probability w in configuration space. To find it, we multiply (14.10) by ψ^* and subtract the corresponding complex-conjugated equation. Taking into account the structure of the Hamiltonian (14.11), we get

$$\mathrm{i}\hbar\frac{\partial}{\partial t}(\psi^*\psi) = -\frac{\hbar^2}{2}\sum_{i=1}^{N}\frac{1}{m_i}(\psi^*\nabla_i^2\psi - \psi\nabla_i^2\psi^*) \ .$$

Setting

$$j_i = \frac{\mathrm{i}\hbar}{2m_i}(\psi\nabla_i\psi^* - \psi^*\nabla_i\psi) \ , \tag{14.12}$$

where ∇_i is the operator with components $\nabla_i = (\partial/\partial x_i, \partial/\partial y_i, \partial/\partial z_i)$, we can thus write (14.12) as

$$\frac{\partial w(\boldsymbol{r}_1, \ldots, \boldsymbol{r}_i, \ldots, \boldsymbol{r}_N, t)}{\partial t} + \sum_{i=1}^{N} \mathrm{div}_i\, \boldsymbol{j}_i(\boldsymbol{r}_1, \ldots, \boldsymbol{r}_i, \ldots, \boldsymbol{r}_N, t) = 0 \ . \tag{14.13}$$

This equation shows that the change in the configuration probability w is determined by the current of that probability. Hence, $\boldsymbol{j}_i$ is a function of the coordinates of all particles (and of time) and represents the current density caused by the motion of the particle i if the coordinates of all other $(n-1)$ particles are fixed. To obtain the current density of the ith particle with the other particles in arbitrary positions, (14.12) has to be integrated over all coordinates, except those of the particle i, i.e.

$$\boldsymbol{J}_i(x_i, y_i, z_i, t) = \int \boldsymbol{j}_i(x_1, \ldots, x_i, y_i, z_i, \ldots, z_N, t)\, \mathrm{d}\Omega_i \ . \tag{14.14}$$

This current density also satisfies the equation of continuity, but now in three-dimensional space, i.e. if we integrate (14.13) over $\mathrm{d}\Omega_i$, we get

$$\begin{aligned}\int \frac{\partial}{\partial t} w(x_1, \ldots, z_N, t)\, \mathrm{d}\Omega_i &= \frac{\partial}{\partial t}\int w(x_1, \ldots, z_N, t)\, \mathrm{d}\Omega_i \\ &= \frac{\partial}{\partial t} W(x_i, y_i, z_i, t) \ .\end{aligned}$$

Moreover,

$$\sum_{i'=1}^{N} \int \mathrm{div}_{i'}\, \boldsymbol{j}\, \mathrm{d}\Omega_i = \int \mathrm{div}_i\, \boldsymbol{j}_i\, \mathrm{d}\Omega_i + \sum_{i' \neq i}^{N} \mathrm{div}_{i'}\, \boldsymbol{j}_{i'}\, \mathrm{d}\Omega_i \ .$$

The volume element $\mathrm{d}\Omega_k$ [see (14.7)] contains the coordinates of all particles with the exception of particle k. The integrals of the form $\int \mathrm{div}_{i'}\, \boldsymbol{j}_{i'}\, \mathrm{d}\Omega_i$ can be transformed into surface integrals and they are, if ψ vanishes at infinity, equal to zero. In the integral $\int \mathrm{div}_i\, \boldsymbol{j}_i\, \mathrm{d}\Omega_i$, we are differentiating and integrating with respect to different variables. Therefore we have

$$\int \mathrm{div}_i\, \boldsymbol{j}_i\, \mathrm{d}\Omega_i = \mathrm{div}_i \int \boldsymbol{j}_i\, \mathrm{d}\Omega_i = \mathrm{div}_i\, \boldsymbol{J}_i(x_i, y_i, z_i, t) \ ,$$

where (14.14) has been used. We thus obtain the continuity equation for each individual particle:

$$\frac{\partial W(x_i, y_i, z_i, t)}{\partial t} + \mathrm{div}\, \boldsymbol{J}_i(x_i, y_i, z_i, t) = 0 \tag{14.15}$$

in three-dimensional space (x_i, y_i, z_i).

14.1 The Conservation of the Total Momentum of a Particle System

In classical mechanics, only the total momentum of a particle system under the influence of internal forces remains constant. Thereby the centre of mass moves in a straight line with constant velocity according to Newton's law. But if there are external forces, then the variation of the total momentum within a time unit is equal to the sum of all forces acting on the particles of the system. We will show that these principles of classical mechanics also retain their validity in the domain of quantum phenomena. For this purpose we assume an operator of total momentum for all the particles of the system. Naturally, by operator of total momentum $\hat{\boldsymbol{p}}$ of the entire particle system we mean the sum over the individual momentum operators $\hat{\boldsymbol{p}}_k$ of all particles $k = 1, 2, \dots, N$:

$$\hat{\boldsymbol{p}} = \sum_{k=1}^{N} \hat{\boldsymbol{p}}_k = -\mathrm{i}\hbar \sum_{k=1}^{N} \boldsymbol{\nabla}_k \ . \tag{14.16}$$

Let us calculate the time derivative of the momentum operator $\hat{\boldsymbol{p}}$. According to the general formula of quantum mechanics (see Chap. 8), it is

$$\frac{\mathrm{d}\hat{\boldsymbol{p}}}{\mathrm{d}t} = \frac{\mathrm{i}}{\hbar}(\hat{H}\hat{\boldsymbol{p}} - \hat{\boldsymbol{p}}\hat{H}) \ . \tag{14.17}$$

Inserting $\hat{H}$ from (14.11) and noting that $\hat{\boldsymbol{p}}$ commutes with the operator of the kinetic energy

$$\hat{T} = -\frac{\hbar^2}{2}\sum_{k=1}^{N}\frac{1}{m_k}\boldsymbol{\nabla}_k^2 \ ,$$

we thus get

$$\frac{\mathrm{d}\hat{\boldsymbol{p}}}{\mathrm{d}t} = \left[\left(\sum_{k=1}^{N} V_k + \sum_{k\neq j=1}^{N} V_{kj}\right)\left(\sum_{i=1}^{N}\boldsymbol{\nabla}_i\right) - \left(\sum_{i=1}^{N}\boldsymbol{\nabla}_i\right)\left(\sum_{k=1}^{N} V_k + \sum_{k\neq j=1}^{N} V_{kj}\right)\right] \ . \tag{14.18}$$

Furthermore,

$$V_k\left(\sum_{i=1}^{N}\boldsymbol{\nabla}_i\right) - \left(\sum_{i=1}^{N}\boldsymbol{\nabla}_i\right)V_k = -(\boldsymbol{\nabla}_k V_k(x_k, y_k, z_k)) \ , \tag{14.19}$$

because $V_k(\boldsymbol{r}_k)$ depends only on the coordinates $\boldsymbol{r}_k$ of particle k.

Last of all, we calculate the commutation of the operator $\sum_{i=1}^{N}\nabla_i$ with the interaction energy of the particles $\sum_{k\neq j}V_{kj}$. Thereby we assume that the forces between the particles depend only on the distances between the particles $\boldsymbol{r}_{ki}$, so that $V_{kj}=V_{kj}(r_{kj})$. Then only those operators ∇_k of the sum $\sum_{i=1}^{N}\nabla_i$ act on V_{kj}, for which $i=k$ or $i=j$; i.e. the pair $\nabla_k+\nabla_j$ acts on V_{kj}.

Therefore we only examine

$$V_{kj}(\nabla_k+\nabla_j)-(\nabla_k+\nabla_j)V_{kj}=-(\nabla_k V_{kj})-(\nabla_j V_{kj}) \ . \tag{14.20}$$

But now

$$(\nabla_k V_{kj})=\frac{\mathrm{d}V_{kj}}{\mathrm{d}r_{kj}}\nabla_k r_{kj}=\frac{\mathrm{d}V_{kj}}{\mathrm{d}r_{kj}}\frac{\boldsymbol{r}_{kj}}{r_{kj}} \ , \quad (\nabla_j V_{kj})=\frac{\mathrm{d}V_{kj}}{\mathrm{d}r_{kj}}\nabla_j r_{kj}=-\frac{\mathrm{d}V_{kj}}{\mathrm{d}r_{kj}}\frac{\boldsymbol{r}_{kj}}{r_{kj}} \ .$$

Consequently, we have

$$(\nabla_k V_{kj})+(\nabla_j V_{kj})=0 \ . \tag{14.21}$$

This is simply Newton's law, according to which *action* = – *reaction*. From this it follows that the commutation of the operators (14.19) is identical to zero. We thus get

$$\frac{\mathrm{d}\hat{\boldsymbol{p}}}{\mathrm{d}t}=-\sum_{k=1}^{N}(\nabla_k V_k(x_k,y_k,z_k,t)) \ , \tag{14.22}$$

i.e. the time derivative of the total momentum is equal to the operator of the resulting force, which acts on the system by external fields. This law is analogous to the classical law of momentum conservation. The only difference lies in the fact that in quantum mechanics it is not formulated for the actual mechanical quantities but for the operators representing these quantities, and therefore for the mean values of these quantities (see Ehrenfest's theorem in Chap. 8). If there are no external forces ($V_k=0$), it follows from (14.21) that

$$\frac{\mathrm{d}\hat{\boldsymbol{p}}}{\mathrm{d}t}=0 \ . \tag{14.23}$$

Hence, the total momentum of a system of particles, interacting mutually, is conserved in the absence of external forces (14.23).

We recall that the operator equation (14.23) means (1) the mean value of the total momentum does not change with time, (2) the probabilities $w(\boldsymbol{p})$ of finding a certain value of $\boldsymbol{p}$ remain unchanged too.

14.2 Centre-of-Mass Motion of a System of Particles in Quantum Mechanics

In the following, we show that the centre-of-mass motion of a system of particles does not depend on the relative motion of its constituents. This fact is well known in classical mechanics and is also valid in quantum mechanics.

We consider the Hamiltonian $\hat{H}$, which takes into account only the influence of the inner forces (two-body forces $V_{kj}(r_{kj})$):

$$\hat{H} = -\frac{\hbar^2}{2}\hat{D} + W \ , \tag{14.24}$$

where

$$\hat{D}\sum_{k=1}^{N}\frac{1}{m_k}\nabla_k^2 \ , \quad W = \sum_{\substack{j,k=1\\ j\neq k}}^{N} V_{kj}(r_{kj}) \ . \tag{14.25}$$

We express the Hamiltonian in terms of an adequate coordinate system, consisting of the centre-of-mass coordinates X, Y, Z and the $3N-3$ relative coordinates. The *Jacobi coordinates*, already introduced in Example 9.6, suit this purpose. As we recall, they are defined by

$$\begin{aligned}
\xi_1 &= \frac{m_1 x_1}{m_1} - x_2 \equiv x_1 - x_2 \ , \\
\xi_2 &= \frac{m_1 x_1 + m_2 x_2}{m_1 + m_2} - x_3 \ , \\
&\vdots \\
\xi_j &= \frac{\sum_{k=1}^{j} m_k x_k}{\sum_{k=1}^{j} m_k} - x_{j+1} \ , \\
&\vdots \\
\xi_N &= \frac{\sum_{k=1}^{N} m_k x_k}{M} \equiv X \ ,
\end{aligned} \tag{14.26}$$

where $M = \sum_{k=1}^{N} m_k$ denotes the total mass of the system. Similar expressions can be obtained for the y and z axes:

$$\eta_j = \frac{\sum_{k=1}^{j} m_k y_k}{\sum_{k=1}^{j} m_k} - y_{j+1} \ , \quad \eta_N \equiv Y \ ; \tag{14.27}$$

$$\zeta_j = \frac{\sum_{k=1}^{j} m_k z_k}{\sum_{k=1}^{j} m_k} - z_{j+1} \ , \quad \zeta_N \equiv Z \ . \tag{14.28}$$

These are generalizations of the relations between the centre of mass and relative coordinates of the two-body system. Important is the principle of construction: the Jacobi vector $\boldsymbol{\xi}_j = \{\xi_j, \eta_j, \zeta_j\}$ is the vector from the $(j+1)$th particle to the centre of mass of the first j particles. Figure 14.1 illustrates the situation.

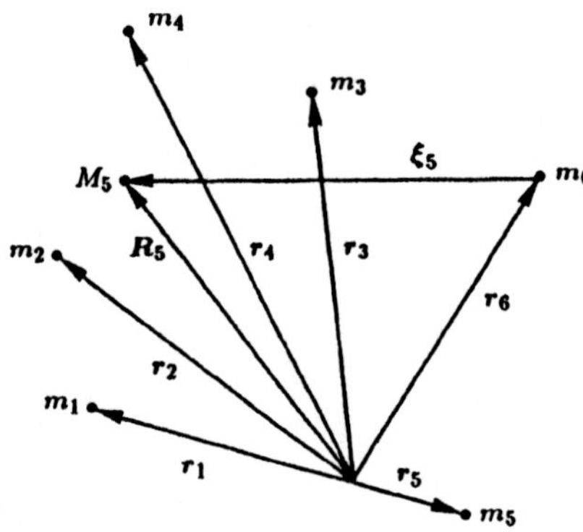

Fig. 14.1. The Jacobi coordinates ξ_j in the case of the vector ξ_5. It points from the centre of mass $\boldsymbol{R}_5$ of the first 5 particles to the position vector $\boldsymbol{r}_6$ of the 6th particle $(M_5 = m_1 + m_2 + m_3 + m_4 + m_5)$

For the kinetic-energy operator (see Example 9.6), we have

$$\hat{D} = \frac{1}{M}\nabla^2 + \sum_{j=1}^{N-1} \frac{1}{\mu_j}\nabla_j^2 \,, \quad \text{where} \tag{14.29}$$

$$\nabla^2 = \frac{\partial^2}{\partial \xi_N^2} + \frac{\partial^2}{\partial \eta_N^2} + \frac{\partial^2}{\partial \zeta_N^2} = \frac{\partial^2}{\partial X^2} + \frac{\partial^2}{\partial Y^2} + \frac{\partial^2}{\partial Z^2} \tag{14.30}$$

denotes the Laplace operator of the centre of mass of all particles, and

$$\nabla_j^2 = \frac{\partial^2}{\partial \xi_j^2} + \frac{\partial^2}{\partial \eta_j^2} + \frac{\partial^2}{\partial \zeta_j^2} \tag{14.31}$$

the Laplace operator of the Jacobi coordinates $\boldsymbol{\xi}_j = \{\xi_j, \eta_j, \zeta_j\}$. The reduced mass μ_j is given by

$$\frac{1}{\mu_j} = \frac{1}{\sum_{k=1}^{j} m_k} + \frac{1}{m_{j+1}} \,, \tag{14.32}$$

where $m_1, \ldots, m_N$ are the masses of the N particles.

The Hamiltonian of (14.24) can be rewritten in the form

$$\begin{aligned} \hat{H} = &-\frac{\hbar^2}{2M}\nabla^2 \\ &- \sum_{j=1}^{N-1} \frac{\hbar^2}{2\mu_j}\nabla_j^2 + W(\xi_1, \ldots, \xi_{N-1}, \eta_1, \ldots, \eta_{N-1}, \zeta_1, \ldots, \zeta_{N-1}) \,. \end{aligned} \tag{14.33}$$

Using (14.30), it follows that

$$\hat{T}_s \equiv -\frac{\hbar^2}{2M}\nabla^2 = -\frac{\hbar^2}{2M}\left(\frac{\partial^2}{\partial X^2}+\frac{\partial^2}{\partial Y^2}+\frac{\partial^2}{\partial Z^2}\right) \tag{14.34}$$

represents the *kinetic-energy operator of the centre of mass* of all particles. The *kinetic-energy operator of the relative (inner) particle motion* is given by

$$\hat{T}_R \equiv -\sum_{j=1}^{N-1}\frac{\hbar^2}{2\mu_j}\nabla_j^2 \ . \tag{14.35}$$

Looking at (14.33), we see that the interaction energy does not depend on the centre-of-mass coordinates $\boldsymbol{\xi}_N = \{\xi_N, \eta_N, \zeta_N\} = \{X, Y, Z\}$. It depends, according to (14.25), on the relative distances between the particles only. But the relative coordinates can be expressed by the first $(N-1)$ Jacobi coordinates $\boldsymbol{\xi}_1, \dots, \boldsymbol{\xi}_{N-1}$, which follows immediately from the relations (14.26). If we transform $\boldsymbol{\xi}_1, \dots, \boldsymbol{\xi}_{N-1}$ (i.e. $\xi_1, \dots, \xi_{N-1}, \eta_1, \dots, \eta_{N-1}, \zeta_1, \dots, \zeta_{N-1}$) by a linear transformation to arbitrary relative coordinates $q_1, q_2, \dots, q_{3N-3}$, the operator $\hat{T}_s$ remains unchanged. Therefore we can generalize (14.33) to

$$\hat{H} = -\frac{\hbar^2}{2M}\nabla^2 + \hat{H}_R(q_1, q_2, \dots, q_{3N-3}) \ . \tag{14.36}$$

The Hamiltonian of the relative motion $\hat{H}_R$ does not depend on the centre-of-mass coordinates; thus the wave function of the system separates into a relative part and a centre-of-mass part. In the next step we introduce the operator of the total momentum,

$$\hat{P}_X = -i\hbar\frac{\partial}{\partial X} \ , \quad \hat{P}_Y = -i\hbar\frac{\partial}{\partial Y} \ , \quad \hat{P}_Z = -i\hbar\frac{\partial}{\partial Z} \ , \tag{14.37}$$

so that we can write the kinetic energy of the centre of mass in the form

$$\begin{aligned}\hat{T}_s &= \frac{\hat{\boldsymbol{P}}^2}{2M} = \frac{\hat{P}_X^2+\hat{P}_Y^2+\hat{P}_Z^2}{2M}\\ &= -\frac{\hbar^2}{2M}\left(\frac{\partial^2}{\partial X^2}+\frac{\partial^2}{\partial Y^2}+\frac{\partial^2}{\partial Z^2}\right) = -\frac{\hbar^2}{2M}\Delta \ .\end{aligned} \tag{14.38}$$

The wave function of the system separates according to (14.36), and we write it as a product of the centre-of-mass part $\phi(X, Y, Z, t)$, with the centre-of-mass coordinates X, Y, Z, and the relative part $\psi(q_1, q_2, \dots, q_{3N-3})$. Thus we have

$$\begin{aligned}&\Psi(X, Y, Z, q_1, q_2, \dots, q_{3N-3}, t)\\ &\quad = \phi(X, Y, Z, t)\psi(q_1, \dots, q_{3N-3}, t) \ .\end{aligned} \tag{14.39}$$

If we insert (14.39) into the Schrödinger equation, we obtain

$$
\begin{aligned}
\mathrm{i}\hbar\frac{\partial}{\partial t}\Psi &= \mathrm{i}\hbar\frac{\partial}{\partial t}(\phi(X,Y,Z,t)\psi(q_1,\dots,q_{3N-3},t)) \\
&= \hat{H}\Psi \\
&= \left(-\frac{\hbar^2}{2M}\nabla^2+\hat{H}_R(q_1,\dots,q_{3N-3})\right) \\
&\quad\times(\phi(X,Y,Z)\psi(q_1,\dots,q_{3N-3},t)) \\
\Leftrightarrow \mathrm{i}\hbar\frac{\partial\phi}{\partial t}\psi+\mathrm{i}\hbar\phi\frac{\partial\psi}{\partial t} &= -\psi\frac{\hbar^2}{2M}\nabla^2\phi+\phi\hat{H}_\mathrm{R}\psi\ .
\end{aligned}
\tag{14.40}
$$

Dividing (14.40) by $\phi\psi$ (for $\phi\psi\neq 0$) gives

$$
\mathrm{i}\hbar\frac{1}{\phi}\frac{\partial\phi}{\partial t}+\mathrm{i}\hbar\frac{1}{\psi}\frac{\partial\psi}{\partial t}=-\frac{1}{\phi}\frac{\hbar^2}{2M}\nabla^2\phi+\frac{1}{\psi}\hat{H}_\mathrm{R}\psi\ .
$$

After reordering, we get

$$
\left(\mathrm{i}\hbar\frac{\partial\phi}{\partial t}+\frac{\hbar^2}{2M}\nabla^2\phi\right)\frac{1}{\phi}=\left(-\mathrm{i}\hbar\frac{\partial\psi}{\partial t}+\hat{H}_\mathrm{R}\psi\right)\frac{1}{\psi}\stackrel{!}{=}E\ .
$$

This equation is fulfilled if both sides are equal to a constant E. Then we have for $\phi(X,Y,Z,t)$ and $\psi(q_1,q_2,\dots,q_{3N-3},t)$

$$
\mathrm{i}\hbar\frac{\partial\phi}{\partial t}=-\frac{\hbar^2}{2M}\nabla^2\phi+E\phi \quad \text{and} \tag{14.41}
$$

$$
\mathrm{i}\hbar\frac{\partial\psi}{\partial t}=\hat{H}_\mathrm{R}\psi-E\psi\ . \tag{14.42}
$$

The first equation describes the centre-of-mass motion of a system of particles with total mass M. If no external forces act on the system, the centre of mass moves like a free particle of mass M. The simplest special solution is given by a plane wave (a de Broglie wave), i.e.

$$
\begin{aligned}
\phi(X,Y,Z,t) &= (2\pi\hbar)^{-3/2}\exp\left[\frac{\mathrm{i}}{\hbar}(E_s t-P_X X-P_Y Y-P_Z Z)\right] \\
&= (2\pi\hbar)^{-3/2}\exp\left[-\frac{\mathrm{i}}{\hbar}(\boldsymbol{P}\cdot\boldsymbol{X}-E_s t)\right]\ .
\end{aligned}
\tag{14.43}
$$

By inserting ϕ in the Schrödinger equation (14.41), we can identify the components of $\boldsymbol{P}=(P_X,P_Y,P_Z)$ as the eigenvalues of the total-momentum operator. For the eigenvalue of the kinetic energy of the centre-of-mass motion E_s, it follows that

$$
E_\mathrm{s}=\frac{1}{2M}(P_X^2+P_Y^2+P_Z^2)+E\ . \tag{14.44}
$$

The additive constant E is unimportant and can be chosen equal to zero ($E=0$).

The wavelength of the de Broglie wave is given by

$$\lambda = \frac{h}{P} = \frac{h}{MV} \ , \quad P = \sqrt{P_X^2 + P_Y^2 + P_Z^2} \ , \tag{14.45}$$

V being the velocity of the centre of mass.

We can now deduce from (14.43)–(14.45) that de Broglie waves (14.44) are not oscillations connected with the internal structure of the particle system, but represent the general quantum-mechanical motion of free particles (or, in our case, a centre-of-mass motion, i.e. the motion of the system as a whole) without external forces.

The essential and interesting aspects of the many-body problem concern the inner degrees of freedom, described by (14.43). The centre-of-mass motion is, as in classical mechanics, a rather trivial aspect. It is only important if all the particles of the system – which, as a result of inner binding forces, are confined relative to each other – are deflected in an external field, or interact with other complex systems. In the latter case we speak of *cluster structure* and mean the splitup of an N-body system into various substructures.

Cluster structure plays an important role in the breakup of a nucleus with A nucleons into two fragments with nucleon numbers A_1 and A_2 $(A_1 + A_2 = A)$ or into three or more fragments. This is called *two-body* (*binary*) or *three-body* (*ternary*) etc. *fission*. If one of the fragments is very big and the other one quite small (e.g. $A \to (A-4)+4$ or $A \to (A-12)+12$ and $A \approx 220$), one speaks of *radioactive decay*. The most famous form of this is α decay, in which an α particle is emitted (^{4}He nucleus).

More recently, so-called *cluster radioactivity* has been discovered, in which ^{12}C nuclei, ^{16}O nuclei, ^{24}Ne nuclei, ^{32}S nuclei, etc. are emitted. It was theoretically predicted[1] and 4 years later, experimentally confirmed.[2]

Returning to our calculations, we finally get for (14.39), after separation of the centre-of-mass motion (14.43) in the general form,

$$\begin{aligned}\Psi(X, Y, Z, q_1, \dots, q_{3N-3}, t)\\ = (2\pi\hbar)^{-3/2} \exp\left[-\frac{\mathrm{i}}{\hbar}(\boldsymbol{P}\cdot\boldsymbol{X} - E_s t)\right] \psi(q_1, \dots, q_{3N-3}, t) \ .\end{aligned} \tag{14.46}$$

14.3 Conservation of Total Angular Momentum in a Quantum-Mechanical Many-Particle System

Again we consider a system of N particles and denote the components of the orbital angular momentum of particle k in terms of Cartesian coordinates by $\hat{\boldsymbol{l}}^k = (\hat{l}_x^k, \hat{l}_y^k, \hat{l}_z^k)$. The position vector of the kth particle is $\boldsymbol{x}_k = (x_k, y_k, z_k)$. We then

[1] A. Sandulescu, D.N. Poenaru, W. Greiner: Sov. J. Part. Nucl. **11**, 528–541 (1980).
[2] H.J. Rose, G.A. Jones: Nature **307**, 245–247 (1984).

have

$$\hat{l}_x^k = -\mathrm{i}\hbar\left(y_k\frac{\partial}{\partial z_k} - z_k\frac{\partial}{\partial y_k}\right) ,$$
$$\hat{l}_y^k = -\mathrm{i}\hbar\left(z_k\frac{\partial}{\partial x_k} - x_k\frac{\partial}{\partial z_k}\right) ,$$
$$\hat{l}_z^k = -\mathrm{i}\hbar\left(x_k\frac{\partial}{\partial y_k} - y_k\frac{\partial}{\partial x_k}\right) . \tag{14.47}$$

The components of the operator for the total orbital angular momentum $\hat{\boldsymbol{l}} = (\hat{l}_x, \hat{l}_y, \hat{l}_z)$ of the system are defined as the sum over the individual angular momenta, i.e.

$$\hat{l}_x = \sum_{k=1}^{N} \hat{l}_x^k , \quad \hat{l}_y = \sum_{k=1}^{N} \hat{l}_y^k , \quad \hat{l}_z = \sum_{k=1}^{N} \hat{l}_z^k . \tag{14.48}$$

In the following we will prove that the derivative of the angular-momentum operator equals the operator of the torque exerted on the system. According to (8.6), the time derivative of a not explicitly time-dependent operator, e.g. $\hat{l}_x$, is

$$\frac{\mathrm{d}\hat{l}_x}{\mathrm{d}t} = \frac{\mathrm{i}}{\hbar}[\hat{H}, \hat{l}_x] . \tag{14.49}$$

The Hamiltonian of the N-particle system with masses $m_1, m_2, \ldots, m_N$ reads

$$\hat{H} = \sum_{k=1}^{N}\left(\frac{\hat{p}_k}{2m_k} + V_k(x_k, y_k, z_k, t)\right)$$
$$+ \sum_{\substack{j,k=1 \\ j\neq k}}^{N} V_{kj}(x_k, y_k, z_k, x_j, y_j, z_j) . \tag{14.50}$$

As before, V_k corresponds to the potential energy of the kth particle in an external field, and V_{kj} is the interaction energy between particles k and j. We know from Sect. 4.8 that every single component of the angular-momentum operator commutes with its square. Because the angular-momentum operators of different particles commute – they act in different coordinate spaces – they are not able to harm each other, e.g. $[\hat{l}_x^k, \hat{l}_y^j] = 0$ for any $k \neq j$. Each component $\hat{l}_i^k$ of a particle's angular-momentum operator commutes with the square $\hat{l}^2$ of the total-angular-momentum operator, i.e.

$$[\hat{l}_i^k, \hat{l}^2] = [\hat{l}_i^k, (\hat{l}^k)^2] = 0 , \quad i = 1, 2, 3 \quad \text{or} \quad x, y, z . \tag{14.51}$$

We also know that $\hat{p}_k^2$ commutes with $\hat{l}_x^k, \hat{l}_y^k, \hat{l}_z^k$, which can be verified, for example, for the x component:

$$\hat{p}_k^2 = -\hbar^2\nabla_k^2 = -\hbar^2\left(\frac{\partial^2}{\partial x_k^2} + \frac{\partial^2}{\partial y_k^2} + \frac{\partial^2}{\partial z_k^2}\right) ,$$

and therefore

$$
\begin{aligned}
[\hat{p}_k^2, \hat{l}_x^k] &= \mathrm{i}\hbar^3 \left[\frac{\partial^2}{\partial x_k^2}, y_k \frac{\partial}{\partial z_k} - z_k \frac{\partial}{\partial y_k} \right] + \mathrm{i}\hbar^3 \left[\frac{\partial^2}{\partial y_k^2} + \frac{\partial^2}{\partial z_k^2}, y_k \frac{\partial}{\partial z_k} - z_k \frac{\partial}{\partial y_k} \right] \\
&= 0 + \mathrm{i}\hbar^3 \left[\frac{\partial}{\partial y_k} \left(\frac{\partial}{\partial z_k} + y_k \frac{\partial}{\partial z_k} \frac{\partial}{\partial y_k} - z_k \frac{\partial^2}{\partial y_k^2} \right) - \left(y_k \frac{\partial}{\partial z_k} - z_k \frac{\partial}{\partial y_k} \right) \frac{\partial^2}{\partial y_k^2} \right. \\
&\qquad + \frac{\partial}{\partial z_k} \left(y_k \frac{\partial^2}{\partial z_k^2} - \frac{\partial}{\partial y_k} - z_k \frac{\partial}{\partial y_k} \frac{\partial}{\partial z_k} \right) \\
&\qquad \left. - \left(y_k \frac{\partial}{\partial z_k} - z_k \frac{\partial}{\partial y_k} \right) \frac{\partial^2}{\partial z_k^2} \right] \\
&= \mathrm{i}\hbar^3 \left(\frac{\partial}{\partial y_k} \frac{\partial}{\partial z_k} + \frac{\partial}{\partial z_k} \frac{\partial}{\partial y_k} + y_k \frac{\partial}{\partial z_k} \frac{\partial^2}{\partial y_k^2} - z_k \frac{\partial^3}{\partial y_k^3} \right. \\
&\qquad - y_k \frac{\partial}{\partial z_k} \frac{\partial^2}{\partial y_k^2} + z_k \frac{\partial^3}{\partial y_k^3} + y_k \frac{\partial^3}{\partial z_k^3} - \frac{\partial}{\partial z_k} \frac{\partial}{\partial y_k} - \frac{\partial}{\partial y_k} \frac{\partial}{\partial z_k} \\
&\qquad \left. - z_k \frac{\partial}{\partial y_k} \frac{\partial^2}{\partial z_k^2} - y_k \frac{\partial^3}{\partial z_k^3} + z_k \frac{\partial}{\partial y_k} \frac{\partial^2}{\partial z_k^2} \right) = 0 ,
\end{aligned}
$$

i.e.

$$
[\nabla_k^2, \hat{l}_x^k] = 0 . \tag{14.52}
$$

Let us now split the kinetic-energy operator in (14.50) into a translational part $\hat{T}_{r_k}$ along the radius vector $\boldsymbol{r}_k$ and a rotational part (see Fig. 14.2 and Sect. 4.9):

$$
-\frac{\hbar^2}{2m_k} \nabla_k^2 = \hat{T}_{r_k} + \frac{(\hat{\boldsymbol{l}}_k)^2}{2m_k r_k^2} . \tag{14.53}
$$

Because each component of the angular-momentum operator of a particle commutes with $(\hat{\boldsymbol{l}}_k)^2$ and with ∇_k^2, it also commutes with $\hat{T}_{r_k}$, according to (14.53):

$$
[\hat{T}_{r_k}, \hat{l}_i^k] = 0 , \quad (i = x, y, z) . \tag{14.54}
$$

In order to evaluate (14.49), we need the commutators of $[V_k, \hat{l}_i^k]$ and $[V_{kj}, \hat{l}_i^k]$ for $(i = x, y, z)$:

$$
\begin{aligned}
V_k \hat{l}_x^k - \hat{l}_x^k V_k &= -\mathrm{i}\hbar \left[V_k \left(y_k \frac{\partial}{\partial z_k} - z_k \frac{\partial}{\partial y_k} \right) - \left(y_k \frac{\partial}{\partial z_k} - z_k \frac{\partial}{\partial y_k} \right) V_k \right] \\
&= \mathrm{i}\hbar \left(y_k \frac{\partial V_k}{\partial z_k} - z_k \frac{\partial V_k}{\partial y_k} \right) ,
\end{aligned} \tag{14.55}
$$

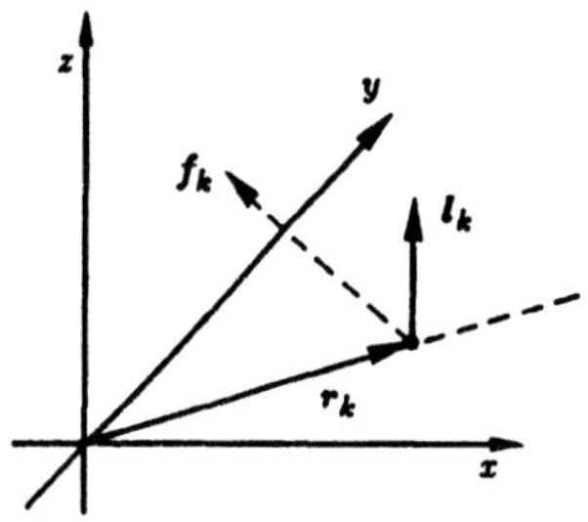

Fig. 14.2. The position vector $\boldsymbol{r}_k$, its angular momentum $\boldsymbol{l}_k$ and the external force $\boldsymbol{f}_k = -\nabla_k V_k(\boldsymbol{r}_k)$ acting on the kth particle

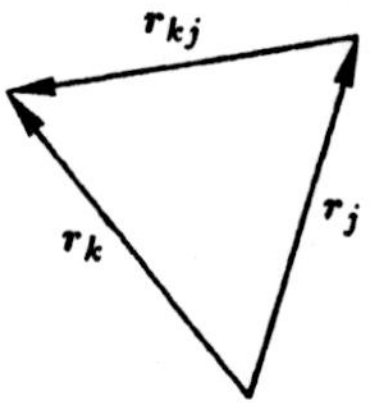

Fig. 14.3. The relative vector $\boldsymbol{r}_{kj}$ between the kth and the jth particle is defined as $\boldsymbol{r}_{kj} = \boldsymbol{r}_k - \boldsymbol{r}_j$

and, analogously, we calculate:

$$[V_{kj}, \hat{l}_x^k] = \mathrm{i}\hbar \left(y_k \frac{\partial V_{kj}}{\partial z_k} - z_k \frac{\partial V_{kj}}{\partial y_k} \right) . \tag{14.56}$$

We express the derivatives in (14.56) in terms of the relative coordinate $\boldsymbol{r}_{kj}$ and have (see Fig. 14.3)

$$\boldsymbol{r}_{kj} = \boldsymbol{r}_k - \boldsymbol{r}_j ,$$

$$\frac{\partial}{\partial z_k} = \frac{\partial r_{kj}}{\partial z_k} \frac{\partial}{\partial r_{kj}} = \frac{\partial}{\partial z_k} \left(\sqrt{(x_k - x_j)^2 + (y_k - y_j)^2 + (z_k - z_j)^2} \right) \frac{\partial}{\partial r_{kj}}$$

$$= \frac{z_k - z_j}{r_{kj}} \frac{\partial}{\partial r_{kj}} .$$

Similarly, we obtain

$$\frac{\partial}{\partial y_k} = \frac{y_k - y_j}{r_{kj}} \frac{\partial}{\partial r_{kj}} , \quad \frac{\partial}{\partial z_k} = \frac{z_k - z_j}{r_{kj}} \frac{\partial}{\partial r_{kj}} .$$

According to the above relations, (14.56) reads:

$$[V_{kj}, \hat{l}_x^k] = \mathrm{i}\hbar \frac{\partial V_{kj}}{\partial r_{kj}} \left(y_k \frac{z_k - z_j}{r_{kj}} - z_k \frac{y_k - y_j}{r_{kj}} \right)$$

$$= \mathrm{i}\hbar (z_k y_j - z_j y_k) \frac{\partial V_{kj}}{\partial r_{kj}} \frac{1}{r_{kj}} . \tag{14.57}$$

Using the relations (14.51)–(14.53), (14.56) and (14.57), the time derivative of a component of the total angular momentum can be evaluated. According to (14.48), we obtain for the x component

$$\frac{\mathrm{d}\hat{l}_x}{\mathrm{d}t} = -\sum_{k=1}^{N} \left(y_k \frac{\partial V_k}{\partial z_k} - z_k \frac{\partial V_k}{\partial y_k} \right) - \sum_{\substack{j,k=1 \\ j \neq k}}^{N} (z_k y_j - z_j y_k) \frac{\partial V_{kj}}{\partial r_{kj}} \frac{1}{r_{kj}} . \tag{14.58}$$

The second part in (14.58) vanishes, since the terms of the sum change their sign by changing their indices, and thus cancel each other. The result is

$$\frac{\mathrm{d}\hat{l}_x}{\mathrm{d}t} = -\sum_{k=1}^{N} \left(y_k \frac{\partial V_k}{\partial z_k} - z_k \frac{\partial V_k}{\partial y_k} \right) . \tag{14.59}$$

Similarly, we obtain for the other two components

$$\frac{\mathrm{d}\hat{l}_y}{\mathrm{d}t} = -\sum_{k=1}^{N} \left(z_k \frac{\partial V_k}{\partial x_k} - x_k \frac{\partial V_k}{\partial z_k} \right) , \tag{14.60}$$

$$\frac{\mathrm{d}\hat{l}_z}{\mathrm{d}t} = -\sum_{k=1}^{N} \left(x_k \frac{\partial V_k}{\partial y_k} - y_k \frac{\partial V_k}{\partial x_k} \right) . \tag{14.61}$$

Thus we have proved the theorem, already known from mechanics, that the time derivative of the orbital angular momentum equals the torque of external forces acting upon the system (the *torque of external forces*). If no external forces are present, or if their torque vanishes, the total angular momentum is conserved:

$$\frac{d\hat{l}_x}{dt} = \frac{d\hat{l}_y}{dt} = \frac{d\hat{l}_z}{dt} = 0 \ . \tag{14.62}$$

Thus for the case of vanishing external forces, the averages $\bar{l}_x$, $\bar{l}_y$, and $\bar{l}_z$ are constant, as are the probabilities $w(l_x)$, $w(l_y)$ and $w(l_z)$ of finding a fixed value for an angular-momentum component.

Including the internal spin $\boldsymbol{s}$ of a particle (see Chap. 12), the relations for the total angular momentum can be straightforwardly modified as follows:

$$\hat{j}_x = \sum_{k=1}^{N}(\hat{l}_x^k + \hat{s}_x^k) \ , \quad \hat{j}_y = \sum_{k=1}^{N}(\hat{l}_y^k + \hat{s}_y^k) \ , \quad \hat{j}_z = \sum_{k=1}^{N}(\hat{l}_z^k + \hat{s}_z^k) \ , \tag{14.63}$$

where $\hat{s}_x^k$, $\hat{s}_y^k$, $\hat{s}_z^k$ denote the projections of the spin of the kth particle on the corresponding coordinate axis. The spin operators are represented by 2×2 Pauli matrices. If no external electromagnetic fields are present, i.e. if no forces are acting on the spin, the conservation law for the angular momentum remains unrestrictedly valid, since in this case the Hamiltonian commutes with each component of $\hat{\boldsymbol{s}}^k$.

The commutation relations of the total angular momentum of a system of particles correspond to those of the orbital angular momentum, because the operators $\hat{l}_x^k$, $\hat{l}_y^k$, $\hat{l}_z^k$, $\hat{s}_x^k$, $\hat{s}_y^k$ and $\hat{s}_z^k$ commute for different particle indices, and, furthermore, $\hat{l}_i^k$ and $\hat{s}_j^k$ commute with each other, because these operators act in different spaces (coordinate space – spin space). Hence,

$$[\hat{j}_x, \hat{j}_y] = \mathrm{i}\hbar\hat{j}_z \ , \quad [\hat{j}_y, \hat{j}_z] = \mathrm{i}\hbar\hat{j}_x \ , \quad [\hat{j}_z, \hat{j}_x] = \mathrm{i}\hbar\hat{j}_y \ , \tag{14.64}$$

$$[\hat{j}^2, \hat{j}_x] = [\hat{j}^2, \hat{j}_y] = [\hat{j}^2, \hat{j}_z] = 0 \ . \tag{14.65}$$

The eigenvalues of $\hat{j}_z^k$ are equal to the sum of the eigenvalues of $\hat{l}_z^k + \hat{s}_z^k$. In Sect. 4.8 we found the eigenvalue of the z component of the orbital momentum to be $\tilde{m}_z^k\hbar$, with $-l^k \le \tilde{m}_z^k \le l^k$ ($l^k = 0, 1, 2, \ldots$ represents the quantum number of the orbital momentum of the kth particle). The eigenvalue of the spin is $\pm\hbar/2$ [see (12.13), (12.39) and (12.40)]; thus we have for the eigenvalues of $\hat{j}_z^k$, the values $\hbar m_z^k$, m_z^k being an integer multiple of $\frac{1}{2}$ for particles with spin $\frac{1}{2}$. For the z component of the total angular momentum, we have

$$j_z = \sum_{k=1}^{N} \hbar m_z^k = \hbar m \ , \quad m = \sum_{k=1}^{N} m_z^k \ . \tag{14.66}$$

Equation (14.66) has to be interpreted as an eigenvalue equation; the index z is omitted. To determine the eigenvalues of $\hat{j}^2$, we introduce the eigenfunctions $|jm\rangle$ of $\hat{j}^2$ and $\hat{j}_z$ with

$$\hat{j}^2\,|jm\rangle = J^2\,|jm\rangle \;, \quad \hat{j}_z\,|jm\rangle = m\,|jm\rangle \;. \tag{14.67}$$

Neither J^2 nor m nor their relation to each other are yet known. To proceed in this direction, it is customary to consider the step operators $\hat{j}_+$ and $\hat{j}_-$, which are defined as

$$\hat{j}_+ = \hat{j}_x + \mathrm{i}\hat{j}_y \;, \quad \hat{j}_- = \hat{j}_x - \mathrm{i}\hat{j}_y \;. \tag{14.68}$$

Using the commutation relations (14.64), the validity of the following relations can easily be verified.

$$\begin{aligned}[\hat{j}_+, \hat{j}_z] &= [\hat{j}_x, \hat{j}_z] + \mathrm{i}[\hat{j}_y, \hat{j}_z] = -\mathrm{i}\hbar\,\hat{j}_y + \mathrm{i}\hbar\mathrm{i}\,\hat{j}_x \\ &= -\hbar(\hat{j}_x + \mathrm{i}\hat{j}_y) = -\hbar\,\hat{j}_+ \;,\end{aligned} \tag{14.69a}$$

and, similarly,

$$[\hat{j}_-, \hat{j}_z] = \hbar\,\hat{j}_- \;. \tag{14.69b}$$

We rewrite the commutation relations (14.68) and (14.69) in matrix form, i.e. we multiply both equations from the left by a bra, and from the right, by a ket vector. We choose a basis in which $\hat{j}_z$ is diagonal: $\hat{j}_z|jm\rangle = \hbar m|jm\rangle$. We calculate $\langle jm'|\ldots|jm''\rangle$ and obtain

$$\begin{aligned}(j_+)_{m'm''}\hbar m'' - \hbar m'(j_+)_{m'm''} &= -\hbar(j_+)_{m'm''} \\ (j_-)_{m'm''}\hbar m'' - \hbar m'(j_-)_{m'm''} &= +\hbar(j_-)_{m'm''}\end{aligned} \tag{14.70}$$

or

$$\begin{aligned}(j_+)_{m'm''}(m'' - m' + 1) &= 0 \\ (j_-)_{m'm''}(m'' - m' - 1) &= 0 \;.\end{aligned} \tag{14.71}$$

Obviously, the only nonvanishing matrix elements of $\hat{j}_+$ and $\hat{j}_-$ are given by $(j_+)_{m,m-1}$ and $(j_-)_{m,m+1}$. The operator of the square of the total angular momentum can be written in terms of $\hat{j}_+\hat{j}_-$ or $\hat{j}_-\hat{j}_+$:

$$\begin{aligned}\hat{j}_+\hat{j}_- &= (\hat{j}_x + \mathrm{i}\hat{j}_y)(\hat{j}_x - \mathrm{i}\hat{j}_y) = \hat{j}_x^2 + \hat{j}_y^2 - \mathrm{i}\hat{j}_x\hat{j}_y + \mathrm{i}\hat{j}_y\hat{j}_x = \hat{j}^2 - \hat{j}_z^2 + \hbar\,\hat{j}_z \\ \hat{j}_-\hat{j}_+ &= \hat{j}^2 - \hat{j}_z^2 - \hbar\,\hat{j}_z \;.\end{aligned} \tag{14.72}$$

Completing the square yields

$$\hat{j}_+\hat{j}_- = \hat{j}^2 + \frac{\hbar^2}{4} - \left(j_z - \frac{\hbar}{2}\right)^2 , \quad \hat{j}_-\hat{j}_+ = \hat{j}^2 + \frac{\hbar^2}{4} - \left(j_z + \frac{\hbar}{2}\right)^2 \;. \tag{14.73}$$

Considering the diagonal matrix elements $\langle jm|\dots|jm\rangle$ yields

$$(j_+j_-)_{mm} = (j_+)_{m,m-1}(j_-)_{m-1,m} = J^2 + \frac{\hbar^2}{4} - \hbar^2(m-\tfrac{1}{2})^2 \ ,$$
$$(j_-j_+)_{mm} = (j_-)_{m,m+1}(j_+)_{m+1,m} = J^2 + \frac{\hbar^2}{4} - \hbar^2(m+\tfrac{1}{2})^2 \ . \qquad (14.74)$$

We assume J^2 to be a given, but still unknown, positive semidefinite quantity. In the following we denote by m' the lowest possible value of m, and by m'', the maximal value of m. From (14.74) it follows that, by making use of

$$(j_+)_{m',m'-1} = 0 = (j_-)_{m'-1,m'} \quad \text{and} \quad (j_-)_{m'',m''+1} = 0 = (j_+)_{m''+1,m''} \ ,$$

we get

$$J^2 + \frac{\hbar^2}{4} = \hbar^2(m'-\tfrac{1}{2})^2 \ , \quad J^2 + \frac{\hbar^2}{4} = \hbar^2(m''+\tfrac{1}{2})^2 \ ,$$

and therefore

$$m' = \frac{1}{2} - \sqrt{\frac{J^2}{\hbar^2} + \frac{1}{4}} \ , \quad m'' = -\frac{1}{2} + \sqrt{\frac{J^2}{\hbar^2} + \frac{1}{4}} \ . \qquad (14.75)$$

In the equation for m', we have chosen the negative value of the root in order to get the smallest possible value for m'. The difference $m - m' + 1$ is an integer, giving the number of possible z projections of the total angular momentum $\boldsymbol{j}$. Setting m'', $-m'+1 = 2j+1$ (in analogy to the orbital momentum), it follows from (14.74) that

$$2j+1 = 2\sqrt{\frac{J^2}{\hbar^2} + \frac{1}{4}} \Leftrightarrow J^2 = \hbar^2 j(j+1) \ . \qquad (14.76)$$

Since for the z projection of the total angular momentum m, both positive and negative values must be equally represented, m'' must be equal to $-m'$. Then, from $m'' - m' + 1 = 2j+1$, we get

$$|m| \le j \quad \text{with} \quad m = 0, \pm 1, \pm 2, \dots, \pm j \ , \quad \text{or}$$
$$m = \pm\frac{1}{2}, \pm\frac{3}{2}, \dots, \pm j \ . \qquad (14.77)$$

Thus the relation for the total angular momentum has been proved to be of the eigenvalue form

$$\hat{j}^2 = \hbar^2 j(j+1) \ , \qquad (14.78)$$
$$\hat{j}_z = \hbar m \ , \quad |m| \le j \ . \qquad (14.79)$$

Depending on the number of particles and on the spin, j has either integral values $0, 1, 2, 3, \dots$, or is an odd multiple of $\frac{1}{2}$, i.e. $\frac{1}{2}, \frac{3}{2}, \frac{5}{2}, \dots$. For the projections of the total angular momentum m we obtain $2j+1$ possible quantum-mechanical orientations with respect to an arbitrary axis (here, the z axis), namely $m = -j, -j+1, \dots, j-1, j$. Electrons have spin $\frac{1}{2}$; therefore, for a system consisting of an even number of electrons, j has integer values, while an odd number leads to an integer multiple of $\frac{1}{2}$.

To prove the eigenvalue equations (14.78) and (14.79) we have made use only of the commutation relations (14.64) and (14.65). Because the total angular momentum $\hat{\boldsymbol{l}}$ and the spin $\hat{\boldsymbol{s}}$ satisfy the same commutation rules, we obtain analogous relations for the eigenvalues of the corresponding operators:

$$\hat{\boldsymbol{l}} = \sum_{k=1}^{N} \hat{\boldsymbol{l}}_k \tag{14.80}$$

$$\hat{\boldsymbol{l}}^2 \Rightarrow \hbar^2 l(l+1) \ , \quad l = 0, 1, 2, \dots \ , \tag{14.81}$$

$$\hat{l}_z \Rightarrow \hbar m_l \ , \quad |m_l| \le l \ , \tag{14.82}$$

$$\hat{\boldsymbol{s}} = \sum_{k=1}^{N} \hat{\boldsymbol{s}}_k \ , \tag{14.83}$$

$$\hat{\boldsymbol{s}}^2 \Rightarrow \hbar^2 s(s+1) \ , \quad s = 0, \frac{1}{2}, 1, \frac{3}{2}, \dots \ , \tag{14.84}$$

$$\hat{s}_z \Rightarrow \hbar m_s \ , \quad |m_s| \le s \ . \tag{14.85}$$

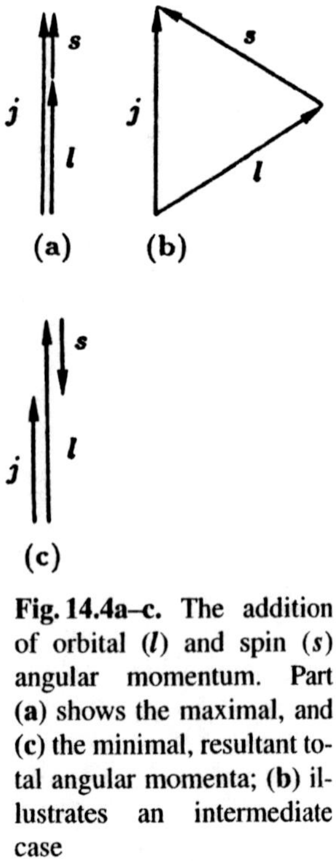

Fig. 14.4a–c. The addition of orbital ($\boldsymbol{l}$) and spin ($\boldsymbol{s}$) angular momentum. Part **(a)** shows the maximal, and **(c)** the minimal, resultant total angular momenta; **(b)** illustrates an intermediate case

For given values of the total angular momentum $\boldsymbol{l}$ and the total spin $\boldsymbol{s}$, $\boldsymbol{j}$ assumes, depending on the relative orientation of $\boldsymbol{l}$ and $\boldsymbol{s}$, all values between $|l-s|$ (*antiparallel orientation*) and $l+s$ (*parallel orientation*):

$$j = l+s, |l+s-1|, \dots, |l-s| \ . \tag{14.86}$$

This is physically reasonable and is illustrated in Fig. 14.4.[3] States with the same l and s form a group of levels called a *multiplet*, which lie close together because of the weak $\boldsymbol{l} \cdot \boldsymbol{s}$ interaction. From (14.85), it follows that for $s \le l$ a multiplet contains $2s+1$ states. In other words, (14.86) tells us that there are $2s+1$ states in a multiplet. Consequently, the specification of j, l and s is essential to characterize the energy of the entire atom. Of course, there will be additional quantum numbers like the principal quantum numbers in the case of the hydrogen atom, but also others typical for the many-body problem. In analogy to the hydrogen atom, terms with $l = 0, 1, 2, \dots$ are marked with the capital letters S, P, D, F. The lower index on the right indicates the j value; the upper index on the left gives the value of the multiplicity of the multiplet. For example, ${}^2P_{1/2}$ marks the term with $l = 1$, $j = \frac{1}{2}$, $s = \frac{1}{2}$, $(2 \times \frac{1}{2} + 1 = 2)$ and ${}^4F_{3/2}$, the term with $l = 3$,

[3] We give a formal derivation of this, using the commutation relations only, in W. Greiner, B. Müller: *Quantum Mechanics – Symmetries*, 2nd ed. (Springer, Berlin, Heidelberg 1994).

$j = \frac{3}{2}$, $s = \frac{3}{2}$, $(2 \times \frac{3}{2} + 1 = 4)$ etc. Strictly speaking, the index at the upper left is redundant, because the multiplicity, always given by $2j + 1$, is immediately deduced from the lower right index.

EXAMPLE

14.1 The Anomalous Zeeman Effect

As an illustrative example of the angular-momentum algebra, we consider level splitting in a complicated atom with several electrons in a weak homogeneous magnetic field (the anomalous Zeeman effect). The interaction of the electrons with the external magnetic field $\boldsymbol{B}$ is given by

$$W = -\hat{\boldsymbol{\mu}} \cdot \hat{\boldsymbol{B}} \ , \tag{1}$$

with the magnetic moment $\boldsymbol{\mu} = e/(2m_e c)(\hat{\boldsymbol{l}} + 2\hat{\boldsymbol{s}})$; m_e is the electron mass. The anomalous g factor of spin, i.e. $g = 2$, is included. The coordinate system is chosen in such a way that the magnetic field $\boldsymbol{B}$ and the z axis are parallel. $\boldsymbol{l}$ and $\boldsymbol{s}$ denote the total orbital angular momentum and total spin. The operator of the magnetic moment can be expressed in terms of the total orbital angular momentum ($\hat{\boldsymbol{j}} = \hat{\boldsymbol{l}} + \hat{\boldsymbol{s}}$):

$$\hat{\boldsymbol{\mu}} = \hat{G}\hat{\boldsymbol{j}} = e/(2m_e c)(\hat{\boldsymbol{l}} + 2\hat{\boldsymbol{s}}) = e/(2m_e c)(\hat{\boldsymbol{j}} + \hat{\boldsymbol{s}}) \ , \tag{2}$$

$$\hat{G} = e/(2m_e c)\left\{1 + \hat{\boldsymbol{j}} \cdot \hat{\boldsymbol{s}}/[j(j+1)]\right\} \ . \tag{3}$$

Here, we have made the assumption that in states $|jm\rangle$ only the component $[\hat{\boldsymbol{s}} \cdot \hat{\boldsymbol{j}}/j(j+1)]\hat{\boldsymbol{j}}$ of the spin vector $\hat{\boldsymbol{s}}$, i.e. the component of $\hat{\boldsymbol{s}}$ parallel to the vector of total angular momentum $\hat{\boldsymbol{j}}$, yields a contribution. The normal component is, on average, zero; this is valid for vector operators (*the projection theorem*).[4]

With $\hat{\boldsymbol{l}}^2 = \hat{\boldsymbol{j}}^2 + \hat{\boldsymbol{s}}^2 - 2\hat{\boldsymbol{s}} \cdot \hat{\boldsymbol{j}}$, we get

$$\hat{G} = \frac{e}{2m_e c}\left[1 + \frac{\hat{\boldsymbol{j}}^2 - \hat{\boldsymbol{l}}^2 + \hat{\boldsymbol{s}}^2}{2j(j+1)}\right] \ . \tag{4}$$

Because of the orientation of the magnetic field $[\boldsymbol{B} = (0, 0, B)]$, we only need $\hat{\mu}_z = \hat{G}\hat{j}_z$.

Let us assume that the magnetic field B is weak enough for first-order perturbation theory to be sufficient to calculate the effect of the magnetic interaction (1). We calculate the matrix element of the interaction in a basis of

[4] See e.g. M.E. Rose: *Elementary Theory of Angular Momentum* (Wiley, New York 1957), Chap. 20.

Example 14.1

eigenfunctions $|jm\rangle$, in which the operators $\hat{G}$ and $\hat{j}_z$ are diagonal, so that

$$\begin{aligned}
&\langle N'j'l'm'|\hat{\mu}_z B|Njlm\rangle \\
&= \left\langle N'j'l'm'\left|\hat{G}\hat{j}_z B\right|Njlm\right\rangle \\
&= mge\hbar B\delta_{mm'}\delta_{ll'}\delta_{jj'}/(2m_e c) \ ,
\end{aligned} \tag{5}$$

with g being the so-called ***Landé*** factor. From (4) it follows that

$$g = 1 + \frac{j(j+1) - l(l+1) + s(s+1)}{2j(j+1)} \ . \tag{6}$$

From Exercise 12.1 we know the Larmor frequency $\omega_L = eB/(2mc)$, and hence we obtain with (1), (5) and (6) the interaction energy of a particle system in a magnetic field as

$$W = \hbar m_e \omega_L \left[1 + \frac{j(j+1) - l(l+1) + s(s+1)}{2j(j+1)}\right] \ . \tag{7}$$

This means that in first-order perturbation theory the modification of the energy levels is

$$E_{Njlm} = E_{Nj} - (e\hbar B/2m_e c)gm \ , \tag{8}$$

so that the shift between two neighbouring levels ($\Delta m = 1$) is

$$\Delta E = (e\hbar B/2m_e c)g \ ; \tag{9}$$

ΔE depends on the Landé factor (i.e. on j, l, and s) and on the intensity of magnetic field. For states with total spin $s = 0$ and therefore $j = l$ (singlet terms of atoms with an even number of electrons), we get $g = 1$ and $\Delta E = e\hbar B/2m_e c$, which is the *normal Zeeman effect.*

Equation (9) is valid only for weak and homogeneous magnetic fields, i.e. for field strengths B, which cause a splitting that is smaller than the energy difference of the unperturbed levels (without fields). This yields the condition

$$|e\hbar B/2m_e c| \ll |E_{Nj} - E_{N'j'}| \ . \tag{10}$$

EXERCISE

14.2 Centre-of-Mass Motion in Atoms

Problem. (a) Take into account the motion of the nucleus in atoms; make use of the results obtained in the section on the centre-of-mass motion of a particle system.

(b) What are the modifications of the transition frequencies of the hydrogen atom discussed in Chap. 9? In other words: what is the true value of the Rydberg constant in the hydrogen atom?

(c) Determine the electron mass, making use of the relations between transition frequency and reduced mass in atoms with one electron.

Solution. (a) Taking into account the motion of the nucleus, the stationary Schrödinger equation reads

$$\left[-\sum_{i=1}^{2}\frac{\hbar^2}{2m_i}\left(\frac{\partial^2}{\partial x_i^2}+\frac{\partial^2}{\partial y_i^2}+\frac{\partial^2}{\partial z_i^2}\right)\right]\psi+V(r)\psi$$
$$=E\psi(x_1,y_1,z_1,x_2,y_2,z_2)\ , \qquad (1)$$

where m_1 is the mass of the nucleus with the coordinates (x_1, y_1, z_1), and m_2 the electron mass located at (x_2, y_2, z_2). The relative distance between the nucleus and the electron is

$$r=\sqrt{(x_1-x_2)^2+(y_1-y_2)^2+(z_1-z_2)^2} \qquad (2)$$

(see next figure).

We introduce Jacobi coordinates, corresponding to our general considerations of the centre-of-mass motion of a particle system (in the Sect. 14.2),

$$\xi_1=x_1-x_2\equiv x\ ,\quad \xi_2=\xi_N=\frac{m_1x_1+m_2x_2}{m_1+m_2}\equiv X\ ,$$
$$\eta_1=y_1-y_2\equiv y\ ,\quad \eta_2=\eta_N=\frac{m_1y_1+m_2y_2}{m_1+m_2}\equiv Y\ ,$$
$$\zeta_1=z_1-z_2\equiv z\ ,\quad \zeta_2=\zeta_N=\frac{m_1z_1+m_2z_2}{m_1+m_2}\equiv Z\ , \qquad (3)$$

applying the results for $N=2$ on [see (14.25)].

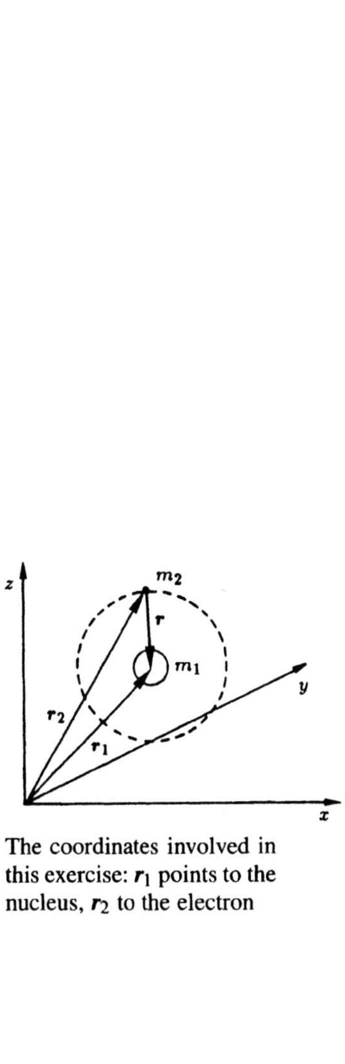

The coordinates involved in this exercise: $\boldsymbol{r}_1$ points to the nucleus, $\boldsymbol{r}_2$ to the electron

We see that the Jacobi coordinates with index 1 represent the relative coordinates; those with index 2, the centre-of-mass coordinates of nucleus and electron.

The transformed Schrödinger equation and its solution follow immediately from (14.32):

$$-\frac{\hbar^2}{2M}\left(\frac{\partial^2\psi}{\partial X^2}+\frac{\partial^2\psi}{\partial Y^2}+\frac{\partial^2\psi}{\partial Z^2}\right)-\frac{\hbar^2}{2\mu}\left(\frac{\partial^2\psi}{\partial x^2}+\frac{\partial^2\psi}{\partial y^2}+\frac{\partial^2\psi}{\partial z^2}\right)$$
$$+V(r)\psi=E\psi(X,Y,Z;x,y,z)\ , \qquad (4)$$

Exercise 14.2

with

$$M = m_1 + m_2 \ , \qquad \mu = m_1 m_2/(m_1 + m_2) \ . \tag{5}$$

For ψ we choose a separation of variables by exploiting our knowledge of a freely moving centre of mass, i.e. we use a plane-wave for the centre-of-mass motion:

$$\psi(X, Y, Z; x, y, z) = N \exp\left[-\frac{\mathrm{i}}{\hbar}(P_x X + P_y Y + P_z Z)\right] \varphi(x, y, z) \ , \tag{6}$$

N being a normalization factor. Inserting (6) into (4) yields the Schrödinger equation for the relative motion:

$$-\frac{\hbar^2}{2\mu}\left(\frac{\partial^2 \varphi}{\partial x^2} + \frac{\partial^2 \varphi}{\partial y^2} + \frac{\partial^2 \varphi}{\partial z^2}\right) + V(r)\varphi = \varepsilon\varphi \ , \tag{7}$$

with

$$\varepsilon = E - P^2/2M \ . \tag{8}$$

Equation (7) describes the motion of a particle with mass μ (the reduced mass) in a given force field $V(r)$. The quantity ε stands for the internal energy of the relative motion. E is the total energy, which contains the energy of the relative motion ε and the kinetic energy $P^2/2M$ of the centre of mass. In Sect. 9.4, we solved an equation analogous to (7), but on the assumption that the mass of the nucleus is very large compared to the electron mass $m_1 \gg m_2$. Indeed, using (5), we see that here $\mu \simeq m_2$. In the case of the hydrogen atom, we found for the transition frequencies between the principal quantum numbers n and n' that

$$\nu_{nn'} = R_\infty \left(\frac{1}{n'^2} - \frac{1}{n^2}\right) \ , \quad \text{with} \tag{9}$$

$$R_\infty = \frac{m_2 e^4}{4\pi\hbar^3} \ .$$

The desired values for ε and $\phi(x, y, z)$ correspond exactly to the quantities derived for the hydrogen atom, if we write μ for m_2.

(b) We have to replace m_2 by μ in order to obtain an accurate value for the Rydberg constant. We then get for the hydrogen atom

$$R_\mathrm{H} = \frac{\mu e^4}{4\pi\hbar^3} \ , \qquad \mu = \frac{m_\mathrm{p} m_\mathrm{e}}{m_\mathrm{p} + m_\mathrm{e}} \ . \tag{10}$$

For a nucleus A of charge Z and mass m_A, we have to replace the Coulomb interaction in the hydrogen atom $-e^2/r$, by $-Ze^2/r$; and we consequently obtain for the modified transition frequencies

$$\begin{aligned}\nu_{nn'} &= \frac{\mu(Ze^2)^2}{4\pi\hbar^3}\left(\frac{1}{n'^2} - \frac{1}{n^2}\right) \\ &= Z^2 R_\mathrm{A}\left(\frac{1}{n'^2} - \frac{1}{n^2}\right) \ ,\end{aligned} \tag{11}$$

where

Exercise 14.2

$$R_\mathrm{A} = \frac{\mu e^4}{4\pi\hbar^3} \ , \quad \mu = \frac{m_\mathrm{A} m_\mathrm{e}}{m_\mathrm{A} + m_\mathrm{e}} \ .$$

(c) The fact that μ assumes different values in different atoms was used by Houston to determine the electron mass through precise measurements of the H_α and H_β lines of the hydrogen atom. The H_α and H_β lines belong to the Balmer series, i.e. transitions that end in the $n = 2$ energy level (see Chap. 9). H_α describes the transition from $n = 3$ to $n = 2$, H_β the transition from $n = 4$ to $n = 2$.

The transition frequencies of the H_α lines can be determined in hydrogen and He^+ (i.e. singly ionized helium with only one electron circling the He nucleus):

$$\nu_\mathrm{H} = R_\mathrm{H}\left(\frac{1}{2^2} - \frac{1}{3^2}\right) = \frac{5}{36} R_\mathrm{H} \ ,$$
$$\nu_\mathrm{He} = 2^2 R_\mathrm{He}\left(\frac{1}{2^2} - \frac{1}{3^2}\right) = \frac{20}{36} R_\mathrm{He} \ . \tag{12}$$

From (12), a relation can be established between the frequencies, depending on the reduced masses:

$$\gamma = \frac{\frac{1}{4}\nu_\mathrm{He} - \nu_\mathrm{H}}{\nu_\mathrm{H}} = \frac{\mu_\mathrm{He} - \mu_\mathrm{H}}{\mu_\mathrm{H}} \ . \tag{13}$$

Expressing μ_He and μ_H by the masses m_H and m_He of the hydrogen and helium nuclei,

$$\mu_\mathrm{H} = \frac{m_\mathrm{H} m_\mathrm{e}}{m_\mathrm{H} + m_\mathrm{e}} \ , \quad \mu_\mathrm{He} = \frac{m_\mathrm{He} m_\mathrm{e}}{m_\mathrm{He} + m_\mathrm{e}} \ , \tag{14}$$

we obtain for γ

$$\gamma = \left(\frac{m_\mathrm{He} - m_\mathrm{H}}{m_\mathrm{He} + m_\mathrm{e}}\right)\frac{m_\mathrm{e}}{m_\mathrm{H}} \ . \tag{15}$$

The spectroscopic determination of γ allows us to compute the ratio $m_\mathrm{e}/m_\mathrm{H}$ according to (15), i.e. the atomic mass of the electron, for given values of m_He and m_H.

Houston found the value

$$m_\mathrm{H}/m_\mathrm{e} = 1838.2 \pm 1.8 \ . \tag{16}$$

This method is also suitable for determining isotopic masses, because the different reduced masses cause a line shift in a quantum transition. The mass of the deuteron $m_\mathrm{D} = 2m_\mathrm{H}$, which contains one proton and one neutron, was determined by this effect.

14.4 Small Oscillations in a Many-Particle System

As the simplest many-particle system, we first consider two identical particles oscillating with small amplitudes about their equilibrium position. In this case we can expand the potential energy in a power series:

$$
\begin{aligned}
V(x_1, x_2) = V(0,0) &+ x_1 \left(\left.\frac{\partial V(x_1, x_2)}{\partial x_1}\right|_{x_1=0} \right) + x_2 \left(\left.\frac{\partial V(x_1, x_2)}{\partial x_2}\right|_{x_2=0} \right) \\
&+ \frac{1}{2} x_1^2 \left(\left.\frac{\partial^2 V}{\partial x_1^2}\right|_{x_1=0} \right) + \frac{1}{2} x_2^2 \left(\left.\frac{\partial^2 V}{\partial x_2^2}\right|_{x_2=0} \right) \\
&+ x_1 x_2 \left(\left.\frac{\partial^2 V}{\partial x_1 \partial x_2}\right|_{x_1=x_2=0} \right) + \dots \, . \qquad (14.87)
\end{aligned}
$$

For a vanishing elongation, the potential energy is minimal (no forces exist in the equilibrium position) and can be set equal to zero. It then follows that

$$
V(0,0) = 0 \, , \quad \left.\frac{\partial V}{\partial x_1}\right|_{x_1=0} = 0 \, , \quad \left.\frac{\partial V}{\partial x_2}\right|_{x_2=0} = 0 \, . \qquad (14.88)
$$

The knowledge of the one-particle oscillator potential (see Chap. 7) suggests setting

$$
\left.\frac{\partial^2 V}{\partial x_1^2}\right|_{x_1=0} = \left.\frac{\partial^2 V}{\partial x_2^2}\right|_{x_2=0} = \mu\omega_0^2 \, , \qquad (14.89)
$$

if we assume equal spring constants, masses μ and frequencies for the two particles. The interaction energy of the particles is taken to be constant in lowest order:

$$
\left.\frac{\partial^2 V}{\partial x_1 \partial x_2}\right|_{\substack{x_1=0 \\ x_2=0}} \equiv \lambda \, . \qquad (14.90)
$$

We thus obtain in the case of small oscillations the following expression for the potential:

$$
V(x_1, x_2) = \frac{\mu\omega_0^2}{2} x_1^2 + \frac{\mu\omega_0^2}{2} x_2^2 + \lambda x_1 x_2 \, . \qquad (14.91)
$$

The Hamiltonian of the system follows immediately as

$$
\hat{H} = -\frac{\hbar^2}{2\mu} \left(\frac{\partial^2}{\partial x_1^2} + \frac{\partial^2}{\partial x_2^2} \right) + \frac{\mu\omega_0^2}{2} (x_1^2 + x_2^2) + \lambda x_1 x_2 \, . \qquad (14.92)
$$

In analogy to classical mechanics, we introduce normal coordinates q_1 and q_2, so that the potential energy $V(x_1, x_2)$ can be represented by a sum of equal terms quadratic in q_1 and q_2. The kinetic energy can be expressed by the squares of the momenta $-i\hbar(\partial/\partial q_1)$, $-i\hbar(\partial/\partial q_2)$. In general, normal coordinates are suitable for describing the eigenoscillations (normal modes) of a system, in which the restoring forces are proportional to the elongation of all particles, and the potential energy is thus a quadratic form of the elongations.

For the system under consideration we set

$$x_1 = \frac{1}{\sqrt{2}}(q_1 + q_2) \ , \quad x_2 = \frac{1}{\sqrt{2}}(q_1 - q_2) \ . \tag{14.93}$$

Now x_1 and x_2 can be expressed in terms of the normal coordinates. For that purpose the derivatives

$$\begin{aligned} \frac{\partial\psi}{\partial q_1} &= \frac{\partial\psi}{\partial x_1}\frac{\partial x_1}{\partial q_1} + \frac{\partial\psi}{\partial x_2}\frac{\partial x_2}{\partial q_1} = \frac{1}{\sqrt{2}}\left(\frac{\partial\psi}{\partial x_1} + \frac{\partial\psi}{\partial x_2}\right) \ , \\ \frac{\partial^2\psi}{\partial q_1^2} &= \frac{1}{2}\left(\frac{\partial^2\psi}{\partial x_1^2} + 2\frac{\partial^2\psi}{\partial x_1\partial x_2} + \frac{\partial^2\psi}{\partial x_2^2}\right) \quad \text{and} \\ \frac{\partial^2\psi}{\partial q_2^2} &= \frac{1}{2}\left(\frac{\partial^2\psi}{\partial x_1^2} - 2\frac{\partial^2\psi}{\partial x_1\partial x_2} + \frac{\partial^2\psi}{\partial x_2^2}\right) \end{aligned} \tag{14.94}$$

are needed. The potential energy becomes

$$\frac{\mu\omega_0^2}{2}(x_1^2 + x_2^2) + \lambda x_1 x_2 = \frac{\mu\omega_0^2}{2}(q_1^2 + q_2^2) + \frac{\lambda}{2}(q_1^2 - q_2^2) \ , \tag{14.95}$$

with

$$\mu\omega_1^2 \equiv \mu\omega_0^2 + \lambda \ , \quad \mu\omega_2^2 \equiv \mu\omega_0^2 - \lambda \ . \tag{14.96}$$

We obtain the Hamiltonian in normal coordinates as

$$\hat{H} = -\frac{\hbar^2}{2\mu}\left(\frac{\partial^2}{\partial q_1^2} + \frac{\partial^2}{\partial q_2^2}\right) + \frac{\mu\omega_1^2}{2}q_1^2 + \frac{\mu\omega_2^2}{2}q_2^2 \ . \tag{14.97}$$

Obviously the Hamiltonian of two coupled oscillators (14.92) is transformed into a sum of the Hamiltonians of uncoupled oscillators with frequencies ω_1 and ω_2. The wave functions and the energies of the system are obtained by solving the associated Schrödinger equation, which is

$$-\frac{\hbar^2}{2\mu}\frac{\partial^2\psi}{\partial q_1^2} + \frac{\mu\omega_1^2}{2}q_1^2\psi - \frac{\hbar^2}{2\mu}\frac{\partial^2\psi}{\partial q_2^2} + \frac{\mu\omega_2^2}{2}q_2^2\psi = E\psi \ . \tag{14.98}$$

Decoupling of this equation is achieved by the separation

$$\psi(q_1, q_2) = \psi_1(q_1)\psi_2(q_2) \ , \quad E = E_1 + E_2 \ . \tag{14.99}$$

By introducing (14.99) into (14.98) and after dividing by $\psi_1(q_1)\psi_2(q_2)$, separate terms depending solely on q_1 or on q_2 are obtained:

$$-\frac{\hbar^2}{2\mu}\frac{\partial^2\psi_1}{\partial q_1^2}+\frac{\mu\omega_1^2}{2}q_1^2\psi_1=E_1\psi_1 , \tag{14.100}$$

$$-\frac{\hbar^2}{2\mu}\frac{\partial^2\psi_2}{\partial q_2^2}+\frac{\mu\omega_2^2}{2}q_2^2\psi_2=E_2\psi_2 . \tag{14.101}$$

We already know the solution of (14.100) and (14.101) from Chap. 7. These two equations describe harmonic oscillators with frequencies ω_1 and ω_2, respectively, and the wave functions are given by Hermite polynomials:

$$\psi_{n_1}=\sqrt{\frac{1}{2^{n_1}n_1!}\sqrt{\frac{\lambda_1}{\pi}}}\exp\left(-\frac{\lambda_1 q_1^2}{2}\right)H_{n_1}(\sqrt{\lambda_1}q_1) , \tag{14.102}$$

with $\lambda_1=\mu\omega_1/h$ and the eigenvalues

$$E_{n_1}=\hbar\omega_1(n_1+\tfrac{1}{2}) , \quad n_1=0,1,2,\dots , \tag{14.103}$$

and, analogously,

$$\psi_{n_2}=\sqrt{\frac{1}{2^{n_2}n_2!}\sqrt{\frac{\lambda_2}{\pi}}}\exp\left(-\frac{\lambda_2 q_2^2}{2}\right)H_{n_2}(\sqrt{\lambda_2}q_2) , \tag{14.104}$$

with the energies

$$E_{n_2}=\hbar\omega_2(n_2+\tfrac{1}{2}) , \quad n_2=0,1,2,\dots . \tag{14.105}$$

The eigenfunctions and energy eigenvalues of the whole system follow by inserting the last results into (14.99):

$$\psi_{n_1n_2}(q_1,q_2)=\psi_{n_1}(q_1)\psi_{n_2}(q_2) , \tag{14.106}$$

or

$$E_{n_1n_2}=\hbar\omega_1(n_1+\tfrac{1}{2})+\hbar\omega_2(n_2+\tfrac{1}{2}) , \tag{14.107}$$

from which we deduce the ground-state energy of the system as

$$E_{00}=\frac{\hbar\omega_1}{2}+\frac{\hbar\omega_2}{2} . \tag{14.108}$$

Now we consider the probability of finding the normal coordinates q_1 and q_2 in the intervals (q_1,q_1+dq_1) and (q_2,q_2+dq_2), with the aim of making a statement concerning the coordinates x_1,x_2 of configuration space. The probability mentioned is described by

$$w(q_1,q_2)\,dq_1\,dq_2=\left|\psi_{n_1n_2}(q_1,q_2)\right|^2 dq_1\,dq_2 . \tag{14.109}$$

Correspondingly, the probability of finding the system in the coordinate space x_1, x_2 in the intervals $(x_1, x_1 + dx_1)$ and $(x_2, x_2 + dx_2)$ follows by using

$$dq_1\, dq_2 = \begin{vmatrix} \dfrac{\partial q_1}{\partial x_1} & \dfrac{\partial q_1}{\partial x_2} \\ \dfrac{\partial q_2}{\partial x_1} & \dfrac{\partial q_2}{\partial x_2} \end{vmatrix} dx_1\, dx_2 = -dx_1\, dx_2 \ . \tag{14.110}$$

This implies that we have to reverse the direction of revolution of the region $G^*(q_1, q_2)$ with respect to that of the region $G(x_1, x_2)$ if a transformation of a surface integral from $G(x_1, x_2)$ to $G^*(q_1, q_2)$ is carried out (negative sign of the functional determinant!). The surface element, of course, remains positive,[5] so that the probability is

$$\begin{aligned} &w(x_1, x_2)\, dx_1\, dx_2 \\ &\quad = \left| \psi_{n_1 n_2}\left(\frac{1}{\sqrt{2}}(x_1 + x_2), \frac{1}{\sqrt{2}}(x_1 - x_2) \right) \right|^2 dx_1\, dx_2 \ . \end{aligned} \tag{14.111}$$

The generalization of these results to an N-particle system performing small oscillations about its equilibrium is straightforward.

We denote the elongation of the kth particle by x_k, y_k, z_k and obtain the potential energy

$$V = \frac{1}{2} \sum_{i,j=1}^{3N} C_{ij} w_i w_j = \frac{1}{2} w^{\mathrm{T}} \hat{C} w \ , \tag{14.112}$$

where

$$w^{\mathrm{T}} = \{w_i\} = (x_1, x_2, \dots, x_N, y_1, y_2, \dots, y_N, z_1, z_2, \dots, z_N) \tag{14.113}$$

stands for the position vector in the configuration space of all N particles. In analogy to (14.87) and (14.90), the coefficients $\hat{C} = (C_{ij})$ are the second-order derivatives of the potential energy:

$$C_{ij} = \left. \frac{\partial^2 V}{\partial w_i \partial w_j} \right|_{\substack{w_i=0 \\ w_j=0}} \ , \quad \text{for} \quad i \neq j \ , \tag{14.114}$$

$$C_{ii} = \left. \frac{\partial^2 V}{\partial w_i^2} \right|_{w_i=0} \ , \quad \text{for} \quad i = j \ . \tag{14.115}$$

As in the simple example previously discussed, we can now introduce $3N$ normal coordinates q_s, $s = 1, 2, \dots, 3N$, which are related to the Cartesian coordinates

[5] Since volume elements are required to be positive, we should define the transformation (14.110) from one volume element to another one using the absolute value of the transformation determinant.

by an orthogonal transformation:

$$q_s = \sum_{k=1}^{3N} a_{sk} w_k \; , \qquad s = 1, 2, \dots, 3N \; , \tag{14.116}$$

and

$$\sum_k a_{ik} a_{jk} = \delta_{ij} = \sum_k a_{ki} a_{kj} \; ; \tag{14.117}$$

$\hat{a} = (a_{ik})$ is a matrix and its inverse $\hat{a}^{-1}$ is equal to its transpose $\hat{a}^{-1} = \hat{a}^{\mathrm{T}}$ or to its Hermitian conjugate, if $\hat{a}$ contains complex elements:

$$(\hat{a}^{-1}\hat{a})_{ij} = \sum_k a_{ik}^{-1} a_{kj} = \sum_k a_{ki} a_{kj} = \delta_{ij} \; , \tag{14.118}$$

from which it follows, together with (14.116), that

$$w_l = \sum_{k=1}^{3N} a_{kl} q_k \; . \tag{14.119}$$

Since $\hat{a}$ is an orthogonal matrix, the terms of the operator of the kinetic energy also decouple in normal coordinates, if all particles have the same mass μ:

$$\begin{aligned}
\frac{\partial \Psi}{\partial q_s} &= \sum_{l=1}^{3N} \frac{\partial \Psi}{\partial w_l} \frac{\partial w_l}{\partial q_s} \; , \qquad \frac{\partial w_l}{\partial q_s} = \sum_{k=1}^{3N} a_{kl} \delta_{sk} = a_{sl} \; , \\
\frac{\partial^2 \Psi}{\partial q_s^2} &= \sum_{m=1}^{3N} \frac{\partial}{\partial w_m} \left(\sum_{l=1}^{3N} \frac{\partial \Psi}{\partial w_l} a_{sl} \right) a_{sm} \; , \\
\frac{\partial^2 \Psi}{\partial q_s^2} &= \sum_{m,l=1}^{3N} \frac{\partial^2 \Psi}{\partial w_m \partial w_l} a_{sl} a_{sm} \; .
\end{aligned} \tag{14.120}$$

Now, using the orthogonality of $\hat{a}$ expressed in (14.117) and (14.118), the kinetic energy is calculated as follows:

$$\begin{aligned}
-\frac{\hbar^2}{2\mu} \sum_{s=1}^{3N} \frac{\partial^2 \Psi}{\partial q_s^2} &= -\frac{\hbar^2}{2\mu} \sum_{s,m,l=1}^{3N} \frac{\partial^2 \Psi}{\partial w_m \partial w_l} a_{sl} a_{sm} \\
&= -\frac{\hbar^2}{2\mu} \sum_{l=1}^{3N} \frac{\partial^2 \Psi}{\partial w_l^2} = -\frac{\hbar^2}{2\mu} \sum_{k=1}^{N} \nabla_k^2 \Psi \; .
\end{aligned} \tag{14.121}$$

The index k at the gradient operator

$$\nabla_k = \left(\frac{\partial}{\partial x_k}, \frac{\partial}{\partial y_k}, \frac{\partial}{\partial z_k} \right)$$

refers to the particle number k, as in (14.113). The potential energy is assumed to be a bilinear form of the coordinates w_i and w_j:

$$V=\frac{1}{2}\sum_{i,j=1}^{3N} C_{ij}w_iw_j=\frac{1}{2}w^{\mathrm{T}}\hat{C}w \ ,$$

$$w=\hat{a}^{\mathrm{T}}q \ , \quad w^{\mathrm{T}}=q^{\mathrm{T}}\hat{a} \ , \tag{14.122}$$

so that

$$V=\frac{1}{2}q^{\mathrm{T}}\hat{a}\hat{C}\hat{a}^{\mathrm{T}}q \ . \tag{14.123}$$

To decouple the potential energy in normal coordinates, we require that

$$\hat{a}\hat{C}\hat{a}^{\mathrm{T}}=\hat{\Lambda} \ , \tag{14.124}$$

where Λ is a diagonal matrix of the form

$$\begin{pmatrix} \Lambda_{11} & & & \\ & \Lambda_{22} & & \\ & & \ddots & \\ & & & \Lambda_{3N3N} \end{pmatrix} = \Lambda_{ii}\delta_{ij} \ .$$

Since $\hat{C}$ is a symmetric matrix according to (14.114) and (14.115), it is possible to construct an orthogonal matrix $\hat{a}$ in such a way that Λ is a real diagonal matrix.

With (14.121) and (14.123), the Hamiltonian of the coupled system splits up into a sum of harmonic oscillator Hamiltonians, namely

$$\begin{aligned}\hat{H}&=-\frac{\hbar^2}{2\mu}\sum_{s=1}^{3N}\frac{\partial^2}{\partial w_s^2}+\frac{1}{2}\sum_{i,j=1}^{3N}C_{ij}w_iw_j\\ &=-\frac{\hbar^2}{2\mu}\sum_{s=1}^{3N}\frac{\partial^2}{\partial q_s^2}+\frac{1}{2}\mu\sum_{s=1}^{3N}\omega_s^2q_s^2 \ ,\end{aligned} \tag{14.125}$$

where we have renamed the diagonal elements of $\hat{\Lambda}$:

$$\Lambda_{ss}=\mu\omega_s^2 \ . \tag{14.126}$$

Now the Schrödinger equation for stationary states reads

$$\begin{aligned}&\sum_{s=1}^{3N}\left[-\frac{\hbar^2}{2\mu}\frac{\partial^2}{\partial q_s^2}+\frac{1}{2}\mu(\omega_sq_s)^2\right]\Psi(q_1,q_2,\dots,q_{3N})\\ &\quad =E\Psi(q_1,q_2,\dots,q_{3N}) \ .\end{aligned} \tag{14.127}$$

As an expression for Ψ, we choose in analogy to similar separation problems

$$\Psi=\Phi_1(q_1)\Phi_2(q_2)\Phi_3(q_3)\dots\Phi_{3N}(q_{3N}) \ , \tag{14.128}$$

so that (14.127) decouples into $3N$ equations, which describe the same number of independent oscillators. The equation for the oscillator with the sth normal reads

$$-\frac{\hbar^2}{2\mu}\frac{\partial^2\Phi_s(q_s)}{\partial q_s^2}+\frac{1}{2}\mu(\omega_s q_s)^2\Phi_s(q_s)=E\Phi_s(q_s)\ . \tag{14.129}$$

The solution of (14.127) is, in analogy to (14.102), of the form

$$\Phi_{ns}(q_s)=[(2^{n_s}n_s!)^{-1}(\lambda_s/\pi)^{1/2}]^{1/2}\exp(-\tfrac{1}{2}\lambda_s q_s^2)H_{ns}(\sqrt{\lambda_s}q_s)\ , \tag{14.130}$$

where $\lambda_s=\mu\omega_s/\hbar$. The energy eigenvalues are

$$E_{ns}=\hbar\omega_s(n_s+\tfrac{1}{2})\ ,\qquad n_s=0,1,2,\dots\ , \tag{14.131}$$

so that the total wave function can be written as

$$\begin{aligned}\Psi&=\Psi_{n_1,n_2,\dots,n_{3N}}(q_1,q_2,\dots,q_{3N})\\&=\Phi_{n_1}(q_1)\Phi_{n_2}(q_2)\Phi_{n_3}(q_3)\cdots\Phi_{n_{3N}}(q_{3N})\ ,\end{aligned} \tag{14.132}$$

$$\begin{aligned}E_{n_1,n_2,\dots,n_{3N}}&=\hbar\omega_1\left(n_1+\tfrac{1}{2}\right)+\hbar\omega_2\left(n_2+\frac{1}{2}\right)+\cdots+\hbar\omega_s\left(n_s+\frac{1}{2}\right)\\&\quad+\cdots+\hbar\omega_{3N}\left(n_{3N}+\frac{1}{2}\right)\ .\end{aligned} \tag{14.133}$$

The range of the quantum numbers $n_1,\dots,n_{3N}$ is over all integers, including zero. As the zero-point energy of the system, we obviously get

$$E_0=\frac{1}{2}\hbar\sum_{s=1}^{3N}\omega_s\ . \tag{14.134}$$

The energy levels of the oscillating particle system are obtained by inserting all allowed combinations of oscillator quantum numbers $n_1,\dots,n_{3N}$. In this case it is sufficient to know the frequencies ω_s of the normal oscillations. As these results were obtained for oscillations with small amplitudes, (14.133) is only valid for the low-energy range of the energy spectrum, i.e. for small quantum numbers n_s.

Such a physical situation can be found, for example, in molecules and solids where the atoms oscillate with small amplitudes about their equilibrium position so that an energy spectrum of the form (14.133) is obtained.

For larger amplitudes of oscillation, we have to take higher-order terms in the Taylor series of the potential into account, such as

$$\left.\frac{\partial^3 V}{\partial x_i\partial y_j\partial z_k}\right|_0 x_i y_j z_k\ . \tag{14.135}$$

In this case, a linear force law no longer applies, i.e. the potential energy is not a quadratic form of the displacements, and hence the oscillations will no longer decouple when normal coordinates are introduced. Under these circumstances, our results are only approximately valid.

EXERCISE

14.3 Two Particles in an External Field

Problem. Calculate the influence of an external field on the motion of an interacting two-particle system with the masses m_1 and m_2. Let the potential energies of the first and second particles in the external field be $V_1(x_1, y_1, z_1)$ and $V_2(x_2, y_2, z_2)$, respectively; let the interaction between the particles be $W(x_1 - x_2, y_1 - y_2, z_1 - z_2)$.

Hints: (a) Determine the time-dependent Schrödinger equation of the system in centre-of-mass and relative coordinates.

(b) Assume the dimension of the system to be small, so that the external potentials can be expanded around the centre of mass in terms of the internal (relative) coordinates.

(c) Expand the total wave function in a basis of wave functions Φ_n, which are undisturbed by the external fields. This basis shall describe the relative motion. The coupling of the basis to the centre-of-mass coordinates is assumed to be weak, or, equivalently, it can be treated as a perturbation.

Solution. (a) Let the mass and the coordinates of the first particle be m_1 and (x_1, y_1, z_1), those of the second particle m_2 and (x_2, y_2, z_2). The interaction energy between the particles is of the form $W(x_1 - x_2, y_1 - y_2, z_1 - z_2)$; the potential energy of the single particles in the external field is $V_1(x_1, y_1, z_1)$ and $V_2(x_2, y_2, z_2)$, respectively. Hence, the Schrödinger equation of the system is

$$i\hbar\frac{\partial\Psi}{\partial t} = -\frac{\hbar^2}{2m_1}\nabla_1^2\Psi + V_1\Psi - \frac{\hbar^2}{2m_2}\nabla_2^2\Psi + V_2\Psi + W\Psi \;, \tag{1}$$

where $\Psi = \Psi(x_1, y_1, z_1, x_2, y_2, z_2, t)$. Instead of the particle coordinates x_1, y_1, z_1 and x_2, y_2, z_2, we introduce centre-of-mass and relative coordinates

$$\begin{aligned} X &= \frac{m_1x_1 + m_2x_2}{m_1 + m_2} \;, \quad x = x_1 - x_2 \;, \\ Y &= \frac{m_1y_1 + m_2y_2}{m_1 + m_2} \;, \quad y = y_1 - y_2 \;, \\ Z &= \frac{m_1z_1 + m_2z_2}{m_1 + m_2} \;, \quad z = z_1 - z_2 \;. \end{aligned} \tag{2}$$

The coordinates of the particles x_1, y_1, z_1 and x_2, y_2, z_2 can be expressed in terms of these new coordinates:

$$\begin{aligned} x_1 &= X + \gamma x \;, \quad x_2 = X - \delta x \;, \\ y_1 &= Y + \gamma y \;, \quad y_2 = Y - \delta y \;, \\ z_1 &= Z + \gamma z \;, \quad z_2 = Z - \delta z \;. \end{aligned} \tag{3}$$

$$\gamma = \frac{m_2}{m_1 + m_2} \;, \quad \delta = \frac{m_1}{m_1 + m_2} \;. \tag{4}$$

Exercise 14.3

Now the Laplace operators are expressed in terms of relative and centre-of-mass coordinates. Therefore we need

$$\frac{\partial\Psi}{\partial x_1}=\frac{\partial\Psi}{\partial X}\frac{\partial X}{\partial x_1}+\frac{\partial\Psi}{\partial x}\frac{\partial x}{\partial x_1}=\delta\frac{\partial\Psi}{\partial X}+\frac{\partial\Psi}{\partial x}\ ,$$

$$\frac{\partial^2\Psi}{\partial x_1^2}=\delta\left(\delta\frac{\partial^2\Psi}{\partial X^2}+\frac{\partial^2\Psi}{\partial x\partial X}\right)+\delta\frac{\partial^2\Psi}{\partial X\partial x}+\frac{\partial^2\Psi}{\partial x^2}$$

$$=\delta^2\frac{\partial^2\Psi}{\partial X^2}+2\delta\frac{\partial^2\Psi}{\partial x\partial X}+\frac{\partial^2\Psi}{\partial x^2}\ . \tag{5}$$

Analogously, we find that

$$\frac{\partial^2\Psi}{\partial x_2^2}=\gamma^2\frac{\partial^2\Psi}{\partial X^2}-2\gamma\frac{\partial^2\Psi}{\partial x\partial X}+\frac{\partial^2\Psi}{\partial x^2}\ , \tag{6}$$

from which we get

$$-\frac{\hbar^2}{2m_1}\frac{\partial^2\Psi}{\partial x_1^2}-\frac{\hbar^2}{2m_2}\frac{\partial^2\Psi}{\partial x_2^2}$$

$$=-\frac{\hbar^2}{2(m_1+m_2)}\frac{\partial^2\Psi}{\partial X^2}-\frac{\hbar^2}{2\left(\dfrac{m_1m_2}{m_1+m_2}\right)}\frac{\partial^2\Psi}{\partial x^2}\ . \tag{7}$$

Using the analogous expressions for the y and z components and (3), we find for the Schrödinger equation (1)

$$\begin{aligned}\mathrm{i}\hbar\frac{\partial\Psi}{\partial t}=&-(\hbar^2/2M)\nabla_X^2\Psi\\&+V_1(X+\gamma x,Y+\gamma y,Z+\gamma z)\Psi-(\hbar^2/2\mu)\nabla_x^2\Psi\\&+V_2(X-\delta x,Y-\delta y,Z-\delta z)\Psi+W(x,y,z)\Psi\ ,\end{aligned} \tag{8}$$

with the Laplace operators

$$\nabla_X^2=\frac{\partial^2}{\partial X^2}+\frac{\partial^2}{\partial Y^2}+\frac{\partial^2}{\partial Z^2}\ ,$$

$$\nabla_x^2=\frac{\partial^2}{\partial x^2}+\frac{\partial^2}{\partial y^2}+\frac{\partial^2}{\partial z^2}\ , \tag{9}$$

and the total mass $M=m_1+m_2$, as well as the reduced mass $\mu=(m_1m_2)/(m_1+m_2)$.

Separation, meanwhile familiar to us, does not work in this particular case because the potentials V_1 and V_2 prevent a decoupling of the centre-of-mass coordinates, making our considerations more complicated.

(b) To proceed analytically with our problem, we assume the extensions of the system to be very small. This implies a restriction on systems and states for

Exercise 14.3

which the wave function Ψ decreases sufficiently fast with increasing relative distance $r = (x^2 + y^2 + z^2)^{1/2}$. A typical distance a, at which the particle probability should approximately be zero, is the spatial extension of the system, for example the expectation value of radius of the valence electron in an atom or the longitudinal extension of a molecule.

Under this assumption, consideration of (8) within the range $r \leq a$ is sufficient and we expand the potentials V_1 and V_2 with respect to powers of x, y, z. This gives

$$\begin{aligned} &V_1(X+\gamma x, Y+\gamma y, Z+\gamma z) + V_2(X-\delta x, Y-\delta y, Z-\delta z) \\ &\quad = V_1(X, Y, Z) + V_2(X, Y, Z) + (\partial V_1/\partial x)\gamma x + \cdots - (\partial V_2/\partial z)\delta z + \cdots \\ &\quad = V(X, Y, Z) + w(X, Y, Z, x, y, z) \ . \end{aligned} \tag{10}$$

The term $V(X, Y, Z)$ denotes the potential energy of the centre of mass; $w(X, Y, Z, x, y, z)$ couples the centre-of-mass motion to the relative motion. With (10), the Schrödinger equation (8) can be cast in the form

$$\begin{aligned} \mathrm{i}\hbar \frac{\partial \Psi}{\partial t} &= [-(\hbar^2/2M)\nabla_X^2 + V(X, Y, Z)]\Psi \\ &\quad + [-(\hbar^2/2\mu)\nabla_x^2 + W(x, y, z)]\Psi + w(X, Y, Z, x, y, z)\Psi \ . \end{aligned} \tag{11}$$

(c) If the external field is absent, the eigenfunctions of the internal motion are denoted by $\Phi_n^0(x, y, z)$, with the energy eigenvalues E_n^0. The following equation is valid for these eigenfunctions Φ_n^0:

$$-(\hbar^2/2\mu)\nabla_x^2 \Phi_n^0 + W(x, y, z)\Phi_n^0 = E_n^0 \Phi_n^0 . \tag{12}$$

The influence of the external field on the inner degrees of freedom of the system is taken into account by the term $w(X, Y, Z, x, y, z)$, so that

$$-(\hbar^2/2\mu)\nabla_x^2 \Phi_n + W(x, y, z)\Phi_n + w(X, Y, Z, x, y, z)\Phi_n = E_n \Phi_n \ . \tag{13}$$

The *centre-of-mass coordinates appear as parameters* in the coupling potential of (13). Hence, the wave functions and energy eigenvalues will also depend on the centre-of-mass coordinates.

If $w(X, Y, Z, x, y, z) \ll W(x, y, z)$, then the coupling potential can be considered as a disturbance. If the solutions Φ_n^0 of the free interacting system (12) are known, (11) can be solved. The eigenfunctions and eigenenergies of (13) then are

$$\begin{aligned} \Phi_n &= \Phi_n(x, y, z, X, Y, Z) \ , \\ E_n &= E_n(X, Y, Z) \ . \end{aligned} \tag{14}$$

As already mentioned, the centre-of-mass coordinates X, Y, Z are only parameters here. The total wave function Ψ in (11) is now expanded with respect to the stationary states Φ_n:

$$\begin{aligned} &\Psi(x, y, z, X, Y, Z, t) \\ &\quad = \sum_n a_n(X, Y, Z, t)\Phi_n(x, y, z, X, Y, Z) \ . \end{aligned} \tag{15}$$

Exercise 14.3

Inserting this into (11), we obtain a system of coupled differential equations with respect to the expansion coefficients $a_n(t)$:

$$
\begin{aligned}
\mathrm{i}\hbar\frac{\partial}{\partial t}\left(\sum_n a_n\Phi_n\right) &= [-(\hbar^2/2M)\nabla_X^2 + V(X,Y,Z)]\sum_n a_n\Phi_n \\
&\quad + [-(\hbar^2/2\mu)\nabla_x^2 + W(x,y,z)]\sum_n a_n\Phi_n \\
&\quad + w(X,Y,Z,x,y,z)\sum_n a_n\Phi_n \; . \qquad (16)
\end{aligned}
$$

It can also be written as

$$
\begin{aligned}
\mathrm{i}\hbar\sum_n \dot{a}_n\Phi_n &= -(\hbar^2/2M)\nabla_X\left[\sum_n(\nabla_X a_n)\Phi_n + \sum_n a_n(\nabla_X\Phi_n)\right] \\
&\quad + V\sum_n a_n\Phi_n - (\hbar^2/2\mu)\sum_n a_n(\nabla_X^2\Phi_n) \\
&\quad + W\sum_n a_n\Phi_n + w\sum_n a_n\Phi_n \; . \qquad (17)
\end{aligned}
$$

With the help of (13), the last three terms in (17) can be seen to be identical with $\sum_n a_n E_n\phi_n$. Multiplying (17) from the left by Φ_m^* and integrating over x, y, z yields

$$
\begin{aligned}
\mathrm{i}\hbar\dot{a}_m &= -(\hbar^2/2M)\sum_n 2\langle\Phi_m|\nabla_X|\Phi_n\rangle\nabla_X a_n \\
&\quad - (\hbar^2/2M)\Delta_X a_m - (\hbar^2/2M)\sum_n\langle\Phi_m|\Delta_X|\Phi_n\rangle a_n \\
&\quad + (V+E_m)a_m \; . \qquad (18)
\end{aligned}
$$

The matrix elements $\langle\Phi_m|\nabla_X|\Phi_n\rangle$ and $\langle\Phi_m|\Delta_X|\Phi_n\rangle$ are nonzero only if the wave function Φ_n depends on the centre-of-mass coordinates. In this case, a transition of the system from the state n to another state m is possible according to the transition matrix elements.

If the system is prepared in the state i at time $t=0$, i.e. $a_i(t=0)\neq 0$ and $a_n(t=0)=0$ for all $n\neq i$, then, according to (18), $\dot{a}_i(t=0)\neq 0$ as well.

Time evolution causes the pure state

$$
\Psi_{t=0} = a_i\Phi_i(x,y,z,X,Y,Z)
$$

to become a superposition according to (15).

If the basis wave functions depend only weakly on the centre-of-mass coordinates X, Y, Z, we can, as an approximation, neglect the transition matrix elements and find

$$
\mathrm{i}\hbar\frac{\partial a_n}{\partial t} = -(\hbar^2/2M)\nabla_X^2 a_n + (V+E)a_n \; . \qquad (19)
$$

Exercise 14.3

So the expansion amplitudes $a_n(t)$ follow the equations of motion of the centre of mass in a potential field of the form

$$V_n = V(X, Y, Z) + E_n(X, Y, Z) \; , \tag{20}$$

which depends on E_n. This corresponds to the condition that the inner state of the system be the nth quantum state. For each n, (19), within the approximation chosen, can be interpreted as the motion of a massive point particle. In other words, the whole system propagates for each internal state Φ_n in a slightly modified potential field (see figure). This is quite reasonable.

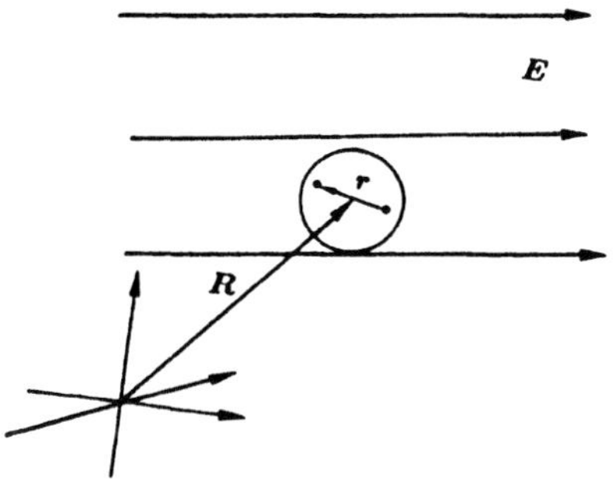

A small cluster moving through an external field. The cluster acts like an elementary particle in any internal state Φ_n as long as the polarization effects (interaction of the internal degrees of freedom with the centre-of-mass motion) are negligible

14.5 Biographical Notes

LANDÉ, Alfred, German-American physicist, *Elberfeld 13.12.1888, †30.10.1975. L. was professor at Tübingen from 1922–1931, and thereafter at Capital University, Columbus, Ohio. He developed the systematics of the multiplet spectra (1921–1923) and the Zeeman effect ("Landésches Vektormodell"). He also introduced the g factor named after him.

15. Identical Particles

One characteristic of quantum mechanics is the indistinguishability of identical particles in the subatomic region. We designate as *identical particles* those particles that have the same mass, charge, spin etc. and behave in the same manner under equal physical conditions. Therefore, in contrast with macroscopic objects, it is not possible to distinguish between particles like electrons (protons, pions, α particles) on the basis of their characteristics or their trajectory. The spreading of the wave packets that describe the particles leads to an overlapping of the probability densities in time (Fig. 15.1); thus we will not be able to establish later on whether particle no. 1 or no. 2 or another particle can be found at the point in space $\boldsymbol{r}$. Because of the possible interaction (momentum exchange etc.), dynamical properties cannot be used to distinguish between them, either.

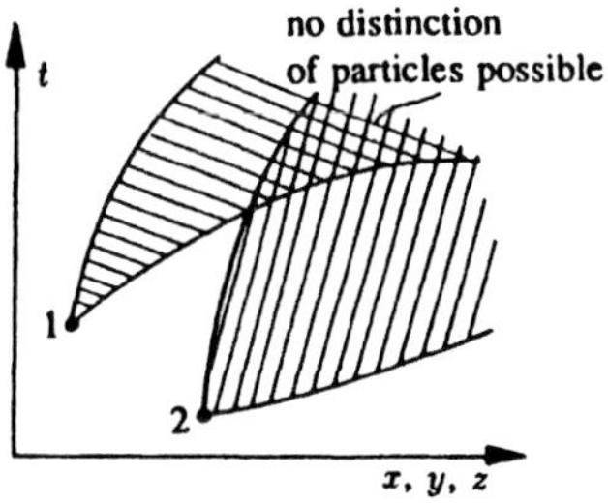

Fig. 15.1. Overlapping of probability densities (schematic). Originally the wave packets for particles no. 1 and no. 2 are prepared separately. As time evolution of the packets proceeds, they overlap (*doubly hatched area*) and it is no longer possible to distinguish the particles

If we regard a quantum-mechanical system of identical particles, we will not be able to relate a state ψ_n to particle no. n; we will only be able to determine the state of the totality of all particles.

In the case of a system of N particles with spin, the wave function of the system is a function of these $4N$ coordinates ($3N$ space and N spin coordinates):

$$\psi = \psi(\boldsymbol{r}_1 s_1, \boldsymbol{r}_2 s_2, \dots, \boldsymbol{r}_N s_N, t) \ . \tag{15.1}$$

Since the system consists of identical particles, the physical state remains the same if the particles j and i are exchanged. This operation is carried out by the operator $\hat{P}_{ij}$:

$$\begin{aligned} &\hat{P}_{ij}\psi(\boldsymbol{r}_1 s_1, \dots, \boldsymbol{r}_i s_i, \dots, \boldsymbol{r}_j s_j, \dots, \boldsymbol{r}_N s_N, t) \\ &\quad = \lambda\psi(\boldsymbol{r}_1 s_1, \dots, \boldsymbol{r}_j s_j, \dots, \boldsymbol{r}_i s_i, \dots, \boldsymbol{r}_N s_N, t) \ , \end{aligned} \tag{15.2}$$

where, for the present, λ is an arbitrary constant factor. A second exchange of the two particles recreates the original state. Hence,

$$\hat{P}_{ij}^2\psi = \lambda^2\psi = \psi \ , \tag{15.3}$$

yielding two values for λ:

$$\lambda = \pm 1 \ . \tag{15.4}$$

Since we are examining systems of identical particles, the exchange of particles always acts in the same way on the wave function. This means that two systems of particles may exist; systems with wave functions that change sign upon the exchange of two particles, and systems whose wave functions remain unchanged.

Therefore either

$$\hat{P}_{ij}\psi_{\mathrm{s}} = \psi_{\mathrm{s}} \quad \text{or} \quad \hat{P}_{ij}\psi_{\mathrm{a}} = -\psi_{\mathrm{a}} \quad \text{holds} \ . \tag{15.5}$$

We call the wave function ψ_{s} with the eigenvalue $+1$ *symmetric* and ψ_{a} with the eigenvalue -1 *antisymmetric* with respect to the exchange of two particles. This is the origin of the indices "s" and "a" in (15.5). Whether particles are described by a symmetric or an antisymmetric wave function will depend on their nature. A transition between symmetric and antisymmetric states is impossible. This is because the interaction between particles is symmetric under their exchange; hence, e.g.,

$$\begin{aligned} &V(\boldsymbol{r}_1, \boldsymbol{r}_2, \dots, \boldsymbol{r}_i, \dots, \boldsymbol{r}_j, \dots, \boldsymbol{r}_N) \\ &\quad = V(\boldsymbol{r}_1, \boldsymbol{r}_2, \dots, \boldsymbol{r}_j, \dots, \boldsymbol{r}_i, \dots, \boldsymbol{r}_N) \ . \end{aligned} \tag{15.6}$$

For this reason, the matrix elements between symmetric and antisymmetric states vanish:

$$\begin{aligned} &\big\langle \psi_{\mathrm{s}}(\boldsymbol{r}_1, \boldsymbol{r}_2, \dots, \boldsymbol{r}_i, \dots, \boldsymbol{r}_j, \dots, \boldsymbol{r}_N) \big| V(\boldsymbol{r}_1, \boldsymbol{r}_2, \dots, \boldsymbol{r}_i, \dots, \boldsymbol{r}_j, \dots, \boldsymbol{r}_N) \\ &\big| \psi_{\mathrm{a}}(\boldsymbol{r}_1, \boldsymbol{r}_2, \dots, \boldsymbol{r}_i, \dots, \boldsymbol{r}_j, \dots, \boldsymbol{r}_N) \big\rangle = 0 \ , \end{aligned} \tag{15.7}$$

and therefore no transitions take place between them. Both kinds of particle occur naturally. The particles described by an antisymmetric wave function are called *fermions* (named after E. Fermi); those particles described by a symmetric wave function are called *bosons* (named after S.N. Bose).

The physical criterion that distinguishes between the two kinds of particles is their spin: *Fermions have half-integer spin; bosons have integer spin.* This relation between spin and symmetry properties of the wave function or – as it is also called – between *spin and statistics*, was first found empirically. Later on, when concerned with quantum field theory (quantum electrodynamics), we will understand why this must be so.

Examples of fermions are electrons, protons, neutrons, neutrinos, C^{13} nuclei, etc. (all spin $\frac{1}{2}$); examples of bosons are π mesons (spin 0), photons (spin 1), deuterons (spin 1), α particles (spin 0), oxygen nuclei (spin 0). For particles that are composed of several elementary particles, the spin also determines the character of the statistics, as has already been mentioned. The α particle that consists of four nucleons with spin $\frac{1}{2}$ has spin 0 and is a boson. We get the same result when considering that the exchange of an α particle requires the exchange of two protons and two neutrons; the signs that result from the two-fermion exchange compensate for each other in this case.

15.1 The Pauli Principle

The antisymmetry of the fermion wave function is equivalent to *Pauli's exclusion principle*, empirically formulated by *Wolfgang Pauli* in 1925 when he was investigating atomic spectra. It states that there can be only one electron in a particular quantum-mechanical state. This simple formulation of Pauli's principle, however, has to be specified in somewhat more detail. We have just stressed that in a system of electrons only the state of the whole system and not that of the single particles is defined. Accordingly, the state of an electron in an atom will certainly change if another electron is put into the electron shell or if the atom is ionized.

We will be able to avoid these difficulties if we refer to the measuring process used on an electron. Taking into account the degree of freedom resulting from the spin, the electron has four degrees of freedom. Therefore, its state is characterized by four independent numbers. As usual, the appropriate quantities to be chosen are energy, angular momentum, the z component of the angular momentum and the z component of the spin. This set of quantities corresponds to the quantum numbers n, l, m_l, m_s. The choice of another set of quantities, e.g. the three momentum components and the spin component, is also possible. According to the choice we make, the wave function is then determined by four quantum numbers:

$$\psi = \psi_{nlm_lm_s} \ . \tag{15.8}$$

Now we can formulate the Pauli principle in a more precise form: In a system of electrons, the measurement of four quantities that are typical of the electron (e.g. the quantum numbers n, l, m_l, m_s) can have a well-defined (fixed) value for one electron only at any one moment. Two electrons can never simultaneously occupy the state (15.8). Soon we will understand the thus formulated Pauli principle as a consequence of the antisymmetry of the wave function described in (15.5).

Since this empirically ascertained principle is a consequence of the antisymmetry of the fermion wave function, the Pauli principle is not only valid for electrons, but for all fermions.

15.2 Exchange Degeneracy

We consider a system of N identical particles without any interaction; the inclusion of interactions would not change any of the following fundamental considerations. The Schrödinger equation for such a system is

$$\begin{aligned}(\hat{H}_1 + \hat{H}_2 + \cdots + \hat{H}_N)\psi(\boldsymbol{r}_1 s_1, \boldsymbol{r}_2 s_2, \ldots, \boldsymbol{r}_N s_N)\\ = E\psi(\boldsymbol{r}_1 s_1, \boldsymbol{r}_2 s_2, \ldots, \boldsymbol{r}_N s_N) \ .\end{aligned} \tag{15.9}$$

The single-particle Hamiltonians $\hat{H}_i(\boldsymbol{r}_i, s_i)$ can be distinguished from each other by the fact that they act on different particles. If we designate the ith eigenfunction of the particle k by $\varphi_i(\boldsymbol{r}_k, s_k)$, we will have for the eigenvalue problem of the single particle

$$\hat{H}_k(\boldsymbol{r}_k, s_k)\varphi_i(\boldsymbol{r}_k, s_k) = E_i\varphi_i(\boldsymbol{r}_k, s_k) \ ,$$
$$k = 1, 2, \dots, N \ ; \quad i = 1, 2, \dots \ . \tag{15.10}$$

The Schrödinger equation (15.9) is then solved by the product of single-particle wave functions:

$$\psi(\boldsymbol{r}_1 s_1, \boldsymbol{r}_2 s_2, \dots, \boldsymbol{r}_N s_N) = \varphi_{i_1}(\boldsymbol{r}_1, s_1)\varphi_{i_2}(\boldsymbol{r}_2, s_2)\dots\varphi_{i_N}(\boldsymbol{r}_N, s_N) \ . \tag{15.11}$$

The $i_j (j = 1, 2, \dots)$ are special numbers that characterize the eigenfunctions. If there are n_i particles in the state φ_i, then we will get for the eigenvalue of the total energy

$$E = \sum_i n_i E_i \quad \text{with} \quad \sum_i n_i = N \ . \tag{15.12}$$

Because of the indistinguishability of the particles, we are not able to say which particle is in which state. This means that there are $N!/(n_1!n_2!n_3!\dots)$ combinations of single-particle wave functions in (15.11) that give the same energy eigenvalue E. This is called *exchange degeneracy*.

Exchange degeneracy is lifted by the requirement of wave-function symmetry for the bosons and fermions. In fact, the entire space of functions spanned by the eigenfunctions to the energy E contains only *one symmetric* and *one antisymmetric* wave function. The symmetric wave function for bosons results from the sum of all possible $N!$ permutations of the arguments of the single-particle wave functions in (15.11). If we designate the permutations by P, then the wave function of a system of bosons together with its normalization factor $(N!n_1!n_2!\dots)^{-1/2}$ reads

$$\psi_{\text{boson}} = \frac{1}{\sqrt{N!n_1!n_2!\dots}}$$
$$\times \sum_{P=1}^{N!} P\varphi_{i_1}(\boldsymbol{r}_1, s_1)\varphi_{i_2}(\boldsymbol{r}_2, s_2)\dots\varphi_{i_N}(\boldsymbol{r}_N, s_N) \ . \tag{15.13}$$

Here, we have assumed that the single-particle wave functions are orthonormalized.

15.3 The Slater Determinant

The antisymmetric wave function is generally accepted to be best expressed in the form of a determinant. ***Slater***'s determinant is an $N \times N$ determinant consisting of a single-particle wave function (15.10) arranged in the following way:

$$\psi_{\text{fermion}} = \frac{1}{\sqrt{N!}} \begin{vmatrix} \varphi_{i_1}(\boldsymbol{r}_1, s_1) & \varphi_{i_2}(\boldsymbol{r}_1, s_1) & \dots & \varphi_{i_N}(\boldsymbol{r}_1, s_1) \\ \varphi_{i_1}(\boldsymbol{r}_2, s_2) & \varphi_{i_2}(\boldsymbol{r}_2, s_2) & \dots & \varphi_{i_N}(\boldsymbol{r}_2, s_2) \\ \vdots & & & \vdots \\ \varphi_{i_1}(\boldsymbol{r}_N, s_N) & \varphi_{i_2}(\boldsymbol{r}_N, s_N) & \dots & \varphi_{i_N}(\boldsymbol{r}_N, s_N) \end{vmatrix} . \tag{15.14}$$

Note that each column always contains the same single-particle wave function, while each row contains the same argument in the single-particle wave function.

The determinant form easily ensures the required properties of the fermion wave function in an elegant way. By interchanging two particles (two rows), the sign changes. The function will vanish if two particles occupy the same state (two columns are equal). This is Pauli's principle! It is thus a consequence of the antisymmetry of the wave function, an indeed most interesting and fundamental result.

We shall now illustrate these features in the following example.

EXAMPLE

15.1 The Helium Atom

Some phenomena occurring in many-body problems may be demonstrated by the helium atom. It consists of the He nucleus surrounded by two orbiting electrons. For the mathematical description of the helium atom, we start with the independent motion of two electrons in the Coulomb field and treat their mutual electrostatic interaction as a perturbation. The Hamiltonian of the systems is

$$\hat{H}\psi = (\hat{H}(1) + \hat{H}(2) + \hat{W}(12))\psi = E\psi(\boldsymbol{r}_1, \boldsymbol{r}_2) . \tag{1}$$

The Hamiltonians $\hat{H}(1)$ and $\hat{H}(2)$ are the operators for the single-particle problem, i.e. the individual electrons. They are

$$\hat{H}(1) = -\frac{\hbar^2}{2m}\Delta_1 + V(\boldsymbol{r}_1) \quad \text{and}$$
$$\hat{H}(2) = -\frac{\hbar^2}{2m}\Delta_2 + V(\boldsymbol{r}_2) , \tag{2}$$

with $V(r) = -Ze^2/r$, $Z = 2$.

Example 15.1

The solutions are obviously obtained from the hydrogen wave functions if $Z = 1$ is replaced by $Z = 2$. Hence,

$$\hat{H}(1)\psi_r(\boldsymbol{r}_1) = E_r\psi_r(\boldsymbol{r}_1) \quad \text{and}$$
$$\hat{H}(2)\psi_s(\boldsymbol{r}_2) = E_s\psi_s(\boldsymbol{r}_2) \ . \tag{3}$$

The indices r, s represent the set of quantum numbers n, l, m. The degeneracy of the hydrogen wave function will not be taken into account to prevent the problem from becoming too complicated. The two-electron Schrödinger equation without interactions now reads

$$\hat{H}_0\psi = (\hat{H}(1) + \hat{H}(2))\psi = E\psi \ . \tag{4}$$

According to the separation of the problem expressed in (2) and (3), we immediately get the product wave function

$$\psi(\boldsymbol{r}_1, \boldsymbol{r}_2) = \psi_r(\boldsymbol{r}_1)\psi_s(\boldsymbol{r}_2) \ , \tag{5}$$

which obeys the eigenvalue equation

$$\hat{H}_0\psi(\boldsymbol{r}_1, \boldsymbol{r}_2) = (E_r + E_s)\psi(\boldsymbol{r}_1, \boldsymbol{r}_2) \ . \tag{6}$$

Obviously, not only the state ψ belongs to the eigenvalue $E_r + E_s$, but also the state

$$\psi'(\boldsymbol{r}_1, \boldsymbol{r}_2) = \psi_s(\boldsymbol{r}_1)\psi_r(\boldsymbol{r}_2) \ , \tag{7}$$

where the first particle is in the state ψ_s and the second particle is in the state ψ_r. This is the exchange degeneracy mentioned earlier. The two states ψ and ψ' emerge from each other by interchanging the coordinates of the particles. As a solution of the Schrödinger equation (4), we therefore have to consider a linear combination of two states:

$$\psi(\boldsymbol{r}_1, \boldsymbol{r}_2) = a\psi_r(\boldsymbol{r}_1)\psi_s(\boldsymbol{r}_2) + b\psi_s(\boldsymbol{r}_1)\psi_r(\boldsymbol{r}_2) \ . \tag{8}$$

Because of the normalization of the states,

$$a^2 + b^2 = 1 \tag{9}$$

holds.

Now we take into account the influence of the interaction by using perturbation theory. Therefore we start with (11.25). (Here, we have named the perturbation energy "W" instead of "εW".) The unperturbed energy is $E^0 = E_r + E_s$; the expansion coefficients $a_{k\alpha}$ are a and b. With the abbreviation

$$\varepsilon = E - E_r - E_s \ , \tag{10}$$

we get the two equations

$$(W_{11} - \varepsilon)a + W_{12}b = 0 \ ,$$
$$W_{21}a + (W_{22} - \varepsilon)b = 0 \ , \tag{11}$$

Example 15.1

with the secular determinant

$$D=\begin{vmatrix} W_{11}-\varepsilon & W_{12} \\ W_{21} & W_{22}-\varepsilon \end{vmatrix} \ . \tag{12}$$

The matrix elements of the perturbation are given, together with the interaction

$$W(1,2)=\frac{e^2}{|\boldsymbol{r}_1-\boldsymbol{r}_2|}=\frac{e^2}{r_{12}} \ , \tag{13}$$

by the integrals

$$W_{11}=W_{22}=e^2\int\frac{|\psi_r(\boldsymbol{r}_1)|^2|\psi_s(\boldsymbol{r}_2)|^2}{r_{12}}\,\mathrm{d}V_1\,\mathrm{d}V_2 \tag{14}$$

and

$$W_{12}=W_{21}=e^2\int\frac{\psi_r^*(\boldsymbol{r}_1)\psi_s(\boldsymbol{r}_1)\psi_r(\boldsymbol{r}_2)\psi_s^*(\boldsymbol{r}_2)}{r_{12}}\,\mathrm{d}V_1\,\mathrm{d}V_2 \ . \tag{15}$$

Usually the matrix elements are denoted by the letters

$$W_{11}=W_{22}=K \quad \text{and} \quad W_{12}=W_{21}=A \ . \tag{16}$$

The quantity K is the *Coulomb interaction* of the two charge densities $e|\psi_r(\boldsymbol{r}_1)|^2$ and $e|\psi_s(\boldsymbol{r}_2)|^2$. The quantity A is called the *exchange energy*; it has no classical analogue. The exchange integral is due to the fact that an electron may be in the state ψ_r as well as in the state ψ_s. The magnitude of the exchange integral depends on the product $\psi_r\psi_s$, i.e. on the overlapping of the two wave functions. Thus, for example, the exchange energy between the ground state and a highly excited state is very small.

From the requirement for a nontrivial solution of system (11), it follows that the secular determinant has to vanish:

$$D=0 \ , \tag{17}$$

so that

$$(K-\varepsilon)^2=A^2 \tag{18}$$

has to hold. Therefore we get for energy splitting by the perturbation

$$\varepsilon=K\pm A \ . \tag{19}$$

For $\varepsilon=K+A$, (9) and (11) yield

$$a=b=\frac{1}{\sqrt{2}} \ , \tag{20}$$

Example 15.1

and for $\varepsilon = K - A$, analogously

$$a = -b = +\frac{1}{\sqrt{2}} \; . \tag{21}$$

The exchange degeneracy is broken by the interaction; the state splits up into a *symmetric* and an *antisymmetric* state:

$$\psi_{\mathrm{s}}(\boldsymbol{r}_1, \boldsymbol{r}_2) = \frac{1}{\sqrt{2}}(\psi_r(\boldsymbol{r}_1)\psi_s(\boldsymbol{r}_2) + \psi_s(\boldsymbol{r}_1)\psi_r(\boldsymbol{r}_2)) \quad \text{and}$$

$$\psi_{\mathrm{a}}(\boldsymbol{r}_1, \boldsymbol{r}_2) = \frac{1}{\sqrt{2}}(\psi_r(\boldsymbol{r}_1)\psi_s(\boldsymbol{r}_2) - \psi_s(\boldsymbol{r}_1)\psi_r(\boldsymbol{r}_2)) \; . \tag{22}$$

Until now, we have regarded the electrons as spinless particles. Since the electrons have a spin, (see Chaps. 12, 13) they are fermions and their total wave function has to be antisymmetric. The interactions involving spin (spin–orbit, spin–spin) are neglected; then we can write the total wave function as the product of the space (ψ) and spin (χ) wave functions:

$$\psi = \psi(\boldsymbol{r}_1, \boldsymbol{r}_2)\chi \; . \tag{23}$$

Since *the total wave function has to be antisymmetric*, the product functions (23) always consist of an antisymmetric and a symmetric function. Either the spatial part ψ is symmetric and the spin function χ antisymmetric, or vice versa.

We denote the spin function of the particle 1 with spin up by χ_1^+, etc. Three symmetric and one antisymmetric states can be constructed from the spin functions:

$$\chi_{\mathrm{s}}^+ = \chi_1^+\chi_2^+ \; ,$$

$$\chi_{\mathrm{s}}^0 = \frac{1}{\sqrt{2}}(\chi_1^+\chi_2^- + \chi_1^-\chi_2^+) \; ,$$

$$\chi_{\mathrm{s}}^- = \chi_1^-\chi_2^- \; ,$$

$$\chi_{\mathrm{a}}^0 = \frac{1}{\sqrt{2}}(\chi_1^+\chi_2^- - \chi_1^-\chi_2^+) \; . \tag{24}$$

The factors $1/\sqrt{2}$ are necessary for normalization. The helium atoms with symmetric spin function are called *orthohelium*; those with an antisymmetric spin function are called *parahelium*. The properties are summarized in the following figure. E_1 is the ground state energy of the hydrogen atom for $Z = 2$; E_2 is the corresponding energy for the first excited state.

Example 15.1

Wave function	Orthohelium	Spin	Energy
$\frac{1}{\sqrt{2}}(\psi_1(\boldsymbol{r}_1)\psi_2(\boldsymbol{r}_2) - \psi_2(\boldsymbol{r}_1)\psi_1(\boldsymbol{r}_2))$	$\begin{cases} \chi_1^+\chi_2^+ \\ \frac{1}{\sqrt{2}}(\chi_1^+\chi_2^- + \chi_2^+\chi_1^-) \\ \chi_1^-\chi_2^- \end{cases}$	↑↑ →← ↓↓	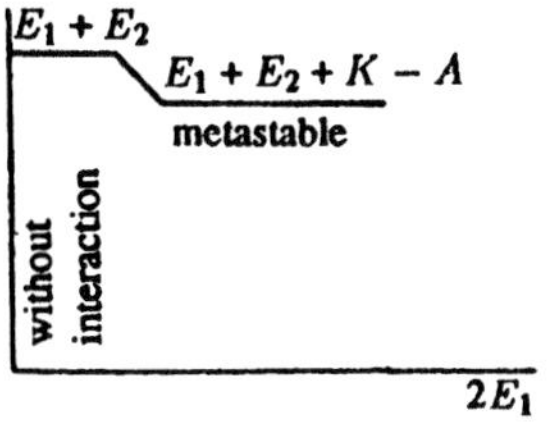
	Parahelium		
$\frac{1}{\sqrt{2}}(\psi_1(\boldsymbol{r}_1)\psi_2(\boldsymbol{r}_2) + \psi_2(\boldsymbol{r}_1)\psi_1(\boldsymbol{r}_2))$	$\frac{1}{\sqrt{2}}(\chi_1^+\chi_2^- - \chi_2^+\chi_1^-)$	↑↓	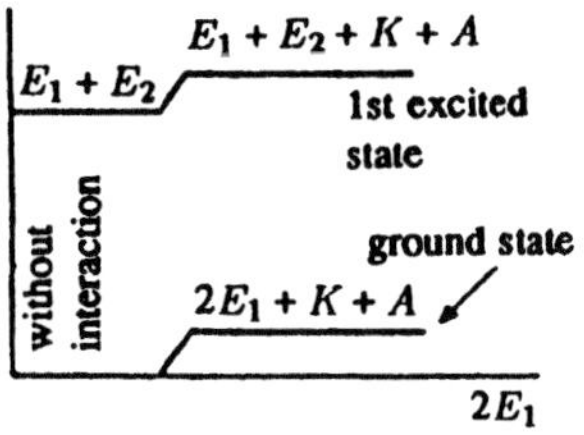

Orthohelium and parahelium wave functions and energies

Parahelium is energetically the lowest state of helium. Its spatial wave function is symmetric; both electrons may simultaneously occupy the ground state. Then the Pauli principle requires an antisymmetric spin function. For orthohelium, *the spin function is symmetric*; the Pauli principle prevents both particles from simultaneously being in the ground state. Because of the small *spin–spin interaction* (interaction of the respective magnetic dipole moments) *the possibility of a spin flip in orthohelium is very small*; therefore orthohelium represents a *metastable state* of helium.

EXAMPLE

15.2 The Hydrogen Molecule

The exchange energy is the reason for the *homeopolar binding of molecules*. To get a better understanding of this kind of binding, we examine the simplest example: the hydrogen molecule. Here, we use perturbation theory, as we did for the helium atom in the preceding example. As an approximation of zeroth order, we use products of the hydrogen eigenfunctions for the molecular wave function. We begin with two hydrogen atoms that are very far away from each other and regard the forces occurring during their approach as a perturbation. It is clear that this way of treating the problem is not very accurate, because the forces appearing in the molecular region of interest will not be small any more. With the notation given in the figure, the Schrödinger equation for the two electrons in the potential of the two protons is

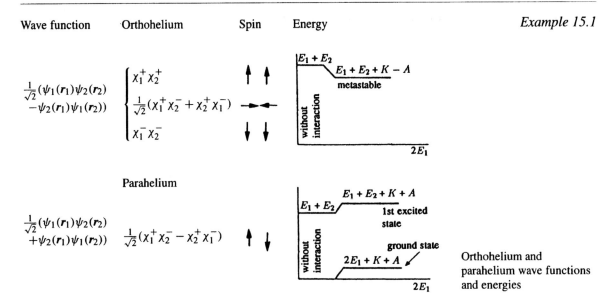

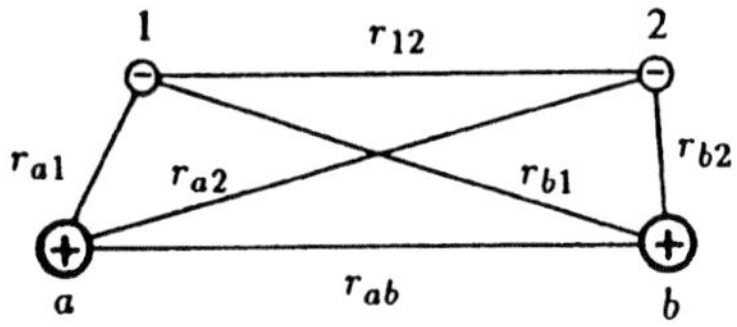

Atomic nuclei (protons) a and b and electrons 1 and 2 in the hydrogen molecule

$$-\frac{\hbar^2}{2m}(\Delta_1+\Delta_2)\psi + \left[\frac{e^2}{r_{12}} - \frac{e^2}{r_{a1}} - \frac{e^2}{r_{a2}} - \frac{e^2}{r_{b1}} - \frac{e^2}{r_{b2}}\right]\psi = E\psi \ . \qquad (1)$$

Example 15.2

The distance r_{ab} is used as a parameter in the following calculation. The Hamiltonian $\hat{H}$ of equation (1) is now split up in two different ways. With the abbreviations

$$\hat{H}_{a1} = -\frac{\hbar^2}{2m}\Delta_1 - \frac{e^2}{r_{a1}} ,$$

$$\hat{H}_{b2} = -\frac{\hbar^2}{2m}\Delta_2 - \frac{e^2}{r_{b2}} \quad \text{and}$$

$$\hat{W}_{a2,b1} = -\frac{e^2}{r_{a2}} - \frac{e^2}{r_{b1}} + \frac{e^2}{r_{12}} , \tag{2}$$

we can write

$$\hat{H} = \hat{H}_{a1} + \hat{H}_{b2} + \hat{W}_{a2,b1} .$$

This decomposition corresponds to associating electron 1 to nucleus a and electron 2 to nucleus b. Consequently, for a large distance between the two nuclei, the $W_{a2,b1}$ part of the Hamiltonian, which is treated as a perturbation in our approximation, vanishes. With this association of electrons to the nuclei, (1) becomes for $r_{ab} \to \infty$

$$(\hat{H}_{a1} + \hat{H}_{b2})u = Eu , \tag{3}$$

which is an equation that describes two noninteracting hydrogen atoms. It is solved by the product wave function

$$u(\boldsymbol{r}_1, \boldsymbol{r}_2) = \psi_a(r_{a1})\psi_b(r_{b2}) , \tag{4}$$

where ψ symbolizes the hydrogen wave functions that follow from the equation

$$\hat{H}_{a1}\psi_a = E_1\psi_a \tag{5}$$

and the corresponding equation for ψ_b. Here, we set $E = 2E_1$, since we like to assume that both hydrogen atoms are in the ground state with energy E_1.

Immediately, we see that we have an exchange degeneracy for this problem, since associating electron 2 with nucleus a, and electron 1 with nucleus b leads to equations that differ from the former one only in the indices. Instead of the operators (2) we then have

$$\hat{H}_{a2} = -\frac{\hbar^2}{2m}\Delta_2 - \frac{e^2}{r_{a2}} , \quad \hat{H}_{b1} = -\frac{\hbar^2}{2m}\Delta_1 - \frac{e^2}{r_{b1}} ,$$

$$W_{a1,b2} = -\frac{e^2}{r_{a1}} - \frac{e^2}{r_{b2}} + \frac{e^2}{r_{12}} . \tag{6}$$

Again, $W_{a1,b2}$ vanishes for $r_{ab} \to \infty$, and the remaining part of (1) is solved by

$$v(\boldsymbol{r}_1, \boldsymbol{r}_2) = \psi_b(r_{b1})\psi_a(r_{a2}) . \tag{7}$$

Example 15.2

The hydrogen wave functions ψ fulfil a correspondingly changed form of (5). Equation (1) is solved for $r_{ab} \to \infty$ by a linear combination of the functions (4) and (7), namely

$$au + bv = a\psi_a(r_{a1})\psi_b(r_{b2}) + b\psi_b(r_{b1})\psi_a(r_{a2}) \ , \tag{8}$$

where a and b are constants of the nuclei a and b, respectively. We use this linear combination (8) of the functions u and v as the zeroth approximation for the solution of our problem. Note that the functions u and v are only orthogonal in the limit $r_{ab} \to \infty$. When the two nuclei come closer to each other, the electron wave functions overlap and the integral

$$\begin{aligned} |S|^2 &= \int u^* v \, dV_1 \, dV_2 \\ &= \underbrace{\int \psi_a^*(r_{a1})\psi_b(r_{b1}) \, dV_1}_{S} \underbrace{\int \psi_a(r_{a2})\psi_b^*(r_{b2}) \, dV_2}_{S^*} \end{aligned} \tag{9}$$

will be nonzero.

As a result of the perturbation W, the energy of the system changes, as does the wave function. We write

$$E = 2E_1 + \varepsilon \quad \text{and} \quad \Psi = au + bv + \varphi \ . \tag{10}$$

The various terms $W_{a1,b2}$, $W_{a2,b1}$, ε and φ are assumed to be small; products of these quantities are neglected in the following. Now we insert (10) into (1) and get, by neglecting products which are small in second order,

$$\begin{aligned} &a(\hat{H}_{a1} + \hat{H}_{b2} + W_{a2,b1})u + b(\hat{H}_{a2} + \hat{H}_{b1} + W_{a1,b2})v + (\hat{H}_{a1} + \hat{H}_{b2})\varphi \\ &\quad = 2E_1(au + bv) + \varepsilon(au + bv) + 2E_1\varphi \ . \end{aligned} \tag{11}$$

The parts of the unperturbed system cancel and after reordering according to the different functions, we have

$$a(W_{a2,b1} - \varepsilon)u + b(W_{a1,b2} - \varepsilon)v + (\hat{H}_{a1} + \hat{H}_{b2} - 2E_1)\varphi = 0 \ . \tag{12}$$

For $a = b = 0$, (12) is a homogenous differential equation for φ with the solution $\varphi = u$, as a comparison with (3) shows. We use the theorem that the solution of a homogenous differential equation is orthogonal to the inhomogeneous part of the differential equation. Hence,

$$\int [a(W_{a2,b1} - \varepsilon)u + b(W_{a1,b2} - \varepsilon)v]^* u \, dV_1 \, dV_2 = 0 \ . \tag{13}$$

In formulating (11), we have expressed the Hamiltonian $\hat{H}$ that acts on φ via the decomposition (2). If we use the decomposition (6) as the next step, we will get instead of (12) the differential equation

$$a(W_{a2,b1} - \varepsilon)u + b(W_{a1,b2} - \varepsilon)v + (\hat{H}_{a2} + \hat{H}_{b1} - 2E_1)\varphi = 0 \ , \tag{14}$$

Example 15.2

which is solved by the function $\varphi = v$ for the homogenous part. Then the same argument as above in (13) leads to the integral

$$\int [a(W_{\mathrm{a2,b1}} - \varepsilon)u + b(W_{\mathrm{a1,b2}} - \varepsilon)v]^* v \, \mathrm{d}V_1 \, \mathrm{d}V_2 = 0 \ . \tag{15}$$

It holds that

$$\int |u|^2 W_{\mathrm{a2,b1}} \, \mathrm{d}V_1 \, \mathrm{d}V_2 = \int |v|^2 W_{\mathrm{a1,b2}} \, \mathrm{d}V_1 \, \mathrm{d}V_2 = K \ , \tag{16}$$

and

$$\int u^* v W_{\mathrm{a2,b1}} \, \mathrm{d}V_1 \, \mathrm{d}V_2 = \int v^* u W_{\mathrm{a1,b2}} \, \mathrm{d}V_1 \, \mathrm{d}V_2 = A \ . \tag{17}$$

The equality of the integrals is due to the fact that the arguments differ by their indices only. Here, K is the *Coulomb energy* of the perturbation; A is the *exchange energy*. If, for example, we insert the perturbation W in the form (2), it will follow that

$$\begin{aligned} K = & -e^2 \int \frac{|\psi_{\mathrm{a}}(r_{\mathrm{a1}})|^2}{r_{\mathrm{b1}}} \mathrm{d}V_1 - e^2 \int \frac{|\psi_{\mathrm{b}}(r_{\mathrm{b2}})|^2}{r_{\mathrm{a2}}} \mathrm{d}V_2 \\ & + e^2 \int \frac{|\psi_{\mathrm{a}}(r_{\mathrm{a1}})|^2 |\psi_{\mathrm{b}}(r_{\mathrm{b2}})|^2}{r_{12}} \mathrm{d}V_1 \, \mathrm{d}V_2 \ , \end{aligned} \tag{18}$$

and

$$\begin{aligned} A = & -e^2 \int \frac{\psi_{\mathrm{a}}^*(r_{\mathrm{a1}}) \psi_{\mathrm{b}}(r_{\mathrm{b1}})}{r_{\mathrm{b1}}} \mathrm{d}V_1 \times S^* - e^2 \int \frac{\psi_{\mathrm{a}}(r_{\mathrm{a2}}) \psi_{\mathrm{b}}^*(r_{\mathrm{b2}})}{r_{\mathrm{a2}}} \mathrm{d}V_2 \times S \\ & + e^2 \int \frac{\psi_{\mathrm{a}}^*(r_{\mathrm{a1}}) \psi_{\mathrm{b}}^*(r_{\mathrm{b2}}) \psi_{\mathrm{a}}(r_{\mathrm{a2}}) \psi_{\mathrm{b}}(r_{\mathrm{b1}})}{r_{12}} \mathrm{d}V_1 \, \mathrm{d}V_2 \ . \end{aligned}$$

In the case of the Coulomb energy, the various terms express the energy of the interaction of the various electron charge distributions with the other nucleus and the mutual interaction of the two electron charge distributions.

In the case of the exchange energy, the mixed densities appear. The quantity S, defined in (9), expresses the overlap of the nonorthogonal electron wave functions. We have

$$S(r_{\mathrm{ab}} \to \infty) = 0 \quad \text{and} \quad S(r_{\mathrm{ab}} \to 0) = 1 \ .$$

By using the abbreviations introduced for the different integrals, we can write (13) and (15) in compact form:

$$(\varepsilon - K)a + (\varepsilon S^2 - A)b = 0 \ , \tag{19}$$

$$(\varepsilon S^2 - A)a + (\varepsilon - K)b = 0 \ . \tag{20}$$

Example 15.2

Thus we have two equations for determining the coefficients of the linear combination (8). Assuming a nontrivial solution for the system of equations, its determinant has to vanish, yielding the relation

$$(\varepsilon - K)^2 = (\varepsilon S^2 - A)^2 \ .$$

The solutions of these equations give the energy shifts

$$\varepsilon_1 = \frac{K-A}{1-S^2} = \varepsilon_{\text{anti}} \ , \tag{21}$$

$$\varepsilon_2 = \frac{K+A}{1+S^2} = \varepsilon_{\text{sym}} \ . \tag{22}$$

The insertion of these two solutions into (19) and (20) yields for the coefficients

$$a = -b \quad \text{for} \quad \varepsilon_1 \quad \text{and} \tag{23}$$

$$a = b \quad \text{for} \quad \varepsilon_2 \ . \tag{24}$$

Therefore we get a symmetric solution, with the energy

$$E_{\text{sym}} = 2E_1 + \frac{K+A}{1+S^2} \ , \tag{25}$$

and an antisymmetric solution, with

$$E_{\text{anti}} = 2E_1 + \frac{K-A}{1-S^2} \ . \tag{26}$$

To ascertain the energies, we have to calculate the integrals K, A and S with the wave function of the ground state of the hydrogen atom. Owing to the extensive calculation necessary for this task, we give only a graphical representation here. We treat the protons as classical point particles. Then the energy

$$\varepsilon' = \varepsilon + \frac{e^2}{r_{\text{ab}}} \tag{27}$$

is the *binding energy of the molecule.* In the following figure, the Coulomb energy $K + e^2/r_{\text{ab}}$ is given as a function of the distance of the nuclei (in units of Bohr's radius). The result is a very weak binding. The *exchange energy is negative* and, except for the case of very small distances, it is greater than the Coulomb energy. This causes a *stronger binding in the symmetric state* (22), *and repulsion in the antisymmetric case* (21). Therefore the sign of the exchange energy is responsible for the binding of the H_2 molecule. The real binding energy of the hydrogen molecule is much smaller than the value of this calculation (-4.4 eV). In spite of this quantitative failing, the calculation gives an idea of how homeopolar binding comes about.

Thus we have found a symmetric local wave function of the ground state of hydrogen. Because of the Pauli principle, the spin function has to be antisymmetric, i.e. the electron spins are oriented in an antiparallel manner [see (24) of

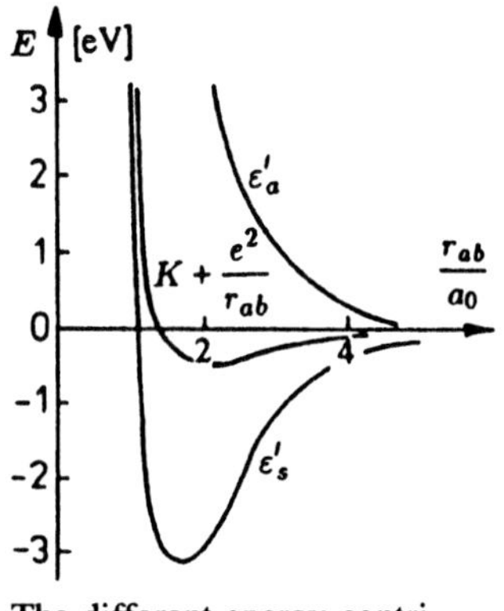

The different energy contributions to the binding of the hydrogen molecule

Example 15.2

Example 15.1]. We distinguish, as with helium, between *parahydrogen (singlet state)* and *metastable orthohydrogen (triplet state)*.

Our presentation of a solution to the problem by perturbation theory is based on considerations by W. ***Heitler*** and F. ***London***. More precise methods first solve the two-centre problem for the electrons, i.e. the Schrödinger equation with the Hamiltonian (2) without the electron-electron interaction e^2/r_{12}. In 1930, E. ***Teller*** and E.A. ***Hylleraas*** applied this new method.

Recently this kind of problem has again attracted attention, particularly in connection with the creation of *very heavy quasi-molecules* in the collision of very heavy ions. Since this scenario is a two-centre problem of very heavy nuclei and one electron (e.g. uranium–uranium molecule) with small distances between the nuclei, the two-centre Dirac equation[1] has to be solved, because the inner electrons in heavy and very heavy elements are relativistic. We shall treat this topic in more detail when discussing relativistic quantum theory.

EXAMPLE

15.3 The van der Waals Interaction

As an example of an application for the variational method (see Chap. 11) we calculate the long-range (***van der Waals***) interaction between two hydrogen atoms in their ground states. To this end, it is useful initially to treat this problem using perturbation theory, because afterwards it is easier to see that the leading term of the interaction energy is inversely proportional to the sixth power ($\sim 1/R^6$) of the varying distance R between the two atoms. It will also become apparent that perturbation theory and the variational calculation represent opposite limits for the determination of the coefficients of the $1/R^6$ term.

The two atomic nuclei A and B of the hydrogen atoms are separated by a distance (see figure) and the z axis is given by the connecting line between A and B. We denote the local vector of electron 1 relative to nucleus A by $\boldsymbol{r}_1$, and the local vector of electron 2 relative to nucleus B by $\boldsymbol{r}_2$.

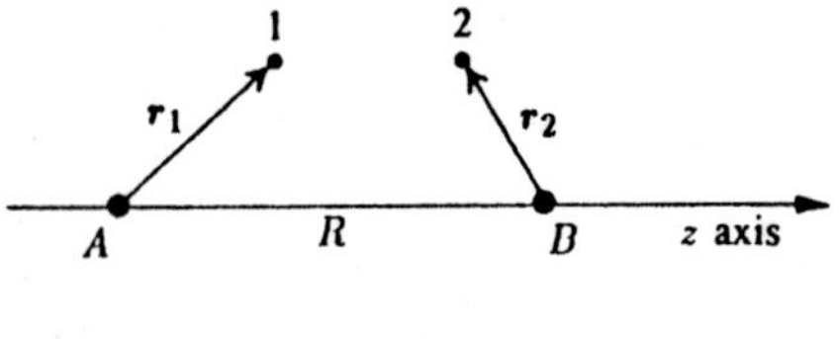

Two hydrogen atoms whose nuclei are separated by a distance R

[1] See B. Müller, W. Greiner: Z. Naturforsch. **31**A, 1 (1976). An extensive discussion of this exciting physics can be found in: W. Greiner, B. Müller, J. Rafelski: *Quantum Electrodynamics of Strong Fields* (Springer, Berlin, Heidelberg 1985).

Example 15.3

The Hamiltonian for the two electrons reads (with spin–orbit coupling neglected)

$$\hat{H} = \hat{H}_0 + \hat{H}' \quad , \quad \text{with}$$

$$\hat{H}_0 = -\frac{\hbar^2}{2m}(\nabla_1^2 + \nabla_2^2) - \frac{e^2}{r_1} - \frac{e^2}{r_2}$$

$$\hat{H}' = \frac{e^2}{R} + \frac{e^2}{r_{12}} - \frac{e^2}{r_{1B}} - \frac{e^2}{r_{2A}} \ , \tag{1}$$

where, in the ground state, the unperturbed Hamiltonian $\hat{H}_0$ has the stationary solution

$$\Psi_0(\boldsymbol{r}_1, \boldsymbol{r}_2) = \psi_{100}(\boldsymbol{r}_1)\psi_{100}(\boldsymbol{r}_2) \ . \tag{2}$$

Here $\Psi_{nlm}(r)$ denotes the known hydrogen eigenfunctions [see (9.45)]. We regard $\hat{H}'$ as a perturbation, which is surely approximately valid for a large distance between the two atoms $R \gg a_0$, a_0 being Bohr's radius.

Since we are interested in the leading term of the interaction energy, we expand $\hat{H}'$ according to powers of $1/R$ and pay attention only to the terms of lowest order:

$$\begin{aligned}\hat{H}' = \frac{e^2}{R}\Bigg\{1 + \Bigg[1 + \frac{2(z_2 - z_1)}{R} \\ + \frac{(x_2 - x_1)^2 + (y_2 - y_1)^2 + (z_2 - z_1)^2}{R^2}\Bigg]^{-1/2} \\ - \left(1 - \frac{2z_1}{R} + \frac{r_1^2}{R^2}\right)^{-1/2} - \left(1 + \frac{2z_2}{R} + \frac{r_2^2}{R^2}\right)^{-1/2}\Bigg\} \\ \approx \frac{e^2}{R^3}(x_1x_2 + y_1y_2 - 2z_1z_2) \ .\end{aligned} \tag{3}$$

The leading term apparently describes the interaction energy of two electric dipoles, which are given by the momentary configuration of the two atoms. The neglected terms of the order $1/R^4$ and $1/R^5$ correspond to the instantaneous dipole–quadrupole interaction and the quadrupole–quadrupole interaction, respectively.

Now, it is clear that the expectation value of the leading term of $\hat{H}'$ (3) in the state $\Psi_0(\boldsymbol{r}_1, \boldsymbol{r}_2)$ (2) vanishes, because Ψ_0 is an even function of $\boldsymbol{r}_1$ and $\boldsymbol{r}_2$, while $\hat{H}'$, on the contrary, is an odd function of $\boldsymbol{r}_1$ and $\boldsymbol{r}_2$. One can also show that all neglected terms of higher order in $\hat{H}'$ have a vanishing expectation value in the state $\Psi_0(\boldsymbol{r}_1, \boldsymbol{r}_2)$, too, because these terms may be expressed by spherical harmonics Y_{lm} with $l \neq 0$. Therefore the leading term of the interaction energy of the two atoms has to be of second order in perturbation theory when the dipole part [see (3)] is taken into account, and so it has to depend on the distance like $1/R^6$.

Example 15.3

Perturbation Theory

In second-order perturbation theory, the interaction energy of the two hydrogen atoms is

$$W(R) = -\sum_{n\neq 0} \frac{|\langle n|\hat{H}'|0\rangle|^2}{E_n - E_0} . \tag{4}$$

Here, the index n stands for all states of the pair of electrons of both hydrogen atoms, including the dissociated states. Because of (4), it is clear that $W(R)$ is negative, since $E_n > E_0$ and therefore nominator and denominator are always positive. Hence, we may conclude that the interaction of the two hydrogen atoms is attractive and proportional to $1/R^6$ if R is large (i.e. $R \gg a_0$). We realize that this conclusion is valid for any pair of atoms which have nondegenerate spherically symmetric ground states; this is generally the case.

We can get an upper bound for the positive quantity $-W(R)$ as follows. Let us substitute all E_n in the nominator by the lowest value E_1. Here, E_1 is the first excited state ($2p$ state) of the hydrogen atoms. Then all denominators in the sum (4) are equal, and the summation is performed in the following way:

$$\begin{aligned}
-W(R) &= \sum_{n\neq 0} \frac{\langle n|\hat{H}'|0\rangle^2}{E_n - E_0} \leq \sum_{n\neq 0} \frac{|\langle n|\hat{H}'|0\rangle|^2}{E_1 - E_0} \\
&= \frac{1}{E_1 - E_0} \sum_{n\neq 0} \left|\left\langle n\left|\hat{H}'\right|0\right\rangle\right|^2 \\
&= \frac{1}{E_1 - E_0} \left(\sum_n \left\langle 0\left|\hat{H}'\right|n\right\rangle\left\langle n\left|\hat{H}'\right|0\right\rangle - \left|\left\langle 0\left|\hat{H}'\right|0\right\rangle\right|^2\right) \\
&= \frac{1}{E_1 - E_0} \left(\left\langle 0\left|\hat{H}'^2\right|0\right\rangle - \left\langle 0\left|\hat{H}'\right|0\right\rangle^2\right) \\
&= \frac{\langle 0|\hat{H}'^2|0\rangle}{E_1 - E_0} .
\end{aligned} \tag{5}$$

Here, we have used the *completeness relation* and the well-known fact that

$$\left\langle 0\left|\hat{H}'\right|0\right\rangle = 0 .$$

Thus

$$W(R) \geq -\frac{\langle 0|\hat{H}'^2|0\rangle}{E_1 - E_0} . \tag{6}$$

Now, we have (see Chap. 9)

$$E_0 = -2\left(\frac{e^2}{2a_0}\right) \quad \text{(both atoms in the ground state)}$$

$$E_1 = -2\left(\frac{e^2}{8a_0}\right) \quad \text{(both atoms in the first excited state)} .$$

Consequently, *Example 15.3*

$$E_1 - E_0 = \frac{3e^2}{4a_0} \ . \tag{7}$$

Furthermore, it follows from (3) that

$$\hat{H}'^2 = \frac{e^2}{R^6}(x_1^2x_2^2 + y_1^2y_2^2 + 4z_1^2z_2^2 + 2x_1x_2y_1y_2 - \dots) \ . \tag{8}$$

The expectation value of the mixed terms (like $2x_1x_2y_1y_2$) vanishes for the same reason as in the former discussion (an odd function of the components of $\boldsymbol{r}_1$ and/or $\boldsymbol{r}_2$). In addition, each of the first three terms of (8) yields a product of identical factors. For instance,

$$\begin{aligned}\int x^2\,|\psi_{100}(\boldsymbol{r})|^2\,\mathrm{d}^3r &= \frac{1}{3}\int r^2\,|\psi_{100}(\boldsymbol{r})|^2\,\mathrm{d}^3r \\ &= \frac{1}{3\pi a_0^3}\int\limits_0^\infty r^2\,\mathrm{e}^{-2r/a_0}4\pi r^2\,\mathrm{d}r = a_0^2 \ .\end{aligned} \tag{9}$$

Then

$$\left\langle 0\left|\hat{H}'^2\right|0\right\rangle = \frac{6e^2a_0^4}{R^6} \ .$$

With this equation, (6) reads

$$W(R) \geq -\frac{8\,\mathrm{e}^2a_0^5}{R^6} \ . \tag{10}$$

Variational Method

Equation (10) represents a lower bound for $W(R)$. An upper bound may always be computed by a variational procedure. We have to consider the problem of choosing a reasonable test wave function Ψ. If we choose Ψ to be independent of R, then the expectation value $\langle\Psi|\hat{H}'|\Psi\rangle$ is proportional to $1/R^3$, which is not useful for our case, since we want to know the coefficient of the $1/R^6$ term. Therefore we have to take into account the *polarization effects* in the wave function. Since we assume that the polarization is proportional to $\hat{H}'$, we write the test function in the following way:

$$\begin{aligned}\Psi(\boldsymbol{r}_1,\boldsymbol{r}_2) &= \psi_{100}(\boldsymbol{r}_1)\psi_{100}(\boldsymbol{r}_2)(1 + A\hat{H}') \\ &= \Psi_0(\boldsymbol{r}_1\boldsymbol{r}_2)(1 + A\hat{H}') \ ,\end{aligned} \tag{11}$$

Example 15.3

where A represents the variational parameter. Then the variational problem [see (11.32), (11.37)] gives

$$E_0+W(R) \leq \frac{\iint \Psi_0^*(\boldsymbol{r}_1\boldsymbol{r}_2)(1+A\hat{H}')(\hat{H}_0+\hat{H}')\Psi_0(\boldsymbol{r}_1\boldsymbol{r}_2)(1+A\hat{H}')\,d^3r_1\,d^3r_2}{\iint |\psi_0(\boldsymbol{r}_1,\boldsymbol{r}_2)|^2(1+A\hat{H}')^2\,d^3r_1\,d^3r_2} \;. \qquad (12)$$

A is assumed to be real. The right-hand side of (12) may be rewritten as

$$\frac{E_0+2A\langle 0|\hat{H}'^2|0\rangle + A^2\langle 0|\hat{H}'\hat{H}_0\hat{H}'|0\rangle}{1+A^2\langle 0|\hat{H}'^2|0\rangle} \;, \qquad (13)$$

since $\Psi_0(\boldsymbol{r}_1,\boldsymbol{r}_2)$ is a normalized eigenfunction of $\hat{H}_0$, with the eigenvalue $E_0 = e^2/a_0$, and, furthermore,

$$\left\langle 0\left|\hat{H}'\right|0\right\rangle = \left\langle 0\left|\hat{H}'^3\right|0\right\rangle = 0 \;.$$

The matrix element $\langle 0|\hat{H}'\hat{H}_0\hat{H}'|0\rangle$ gives a negative contribution. This can be shown by inserting complete sets of eigenfunctions of $\hat{H}_0$:

$$\begin{aligned}\left\langle 0\left|\hat{H}'\hat{H}_0\hat{H}'\right|0\right\rangle &= \sum_{n,m}\left\langle 0\left|\hat{H}'\right|n\right\rangle\left\langle n\left|\hat{H}_0\right|m\right\rangle\left\langle m\left|\hat{H}'\right|0\right\rangle \\ &= \sum_n E_n\left|\left\langle 0\left|\hat{H}'\right|n\right\rangle\right|^2 < 0 \;,\end{aligned}$$

because all E_n are negative. Thus, by

$$\frac{E_0+2A\langle 0|\hat{H}'^2|0\rangle + A^2\langle 0|\hat{H}'\hat{H}_0\hat{H}'|0\rangle}{1+A^2\langle 0|\hat{H}'^2|0\rangle} \leq \frac{E_0+2A\langle 0|\hat{H}'^2|0\rangle}{1+A^2\langle 0|\hat{H}'^2|0\rangle} \;, \qquad (14)$$

we can give an upper bound for (13).

As we are interested only in terms up to the order $\hat{H}'^2$, we expand the denominator of (14) and obtain

$$\begin{aligned}&\left(E_0+2A\left\langle 0\left|\hat{H}'^2\right|0\right\rangle\right)\left(1+A^2\left\langle 0\left|\hat{H}'^2\right|0\right\rangle\right)^{-1} \\ &\approx E_0+(2A-E_0A^2)\left\langle 0\left|\hat{H}'^2\right|0\right\rangle \;.\end{aligned} \qquad (15)$$

E_0, the energy of the ground state of the two hydrogen atoms, is negative. Therefore (15) has a minimum at $A=1/E_0$, and thus (12) takes the form

$$E_0+W(R) \leq E_0 + \frac{\langle 0|\hat{H}'^2|0\rangle}{E_0} = E_0 - \frac{6e^2a_0^5}{R^6} \;. \qquad (16)$$

Example 15.3

Together with (10) we have both an upper and lower bound for the interaction energy, which can be expressed by the inequality

$$-\frac{8e^2a_0^5}{R^6} \leq W(R) \leq -\frac{6e^2a_0^5}{R^6} \; . \tag{17}$$

Finally, we should note that careful variational calculations have shown that the numerical coefficient in $W(R)$ is close to 6.50.[2] The result achieved in this way is not absolutely correct, because only the static dipole–dipole interaction has been considered. If we also take into account the retardation caused by the finite speed of propagation of the electromagnetic interaction of the two dipoles, we find that $W(R) \simeq -1/R^7$, if R is large compared to the wavelength of the electromagnetic radiation of the atomic transition:

$$\left(R \gg \frac{\hbar c a_0}{e^2} = 137 a_0\right) \; . \tag{18}$$

But the interaction energy at these large distances is so small that it is physically uninteresting (insignificant). Therefore we may proceed from the assumption that expression (17) is a useful and reasonable approximation for the interaction of two spherical atoms.[3]

15.4 Biographical Notes

HEITLER, Walter Heinrich, German physicist, *Karlsruhe, 2.1.1904, † 15.11.1981. H. was in Göttingen from 1929–33; thereafter, he worked in Great Britain. H. was a professor at the Institute for Advanced Studies in Dublin from 1941–49 and at the Universität Zürich from 1949. After the development of the fundamentals of quantum mechanics in 1927, H. and F. London were able to explain the homeopolar chemical binding within the framework of quantum mechanics. Then H. applied quantum-mechanical methods to radiation theory and cosmic radiation; in particular, together with H.J. Bhabha, he was able to explain the origin of the showers of cosmic rays by his *cascade theory.* Further research dealt with the theory of nuclear forces and meson theory. In 1968 he was awarded the Max-Planck Medal of the Deutsche Physikalische Gesellschaft.

LONDON, Fritz, German-American physicist, *Breslau, 7.3.1900, † Durham, North Carolina, 30.3.1954. L. grew up in a cultivated, liberal German-Jewish family. He studied at the universities of Bonn, Frankfurt and Munich, and wrote his doctoral thesis in

[2] See, e.g., L. Pauling, E.B. Wilson, Jr.: *Introduction to Quantum Mechanics* (McGraw Hill, New York 1935), Chap. 47a.

[3] See, e.g., H.B.G. Casimir, D. Polder: Phys. Rev. **73**, 360 (1948).

Munich entitled "Über die symbolischen Methoden von Peano, Russell and Whitehead". In 1939 he published, together with Ernst Bauer, a short monograph on the theory of measurement in quantum mechanics. In 1925 he began to work in theoretical physics under the direction of Sommerfeld in Zürich and Berlin. In 1933, L. and his brother, Heinz, left Germany because of the political situation. For two years L. worked in Oxford and spent another two in Paris at the Institut Henri Poincaré. In 1939 he became a professor of theoretical chemistry at Duke University, in North Carolina. In 1927 he and W. Heitler solved the quantum-mechanical many-body problem of the hydrogen molecule. To do so, they took advantage of an analytical technique that was formulated by Lord Rayleigh in his "theory of sound". Thereafter L. worked mainly in the field of molecular theory.

TELLER, Edward, Hungarian-American physicist, *Budapest, 15.1.1908. Since 1935, T. has been a professor in the United States (New York, Chicago, Los Angeles, Livermore, Berkeley). T. took part in the development of the atomic bomb and very early on promoted the construction of the hydrogen bomb. He was one of the founders of Livermore National Laboratory and has been scientific advisor to several presidents of the U.S.A.

HYLLERAAS, Egil Andersen, Norwegian physicist, *Engerdal (Norway), 15.5.1898, † Oslo, 28.10.1965. H. studied at the universities of Oslo and Göttingen and graduated as Dr. phil. in 1924 as a student of L. Vergard. H. became a member of the Chr. Michelsons Institute (Bergen) in 1931 and won the Gunnerusmedalje (Kgl. Norske Videnskalero Selskab, Trondheim).

van der WAALS, Johannes Diderik, Dutch physicist, *Leiden, 23.11.1837, † Amsterdam, 8.3.1923. After years of teaching, W. studied physics at the University of Leiden. On the basis of his knowledge of the work of Clausius and other molecular theorists, he wrote his doctoral thesis "Over de continuiteit van den gasen en Vloeistoftestand" (1873). By applying simple mathematical equations in this thesis, he gave a satisfactory explanation of the properties of gases and fluids in the framework of molecular theory. Thomas Andrews and other experimental physicists later confirmed W.'s thesis, in particular the existence of critical temperature. In 1875 he was appointed a member of the Royal Dutch Academy of Sciences and two years later became a tenured professor at the University of Amsterdam. W. was a much-admired teacher and inspired his students to do both experimental and theoretical work. His scientific papers were primarily concerned with topics in molecular physics and thermodynamics. In 1910 W. was awarded the Nobel Prize in Physics.

SLATER, John Clarke, American physicist, *Oak Park, Illinois, 22.12.1900. S.'s most important contributions were to quantum theory, but he also worked on the theory of solids, thermodynamics and microwave physics at the Massachusetts Institute of Technology (MIT) during the years 1930–1951.

16. The Formal Framework of Quantum Mechanics

In this chapter we summarize the mathematical principles of quantum mechanics, using a more abstract mathematical formulation than before. Many of the relations which will be considered here have already been discussed in the preceding chapters in a more "physical" way and most have been proved in detail. Some of the explanations and proofs are supplemented or demonstrated once again in a more compact manner in additional exercises.

16.1 The Mathematical Foundation of Quantum Mechanics: Hilbert Space

By a *Hilbert space* H we mean an abstract number of elements, which are called vectors $|a\rangle$, $|b\rangle$, $|c\rangle$ etc. H has the following properties:

1. The space H is a *linear vector space* above the body of the complex numbers μ and ν. It has three properties:

(a) To every pair of vectors $|a\rangle$, $|b\rangle$, a new vector $|c\rangle$ is related, which is called the *sum vector*. It holds that

$$
\begin{aligned}
&|a\rangle + |b\rangle = |b\rangle + |a\rangle \quad \text{(commutative law)} \\
&(|a\rangle + |b\rangle) + |c\rangle = |a\rangle + (|b\rangle + |c\rangle) \quad \text{(associative law)}\ .
\end{aligned} \tag{16.1}
$$

(b) A *zero vector* $|0\rangle$ exists, with the property

$$
|a\rangle + |0\rangle = |a\rangle\ . \tag{16.2}
$$

(c) To each vector $|a\rangle$ of H, an antivector $|-a\rangle$ exists, fulfilling the relation

$$
|a\rangle + |-a\rangle = |0\rangle\ ; \tag{16.3}
$$

for arbitrary complex numbers μ and ν, we have

$$
\begin{aligned}
&\mu(|a\rangle + |b\rangle) = \mu\,|a\rangle + \mu\,|b\rangle\ , \\
&(\mu+\nu)\,|a\rangle = \mu\,|a\rangle + \nu\,|a\rangle\ , \\
&\mu\nu\,|a\rangle = \mu(\nu\,|a\rangle)\ , \\
&1\,|a\rangle = |a\rangle\ .
\end{aligned} \tag{16.4}
$$

2. A *scalar product* is defined in the space H. It is denoted by

$$(|a\rangle\,,|b\rangle) \quad \text{or} \quad \langle a|b\rangle\ , \tag{16.5}$$

yielding a complex number. The scalar product has to fulfil the relations

$$\begin{aligned}&(|a\rangle\,,\lambda\,|b\rangle) = \lambda(|a\rangle\,,|b\rangle)\ ,\\&(|a\rangle\,,|b\rangle + |c\rangle) = (|a\rangle\,,|b\rangle) + (|a\rangle\,,|c\rangle)\ ,\\&(|a\rangle\,,|b\rangle) = (|b\rangle\,,|a\rangle)^*\ .\end{aligned} \tag{16.6}$$

The last equation may also be written as

$$\langle a|b\rangle = \langle b|a\rangle^*\ .$$

It is easily shown from this, that

$$(\lambda\,|a\rangle\,,|b\rangle) = \lambda^*(|a\rangle\,,|b\rangle) = \lambda^*\,\langle a|b\rangle\ , \tag{16.7a}$$

and

$$(|a_1\rangle + |a_2\rangle\,,|b\rangle) = (|a_1\rangle\,,|b\rangle) + (|a_2\rangle\,,|b\rangle) = \langle a_1|b\rangle + \langle a_2|b\rangle \tag{16.7b}$$

follow. The *norm of the vectors* is defined by

$$\|\,|a\rangle\,\| = \sqrt{\langle a|a\rangle}$$

(read: norm of vector $|a\rangle = \sqrt{\langle a|a\rangle}$).

It can be shown that *Schwartz's inequality*,

$$\|\,|a\rangle\,\|\ \|\,|b\rangle\,\| \geq |\langle a|b\rangle|\ , \qquad 6.8)$$

is valid and that the equality is only valid for the case

$$|a\rangle = \lambda\,|b\rangle$$

(parallelism of the vectors).

3. For every vector $|a\rangle$ of H, a series $|a_n\rangle$ of vectors exists, with the property that for every $\varepsilon > 0$, there is at least one vector $|a_n\rangle$ of the series with

$$\|\,|a\rangle - |a_n\rangle\,\| < \varepsilon\ . \tag{16.9}$$

A series with this property is called *compact*, or we may say $|a_n\rangle$ of the space H is separable.

4. The Hilbert space is *complete*. This means that every vector $|a\rangle$ of H can be arbitrarily exactly approximated by a series $|a_n\rangle$:

$$\lim_{n\to\infty} \| \, |a\rangle - |a_n\rangle \, \| = 0 \ . \tag{16.10}$$

Then the series $|a_n\rangle$ has a unique limiting value $|a\rangle$.

For Hilbert spaces with finite dimensions, axioms 3 and 4 follow from axioms 1 and 2; then 3 and 4 are superfluous. But they are necessary for spaces of dimension ∞ that occur in quantum mechanics in most cases. In the following, we discuss once again some definitions that are used very often.

1. *Orthogonality of Vectors*:

   Two vectors $|f\rangle$ and $|g\rangle$ are orthogonal if

   $$\langle f|g\rangle = 0 \ . \tag{16.11}$$

2. *Orthonormal System*:

   The set $\{|f_n\rangle\}$ of vectors is an orthonormal system if

   $$\langle f_n|f_m\rangle = \delta_{nm} \ . \tag{16.12}$$

3. *Complete Orthonormal System*:

   The orthonormal system $\{|f_n\rangle\}$ is complete in H if an arbitrary vector $|f\rangle$ of H can be expressed by

   $$|f\rangle = \sum_n \alpha_n \, |f_n\rangle \ . \tag{16.13}$$

   In general, α_n are complex numbers:

   $$\begin{aligned} \alpha_m = \langle f_m|f\rangle &= \left\langle f_m \middle| \sum_n \alpha_n f_n \right\rangle \\ &= \sum_n \alpha_n \, \langle f_m|f_n\rangle \\ &= \sum_n \alpha_n \delta_{mn} \\ &= \alpha_m \ , \end{aligned} \tag{16.14}$$

   so that we can write

   $$|f\rangle = \sum_n |f_n\rangle \, \langle f_n|f\rangle \ . \tag{16.15}$$

   The complex numbers α_n are called the *f_n representation* of $|f\rangle$; they represent, so to say, the vector $|f\rangle$; they are the components of $|f\rangle$ with respect to the basis $\{|f_n\rangle\}$. If the sum in the last equation encloses an infinite number of terms, then we speak of a *Hilbert space of infinite dimensions*. In quantum mechanics, this is usually the case.

16.2 Operators in Hilbert Space

A *linear operator* $\hat{A}$ induces a mapping of H upon itself or upon a subspace of H. Here,

$$\hat{A}(\alpha\,|f\rangle+\beta\,|g\rangle)=\alpha\hat{A}\,|f\rangle+\beta\hat{A}\,|g\rangle\ . \tag{16.16}$$

The operator A is *bounded*, if

$$\left\|\hat{A}\,|f\rangle\right\|\leq C\,\||f\rangle\| \tag{16.17}$$

for all $|f\rangle$ of H, C being the same constant for all $|f\rangle$. *Bounded linear operators are continuous.* This means that for

$$|f_n\rangle\rightarrow|f\rangle\ , \tag{16.18a}$$

$$\hat{A}\,|f_n\rangle\rightarrow\hat{A}\,|f\rangle \tag{16.18b}$$

also follows. Two operators $\hat{A}$ and $\hat{B}$ are *equal* ($\hat{A}=\hat{B}$) if, for all vectors $|f\rangle$ of H,

$$\hat{A}\,|f\rangle=\hat{B}\,|f\rangle\ . \tag{16.19}$$

The following definitions are often used:

$$\begin{array}{llll}
\text{(a)} & \text{unity operator} & \hat{1}: & \hat{1}\,|f\rangle=|f\rangle\ ;\\
\text{(b)} & \text{zero operator} & \hat{0}: & \hat{0}\,|f\rangle=|0\rangle\ ;\\
\text{(c)} & \text{sum operator} & \hat{A}+\hat{B}: & (\hat{A}+\hat{B})\,|f\rangle=\hat{A}\,|f\rangle+\hat{B}\,|f\rangle\ ;\\
\text{(d)} & \text{product operator} & \hat{A}\hat{B}: & (\hat{A}\hat{B})\,|f\rangle=\hat{A}(\hat{B}\,|f\rangle)\ .
\end{array} \tag{16.20}$$

The relations shown here have to be valid for all $|f\rangle$ of H. With respect to the product operator, we have to add that, in general,

$$\hat{A}\hat{B}\neq\hat{B}\hat{A}\ .$$

The *commutator of $\hat{A}$ and $\hat{B}$* is defined by

$$[\hat{A},\hat{B}]_-=\hat{A}\hat{B}-\hat{B}\hat{A}\ . \tag{16.21}$$

Now we explain the very important concept of the *adjoint of a restricted operator*. If an operator $\hat{A}^+$ exists for the operator $\hat{A}$ for all $|f\rangle$ and $|g\rangle$ of H in such a way that

$$(|g\rangle\,,\hat{A}\,|f\rangle)=(\hat{A}^+\,|g\rangle\,,|f\rangle)\ , \tag{16.22}$$

then $\hat{A}^+$ is called the *adjoint operator* of $\hat{A}$. This relation can also be expressed by

$$\langle g|\hat{A}|f\rangle=\langle f|\hat{A}^+|g^*\rangle\ . \tag{16.23}$$

The adjoint of an operator (16.22) possesses the following properties, which are easily derived:

$$\begin{aligned}
&(1)\quad (\alpha\hat{A})^{+}=\alpha^{*}\hat{A}^{+}\ ;\\
&(2)\quad (\hat{A}+\hat{B})^{+}=\hat{A}^{+}+\hat{B}^{+}\ ;\\
&(3)\quad (\hat{A}\hat{B})^{+}=\hat{B}^{+}\hat{A}^{+}\ ;\\
&(4)\quad (\hat{A}^{+})^{+}=\hat{A}\ .
\end{aligned}\tag{16.24}$$

All these properties were discussed and proved in Chaps. 4 and 10. On the basis of the definition given above, the properties may immediately be confirmed.

An operator $\hat{A}$ fulfilling the relation

$$\hat{A}=\hat{A}^{+}\tag{16.25}$$

is called a *Hermitian operator*. From this it follows that the expectation values are real:

$$\langle f|\hat{A}|f\rangle=\langle f|\hat{A}^{+}|f\rangle^{*}=\langle f|\hat{A}|f\rangle^{*}=\text{real}\ .\tag{16.26}$$

16.3 Eigenvalues and Eigenvectors

We speak of an *eigenvector* $|a\rangle$ of the operator $\hat{A}$ belonging to the *eigenvalue a* in the case

$$\hat{A}\,|a\rangle=a\,|a\rangle\ .\tag{16.27}$$

Here, the eigenvalue a is, in general, a complex number. Especially for Hermitian operators $\hat{A}(\hat{A}^{+}=\hat{A})$, the following is true.

(a) The eigenvalues of Hermitian operators are real.
(b) If $|a'\rangle$ and $|a''\rangle$, are two eigenvectors of a Hermitian operator $\hat{A}$ with two different eigenvalues $a'\neq a''$, then

$$\langle a'|a''\rangle=0\ .$$

(c) The normalized eigenvectors of a bounded Hermitian operator $\hat{A}$ create a *countable, complete orthonormal system*. In this case the eigenvalues are discrete. Then we speak of a *discrete spectrum*.

We therefore can conclude that an arbitrary vector $|\psi\rangle$ may be expanded in terms of the complete orthonormal system $|a\rangle$ of the Hermitian, restricted operator $\hat{A}$:

$$|\psi\rangle=\sum_{a}|a\rangle\,\langle a|\psi\rangle\ .\tag{16.28}$$

As noted above, we have

$$\left\langle a' \middle| a'' \right\rangle = \delta_{a'a''} \ . \tag{16.29}$$

The scalar product of two vectors $|\varphi\rangle$ and $|\psi\rangle$ may be expressed in the A representation; also,

$$\langle\varphi|\psi\rangle = \sum_a \langle\varphi|a\rangle \, \langle a|\psi\rangle \ . \tag{16.30}$$

Here, a helpful trick has been used. If we introduce the unity operator $\mathbb{1}$, known as completeness, by

$$\hat{\mathbb{1}} = \sum_a |a\rangle \, \langle a| \ , \tag{16.31}$$

we get

$$|\psi\rangle = \hat{\mathbb{1}} \, |\psi\rangle = \sum_a |a\rangle \, \langle a|\psi\rangle \ , \tag{16.32}$$

and, further,

$$\langle\varphi|\psi\rangle = \left\langle\varphi\middle|\hat{\mathbb{1}}\middle|\psi\right\rangle = \sum_a \langle\varphi|a\rangle \, \langle a|\psi\rangle \ , \tag{16.33}$$

which is consistent with (16.28) and (16.30). The expansion (16.32) implies that

$$\sum_a |\langle a|\psi\rangle|^2 = 1 \ . \tag{16.34}$$

Therefore we may also say that $\langle a|\psi\rangle$ is square-summable. Apparently the abstract Hilbert space is mapped onto the *space of the square-summable functions (eigenfunctions of the operator* $\hat{A}$). This we call the *A representation of* ψ and mean the infinite set of numbers $\langle a|\psi\rangle$ in (16.32). Applying an operator $\hat{B}$ to $|\psi\rangle$ yields

$$\left\langle a'\middle|\hat{B}\middle|\psi\right\rangle = \sum_{a''} \left\langle a'\middle|\hat{B}\middle|a''\right\rangle \left\langle a''\middle|\psi\right\rangle \ . \tag{16.35}$$

Thus the *operator* $\hat{B}$ *can be written in the A representation* as the matrix

$$\hat{B} \rightarrow \left\langle a'\middle|\hat{B}\middle|a''\right\rangle = \begin{pmatrix} \langle a_1|\hat{B}|a_1\rangle & \langle a_1|\hat{B}|a_2\rangle & \cdots \\ \langle a_2|\hat{B}|a_1\rangle & \langle a_2|\hat{B}|a_2\rangle & \cdot \\ \vdots & \vdots & \vdots \\ \vdots & \vdots & \vdots \end{pmatrix} \ , \tag{16.36}$$

and the *vector* ψ in *A representation* as

$$|\psi\rangle \to \langle a'|\psi\rangle = \begin{pmatrix} \langle a_1|\psi\rangle \\ \langle a_2|\psi\rangle \\ \vdots \\ \langle a_n|\psi\rangle \\ \vdots \end{pmatrix} . \tag{16.37}$$

Therefore the operator $\hat{B}$ in A representation is a quadratic matrix; the vector $|\psi\rangle$, a column matrix. The operator $\hat{A}$ itself is given in the A representation of its *eigenrepresentation* as

$$\langle a'|\hat{A}|a''\rangle = a'\delta_{a'a''} . \tag{16.38}$$

Sometimes it is advantageous to write the (arbitrary) operator $\hat{B}$ in the form

$$\hat{B} = \hat{\mathbf{1}}\hat{B}\hat{\mathbf{1}} = \sum_{a',a''} |a'\rangle\langle a'|\hat{B}|a''\rangle\langle a''| . \tag{16.39}$$

The analogy of the representation of a vector in a Hilbert space to the components of a vector in vector space is evident. The choice of the representation coincides with the choice of the coordinate system in the Hilbert space.

Now we proceed to the *transformation of the A representation into B representation*. Here, the so-called transformation matrix

$$\langle a|b\rangle \tag{16.40}$$

plays an important role. In analogy to (16.38) it follows that

$$\langle b'|\hat{B}|b''\rangle = b'\delta_{b'b''} . \tag{16.41}$$

It is convenient to start from the unity operator

$$\hat{\mathbf{1}} = \sum_{a'} |a'\rangle\langle a'| = \sum_{b'} |b'\rangle\langle b'| . \tag{16.42}$$

The following relations can be understood immediately:

$$\begin{aligned} \langle b'|\psi\rangle &= \langle b'|\hat{\mathbf{1}}|\psi\rangle = \sum_{a'} \langle b'|a'\rangle\langle a'|\psi\rangle , \\ \langle a'|\psi\rangle &= \langle a'|\hat{\mathbf{1}}|\psi\rangle = \sum_{b'} \langle a'|b'\rangle\langle b'|\psi\rangle , \\ \langle b'|\hat{C}|b''\rangle &= \langle b'|\hat{\mathbf{1}}\hat{C}\hat{\mathbf{1}}|b''\rangle = \sum_{a',a''} \langle b'|a'\rangle\langle a'|\hat{C}|a''\rangle\langle a''|b''\rangle . \end{aligned} \tag{16.43}$$

Similarly to (16.42), we get

$$\langle a'|\hat{B}\hat{C}|a''\rangle = \langle a'|\hat{B}\hat{\mathbf{1}}\hat{C}|a''\rangle = \sum_{a'''} \langle a'|\hat{B}|a'''\rangle\langle a'''|\hat{C}|a''\rangle . \tag{16.44}$$

This means that for the matrix element of the product of two operators $\hat{B}\hat{C}$, the customary rules for matrix multiplication are valid.

EXERCISE

16.1 The Trace of an Operator

Problem. Show that the trace of an operator is independent of its representation.

Solution. The trace[1] of the operator $\hat{C}$ in the A representation is

$$\mathrm{tr}\hat{C} = \sum_{a'} \langle a'|\hat{C}|a'\rangle \,.$$

Then we write

$$\begin{aligned}
\mathrm{tr}\hat{C} &= \sum_{a'} \langle a'|\hat{C}|a'\rangle \ldots = \mathrm{tr}\hat{\mathbf{1}}\hat{C}\hat{\mathbf{1}} \\
&= \sum_{a'}\sum_{b'}\sum_{b''} \langle a'|b'\rangle\langle b'|\hat{C}|b''\rangle\langle b''|a'\rangle \\
&= \sum_{a'}\sum_{b'}\sum_{b''} \langle b''|a'\rangle\langle a'|b'\rangle\langle b'|\hat{C}|b''\rangle \\
&= \sum_{b'}\sum_{b''} b''b'\langle b'|\hat{C}|b''\rangle \\
&= \sum_{b''} \langle b''|\hat{\mathbf{1}}\hat{C}|b''\rangle = \sum_{b''} \langle b''|\hat{C}|b''\rangle \,.
\end{aligned}$$

Since $\hat{C}|c\rangle = c|c\rangle$, we have in the eigenrepresentation of $\hat{C}$

$$\begin{aligned}
\mathrm{tr}\hat{C} &= \sum_{c'} \langle c'|\hat{C}|c'\rangle \\
&= \sum_{c'} c'\langle c'|c'\rangle = \sum_{c'} c' \,.
\end{aligned}$$

[1] The German name for trace, "Spur", is often used in the literature.

EXERCISE

16.2 A Proof

Problem. Show that

$$\sum_{a'}\sum_{a''}\left|\langle a'|\hat{C}|a''\rangle\right|^2 = \mathrm{tr}\hat{C}\hat{C}^+ \ .$$

Solution. It can easily be seen that

$$\begin{aligned}\sum_{a'}\sum_{a''}\left|\langle a'|\hat{C}|a''\rangle\right|^2 &= \sum_{a'}\sum_{a''}\langle a'|\hat{C}|a''\rangle\langle a'|\hat{C}|a''\rangle^* \\ &= \sum_{a'}\sum_{a''}\langle a'|\hat{C}|a''\rangle\langle a''|\hat{C}^+|a'\rangle \\ &= \sum_{a'}\langle a'|\hat{C}\hat{C}^+|a'\rangle = \mathrm{tr}\hat{C}\hat{C}^+ \ .\end{aligned}$$

Here we have used (16.23) and (16.44).

16.4 Operators with Continuous or Discrete-Continuous (Mixed) Spectra

Many operators occurring in quantum mechanics do not have a discrete, but a *continuous* or a *mixed* (*discrete-continuous*), *spectrum.* An example of an operator with a mixed spectrum is the well-known Hamiltonian of the hydrogen atom. Actually all Hamiltonians for atoms and nuclei have discrete and continuous spectral ranges; therefore they have mixed spectra. Usually the discrete eigenvalues are connected with bound states and the continuous eigenvalues are connected with free, unbound states. The representations related to such operators cause some difficulties because, for continuous spectra, the eigenvectors are not normalizable to unity (cf. our discussion of Weyl's eigendifferentials in Chaps. 4 and 5).

1. Operators with a Continuous Spectrum

The operator $\hat{A}$ has a *continuous spectrum* if the eigenvalue a in

$$\hat{A}\,|a\rangle = a\,|a\rangle \tag{16.45}$$

is continuous. The states $|a\rangle$ can no longer be normalized to unity, but must be normalized to Dirac's delta function:

$$\langle a'|a''\rangle = \delta(a'-a'') \; . \tag{16.46}$$

Here, the delta function replaces, so to speak, Kronecker's δ of the discrete spectrum [cf. (16.29)]. In the expansion of a state $|\psi\rangle$ in terms of a complete set $|a\rangle$, the sums [cf. (16.28)] are replaced by integrals:

$$|\psi\rangle = \int |a'\rangle\langle a'|\psi\rangle \, da' \; . \tag{16.47}$$

$\langle a'|\psi\rangle$ represents the wave function in the A representation. The inner product of two vectors $|\varphi\rangle$ and $|\psi\rangle$ changes analogously to (16.30) into

$$\langle\varphi|\psi\rangle = \int \langle\varphi|a'\rangle\langle a'|\psi\rangle \, da' \; , \tag{16.48a}$$

which is sometimes written as

$$\langle\varphi|\psi\rangle = \int \varphi^*(a')\psi(a') \, da' \; . \tag{16.48b}$$

Here, $\psi(a) = \langle a|\psi\rangle$ may be understood (somewhat imprecisely) as a "wave function in A space". Of course, it is just the A representation of $|\psi\rangle$.

2. Operators with a Mixed Spectrum

If the equation

$$\hat{A}\,|a\rangle = a\,|a\rangle$$

yields discrete as well as continuous eigenvalues a, we are dealing with a *mixed spectrum* (cf. Fig. 16.1).

In these cases, the expansion of $|\psi\rangle$ in terms of $|a\rangle$ reads

$$|\psi\rangle = \sum_{a'} |a'\rangle\langle a'|\psi\rangle + \int |a'\rangle\langle a'|\psi\rangle \, da' \; , \tag{16.49}$$

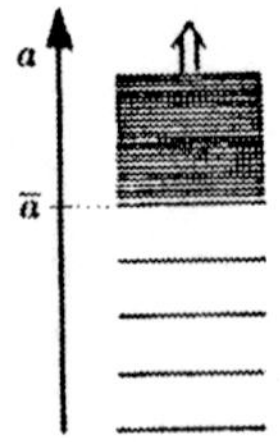

Fig. 16.1. A mixed spectrum. For $a < \bar{a}$, the spectrum is discrete; for $a > \bar{a}$ it is continuous

where the sum extends over the discrete, and the integral over the continuous, eigenstates $|a\rangle$.

In order to make the notation more compact, it is understood that $\sum_{a'}$ or $\int \ldots da'$ is split into the discrete and the continuous parts of the spectrum, if there are any, according to (16.49).

16.5 Operator Functions

Operator functions $f(\hat{A})$ may be defined as a power series if the function $f(x)$ can be expanded in this way. Thus if

$$f(x) = \sum_{n=0}^{\infty} C_n x^n \ ,$$

then the *operator function* $f(\hat{A})$ is defined by

$$f(\hat{A}) = \sum_{n=0}^{\infty} C_n \hat{A}^n \ . \qquad (16.50)$$

For example, $e^{\hat{A}}$, $\cos \hat{A}$, etc. may be defined in this way. Another possibility of defining operator functions is obtained via their eigenvalues: if

$$\hat{A}\,|a'\rangle = a'\,|a'\rangle \ ,$$

then we have

$$f(\hat{A})\,|a'\rangle = f(a')\,|a'\rangle \ . \qquad (16.51)$$

For operator functions of the form (16.50), (16.51) follows immediately. Two exercises will illustrate these points.

EXERCISE

16.3 Operator Functions

Problem. Derive the relation

$$\langle b'|f(\hat{A})|b''\rangle = \sum_{a'} \langle b'|a'\rangle\, f(a')\,\langle a'|b''\rangle \ .$$

Solution. We calculate:

$$\begin{aligned}
\langle b'|f(\hat{A})|b''\rangle &= \langle b'|\hat{\mathbf{1}}\, f(\hat{A})\,\hat{\mathbf{1}}\,|b''\rangle \\
&= \sum_{a',a''} \langle b'|a'\rangle\langle a'|f(\hat{A})|a''\rangle\langle a''|b''\rangle \\
&= \sum_{a',a''} \langle b'|a'\rangle\, f(a')\delta_{a',a''}\,\langle a''|b''\rangle \\
&= \sum_{a'} \langle b'|a'\rangle\, f(a')\,\langle a'|b''\rangle \ . \qquad (1)
\end{aligned}$$

EXERCISE

16.4 Power-Series and Eigenvalue Methods

Problem. Show by the method of power-series expansion (16.50) and by the method of eigenvalues (16.51) that

$$\mathrm{e}^{\mathrm{i}(\beta/2)\sigma_x} = \begin{pmatrix} \cos\frac{\beta}{2} & \mathrm{i}\sin\frac{\beta}{2} \\ \mathrm{i}\sin\frac{\beta}{2} & \cos\frac{\beta}{2} \end{pmatrix}$$

if

$$\sigma_x = \begin{pmatrix} 0 & 1 \\ 1 & 0 \end{pmatrix} .$$

Solution. (a) We use the power series of the exponential function and get

$$\mathrm{e}^{\mathrm{i}(\beta/2)\sigma_x} = \sum_{n=0}^{\infty} \frac{1}{n!}\left(\frac{\mathrm{i}\beta}{2}\right)^n \sigma_x^n . \tag{1}$$

We have $\sigma_x^2 = \mathbf{1} = \begin{pmatrix} 1 & 0 \\ 0 & 1 \end{pmatrix}$ and therefore $\sigma_x^3 = \sigma_x$. For this reason, the series (1) splits into even and odd powers. We get

$$\begin{aligned} \mathrm{e}^{\mathrm{i}(\beta/2)\sigma_x} &= \mathbf{1} \sum_{n\ \text{even}}^{\infty} \frac{1}{n!}\left(\frac{\mathrm{i}\beta}{2}\right)^n + \sigma_x \sum_{n\ \text{odd}} \frac{1}{n!}\left(\frac{\mathrm{i}\beta}{2}\right)^n \\ &= \mathbf{1}\cos\frac{\beta}{2} + \mathrm{i}\sigma_x \sin\frac{\beta}{2} . \end{aligned} \tag{2}$$

(b) We use the method of eigenvalues (16.51). It is suitable to introduce the vectors

$$|z, +1\rangle = \begin{pmatrix} 1 \\ 0 \end{pmatrix} \quad \text{and} \quad |z, -1\rangle = \begin{pmatrix} 0 \\ 1 \end{pmatrix} , \tag{3}$$

i.e. the eigenstates of $\sigma_z = \begin{pmatrix} 1 & 0 \\ 0 & -1 \end{pmatrix}$. This property is expressed by the notation $|z, \lambda\rangle$. Now, it can easily be checked that

$$\langle zi\,|\sigma_x|\,zj\rangle = \begin{pmatrix} 0 & 1 \\ 1 & 0 \end{pmatrix} . \tag{4}$$

To use the method of eigenvalues, we need the eigenvalues of σ_x. For this purpose, we solve the eigenvalue problem

$$\sigma_x\,|x, \lambda\rangle = \lambda\,|x, \lambda\rangle , \tag{5}$$

Exercise 16.4

and find $\lambda = \pm 1$ and the normalized eigenvectors,

$$|x,1\rangle = \frac{1}{\sqrt{2}}\begin{pmatrix}1\\1\end{pmatrix} \;, \quad |x,-1\rangle = \frac{1}{\sqrt{2}}\begin{pmatrix}1\\-1\end{pmatrix} \;. \tag{6}$$

Using (1) of Exercise 16.3, we get

$$\langle z,i|\,\mathrm{e}^{\mathrm{i}(\beta/2)\sigma_x}\,|z,j\rangle = \sum_{\lambda=\pm 1} \langle z,i|x,\lambda\rangle\, \mathrm{e}^{\mathrm{i}(\beta/2)\lambda}\, \langle x,\lambda|z,j\rangle \;. \tag{7}$$

From this we are able to construct all matrix elements. For example, for $i = j = 1$, we get

$$\begin{aligned}\langle z,1|\,\mathrm{e}^{\mathrm{i}(\beta/2)\sigma_x}\,|z,1\rangle &= \frac{1}{\sqrt{2}}(1\quad 0)\begin{pmatrix}1\\1\end{pmatrix}\mathrm{e}^{\mathrm{i}(\beta/2)}\frac{1}{\sqrt{2}}(1\quad 1)\begin{pmatrix}1\\0\end{pmatrix}\\ &\quad + \frac{1}{\sqrt{2}}(1\quad 0)\begin{pmatrix}1\\-1\end{pmatrix}\mathrm{e}^{-\mathrm{i}(\beta/2)}\frac{1}{\sqrt{2}}(1\quad -1)\begin{pmatrix}1\\0\end{pmatrix}\\ &= \frac{1}{2}\mathrm{e}^{\mathrm{i}(\beta/2)} + \frac{1}{2}\mathrm{e}^{-\mathrm{i}(\beta/2)} = \cos\frac{\beta}{2} \;.\end{aligned}$$

In a similar manner we derive the other matrix elements, and finally arrive at

$$\langle z,i|\,\mathrm{e}^{\mathrm{i}(\beta/2)\sigma_x}\,|z,j\rangle = \begin{pmatrix}\cos\frac{\beta}{2} & \mathrm{i}\sin\frac{\beta}{2}\\ \mathrm{i}\sin\frac{\beta}{2} & \cos\frac{\beta}{2}\end{pmatrix} \;. \tag{8}$$

Even the inverse operator $\hat{A}^{-1}$ can be defined by the method of eigenvalues (and not only by the inversion of the matrix), namely:

$$\hat{A}^{-1}\,|a'\rangle = \frac{1}{a'}\,|a'\rangle \;. \tag{16.52}$$

With $\hat{A}|a'\rangle = a'|a'\rangle$, we have

$$\hat{A}^{-1}\hat{A} = \hat{A}\hat{A}^{-1} = \mathbb{1} \;.$$

If one of the eigenvalues of $\hat{A}$, i.e. one of the quantities a', vanishes, the inverse operator cannot be defined. *In this case, $\hat{A}^{-1}$ does not exist.*

16.6 Unitary Transformations

An operator $\hat{U}$ is *unitary* if

$$\hat{U}^{-1} = \hat{U}^{+} \,. \tag{16.53}$$

A *unitary transformation* is given by a unitary operator:

$$|a'_{\text{new}}\rangle = \hat{U}\,|a'_{\text{old}}\rangle \,. \tag{16.54}$$

Hence, for an operator, it follows that

$$\left\langle a'_{\text{new}}\left|\hat{A}_{\text{new}}\right|a''_{\text{new}}\right\rangle = \left\langle \hat{U}a'_{\text{old}}\left|\hat{A}_{\text{new}}\right|\hat{U}a''_{\text{old}}\right\rangle = \left\langle a'_{\text{old}}\left|\hat{U}^{+}\hat{A}_{\text{new}}\hat{U}\right|a''_{\text{old}}\right\rangle$$
$$\stackrel{\text{def}}{\equiv} \left\langle a'_{\text{old}}\left|\hat{A}_{\text{old}}\right|a''_{\text{old}}\right\rangle \,.$$

Therefore

$$\hat{A}_{\text{old}} = \hat{U}^{+}\hat{A}_{\text{new}}\hat{U} \,, \quad \text{or}$$
$$\hat{A}_{\text{new}} = (\hat{U}^{+})^{-1}\hat{A}_{\text{old}}\hat{U}^{-1} = \hat{U}\hat{A}_{\text{old}}\hat{U}^{+} \,, \tag{16.55}$$

where we have used (16.53). It can easily be checked that scalar products are invariant under unitary transformations, because

$$\langle b'_{\text{new}}|a'_{\text{new}}\rangle = \left\langle \hat{U}b'_{\text{old}}\middle|\hat{U}a'_{\text{old}}\right\rangle = \left\langle b'_{\text{old}}\left|\hat{U}^{+}\hat{U}\right|a'_{\text{old}}\right\rangle = \langle b'_{\text{old}}|a'_{\text{old}}\rangle \,. \tag{16.56}$$

Also the eigenvalues of $\hat{A}_{\text{new}}$ are the same as those of $\hat{A}_{\text{old}}$ (invariance of the eigenvalues), i.e.

$$\hat{A}_{\text{new}}\,|a'_{\text{new}}\rangle = \hat{U}\hat{A}_{\text{old}}\underbrace{\hat{U}^{+}\hat{U}}_{1}\,|\hat{a}'_{\text{old}}\rangle = \hat{U}\hat{A}_{\text{old}}\,|a'_{\text{old}}\rangle = \hat{U}a'_{\text{old}}\,|a'_{\text{old}}\rangle$$
$$= a'_{\text{old}}\hat{U}\,|a'_{\text{old}}\rangle = a'_{\text{old}}\,|a'_{\text{new}}\rangle \,. \tag{16.57}$$

It can easily be shown that, given

$$\hat{C}_{\text{old}} = \hat{A}_{\text{old}}\hat{B}_{\text{old}} \quad \text{and} \tag{16.58}$$
$$\hat{D}_{\text{old}} = \hat{A}_{\text{old}} + \hat{B}_{\text{old}} \,, \tag{16.59}$$

it also holds that

$$\hat{C}_{\text{new}} = \hat{A}_{\text{new}}\hat{B}_{\text{new}} \quad \text{and} \tag{16.58a}$$
$$\hat{D}_{\text{new}} = \hat{A}_{\text{new}} + \hat{B}_{\text{new}} \,. \tag{16.59a}$$

The generalization of these relations is obvious: *All algebraic operations remain unchanged by unitary transformations.*

16.7 The Direct-Product Space

Frequently the Hilbert space must be expanded, because new degrees of freedom are discovered. One example we have already encountered is the spin of the electron (see Chap. 12). The total wave function consists of the product of the spatial wave function $\psi(x, y, z)$ and the spin wave function $\chi(\sigma)$:

$$\psi(x, y, z)\chi(\sigma) \ .$$

We say the Hilbert space is extended by *direct-product formation*. The following examples explain this further.

A nucleon may be either a neutron or a proton with nearly identical masses: $m_{\mathrm{p}}c^2 = 938.256\,\mathrm{MeV}$, $m_{\mathrm{n}}c^2 = 939.550\,\mathrm{MeV}$. For this reason we consider it as a particle with two states, the proton state $|p\rangle$ and the neutron state $|n\rangle$:

$$|p\rangle = \begin{pmatrix} 1 \\ 0 \end{pmatrix}_{\mathrm{charge}} \ , \quad |n\rangle = \begin{pmatrix} 0 \\ 1 \end{pmatrix}_{\mathrm{charge}} \ . \tag{16.60}$$

The vectors $|p\rangle$ and $|n\rangle$ span the two-dimensional charge space or *isospin space* (in analogy to the spin). Since the nucleon may also occupy two different spin states

$$|\uparrow\rangle = \begin{pmatrix} 1 \\ 0 \end{pmatrix}_{\mathrm{spin}} \quad \text{and} \quad |\downarrow\rangle = \begin{pmatrix} 0 \\ 1 \end{pmatrix}_{\mathrm{spin}} \ , \tag{16.61}$$

the direct product space consisting of spin and isospin space is given by the four-dimensional space with the basis vectors

$$|p\uparrow\rangle = \begin{pmatrix} 1 \\ 0 \end{pmatrix}_{\mathrm{charge}} \times \begin{pmatrix} 1 \\ 0 \end{pmatrix}_{\mathrm{spin}} = \begin{pmatrix} 1 \\ 0 \\ 0 \\ 0 \end{pmatrix} \ ,$$

$$|p\downarrow\rangle = \begin{pmatrix} 1 \\ 0 \end{pmatrix}_{\mathrm{charge}} \times \begin{pmatrix} 0 \\ 1 \end{pmatrix}_{\mathrm{spin}} = \begin{pmatrix} 0 \\ 1 \\ 0 \\ 0 \end{pmatrix} \ ,$$

$$|n\uparrow\rangle = \begin{pmatrix} 0 \\ 1 \end{pmatrix}_{\mathrm{charge}} \times \begin{pmatrix} 1 \\ 0 \end{pmatrix}_{\mathrm{spin}} = \begin{pmatrix} 0 \\ 0 \\ 1 \\ 0 \end{pmatrix} \ ,$$

$$|n\downarrow\rangle = \begin{pmatrix} 0 \\ 1 \end{pmatrix}_{\mathrm{charge}} \times \begin{pmatrix} 0 \\ 1 \end{pmatrix}_{\mathrm{spin}} = \begin{pmatrix} 0 \\ 0 \\ 0 \\ 1 \end{pmatrix} \ . \tag{16.62}$$

Thus, in this four-dimensional space, the charge properties as well as the spin properties of the nucleon can be described. If further "intrinsic" properties (i.e. more inner degrees of freedom) of the nucleon should be discovered, the space will have to be further enlarged. In fact, a situation similar to the one just discussed arises if we consider particles and antiparticles.[2]

16.8 The Axioms of Quantum Mechanics

It is not easy to summarize the axioms or rules of quantum mechanics. Here we will follow E.G. Harris[3] and refer to the extensive discussions of von Neumann[4] and Jauch.[5]

Quantum mechanics is based upon the following correspondence between physical and mathematical quantities.

1. The state of a physical system is characterized by a vector (more precisely, by a vector beam) in Hilbert space. Hence, $|\psi\rangle$ and $\lambda|\psi\rangle$ describe the same state. In general, the state vectors are normalized to unity, to enable the interpretation of probability.
2. The dynamic observable physical quantities (*observables*) are described by operators in the Hilbert space H. *These operators of observables* are Hermitian operators. Their eigenvectors form a basis of H; any vector of H may be expanded in terms of this basis.

These general principles are supplemented by the following fundamental physical axioms.

Axiom 1: As a result of the measurement of an observable, only one of the eigenvalues of the corresponding operator can be found. After the measurement, the system occupies that state which corresponds to the measured eigenvalue.

Axiom 2: If the system occupies the state $|a'\rangle$, the probability of finding the value b' in a measurement for B reads

$$W(A', B') = \left|\langle a'|b'\rangle\right|^2 . \tag{16.63}$$

---

[2] We will encounter this situation in W. Greiner: *Relativistic Quantum Mechanics*, 3rd ed. (Springer, Berlin, Heidelberg 2000), where the Dirac spinor also turns out to have four components: two for the spin and two for the particle–antiparticle degrees of freedom.

[3] E.G. Harris: *A Pedestrian Approach to Quantum Field Theory* (Wiley, New York 1972).

[4] J. von Neumann: *The Mathematical Foundations of Quantum Mechanics* (Princeton, NJ 1955).

[5] J.M. Jauch: *Foundations of Quantum Mechanics* (Addison-Wesley, Reading, MA 1968).

If B has a continuous spectrum,

$$\mathrm{d}W(A', B') = \left| \langle a'|b\rangle' \right|^2 \mathrm{d}b' \tag{16.63a}$$

is the probability that B has a value within the interval between b' and $b' + \mathrm{d}b'$.

Axiom 3: The operators $\hat{A}$ and $\hat{B}$, which correspond to the classical quantities A and B, fulfil the commutation relation

$$[\hat{A}, \hat{B}]_- = \hat{A}\hat{B} - \hat{B}\hat{A} = \mathrm{i}\hbar\{A, B\}_{\mathrm{op}} \;, \tag{16.64}$$

where $\{A, B\}_{\mathrm{op}}$ is the operator which corresponds to the classical Poisson bracket,

$$\{A, B\} = \sum_i \left(\frac{\partial A}{\partial q_i}\frac{\partial B}{\partial p_i} - \frac{\partial A}{\partial p_i}\frac{\partial B}{\partial q_i} \right) \;; \tag{16.65}$$

q_i and p_i are the classical coordinates and momenta of the system. It follows that

$$[\hat{q}_i, \hat{q}_j]_- = [\hat{p}_i, \hat{p}_j]_- = 0 \quad , \quad [\hat{q}_i, \hat{p}_j]_- = \mathrm{i}\hbar\delta_{ij}\mathbf{1} \;, \tag{16.66}$$

and, similarly, for the orbital angular momentum,

$$\begin{aligned}
\hat{\boldsymbol{L}} &= \boldsymbol{r} \times \hat{\boldsymbol{p}} = (y\hat{p}_z - z\hat{p}_y, z\hat{p}_x - x\hat{p}_z, x\hat{p}_y - y\hat{p}_x) \;, \\
[\hat{L}_x, \hat{L}_y]_- &= \mathrm{i}\hbar \sum_i \left(\frac{\partial L_x}{\partial q_i}\frac{\partial L_y}{\partial p_i} - \frac{\partial L_x}{\partial p_i}\frac{\partial L_y}{\partial q_i} \right)_{\mathrm{op}} \\
&= \mathrm{i}\hbar[(-\hat{p}_y)(-x) - (y)(\hat{p}_x)]_{\mathrm{op}} \\
&= \mathrm{i}\hbar(x\hat{p}_y - y\hat{p}_x) = \mathrm{i}\hbar\hat{L}_z \;.
\end{aligned} \tag{16.67}$$

For the other angular-momentum commutation relations we get a similar result and may write

$$\hat{\boldsymbol{L}} \times \hat{\boldsymbol{L}} = \mathrm{i}\hbar\hat{\boldsymbol{L}} \;. \tag{16.68}$$

We should pay attention to the following consequence of this axiom. If we define the *expectation value of an observable A* by

$$\langle\hat{A}\rangle = \langle\psi|\hat{A}|\psi\rangle \tag{16.69}$$

and the uncertainty (the mean variation) by

$$\Delta A = \sqrt{\langle(\hat{A} - \langle\hat{A}\rangle)^2\rangle} = \sqrt{\langle\psi|(\hat{A} - \langle\psi|\hat{A}|\psi\rangle)^2|\psi\rangle} \;, \tag{16.70}$$

it follows that (see Sect. 4.7)

$$(\Delta A)^2(\Delta B)^2 \geq \tfrac{1}{4}\left|\langle|[\hat{A}, \hat{B}]_-|\rangle\right|^2 \;. \tag{16.71}$$

This is the general formulation of *Heisenberg's uncertainty relation*. In particular, for the variables p_i and q_i, using (16.66) we have

$$\Delta p_i \Delta q_j \geq \frac{\hbar}{2}\delta_{ij} \; . \tag{16.72}$$

Hitherto, we have dealt with states (vectors) and observables at one instant of time. The dynamics of a system may be described in different, equivalent ways. The most customary one is the *Schrödinger picture*, in which the state vector is time dependent, but the operators of the observables are independent of time.

Axiom 4: If a system is described by the state $|\psi_{t_0}\rangle$ at time t_0 and by $|\psi_t\rangle$ at time t, both states are connected by the unitary transformation

$$|\psi_t\rangle = \hat{U}(t-t_0)\left|\psi_{t_0}\right\rangle \; , \tag{16.73}$$

where

$$\hat{U}(t-t_0) = \exp\left[-\frac{\mathrm{i}}{\hbar}\hat{H}(t-t_0)\right] \tag{16.74}$$

and $\hat{H}$ is the Hamiltonian of the system.

Schrödinger's equation follows from (16.73) and (16.74). Let

$$\mathrm{d}t = t - t_0 \quad , \quad \mathrm{d}|\psi\rangle = \left|\psi_{t_0+\mathrm{d}t}\right\rangle - \left|\psi_{t_0}\right\rangle \quad \text{and} \quad \hat{U}(\mathrm{d}t) = 1 - \frac{\mathrm{i}}{\hbar}\hat{H}\,\mathrm{d}t \; ;$$

then

$$-\frac{\hbar}{\mathrm{i}}\frac{\partial}{\partial t}|\psi\rangle = \hat{H}\,|\psi\rangle \; . \tag{16.75}$$

Notice that *Schrödinger's equation is generally valid*. In particular, it is valid for time-independent, as well as for time-dependent, Hamiltonians $\hat{H}$. Only in the first case ($\hat{H}$ time independent) may we conclude (16.73) from (16.75) (cf. Chap. 11). Therefore the special form of the time development (16.73) is only valid for time-independent Hamiltonians.[6]

The *Heisenberg picture* is another description of the dynamics of a physical system, which is equivalent to the Schrödinger picture, as mentioned above. We obtain it from (16.73), by applying the unitary transformation

$$|\psi_t\rangle_{\mathrm{H}} = \hat{U}^{-1}\,|\psi_t\rangle_{\mathrm{S}} = \hat{U}^{-1}\hat{U}\left|\psi_{t_0}\right\rangle_{\mathrm{S}} = \left|\psi_{t_0}\right\rangle_{\mathrm{S}} \tag{16.76}$$

[6] See W. Greiner, B. Müller: *Quantum Mechanics – Symmetries*, 2nd ed. (Springer, Berlin, Heidelberg 1994), especially the section on isotropy in time.

to the state vectors. Then the operators transform according to (16.55), and we get

$$\hat{A}_{\mathrm{H}}(t) = \hat{U}_t^{-1} \hat{A}_{\mathrm{S}} \hat{U}_t \ . \tag{16.77}$$

The subscripts H and S stand for "Heisenberg" and "Schrödinger", respectively. In Heisenberg's representation, the state $|\psi_t\rangle_{\mathrm{H}} = |\psi_{t_0}\rangle_{\mathrm{S}}$ is apparently a fixed time-independent state. Compared to this, the operators

$$\hat{A}_{\mathrm{H}}(t) = \exp\left[+\frac{\mathrm{i}}{\hbar}\hat{H}(t-t_0)\right] \hat{A}_{\mathrm{S}} \exp\left[-\frac{\mathrm{i}}{\hbar}\hat{H}(t-t_0)\right] \tag{16.78}$$

are time dependent because of (16.77) and (16.74). By differentiation of (16.78), we find that $\hat{A}_{\mathrm{H}}(t)$ fulfils the equation

$$-\frac{\hbar}{\mathrm{i}}\frac{\partial}{\partial t}\hat{A}_{\mathrm{H}} = \hat{A}_{\mathrm{H}}\hat{H} - \hat{H}\hat{A}_{\mathrm{H}} = [\hat{A}_{\mathrm{H}}, \hat{H}]_- \ . \tag{16.79}$$

It is called *Heisenberg's equation of motion* for the operator $\hat{A}_{\mathrm{H}}$ in the Heisenberg picture and has to be considered in analogy to the classical equation of motion of a dynamic variable A in the form of Poisson brackets,

$$\frac{\mathrm{d}A}{\mathrm{d}t} = \{A, H\} \ . \tag{16.80}$$

Heisenberg's equation leads immediately to the important result that an operator which commutes with the Hamiltonian is a constant of motion.

16.9 Free Particles

It will be useful to study the motion of a free particle more carefully and to systematically summarize the various mathematical operations and tricks once again. First we consider the free motion of a particle in one dimension and later we will devote ourselves to the three-dimensional problem. The dynamic variables are now the coordinate x, the momentum p, and the Hamiltonian is $\hat{H} = \hat{p}^2/2m$. The eigenvalue equations for x and p read

$$\hat{x}\,|x'\rangle = x'\,|x'\rangle \ , \tag{16.81a}$$

$$\hat{p}\,|p'\rangle = p'\,|p'\rangle \ . \tag{16.81b}$$

By definition, a truly free particle may occupy any position x' and also have any momentum p'. Therefore in (16.81) we have to deal with continuous spectra, so that the eigenstates $|x'\rangle$ and $|p'\rangle$ must be normalized to δ functions

$$\langle x'|x''\rangle = \delta(x'-x'') \ , \tag{16.82a}$$

$$\langle p'|p''\rangle = \delta(p'-p'') \ . \tag{16.82b}$$

Using the commutation relation

$$[\hat{x},\hat{p}]_- = \hat{x}\hat{p} - \hat{p}\hat{x} = \mathrm{i}\hbar\mathbf{1} \ , \tag{16.83}$$

we can calculate the matrix elements of $\hat{p}$ in x representation:

$$\begin{aligned}
\langle x'|\hat{x}\hat{p} - \hat{p}\hat{x}|x''\rangle &= \langle x'|\hat{x}\mathbf{1}\hat{p} - \hat{p}\mathbf{1}\hat{x}|x''\rangle \\
&= \int \mathrm{d}x''' [\langle x'|\hat{x}|x'''\rangle\langle x'''|\hat{p}|x''\rangle - \langle x'|\hat{p}|x'''\rangle\langle x'''|\hat{x}|x''\rangle] \\
&= \int \mathrm{d}x''' \Big[x'''\delta(x'-x''')\langle x'''|\hat{p}|x''\rangle \\
&\qquad - \langle x'|\hat{p}|x'''\rangle x''\delta(x''-x''')\Big] \\
&= x'\langle x'|\hat{p}|x''\rangle - x''\langle x'|\hat{p}|x''\rangle \\
&= (x'-x'')\langle x'|\hat{p}|x''\rangle \ .
\end{aligned} \tag{16.84}$$

On the other hand, because of (16.83),

$$\langle x'|\hat{x}\hat{p} - \hat{p}\hat{x}|x''\rangle = \mathrm{i}\hbar\delta(x'-x'') \ , \tag{16.85}$$

so that

$$(x'-x'')\langle x'|\hat{p}|x''\rangle = \mathrm{i}\hbar\delta(x'-x'') \ . \tag{16.86}$$

With the aid of the identity

$$x\frac{\mathrm{d}}{\mathrm{d}x}\delta(x) = -\delta(x) \ , \tag{16.87}$$

we get

$$\begin{aligned}
\mathrm{i}\hbar\delta(x'-x'') &= -\mathrm{i}\hbar(x'-x'')\frac{\partial}{\partial(x'-x'')}\delta(x'-x'') \\
&= -\mathrm{i}\hbar(x'-x'')\frac{\partial\delta(x'-x'')}{\partial x'} \ .
\end{aligned} \tag{16.88}$$

Finally, by using (16.86), we obtain

$$\langle x'|\hat{p}|x''\rangle = -\mathrm{i}\hbar\frac{\partial}{\partial x'}\delta(x'-x'') \ . \tag{16.89}$$

In the following exercise we will recalculate the analogous relation

$$\langle p'|\hat{x}|p''\rangle = \mathrm{i}\hbar\frac{\partial}{\partial p'}\delta(p'-p'') \ , \tag{16.90}$$

which is what we expect, because of the antisymmetric position of $\hat{x}$ and $\hat{p}$ in (16.84).

EXERCISE

16.5 Position Operator in Momentum Space

Problem. Prove the relation

$$\langle p'|\hat{x}|p''\rangle = \mathrm{i}\hbar\frac{\partial}{\partial p'}\delta(p'-p'') \tag{1}$$

in a similar manner to relation (16.89).

Solution.

$$\begin{aligned}\langle p'|\hat{x}\hat{p}-\hat{p}\hat{x}|p''\rangle &= \langle p'|\hat{x}\mathbf{1}\hat{p}-\hat{p}\mathbf{1}\hat{x}|p''\rangle \\ &= \int \mathrm{d}p'''\left[\langle p'|\hat{x}|p'''\rangle\langle p'''|\hat{p}|p''\rangle - \langle p'|\hat{p}|p'''\rangle\langle p'''|\hat{x}|p''\rangle\right] \\ &= \int \mathrm{d}p'''[p''\delta(p''-p''')\langle p'|\hat{x}|p'''\rangle \\ &\quad - p'\delta(p'''-p')\langle p'''|\hat{x}|p''\rangle] = (p''-p')\langle p'|\hat{x}|p''\rangle\end{aligned} \tag{2}$$

and, on the other hand, because of (16.83), this is equal to

$$\mathrm{i}\hbar\delta(p'-p'') \; . \tag{3}$$

So we get

$$\begin{aligned}&-(p'-p'')\langle p'|\hat{x}|p''\rangle \\ &\quad = \mathrm{i}\hbar\delta(p'-p'') \\ &\quad = -\mathrm{i}\hbar(p'-p'')\frac{\partial}{\partial(p'-p'')}\delta(p'-p'') \quad \text{[according to (16.87)]} \\ &\quad = \mathrm{i}\hbar(p'-p'')\frac{\partial}{\partial p'}\delta(p'-p'') \; .\end{aligned} \tag{4}$$

It follows that

$$\langle p'|\hat{x}|p''\rangle = \mathrm{i}\hbar\frac{\partial}{\partial p'}\delta(p'-p'') \; . \tag{5}$$

The matrix elements $\langle x'|\hat{p}^2|x''\rangle$ can also be calculated directly by computing the matrix product. Briefly,

$$\begin{aligned}
\langle x'|\hat{p}^2|x''\rangle &= \langle x'|\hat{p}\,\mathbf{1}\,\hat{p}|x''\rangle = \int \mathrm{d}x'''\langle x'|\hat{p}|x'''\rangle\langle x'''|\hat{p}|x''\rangle \\
&= \int \mathrm{d}x''' \left[-\mathrm{i}\hbar\frac{\partial}{\partial x'}\delta(x'-x''')\left(-\mathrm{i}\hbar\frac{\partial}{\partial x'''}\delta(x'''-x'')\right)\right] \\
&= -\mathrm{i}\hbar\frac{\partial}{\partial x'}\int \mathrm{d}x'''\delta(x'-x''')\left(-\mathrm{i}\hbar\frac{\partial}{\partial x'''}\delta(x'''-x'')\right) \\
&= \left(-\mathrm{i}\hbar\frac{\partial}{\partial x'}\right)^2 \delta(x'-x'') \,. \qquad (16.91)
\end{aligned}$$

Similarly, we get the more general relations,

$$\langle x'|\hat{p}^n|x''\rangle = \left(-\mathrm{i}\hbar\frac{\partial}{\partial x'}\right)^n \delta(x'-x'') \quad \text{and} \qquad (16.92)$$

$$\langle p'|\hat{x}^n|p''\rangle = \left(\mathrm{i}\hbar\frac{\partial}{\partial p'}\right)^n \delta(p'-p'') \,. \qquad (16.93)$$

Now we consider the eigenvalue problem for the momentum in coordinate representation:

$$\hat{p}\,|p'\rangle = p'\,|p'\rangle \,. \qquad (16.94)$$

We have

$$\begin{aligned}
\langle x'|\hat{p}|p'\rangle &= \int \mathrm{d}x''\langle x'|\hat{p}|x''\rangle\langle x''|p'\rangle = \int \mathrm{d}x''\left(-\mathrm{i}\hbar\frac{\partial}{\partial x'}\delta(x'-x'')\right)\langle x''|p'\rangle \\
&= -\mathrm{i}\hbar\frac{\partial}{\partial x'}\int \mathrm{d}x''\delta(x'-x'')\langle x''|p'\rangle \\
&= -\mathrm{i}\hbar\frac{\partial}{\partial x'}\langle x'|p'\rangle \,. \qquad (16.95)
\end{aligned}$$

On the other hand, it follows from (16.94) that

$$\langle x'|\hat{p}|p'\rangle = p'\,\langle x'|p'\rangle \,,$$

so that the differential equation for $\langle x'|p'\rangle$,

$$-\mathrm{i}\hbar\frac{\partial}{\partial x'}\langle x'|p'\rangle = p'\,\langle x'|p'\rangle \,, \qquad (16.96)$$

results. Its solution is

$$\langle x'|p'\rangle \equiv \psi_{p'}(x') = \frac{1}{\sqrt{2\pi\hbar}}\exp\left(\frac{\mathrm{i}}{\hbar}p'x'\right) \,. \qquad (16.97)$$

Here, we have chosen the normalization in such a way that

$$\begin{aligned}\langle p''|p'\rangle &= \int \mathrm{d}x' \langle p''|x'\rangle\langle x'|p'\rangle \\ &= \int \mathrm{d}x' \psi^*_{p''}(x')\psi_{p'}(x') = \delta(p''-p') \ . \end{aligned} \tag{16.98}$$

Now we generalize the above results to three dimensions. According to (16.66), the three space coordinates commute with each other. Hence, they may be combined into the state

$$|\boldsymbol{x}\rangle = |x, y, z\rangle \ . \tag{16.99}$$

By definition $|\boldsymbol{x}\rangle$ is also an eigenstate of the operators $\hat{x}$, $\hat{y}$ and $\hat{z}$:

$$\hat{x}\,|\boldsymbol{x}'\rangle = x'\,|\boldsymbol{x}'\rangle \ , \quad \hat{y}\,|\boldsymbol{x}'\rangle = y'\,|\boldsymbol{x}'\rangle \ , \quad \hat{z}\,|\boldsymbol{x}'\rangle = z'\,|\boldsymbol{x}'\rangle \ ,$$

or, in short,

$$\hat{\boldsymbol{x}}\,|\boldsymbol{x}'\rangle = \boldsymbol{x}'\,|\boldsymbol{x}'\rangle \ . \tag{16.100}$$

As the spectrum is continuous, we may (must) normalize to δ functions:

$$\langle \boldsymbol{x}''|\boldsymbol{x}'\rangle = \delta(\boldsymbol{x}'-\boldsymbol{x}'') = \delta(x'-x'')\delta(y'-y'')\delta(z'-z'') \ . \tag{16.101}$$

The operators $\hat{p}_x$, $\hat{p}_y$, $\hat{p}_z$ commute with each other, too, so that we may form the common eigenvector $|\boldsymbol{p}\rangle$ with

$$\hat{\boldsymbol{p}}\,|\boldsymbol{p}'\rangle = \boldsymbol{p}'\,|\boldsymbol{p}'\rangle \ . \tag{16.102}$$

Again, we have normalization to δ functions:

$$\langle \boldsymbol{p}''|\boldsymbol{p}'\rangle = \delta(\boldsymbol{p}'-\boldsymbol{p}'') = \delta(p'_x-p''_x)\delta(p'_y-p''_y)\delta(p'_z-p''_z) \ . \tag{16.103}$$

Now we want to return to (16.89). Every single step which led to this solution may be repeated for each component $\hat{p}_x$, $\hat{p}_y$, $\hat{p}_z$ with the state vector $|\boldsymbol{x}\rangle$. Thus we get

$$\begin{aligned}\langle \boldsymbol{x}'|\hat{p}_x|\boldsymbol{x}''\rangle &= -\mathrm{i}\hbar\frac{\partial}{\partial x'}\delta(\boldsymbol{x}'-\boldsymbol{x}'') \\ &\vdots \qquad \text{etc.}\end{aligned} \tag{16.104}$$

We may combine this in the form

$$\begin{aligned}\langle \boldsymbol{x}'|\hat{\boldsymbol{p}}|\boldsymbol{x}''\rangle &= -\mathrm{i}\hbar\frac{\partial}{\partial \boldsymbol{x}'}\delta(\boldsymbol{x}'-\boldsymbol{x}'') \\ &\equiv -\mathrm{i}\hbar\left(\frac{\partial}{\partial x'}\delta(\boldsymbol{x}'-\boldsymbol{x}'') \ , \quad \frac{\partial}{\partial y'}\delta(\boldsymbol{x}'-\boldsymbol{x}'') \ , \quad \frac{\partial}{\partial z'}\delta(\boldsymbol{x}'-\boldsymbol{x}'')\right) \ .\end{aligned} \tag{16.105}$$

Similarly, we conclude immediately that

$$\langle \boldsymbol{p}'|\hat{\boldsymbol{x}}|\boldsymbol{p}''\rangle = \mathrm{i}\hbar \frac{\partial}{\partial \boldsymbol{p}'}\delta(\boldsymbol{p}'-\boldsymbol{p}'') \equiv \mathrm{i}\hbar\left(\frac{\partial}{\partial p'_x}\delta(\boldsymbol{p}'-\boldsymbol{p}'')\,,\quad \frac{\partial}{\partial p'_y}\delta(\boldsymbol{p}'-\boldsymbol{p}'')\,,\quad \frac{\partial}{\partial p'_z}\delta(\boldsymbol{p}'-\boldsymbol{p}'')\right)\,, \tag{16.106}$$

which is analogous to (16.90). The differential equation (16.96) can also be generalized to three dimensions without any difficulties:

$$-\mathrm{i}\hbar\frac{\partial}{\partial \boldsymbol{x}'}\langle \boldsymbol{x}'|\boldsymbol{p}'\rangle = \boldsymbol{p}'\langle \boldsymbol{x}'|\boldsymbol{p}'\rangle\,, \tag{16.107}$$

with the solution

$$\langle \boldsymbol{x}'|\boldsymbol{p}'\rangle \equiv \psi_{\boldsymbol{p}'}(\boldsymbol{x}') = \frac{1}{(2\pi\hbar)^{3/2}}\exp\left(\frac{\mathrm{i}}{\hbar}\boldsymbol{p}'\cdot\boldsymbol{x}'\right)\,, \tag{16.108}$$

normalized to δ functions. Using the results (16.91) and (16.92), we get the Hamiltonian of a free particle $\hat{H} = \hat{\boldsymbol{p}}^2/2m$ in x representation:

$$\langle \boldsymbol{x}'|\hat{H}|\boldsymbol{x}''\rangle = \left\langle \boldsymbol{x}'\left|\frac{\hat{\boldsymbol{p}}^2}{2m}\right|\boldsymbol{x}''\right\rangle = -\frac{\hbar^2}{2m}\nabla^2\delta(\boldsymbol{x}'-\boldsymbol{x}'')\,. \tag{16.109}$$

In p representation, this reads

$$\langle \boldsymbol{p}'|\hat{H}|\boldsymbol{p}''\rangle = \left\langle \boldsymbol{p}'\left|\frac{\hat{\boldsymbol{p}}^2}{2m}\right|\boldsymbol{p}''\right\rangle = \frac{(\boldsymbol{p}')^2}{2m}\delta(\boldsymbol{p}'-\boldsymbol{p}'')\,. \tag{16.110}$$

Now we turn to the time-dependent description. In particular, we are interested in the propagation of the wave which describes a free particle; this is called *free propagation*. For this, we use (16.73) and (16.74), and express $\psi(\boldsymbol{x}',t) = \langle \boldsymbol{x}'|\psi_t\rangle$ by $\psi(\boldsymbol{x}',t_0) = \langle \boldsymbol{x}'|\psi_{t_0}\rangle$ as

$$\begin{aligned}
|\psi_t\rangle &= \exp[-\mathrm{i}\hat{H}(t-t_0)/\hbar]\,|\psi_{t_0}\rangle\,,\\
\psi(\boldsymbol{x}',t) &\equiv \langle \boldsymbol{x}'|\psi_t\rangle = \langle \boldsymbol{x}'|\exp[-\mathrm{i}\hat{H}(t-t_0)/\hbar]\,|\psi_{t_0}\rangle\\
&= \int \mathrm{d}^3x''\langle \boldsymbol{x}'|\exp[-\mathrm{i}\hat{H}(t-t_0)/\hbar]|\boldsymbol{x}''\rangle\langle \boldsymbol{x}''|\psi_{t_0}\rangle\\
&= \int \mathrm{d}^3x''G(\boldsymbol{x}',t|\boldsymbol{x}'',t_0)\psi(\boldsymbol{x}'',t_0)\,.
\end{aligned} \tag{16.111}$$

Here,

$$G(\boldsymbol{x}',t|\boldsymbol{x}'',t_0) = \langle \boldsymbol{x}'|\exp[-\mathrm{i}\hat{H}(t-t_0)/\hbar]|\boldsymbol{x}''\rangle \tag{16.112}$$

is called *Green's function* or the *propagator*. It describes the time development of the wave $\psi(\boldsymbol{x}',t)$, starting with the initial waves $\psi(\boldsymbol{x}'',t_0)$. Its explicit calculation can be accomplished immediately in the case of free particles with $\hat{H} = \hat{\boldsymbol{p}}^2/2m$:

$$
\begin{aligned}
G(\boldsymbol{x}', t|\boldsymbol{x}'', t_0) &= \iint \mathrm{d}^3p'\,\mathrm{d}^3p''\,\langle \boldsymbol{x}'|\boldsymbol{p}'\rangle \\
&\quad \times \left\langle \boldsymbol{p}' \left| \exp\left[-\frac{\mathrm{i}}{\hbar}\frac{\hat{\boldsymbol{p}}^2}{2m}(t-t_0)\right] \right| \boldsymbol{p}''\right\rangle \langle \boldsymbol{p}''|\boldsymbol{x}''\rangle \\
&= \iint \mathrm{d}^3p'\,\mathrm{d}^3p''\,\langle \boldsymbol{x}'|\boldsymbol{p}'\rangle \exp\left[-\frac{\mathrm{i}}{\hbar}\frac{\boldsymbol{p}''^2}{2m}(t-t_0)\right] \\
&\quad \times \delta(\boldsymbol{p}'-\boldsymbol{p}'')\,\langle \boldsymbol{p}''|\boldsymbol{x}''\rangle \\
&= \int \mathrm{d}^3p'\,\langle \boldsymbol{x}'|\boldsymbol{p}'\rangle\langle \boldsymbol{p}'|\boldsymbol{x}''\rangle \exp\left[-\frac{\mathrm{i}}{\hbar}\frac{\boldsymbol{p}'^2}{2m}(t-t_0)\right] \qquad (16.113) \\
&= \int \frac{\mathrm{d}^3p'}{(2\pi\hbar)^3} \exp\left\{\frac{\mathrm{i}}{\hbar}\left[\boldsymbol{p}'\cdot(\boldsymbol{x}''-\boldsymbol{x}') - \frac{\boldsymbol{p}'^2}{2m}(t-t_0)\right]\right\} .
\end{aligned}
$$

The integral can be computed analytically (cf. Exercise 16.6), giving

$$
G(\boldsymbol{x}'t|\boldsymbol{x}''t_0) = \left[\frac{m}{2\pi\mathrm{i}\hbar(t-t_0)}\right]^{3/2} \exp\left[\frac{\mathrm{i}m}{2\hbar}\frac{(\boldsymbol{x}''-\boldsymbol{x}')^2}{t-t_0}\right] . \qquad (16.114)
$$

Finally, we want to make some comments on the description of free particles with spin. This is simply done by constructing the direct product of a vector $|\boldsymbol{x}'\rangle$, $|\boldsymbol{p}\rangle$ or $|\psi\rangle$ and a spin vector $|\sigma\rangle$. For particles with spin 1/2, the vector $|\sigma\rangle$ is, for example, given by

$$
|z,\uparrow\rangle = \begin{pmatrix}1\\0\end{pmatrix} , \quad |z,\downarrow\rangle = \begin{pmatrix}0\\1\end{pmatrix} . \qquad (16.115)
$$

The argument z of these spin vectors indicates that we have chosen the representation with $\sigma_z = \begin{pmatrix}1 & 0\\ 0 & -1\end{pmatrix}$ being diagonal. Hence, we have

$$
|\psi,\sigma\rangle = |\psi\rangle\,|\sigma\rangle \quad \text{and} \qquad (16.116)
$$

$$
\langle \boldsymbol{x}|\psi,\sigma\rangle = \psi(\boldsymbol{x})\,|\sigma\rangle = \begin{pmatrix}\psi_1(\boldsymbol{x})\\ \psi_2(\boldsymbol{x})\end{pmatrix} . \qquad (16.117)
$$

Thus a spin-1/2 particle is represented by a wave function with two components (a spinor).

EXERCISE

16.6 Calculating the Propagator Integral

Problem. Calculate the propagator integral (16.113)

$$G(\boldsymbol{x}'t|\boldsymbol{x}''t_0) = \int_{-\infty}^{\infty} \frac{d^3p'}{(2\pi\hbar)^3} \exp\left\{\frac{i}{\hbar}\left[\boldsymbol{p}'\cdot(\boldsymbol{x}''-\boldsymbol{x}') - \frac{p'^2}{2m}(t-t_0)\right]\right\} .$$

Solution. By suitably rearranging the terms in the exponent, we convert the propagator integral to

$$G(\boldsymbol{x}'t|\boldsymbol{x}''t_0) = \frac{1}{(2\pi\hbar)^3}\int_{-\infty}^{\infty} d^3p' \times \exp\left\{-\frac{i}{\hbar}\frac{(t-t_0)}{2m}\left[p'^2 - \frac{2m\boldsymbol{p}'\cdot(\boldsymbol{x}''-\boldsymbol{x}')}{(t-t_0)}\right]\right\} . \quad (1)$$

By adding and subtracting the quadratic complement in the exponent, we find

$$G(\boldsymbol{x}'t|\boldsymbol{x}''t_0) = \frac{1}{(2\pi\hbar)^3}\int_{-\infty}^{\infty} d^3p' \exp\left(\left[-\frac{i}{\hbar}\frac{(t-t_0)}{2m}\right] \times \left\{\left[\boldsymbol{p}' - \frac{m(\boldsymbol{x}''-\boldsymbol{x}')}{(t-t_0)}\right]^2 - \frac{m^2(\boldsymbol{x}''-\boldsymbol{x}')^2}{(t-t_0)^2}\right\}\right) . \quad (2)$$

Now we may put a factor in front of the integral:

$$G(\boldsymbol{x}'t|\boldsymbol{x}''t_0) = \frac{1}{(2\pi\hbar)^3}\exp\left[\frac{im}{2\hbar}\frac{(\boldsymbol{x}''-\boldsymbol{x}')^2}{(t-t_0)}\right] \times \int_{-\infty}^{\infty} d^3p' \exp\left\{-\frac{i}{\hbar}\frac{(t-t_0)}{2m}\left[\boldsymbol{p}' - \frac{m(\boldsymbol{x}''-\boldsymbol{x}')}{(t-t_0)}\right]^2\right\} . \quad (3)$$

We substitute the new variables

$$\boldsymbol{P}' = \boldsymbol{p}' - \frac{m(\boldsymbol{x}''-\boldsymbol{x}')}{(t-t_0)} \quad \text{and} \quad \lambda^2 = \frac{i}{\hbar}\frac{(t-t_0)}{2m} . \quad (4)$$

The integration over d^3p' is transformed into an integration over $d^3p' = 4\pi P'^2 dP'$. Here, attention has to be paid to the fact that the lower limit of the integral becomes zero:

Exercise 16.6

$$G(\boldsymbol{x}'t|\boldsymbol{x}''t_0) = \frac{4\pi}{(2\pi\hbar)^3}\exp\left[\mathrm{i}\frac{m}{2\hbar}\frac{(\boldsymbol{x}''-\boldsymbol{x}')^2}{(t-t_0)}\right] \times \int_0^\infty \mathrm{d}P'\,P'^2\exp(-\lambda^2\boldsymbol{P}'^2)\ . \tag{5}$$

Because

$$\int_0^\infty \mathrm{d}x\,x^2\,\mathrm{e}^{-a^2x^2} = \frac{\sqrt{\pi}}{4a^3}\ , \tag{6}$$

we get immediately

$$G(\boldsymbol{x}'t|\boldsymbol{x}''t_0) = \left(\frac{m}{2\pi\mathrm{i}\hbar(t-t_0)}\right)^{3/2}\exp\left(\frac{\mathrm{i}m}{2\hbar}\frac{(\boldsymbol{x}''-\boldsymbol{x}')^2}{t-t_0}\right)\ . \tag{7}$$

EXAMPLE

16.7 The One-Dimensional Oscillator in Various Representations

The harmonic oscillator plays an important role in many areas of physics; above all, in field theory (e.g. quantization of the electromagnetic field). Thus it is useful to summarize its properties. The Hamiltonian of the one-dimensional oscillator is given by

$$\hat{H}(\hat{x},\hat{p}) = \frac{1}{2m}\hat{p}^2 + \frac{m\omega^2}{2}\hat{x}^2\ , \tag{1}$$

and the corresponding energy eigenvalue problem reads

$$\hat{H}\,|E\rangle = E\,|E\rangle\ . \tag{2}$$

We solve the problem (2) for different representations.

(a) x Representation (Coordinate Representation)

Here, $\hat{H}$ may be written as [cf. (16.91) and (16.92)]

$$\begin{aligned}\langle x'|\hat{H}|x''\rangle &= \hat{H}(x',\hat{p}')\delta(x'-x'')\\ &= \hat{H}\left(x',\frac{\hbar}{\mathrm{i}}\frac{\partial}{\partial x'}\right)\delta(x'-x'')\ ,\end{aligned} \tag{3}$$

where

$$\hat{H}\left(x',\frac{\hbar}{\mathrm{i}}\frac{\partial}{\partial x'}\right) = -\frac{\hbar^2}{2m}\frac{\partial^2}{\partial x'^2} + \frac{m\omega^2}{2}x'^2\ . \tag{4}$$

Example 16.7

Now (2) reads

$$\begin{aligned}\langle x'|\hat{H}|E\rangle &= \int \mathrm{d}x''\langle x'|\hat{H}|x''\rangle\langle x''|E\rangle \\ &= \int \mathrm{d}x''\hat{H}\left(x', \frac{\hbar}{\mathrm{i}}\frac{\partial}{\partial x'}\right)\delta(x'-x'')\langle x''|E\rangle \\ &= \hat{H}\left(x', \frac{\hbar}{\mathrm{i}}\frac{\partial}{\partial x'}\right)\langle x'|E\rangle = E\langle x'|E\rangle \ . \end{aligned} \tag{5}$$

In the last step, we have written down the right-hand side of (2). Using $\psi_E(x') = \langle x'|E\rangle$, we may write

$$\hat{H}\left(x', \frac{\hbar}{\mathrm{i}}\frac{\partial}{\partial x'}\right)\psi_E(x') = E\psi_E(x') \ . \tag{6}$$

This differential equation, familiar from Chap. 7, has usable solutions only if E adopts the eigenvalues

$$E_n = \hbar\omega(n+\tfrac{1}{2}) \ , \quad n = 0, 1, 2, \dots, \infty \ . \tag{7}$$

The solutions belonging to these eigenvalues are

$$\psi_{E_n}(x') = \left(\frac{m\omega}{\pi\hbar}\right)^{1/4}\frac{1}{\sqrt{2^n n!}}H_n(\xi)\,\mathrm{e}^{-(\xi^2/2)} \ , \tag{8}$$

with $\xi = \sqrt{m\omega/\hbar}\,x'$. $H_n(\xi)$ are the well-known *Hermite polynomials.*

(b) p Representation (Momentum Representation)

Because of (16.81b) and (16.93), the Hamiltonian (1) in p representation is given by

$$\begin{aligned}\langle p'|\hat{H}|p''\rangle &= \hat{H}(x, p')\delta(p'-p'') \\ &= \hat{H}\left(-\frac{\hbar}{\mathrm{i}}\frac{\partial}{\partial p'}, p'\right)\delta(p'-p'') \ , \end{aligned} \tag{9}$$

where

$$\hat{H}\left(-\frac{\hbar}{\mathrm{i}}\frac{\partial}{\partial p'}, p'\right) = \frac{1}{2m}p'^2 - \frac{m\omega^2}{2}\hbar^2\frac{\partial^2}{\partial p'^2} \ . \tag{10}$$

Then, with $\langle p'|E\rangle = \psi_E(p')$, (2) reads

$$\begin{aligned}\langle p'|\hat{H}|E\rangle &= \int \mathrm{d}p''\langle p'|\hat{H}|p''\rangle\langle p''|E\rangle \\ &= \int \mathrm{d}p''\hat{H}\left(-\frac{\hbar}{\mathrm{i}}\frac{\partial}{\partial p'}, p'\right)\delta(p'-p'')\langle p''|E\rangle \\ &= \hat{H}\left(-\frac{\hbar}{\mathrm{i}}\frac{\partial}{\partial p'}, p'\right)\psi_E(p') \\ &= E\psi_E(p') \ . \end{aligned} \tag{11}$$

Example 16.7

In the last step we added the right-hand side of (2). The resulting eigenvalue equation in momentum space,

$$\left(-\frac{m\omega^2\hbar^2}{2}\frac{\partial}{\partial p'^2}+\frac{1}{2m}p'^2\right)\psi_E(p')=E\psi_E(p')\ , \tag{12}$$

is easily transformed into

$$\left(-\frac{\hbar^2}{2m}\frac{\partial^2}{\partial p'^2}+\frac{1}{2m^3\omega^2}p'^2\right)\psi_E(p')=\frac{E}{m^2\omega^2}\psi_E(p')\ , \tag{13}$$

$$\left(-\frac{\hbar^2}{2m}\frac{\partial^2}{\partial p'^2}+\frac{m\bar{\omega}^2}{2}p'^2\right)\psi_E(p')=\bar{E}\psi_E(p')\ . \tag{14}$$

Here, we have set $\bar{\omega}^2=1/m^4\omega^2$ and $\bar{E}=E/m^2\omega^2$. Equation (14) is identical to (4) and (6). Thus the eigenvalues are

$$\bar{E}_n=\hbar\bar{\omega}(n+\tfrac{1}{2})\ ,\qquad n=0,1,\dots,\infty$$

or

$$E_n=\hbar\frac{m^2\omega^2}{m^2\omega}(n+\tfrac{1}{2})=\hbar\omega(n+\tfrac{1}{2})\ . \tag{15}$$

Thus they are identical to the values we got in (7). In the same easy way, (8) is transformed into the wave function $\psi_{E_n}(p')$ in momentum space:

$$\psi_{E_n}(p')=\left(\frac{m\bar{\omega}}{\pi\hbar}\right)^{1/2}\frac{1}{2^n n!}H_n(\eta)\,\mathrm{e}^{-(\eta^2/2)}\ ,$$

$$\eta=\sqrt{\frac{m\bar{\omega}}{\hbar}}p'\ . \tag{16}$$

According to Axiom 2 [see (16.63) and (16.63a)], the probability of finding a particle in the energy state $\psi_{E_n}(x')$ in the interval between x' and $x'+\mathrm{d}x'$ is given by

$$\mathrm{d}W(x')=\left|\langle x'|E_n\rangle\right|^2\mathrm{d}x'=\left|\psi_{E_n}(x')\right|^2\mathrm{d}x'\ . \tag{17}$$

In the same manner, the probability of finding the particle with a momentum between p' and $p'+\mathrm{d}p'$ can be calculated:

$$\mathrm{d}W(p')=\left|\langle p'|E_n\rangle\right|^2\mathrm{d}p'=\left|\psi_{E_n}(p')\right|^2\mathrm{d}p'\ . \tag{18}$$

The wave functions in coordinate space $\psi_{E_n}(x')$ and in momentum space $\psi_{E_n}(p')$ are connected by

$$\begin{aligned}\langle x'|E_n\rangle=\psi_{E_n}(x')&=\int\mathrm{d}p'\,\langle x'|p'\rangle\langle p'|E_n\rangle\\&=\int\frac{\mathrm{d}p'}{\sqrt{2\pi\hbar}}\exp\left(\frac{\mathrm{i}}{\hbar}p'x'\right)\psi_{E_n}(p')\end{aligned} \tag{19a}$$

Example 16.7

and

$$\psi_{E_n}(p') = \int \frac{\mathrm{d}x'}{\sqrt{2\pi\hbar}} \exp\left(-\frac{\mathrm{i}}{\hbar}p'x'\right)\psi_{E_n}(x') \ , \tag{19b}$$

respectively.

(c) Algebraic Method (Algebraic Representation)

The algebraic method does not explicitly use any representation to solve the eigenvalue problem (2); it is particularly useful in field theory. We introduce the operators

$$\hat{a} = \sqrt{\frac{m\omega}{2\hbar}}\hat{x} + \frac{\mathrm{i}}{\sqrt{2m\hbar\omega}}\hat{p} \tag{20a}$$

and

$$\hat{a}^+ = \sqrt{\frac{m\omega}{2\hbar}}\hat{x} - \frac{\mathrm{i}}{\sqrt{2m\hbar\omega}}\hat{p} \ . \tag{20b}$$

They are called the *annihilation operator* ($\hat{a}$) and the *creation operator* ($\hat{a}^+$), respectively, of oscillator phonons. It can be seen that $(\hat{a})^+ = \hat{a}^+$; i.e. $\hat{a} = (\hat{a}^+)^+$. Their commutation relations are easily calculated:

$$\begin{aligned}[\hat{a}, \hat{a}^+]_- &= \left[\sqrt{\frac{m\omega}{2\hbar}}\hat{x} + \frac{\mathrm{i}}{\sqrt{2m\hbar\omega}}\hat{p} \ , \ \sqrt{\frac{m\omega}{2\hbar}}\hat{x} - \frac{\mathrm{i}}{\sqrt{2m\hbar\omega}}\hat{p}\right] \\ &= -\frac{\mathrm{i}}{2\hbar}[\hat{x}, \hat{p}] + \frac{\mathrm{i}}{2\hbar}[\hat{p}, \hat{x}] = \frac{\mathrm{i}}{\hbar}[\hat{p}, \hat{x}] = 1\end{aligned} \tag{21}$$

and

$$[\hat{a}, \hat{a}]_- = 0 = [\hat{a}^+, \hat{a}^+]_- \ . \tag{22}$$

In a similar manner, we find

$$\begin{aligned}\hat{a}^+\hat{a} &= \frac{m\omega}{2\hbar}\hat{x}^2 + \frac{p^2}{2m\hbar\omega} + \frac{\mathrm{i}}{2\hbar}(\hat{x}\hat{p} - \hat{p}\hat{x}) \\ &= \frac{1}{\hbar\omega}\hat{H} - \frac{1}{2} \ ; \quad \text{thus}\end{aligned}$$

$$\hat{H} = \hbar\omega(\hat{a}^+\hat{a} + \tfrac{1}{2}) = \hbar\omega(\hat{N} + \tfrac{1}{2}) \ , \quad \text{where} \tag{23}$$

$$\hat{N} = \hat{a}^+\hat{a} \ . \tag{24}$$

We denote the eigenvectors of $\hat{N}$ by $|n\rangle$, so we have

$$\hat{N}\,|n\rangle = n\,|n\rangle \ . \tag{25}$$

We study the vector $|g\rangle$, which results from applying the annihilation operator $\hat{a}$ to $|n\rangle$:

$$|g\rangle = \hat{a}\,|n\rangle \ . \tag{26}$$

Example 16.7

We study $|g\rangle$ by applying $\hat{N}$ and find that

$$\begin{aligned}\hat{N}\,|g\rangle &= \hat{a}^+\hat{a}\hat{a}\,|n\rangle = (\hat{a}\hat{a}^+ - 1)\hat{a}\,|n\rangle \\ &= \hat{a}(\hat{a}^+\hat{a})\,|n\rangle - \hat{a}\,|n\rangle = \hat{a}n\,|n\rangle - \hat{a}\,|n\rangle \\ &= (n-1)\hat{a}\,|n\rangle = (n-1)\,|g\rangle \quad . \end{aligned} \tag{27}$$

Thus $|g\rangle$ is an eigenvector of $\hat{N}$, too, but it belongs to the eigenvalue $(n-1)$. Therefore, assuming that the eigenvalue $(n-1)$ is not degenerate, $|g\rangle$ and $|n-1\rangle$ are identical except for a constant. So we can write

$$|g\rangle = \hat{a}\,|n\rangle = C_n\,|n-1\rangle \quad . \tag{28}$$

The constant C_n can be determined by calculating the norm

$$\begin{aligned}&(\hat{a}\,|n\rangle\,,\hat{a}\,|n\rangle) = (C_n\,|n-1\rangle\,, C_n\,|n-1\rangle) \Leftrightarrow \\ &\langle n|\hat{a}^+\hat{a}|n\rangle = |C_n|^2\,\langle n-1|n-1\rangle \Leftrightarrow n = |C_n|^2 \quad .\end{aligned}$$

Setting $C_n = \sqrt{n}$, a phase factor equal to unity is chosen, we find that (28) reads

$$\hat{a}\,|n\rangle = \sqrt{n}\,|n-1\rangle \quad . \tag{29}$$

Similarly, we may conclude that

$$\hat{a}^+\,|n\rangle = \sqrt{n+1}\,|n+1\rangle \quad . \tag{30}$$

Equations (29) and (30) explain the nomenclature of annihilation and creation operators. Equation (25) suggests interpreting N as a number operator for oscillator quanta. Now $n \geq 0$ always. We see this by multiplying (25) by $\langle n|$:

$$\langle n|\hat{N}|n\rangle = n\,\langle n|n\rangle = \langle n|\hat{a}^+\hat{a}|n\rangle = \langle \hat{a}n|\hat{a}n\rangle \quad ;$$

hence,

$$n = \frac{\langle \hat{a}n|\hat{a}n\rangle}{\langle n|n\rangle} = \frac{\|\hat{a}|n\rangle\|^2}{\||n\rangle\|^2} \geq 0 \ . \tag{31}$$

Starting with the vector $|n\rangle$, the states

$$|n-1\rangle\,,\ |n-2\rangle\,,\ |n-3\rangle\ldots \tag{32}$$

may be generated by successive applications of the annihilation operator $\hat{a}$. Because of (31), only positive eigenvalues are allowed. Thus the series (32) has to terminate, that is with the state $|0\rangle$. For this state we have

$$\hat{a}\,|0\rangle = 0\ , \quad \hat{N}\,|0\rangle = 0\ . \tag{33}$$

$|0\rangle$ is called a *vacuum state* (also ground state) for oscillator quanta. On the basis of (24) and (25), we conclude that the states $|n\rangle$ are eigenstates of $\hat{H}$ with the eigenvalues $E_n = \hbar\omega(n+\frac{1}{2})$:

$$\hat{H}\,|n\rangle = \hbar\omega(\hat{N}+\tfrac{1}{2})|n\rangle = \hbar\omega(n+\tfrac{1}{2})|n\rangle \quad . \tag{34}$$

Example 16.7

Frequently the matrix elements of $\hat{x}$ and $\hat{p}$ have to be determined. They are easily specified by solving (20) for $\hat{x}$ and $\hat{p}$, respectively:

$$\hat{x} = \sqrt{\frac{\hbar}{2m\omega}}(\hat{a}^+ + \hat{a}) \ , \tag{35a}$$

$$\hat{p} = \mathrm{i}\sqrt{\frac{m\hbar\omega}{2}}(\hat{a}^+ - \hat{a}) \ , \tag{35b}$$

and, by using (29) and (30),

$$\begin{aligned} \langle n_1|\hat{x}|n_2\rangle &= \sqrt{\frac{\hbar}{2m\omega}}\left[\sqrt{n_2+1}\,\delta_{n_1,n_2+1} + \sqrt{n_2}\,\delta_{n_1,n_2-1}\right] \ , \\ \langle n_1|\hat{p}|n_2\rangle &= \mathrm{i}\sqrt{\frac{m\hbar\omega}{2}}\left[\sqrt{n_2+1}\,\delta_{n_1,n_2+1} - \sqrt{n_2}\,\delta_{n_1,n_2-1}\right] \ . \end{aligned} \tag{36}$$

In a similar manner we get

$$\begin{aligned} \langle n_1|\hat{x}^2|n_2\rangle &= \sum_n \langle n_1|\hat{x}|n\rangle\langle n|\hat{x}|n_2\rangle \\ &= \frac{\hbar}{2m\omega}\sum_n\left[\sqrt{n+1}\,\delta_{n_1,n+1} + \sqrt{n}\,\delta_{n_1,n-1}\right] \\ &\quad \times\left[\sqrt{n_2+1}\,\delta_{n,n_2+1} + \sqrt{n_2}\,\delta_{n,n_2-1}\right] \\ &= \frac{\hbar}{2m\omega}\left[(2n_1+1)\delta_{n_1,n_2} + \sqrt{n_1}\sqrt{n_2+1}\right. \\ &\quad \left.\times\delta_{n_1,n_2+2} + \sqrt{n_1+1}\sqrt{n_2}\,\delta_{n_1,n_2-2}\right] \end{aligned} \tag{37}$$

and

$$\begin{aligned} \langle n_1|\hat{p}^2|n_2\rangle = \frac{m\hbar\omega}{2}&\left[(2n_1+1)\delta_{n_1,n_2} - \sqrt{n_1}\sqrt{n_2+1}\right. \\ &\left.\times\delta_{n_1,n_2+2} - \sqrt{n_1+1}\sqrt{n_2}\,\delta_{n_1,n_2-2}\right] \ . \end{aligned} \tag{38}$$

From these two equations we obtain

$$\begin{aligned} \langle n_1|\hat{H}|n_2\rangle &= \frac{1}{2m}\langle n_1|\hat{p}^2|n_2\rangle + \frac{m\omega^2}{2}\langle n_1|\hat{x}^2|n_2\rangle \\ &= \hbar\omega(n_1+\tfrac{1}{2})\delta_{n_1,n_2} \ . \end{aligned} \tag{39}$$

16.10 A Summary of Perturbation Theory

Usually it is not possible to give an exact solution of a problem in quantum mechanics; we have to restrict ourselves to approximate solutions of the Schrödinger equation

$$\mathrm{i}\hbar\frac{\partial}{\partial t}\,|\psi\rangle = (\hat{H}_0+\hat{H}')\,|\psi\rangle \ . \tag{16.118}$$

The splitting of $\hat{H}=\hat{H}_0+\hat{H}'$ is chosen in such a manner that the solutions of $\hat{H}_0$ are known:

$$\hat{H}_0\,|\varphi_n\rangle = E_n\,|\varphi_n\rangle \ , \tag{16.119}$$

and $\hat{H}'$ is small enough compared with $\hat{H}_0$, so that its influence may be considered as a perturbation. In Chap. 11, the perturbation $\hat{H}'$ was denoted by εW. We expand $|\psi\rangle$ in terms of $|\varphi_n\rangle$ and write

$$|\psi\rangle = \sum_n C_n(t)\exp\left(-\frac{\mathrm{i}}{\hbar}E_n t\right)|\varphi_n\rangle \ . \tag{16.120}$$

By inserting this into (16.118) and taking (16.119) into account, the following system of coupled differential equations for the expansion coefficients $C_n(t)$ is obtained:

$$\frac{\mathrm{d}}{\mathrm{d}t}C_m(t) = -\frac{\mathrm{i}}{\hbar}\sum_n\langle\varphi_m|\hat{H}'|\varphi_n\rangle\exp\left[\frac{\mathrm{i}}{\hbar}(E_m-E_n)t\right]C_n(t) \ ,$$
$$m=0,1,2,\dots \ . \tag{16.121}$$

These equations may easily be transformed by integration into a system of coupled integral equations:

$$C_m(t) = C_m(0) - \frac{\mathrm{i}}{\hbar}\sum_n\int_0^t \mathrm{d}t'\langle\varphi_m|\hat{H}'|\varphi_n\rangle \times\exp\left[\frac{\mathrm{i}}{\hbar}(E_m-E_n)t'\right]C_n(t') \ . \tag{16.122}$$

Up to this point, everything is exact. Now approximations enter. If we assume the system to be in state $|\varphi_\mathrm{i}\rangle$ at time $t=0$, we have

$$C_m(0) = \delta_{m\mathrm{i}} \ . \tag{16.123}$$

As the perturbation $\hat{H}'$ is assumed to be small and of little influence on the undisturbed states $|\varphi_n\rangle$, it is consistent to assume that none of the $C_n(t)$ coefficients differs considerably from the initial value. Further, assuming that $\hat{H}'$ is independent of time, it follows for $\mathrm{f}\neq\mathrm{i}$ that

$$
\begin{aligned}
C_{\mathrm{f}}(t) &= -\frac{\mathrm{i}}{\hbar}\langle\varphi_{\mathrm{f}}|\hat{H}'|\varphi_{\mathrm{i}}\rangle \int\limits_0^t \mathrm{d}t' \exp\left[\frac{\mathrm{i}}{\hbar}(E_{\mathrm{f}}-E_{\mathrm{i}})t'\right] \\
&= -\frac{\mathrm{i}}{\hbar}\langle\varphi_{\mathrm{f}}|\hat{H}'|\varphi_{\mathrm{i}}\rangle \left[\frac{\exp[\mathrm{i}(E_{\mathrm{f}}-E_{\mathrm{i}})t/\hbar]-1}{\mathrm{i}(E_{\mathrm{f}}-E_{\mathrm{i}})/\hbar}\right] .
\end{aligned}
\tag{16.124}
$$

After time t, the probability of finding the system in the state $|\varphi_{\mathrm{f}}\rangle$ is given by

$$
\left|C_{\mathrm{f}}(t)\right|^2 = \frac{4}{\hbar^2}\left|\langle\varphi_{\mathrm{f}}|\hat{H}'|\varphi_{\mathrm{i}}\rangle\right|^2 \frac{\sin^2(\omega_{\mathrm{fi}}t/2)}{\omega_{\mathrm{fi}}^2} ,
\tag{16.125}
$$

with

$$
\omega_{\mathrm{fi}} = \frac{E_{\mathrm{f}}-E_{\mathrm{i}}}{\hbar} .
$$

The function $\sin^2(\omega t/2)/\omega^2$ has a peak at $\omega=0$ as a function of ω, which becomes sharper if t increases. Moreover, we have

$$
\int\limits_{-\infty}^{\infty} \mathrm{d}\omega \frac{\sin^2(\omega t/2)}{\omega^2} = \frac{\pi t}{2} ,
\tag{16.126}
$$

because

$$
\int\limits_{-\infty}^{\infty} \frac{\sin^2 x}{x^2}\,\mathrm{d}x = \pi .
$$

Clearly this means that

$$
\lim_{t\to\infty} \frac{\sin^2(\omega t/2)}{\omega^2 t} = \frac{\pi}{2}\delta(\omega) .
\tag{16.127}
$$

Inserting this result in (16.125), we obtain immediately the *transition probability per unit time* from state $|\varphi_{\mathrm{i}}\rangle$ to $|\varphi_{\mathrm{f}}\rangle$:

$$
\lim_{t\to\infty} \frac{|C_{\mathrm{f}}(t)|^2}{t} = \frac{2\pi}{\hbar}\left|\langle\varphi_{\mathrm{f}}|\hat{H}'|\varphi_{\mathrm{i}}\rangle\right|^2 \delta(E_{\mathrm{f}}-E_{\mathrm{i}}) .
\tag{16.128}
$$

This is *Fermi's golden rule*. Here, the δ function expresses the conservation of energy. The δ function is a distribution and as such its significance in a physical formula is not quite clear. It disappears from (16.128) if we consider that in all practical cases we have to integrate over a continuum from a lower to an upper energy limit (cf. the extensive discussion in Chap. 11).

Now the general approximation scheme for solving the system of integral equations (16.122) is evident. To obtain higher approximations, we have to iterate as often as necessary. These calculations are tedious; however, their result is simple enough to be presented here without proof. Generally the transition probability per time for the transition $\mathrm{i} \to \mathrm{f}$ is given by

$$\left(\frac{\text{transition probability}}{\text{time}}\right)_{\mathrm{i}\to\mathrm{f}} = \frac{2\pi}{\hbar}|M_{\mathrm{fi}}|^2\delta(E_\mathrm{f}-E_\mathrm{i}) \ . \qquad (16.129)$$

Here, the transition matrix M_{fi} reads:

$$\begin{aligned} M_{\mathrm{fi}} &= \langle\mathrm{f}|\hat{H}|\mathrm{i}\rangle + \sum_I \frac{\langle\mathrm{f}|\hat{H}'|I\rangle\langle I|\hat{H}'|\mathrm{i}\rangle}{E_\mathrm{i}-E_I+\mathrm{i}\eta} \\ &\quad + \sum_I\sum_{II} \frac{\langle\mathrm{f}|\hat{H}'|I\rangle\langle I|\hat{H}'|II\rangle\langle II|\hat{H}'|\mathrm{i}\rangle}{(E_\mathrm{i}-E_I+\mathrm{i}\eta)(E_\mathrm{i}-E_{II}+\mathrm{i}\eta)} + \dots \ . \end{aligned} \qquad (16.130)$$

To simplify the formula, we have abbreviated $\langle\varphi_\mathrm{f}|\hat{H}'|\varphi_\mathrm{i}\rangle \equiv \langle\mathrm{f}|\hat{H}'|\mathrm{i}\rangle$, etc. The states I, II are intermediate states via which the higher-order transitions proceed. The infinitesimal quantity η, which appears in the denominators, is positive and indicates how the singularities in the expression M_{fi}[7] are to be treated.

[7] See the extensive presentation of perturbation theory in Chap. 11 and, e.g., in A.S. Davydov: *Quantum Mechanics* (Pergamon, Oxford 1965), Chap. VII; L.I. Schiff: *Quantum Mechanics*, 3rd ed. (McGraw-Hill, New York 1968), Chap. 8; A. Messiah: *Quantum Mechanics*, Vol. II (North-Holland, Amsterdam 1965).

17. Conceptual and Philosophical Problems of Quantum Mechanics

In the preceding chapters we have tried to develop the conceptual basis of quantum mechanics. In addition, various exercises have demonstrated how quantum theory can be used to solve real physical problems. Of course, these problems represent only a small fraction of the many applications of quantum mechanics; numerous experimental results involving physics and chemistry can be explained successfully by adopting this theory. Up to now, no prediction made by quantum mechanics has been disproved experimentally. In spite of this success, conceptual problems with the theory exist; in fact, there have been many attempts to interpret quantum mechanics in a new way or even to replace it by another theory based on a more obvious conceptual and philosophical foundation.

Therefore, in this chapter,[1] we will try to gain insight into the principal conceptual difficulties of quantum mechanics, which we will illustrate. The most important alternatives will be presented. Many questions occurring during this discussion belong to the realm of opinion rather than to that of fact; therefore, many physicists assign these considerations to the field of philosophy. Nevertheless, the basis of quantum mechanics is so important and fundamental to our perception of microscopic processes in nature, that every physicist should at least understand the nature of the principal conceptual problems and questions arising therefrom.

17.1 Determinism

Quantum mechanics is an *indeterminate* theory which states that there are physical measurements whose results are not definitely determined by the state of the systems prior to the measurement (at least, as far as it is in principal possible to observe that state). If, just before the measurement, the wave function of the system is not an eigenfunction of the operator whose observable is to be meas-

[1] In conceiving this chapter, we were guided by similar discussions by A.I.M Rae in his book: *Quantum Mechanics* (McGraw-Hill, London 1981); M. Jammer: *The Philosophy of Quantum Mechanics: The Interpretations of Quantum Mechanics in Historical Perspective* (Wiley, New York 1974)

ured, then the result of the measurement is not definitely predictable; only the *probability of the various possible results* can be determined.

By way of example, recall the Stern–Gerlach experiment (Chap. 12). If a beam of spin-$\frac{1}{2}$ particles goes through an experimental setup oriented in the z direction, the beam is split up according to the two possible directions of s_z. If the beam consisting of $s_z = +\frac{1}{2}\hbar$ particles then travels through another apparatus measuring the spin component in another direction, then the beam will split up again into the two components $+\frac{1}{2}\hbar$ and $-\frac{1}{2}\hbar$. The probability of these two results can be calculated, but one cannot predict whether a single particle will have spin up or spin down.

The indeterminism of quantum mechanics is opposed to the *determinism* of classical mechanics. For the latter, the evolution of a system is definitely fixed by its initial state and by the forces acting on it. Even the final state of a "purely random" experiment, like the tossing of a coin, can be determined if the initial position, velocity and angular momentum of the coin, as well as the gravitational and frictional forces acting upon it, are known.

The randomness of quantum mechanics, however, has a totally different character: results of measurements are not an unequivocal consequence of the previous state of the system. The fact that spin-$\frac{1}{2}$ particles are eigenstates of $\hat{s}_z$ fully determines the part of the wave functions attributed to the spin, and these particles are identical. Nevertheless, they may act differently if another spin component is measured. The guiding-field interpretation introduced in Chap. 3 permits only statements concerning probabilities of the individual behaviour of the particles moving in the guiding field.

To circumvent the problem of indeterminism, it was suggested that particles in the same state just *seem* to be identical, and that in reality, they have additional, different properties which determine the result of a further experiment. For instance, a particle in an eigenstate of s_z would also have a fixed value of s_x, even if the latter value cannot be measured. Theories based on assumptions like this are called *hidden-variable theories*; the properties of these theories will be discussed below.

17.2 Locality

We now turn our attention to a second problem of quantum mechanics, the so-called *locality* of the theory. When talking about particles, we assume them to be pointlike or at least significantly smaller than other typical dimensions of the system under consideration. Thus it makes good sense to assume that a particle "feels" the value of some field acting upon it only at its momentary position. Now we will discuss two examples in which that assumption proves incompatible with the outcome of the measurement.

First we will look at an experiment with a parallel beam (e.g. a light ray or an electron beam) falling through a double slit (see Fig. 17.1). At some distance

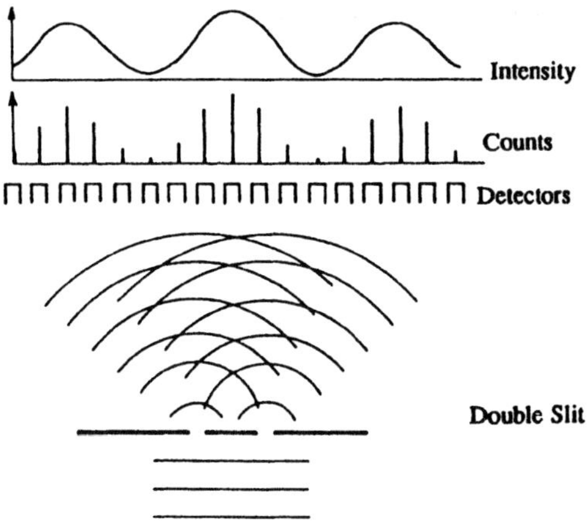

Fig. 17.1. Double-slit experiment (schematic)

from the slit, a row of detectors register the arrival of photons or electrons. After a sufficiently long period of time, the number of particles counted by a single detector is proportional to the intensity of a wave at that point, which has been calculated by assuming that the single waves interfere. The wave model (guiding field) as well as the particle model are necessary to explain this result: the detector registers particles of a definite energy, but the interference pattern can only be explained by the passage of the waves (of the guiding field) through both slits.

Now we alter the experimental setup by placing the detectors directly behind each slit. Then particles will be measured passing through one of the slits, but not through both at the same time. But the appearance of an interference pattern necessarily requires that a wave pass through both slits at once; this can be shown by shutting one of the slits during the previous experiment. Therefore the probability of observing a particle at a certain point depends upon whether only one or both slits are open. We could conclude that the "object" (photon or electron) acts only in some situations as a particle (and in other situations, as a wave), or that it is influenced by a slit through which it does not pass. The second assumption implies a kind of nonlocal interaction.

A further remark: this difficulty in interpreting the experiment can be avoided by applying the concept of a guiding field ψ, leading the single particles statistically at the various space–time points corresponding to $|\psi(x, y, z, t)|^2$.

One further example of nonlocal consequences in quantum mechanics is given by the behaviour of particles separated in space, whose properties are correlated in some way. A system of this kind was first discussed by Einstein, Podolski and Rosen.[2] Here we will develop a *Gedankenexperiment*, conceived

[2] A. Einstein, B. Podolski, N. Rosen: Phys. Rev. **47**, 777 (1935).

of by Bohm[3], and look at a pair of spin-$\frac{1}{2}$ particles with total spin zero, each having zero orbital angular momentum. Pairs with these properties may be created by the scattering of a beam of low-energy protons from hydrogen gas. The incoming and target protons then withdraw to a state with vanishing orbital angular momentum and total spin. When the protons are far apart, we measure, for instance, the z component of the spin of the first particle (s_{z_1}) and, afterwards, the same component of the second particle (s_{z_2}). Since the total spin is zero, the components must have opposite sign: $s_{z_1} = \frac{1}{2}\hbar$, $s_{z_2} = -\frac{1}{2}\hbar$. Thus, in principle, the second measurement is not necessary, as the value of s_{z_2} can be deduced from the value of s_{z_1}. After the measurement of s_{z_1}, the second particle is in an eigenstate with s_{z_2}. This also means a change in the probabilities of the measurement of another spin component of the second particle [see the more detailed discussion below, following (17.1)].

So the result of the measurement of the second particle is influenced by the measurement of the spin of the first particle, although both particles are far apart and do not interact with each other. According to quantum mechanics, two particles in a state of this kind are not independent of each other. If they are nevertheless considered to be independent, then a hidden-variable theory must be introduced.

Attempts have been made to develop a theory conserving determinism and locality that could reproduce all the experimentally proved results of quantum mechanics. Only recently has it become clear that this is not possible. This will be elucidated in the following section.

17.3 Hidden-Variable Theories

The aim of a hidden-variable theory is to consider quantum mechanics as a statistical theory in the sense that it furnishes probabilities of possible events, which in reality are fixed by nonobservable properties. This is in analogy to classical statistical mechanics, which incorporates random fluctuations as a property of thermodynamical systems, although a discussion of individual atoms should yield completely deterministic behaviour. It would be advantageous to find similar substructures for quantum mechanics to preserve locality and determinism. Of course, a theory including hidden variables has to reproduce all the experimentally confirmed results of quantum mechanics. On the other hand, we could decide which theory is better if new experiments were performed, for which the theories predict different results.

One such experiment, for which quantum mechanics and the hidden-variable theory give different results, is Bohm's *Gedankenexperiment*. To prove a discrepancy, we first have to derive the quantitative predictions of quantum mechanics

[3] D. Bohm: *Quantum Theory* (Prentice-Hall, Englewood Cliffs, NJ 1951) pp. 614–622.

for measuring the spin component s_{ϕ_2} of the second particle at an angle ϕ to the z axis, having previously determined the spin component s_{z_1} of the first particle. If the first measurement results in the value $+\frac{1}{2}\hbar$, s_{z_2} necessarily has to be negative. The spin part of the wave function of the second particle therefore is (compare Chap. 12):

$$\chi_- = \begin{pmatrix} 0 \\ 1 \end{pmatrix} . \tag{17.1}$$

The operator $\hat{s}_\phi$, standing for the spin component at an angle ϕ to the z axis, is given by

$$\hat{s}_\phi = \hat{s}_z \cos\phi + \hat{s}_x \sin\phi = \frac{\hbar}{2}\begin{pmatrix} \cos\phi & \sin\phi \\ \sin\phi & -\cos\phi \end{pmatrix} . \tag{17.2}$$

$\hat{s}_\phi$ has the eigenvalues $+\hbar/2$ and $-\hbar/2$, and the corresponding eigenvectors are easily determined as $\binom{\cos(\phi/2)}{\sin(\phi/2)}$ and $\binom{-\sin(\phi/2)}{\cos(\phi/2)}$. We expand the wave function χ_- as a linear combination of these two eigenvectors, which gives

$$\begin{pmatrix} 0 \\ 1 \end{pmatrix} = \sin(\phi/2)\begin{pmatrix} \cos(\phi/2) \\ \sin(\phi/2) \end{pmatrix} + \cos(\phi/2)\begin{pmatrix} -\sin(\phi/2) \\ \cos(\phi/2) \end{pmatrix} . \tag{17.3}$$

The probability that the second measurement will result in a positive value is therefore $P_{++}(\phi) = \sin^2(\phi/2)$. In analogy, we define the probabilities P_{+-}, P_{-+}, P_{--} for the various possible results of both experiments. Similar considerations lead to

$$\begin{aligned} &P_{++}(\phi) = \sin^2(\phi/2) \ , \quad P_{+-}(\phi) = \cos^2(\phi/2) \ , \\ &P_{-+}(\phi) = \cos^2(\phi/2) \ , \quad P_{--}(\phi) = \sin^2(\phi/2) \ . \end{aligned} \tag{17.4}$$

It is useful to introduce the *correlation coefficient* $C(\phi)$, defined as the average value of the product $S_{z1}S_{\phi 2}$, averaged over a great number of measurements of such pairs of particles. Then,

$$\begin{aligned} C(\phi) &= \frac{\hbar^2}{8}(P_{++}(\phi) - P_{+-}(\phi) - P_{-+}(\phi) + P_{--}(\phi)) \\ &= \frac{\hbar^2}{4}[\sin^2(\phi/2) - \cos^2(\phi/2)] = -\frac{\hbar^2}{4}\cos\phi \ . \end{aligned} \tag{17.5}$$

Now we will look at an example of a hidden-variable theory which assumes all spin components (only one being determinable at a given time) to have fixed, but unknown, values. This means that a "*real*" *spin vector*, analogous to a classical angular-momentum vector, should exist. We choose as our example the two-

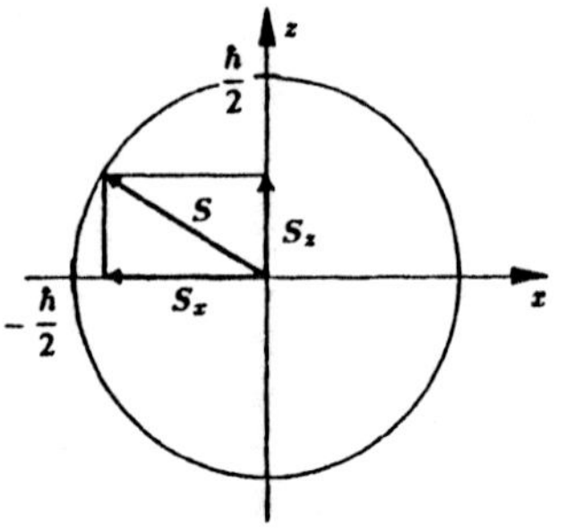

Fig. 17.2. A "real" spin vector $\boldsymbol{s}$. It is postulated that a measurement in the z direction yields the result $\hbar/2$, although the absolute value of the "real" spin component s_z is smaller. A measurement in the x direction yields $-\hbar/2$. This postulate has to be introduced in order to preserve the theoretically established (quantum-mechanical) results of the measurement of a spin component

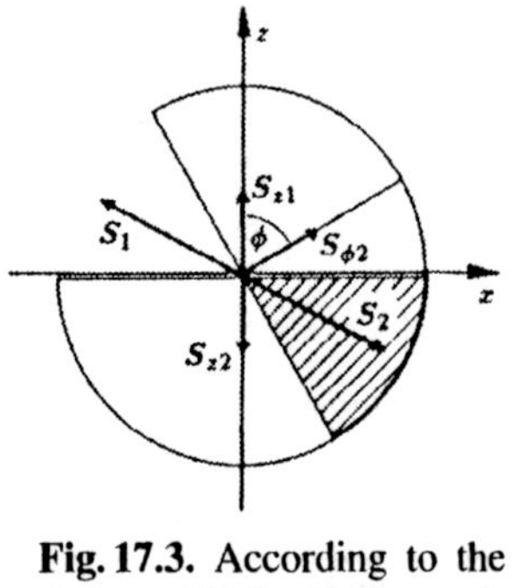

Fig. 17.3. According to the theory with the hidden variables s_x and s_y, based on the concept of a "real" spin vector, the number of particle pairs that have a positive s_{z1} component of the first particle and a positive $s_{\phi 2}$ component of the second particle, is proportional to the hatched area. Two "real" spin vectors $\boldsymbol{s}_1$ and $\boldsymbol{s}_2$, fulfilling these conditions, are shown

particle scattering for a state with vanishing total spin, which has already been explained. For total spin zero, it follows naturally that these "real" spin vectors of the two particles should have the same value but opposite direction, when the particles move away from each other. In quantum mechanics, only one spin component of a particle can be determined at a given time. The only possible results for that component are $\pm\frac{1}{2}\hbar$. The other components, which are not measurable, may be thought of as hidden variables. We assume they exist, but it is not possible to determine them at the same time as the previously chosen component (which can be any of the three components).

We denote the "real" spin vector of a particle by $\boldsymbol{s}$. This spin vector $\boldsymbol{s}$ has to interact with the experimental apparatus in such a way as to ensure that the value of a spin component is always measured as $\pm\hbar/2$, although the corresponding component of $\boldsymbol{s}$ may have another value. This may be achieved by requiring that if $\boldsymbol{s}$ has a positive z component, its measured value must always be $+\hbar/2$, and that for a negative z component, the result must be $-\hbar/2$. In general, this should be valid for any direction.

The point here should be clear: the "real" spin vector of a particle does not have to be parallel to the z axis if we measure $+\hbar/2$ as the result for the z component; it should only be somewhere in the upper hemisphere, with the z axis being the symmetry axis (see Fig. 17.2).

The assumption that each individual particle has a well-defined but not measurable spin vector $\boldsymbol{s}$ is the basic postulate of the theory of hidden variables: if s_z is measured, the components s_x and s_y also exist, constituting together with s_z the total "real" spin. This concept of "real spin" can only be made compatible with experiments by artificially introducing unjustifiable prescriptions for the result of a measurement.

In the following we will show that this leads to a correlation function $C'(\phi)$ of the spin components of both particles, differing from the quantum-mechanical result $C(\phi)$ of (17.5). If we get the value $s_{z1} = +\hbar/2$ for the first particle in our hypothetical experiment, the z component of the second particle's spin has to be negative. The "real" spin vector of the second particle is therefore placed in the lower hemisphere, with the negative z axis as symmetry axis (see Fig. 17.3). If the second particle's spin points in a direction at an angle ϕ to the z axis, i.e. $s_{\phi 2}$ being positive, the spin vector of the second particle has to be in the hemisphere with the symmetry axis inclined by ϕ with respect to the z axis. Assuming the orientation of the spin vectors to be homogeneously distributed in space, the probability $P'_{++}(\phi)$ for s_{z1} and $s_{\phi 2}$ being positive at the same time has to be proportional to the overlapping volume of both hemispheres (hatched part of the Fig. 17.3).

The overlap is proportional to ϕ for $0 \leq \phi \leq \pi$. For angles between π and 2π, the angle ϕ has to be replaced by $2\pi - \phi$ in the following formula. By taking into account that $P'_{++}(\pi) = 1$, it follows that $P'_{++}(\phi) = \phi/\pi$. The remaining probabilities can be derived by analogous considerations:

$$
\begin{aligned}
P'_{++}(\phi) &= \phi/\pi \ , \qquad & P'_{+-}(\phi) &= 1 - \phi/\pi \ , \\
P'_{-+}(\phi) &= 1 - \phi/\pi \ , & P'_{--}(\phi) &= \phi/\pi \ .
\end{aligned}
\tag{17.6}
$$

Thus, for a hidden-variable theory, the correlation coefficient $C'(\phi)$, defined by the average value of the product of both measured spin components, is

$$C'(\phi) = \frac{\hbar^2}{4}(2\phi/\pi - 1) \ . \tag{17.7}$$

This result coincides with the quantum-mechanical formula (17.5) only for $\phi = 0$, $\phi = \frac{1}{2}\pi$, or $\phi = \pi$. For different orientations of the experimental apparatus, the results differ considerably (see Fig. 17.4).

Thus we have shown a discrepancy between the predictions of quantum mechanics and the results of a hidden-variable theory, assuming that the single particles have "real" spin vectors. In the following, we will demonstrate as well that any hidden-variable theory will end up predicting experimental results that contradict quantum mechanics. The first proof of this important statement was given by *Bell*,[4] and is thus called *Bell's theorem.*

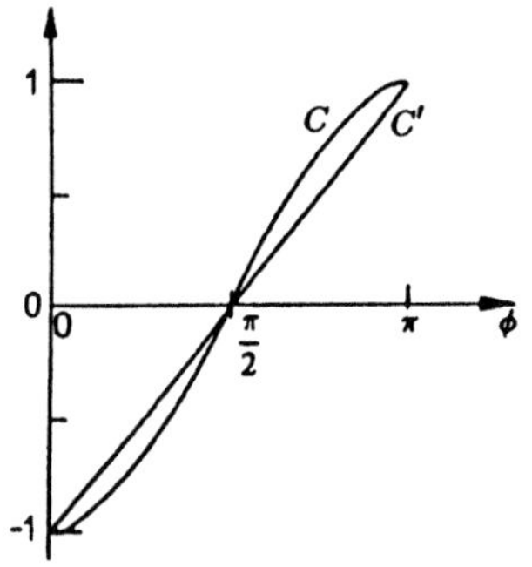

Fig. 17.4. The quantum-mechanical correlation function $C(\phi)$ and the correlation function $C'(\phi)$ of a hidden-variable theory (based on the existence of a "real" spin vector) are considerably different

17.4 Bell's Theorem

To succeed with a general proof, we first have to state the minimum requirements of a local deterministic hidden-variable theory for the case of Bohm's hypothetical experiment. After the two particles have separated, the system should have a property that determines in advance the result of measuring any spin component of any particle. In the previous example, that property was granted by the existence of a "real" spin vector, but for the general case a hidden variable does not have to correspond with a parameter of a special physical model.

The result of a measurement of the first particle's z component of spin s is designated as $s_{z1}(\lambda)$, where the only values allowed for s_{z1} are $\pm\frac{1}{2}\hbar$, and λ represents the hidden variable (of course, λ may be replaced by multiple hidden variables; the following discussion can easily be extended to that case). A theory of this kind is *deterministic*, since the values of s_{z1} and $s_{\phi 2}$ are given by the value of λ; the theory is *local*, because the result of an experiment does not depend on an experiment determining the spin value of the other particle. Every particle pair has a definite value of λ, and we define the probability density $p(\lambda)$, giving the probability of a pair with a value between λ and $\lambda + \mathrm{d}\lambda$, as $p(\lambda)\,\mathrm{d}\lambda$. The normalization condition is

$$\int p(\lambda)\,\mathrm{d}\lambda = 1 \ . \tag{17.8}$$

[4] J.S. Bell: Physics **1**, 195 (1965). The following synoptic article can be recommended: J.F. Clauser, A. Shimony: "Bell's Theorem: Experimental Tests and Implications", Rep. Prog. Phys. **41**, 1881 (1978). We also mention a more popular article by B. d'Espagnat: "The Quantum Theory and Reality", Scientific American **241**(11) 128 (1979).

Now we consider an experiment determining the spin components s_{z1} and $s_{\phi 2}$ of a large number of pairs. The average value $C''(\phi)$ of the products $s_{z1}s_{\phi 2}$ follows as

$$C''(\phi) = \int s_{z1}(\lambda)s_{\phi 2}(\lambda)p(\lambda)\,d\lambda \ . \tag{17.9}$$

Let us look at a second experiment measuring, as before, the z component of the first particle, but where the second apparatus is placed at an angle θ to the z axis. We get an analogous expression of the correlation coefficient $C''(\theta)$. It is

$$C''(\phi) - C''(\theta) = \int (s_{z1}(\lambda)s_{\phi 2}(\lambda) - s_{z1}(\lambda)s_{\theta 2}(\lambda))p(\lambda)\,d\lambda \ . \tag{17.10}$$

We know the spins of both particles to be of equal magnitude but opposite direction. Thus it follows that

$$s_{\theta 1}(\lambda) = -s_{\theta 2}(\lambda) \ , \quad s_{\phi 1}(\lambda) = -s_{\phi 2}(\lambda) \ . \tag{17.11}$$

Inserting (17.11) into (17.10) yields

$$\begin{aligned} C''(\phi) - C''(\theta) &= -\int s_{z1}(\lambda)(s_{\phi 1}(\lambda) - s_{\theta 1}(\lambda))p(\lambda)\,d\lambda \\ &= -\int s_{z1}(\lambda)s_{\phi 1}(\lambda)\left(1 - \frac{4}{\hbar^2}s_{\phi 1}(\lambda)s_{\theta 1}(\lambda)\right)p(\lambda)\,d\lambda \ , \end{aligned} \tag{17.12}$$

where we have used $\langle s_{\phi 1}(\lambda)\rangle^2 = \frac{1}{4}\hbar^2$. The magnitude of (17.12) can be estimated by

$$\begin{aligned} &\left|C''(\phi) - C''(\theta)\right| \\ &\le \int \left|s_{z1}(\lambda)s_{\phi 1}(\lambda)\left(1 - \frac{4}{\hbar^2}s_{\phi 1}(\lambda)s_{\theta 1}(\lambda)\right)p(\lambda)\right|\,d\lambda \ . \end{aligned} \tag{17.13}$$

Since the term $p(\lambda)$ always has to be positive and the spins only have values $\pm\frac{1}{2}\hbar$ (thus $|S_{z1}(\lambda)S_{\phi 1}(\lambda)| = \frac{1}{4}\hbar^2$), it follows by applying (17.8) and (17.11) that

$$\begin{aligned} \left|C''(\phi) - C''(\theta)\right| &\le \int \left(\frac{\hbar^2}{4} - s_{\phi 1}(\lambda)s_{\theta 1}(\lambda)\right)p(\lambda)\,d\lambda \\ &= \frac{\hbar^2}{4} + \int s_{\phi 1}(\lambda)s_{\theta 2}(\lambda)p(\lambda)\,d\lambda \ . \end{aligned} \tag{17.14}$$

We will restrict our consideration to cases where the z axis and the directions given by θ and ϕ are in one plane. The mean values defined by (17.9) only depend on the relative orientation of the single measurements – not on the absolute direction of the z axis (this may be proved by additional experiments). Therefore we can replace the integral in (17.14) by the correlation function $C''(\theta - \phi)$: we get

$$\left|C''(\phi) - C''(\theta)\right| - C''(\theta - \phi) \le \frac{\hbar^2}{4} \ . \tag{17.15}$$

This inequality is called *Bell's theorem*. It is a direct consequence of every local deterministic hidden-variable theory. We now have to establish whether this inequality coincides with the predictions of quantum mechanics. To do so, we look at the special case $\theta = 2\phi$. The two correlation functions can be derived with the help of (17.5):

$$C(\phi) = -\frac{\hbar^2}{4}\cos\phi\ , \quad C(\theta) = -\frac{\hbar^2}{4}\cos 2\phi\ . \tag{17.16}$$

Comparing (17.16) with (17.15), we find that quantum mechanics is only consistent with a hidden-variable theory if $\frac{1}{4}\hbar^2(|\cos\phi - \cos 2\phi| + \cos\phi) \le \frac{1}{4}\hbar^2$.

Figure 17.5 shows Bell's theorem to be satisfied for $\frac{1}{2}\pi \le \phi \le \pi$, but not for $0 \le \phi \le \frac{1}{2}\pi$. The function has a maximum when $\phi = \pi/3$ with a value of $3\hbar^2/8$. In addition, the figure shows the function (dashed lines) calculated by applying the previously discussed theory of "real" spin vectors (17.7). Of course, this is consistent with Bell's theorem and inconsistent with quantum mechanics.

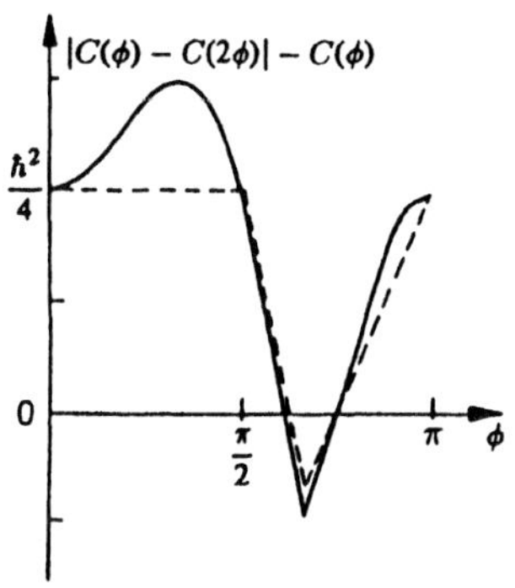

Fig. 17.5. The *solid line* shows the function $|C(\phi) - C(2\phi)| - C(\phi)$ according to quantum mechanics; the *dashed line* represents the result of the same function for the hidden-variable theory containing "real" spin

It is remarkable that Bell's inequality is invalid for that range of parameters in which the difference between the results of a hidden-variable theory and quantum mechanics is maximal.

Owing to these considerations, we have to conclude that *no local deterministic hidden-variable theory can reproduce the results attained by quantum mechanics for experiments of this kind.*

As natural scientists, we have to determine which theory is correct, i.e. which theory is consistent with nature. Thus experiments have to show which of the two theories yields better predictions. At first, one might think this unnecessary, since quantum mechanics has become so well-established on the basis of experimental evidence. But experiments performed before the formulation of Bell's theorem did not check the consistency of quantum mechanics regarding this special point. In recent years, several experiments testing this point have been carried out. Although the first experiments yielded results consistent with Bell's theorem (leading to doubts about the validity of quantum mechanics), the following more accurate experiments yielded results inconsistent with Bell's theorem but in accord with quantum-mechanical predictions. Most of these experiments measured the polarization of correlated photon pairs, in which case the formulation of Bell's theorem is slightly different from (17.15).[5] (For our discussion, these minor differences yield no new information.)

Many of the difficulties concerning the nonlocal indeterministic character of quantum mechanics can be avoided by the *Copenhagen interpretation*, which treats the experimental apparatus as a part of the total quantum-mechanical system (see next section). As a variation of that interpretation, an additional formulation may be made by constructing a hidden-variable theory that also treats both correlated particles and apparatus as one system. In fact it is possible to con-

[5] Further information is given in the article by Clauser and Shimony cited above (see footnote 4). A new experiment was done by W. Perrie, A.J. Duncan, H.J. Beyer, and H. Kleinpoppen, and reported on in Phys. Rev. Lett. **54**, 1790 (1985).

struct such a nonlocal but deterministic theory, which reproduces the results of quantum mechanics. Here, the behaviour of the measuring apparatus at one point has to be influenced by another one, possibly placed far away at another point. This assumption proposes the existence of a new, up to now unknown, mechanism mediating the influence between the two apparatuses. A postulate of this kind seems to be unnatural, and, furthermore, the conceptual difficulties of this hidden-variable theory seem to be at least as extreme as in the case of quantum mechanics. Thus there is no reason to replace quantum mechanics by another theory.

Hence, many physicists believe indeterminism to be a property of physical phenomena. Our common conception of causality results from our experience in the macroscopic world and cannot be translated directly to microscopic processes, where quantum-mechanical effects occur. There is no reason why there should not be some randomness in nature. If *different causes* can lead to the *same effects* (there are many examples of this in physics), it does not seem to be unreasonable that *one cause* could lead to *different effects*. The image of a guiding field leading its quanta according to the resulting probabilities seems to be a convincing concept. But, indeed, it is only a concept allowing predictions of expectation values, i.e. of results of many measurements.

17.5 Measurement Theory

In this section we want to discuss in more detail what the measurement of quantum-mechanical quantities really means. Let us consider a system described by a wave function $\psi(x)$. We want to measure a variable q represented by the Hermitian operator $\hat{Q}$. The eigenvalues q_n of $\hat{Q}$ are the possible results of measuring that variable. The wave function may be expanded in terms of the set of eigenfunctions $\phi_n(x)$ of $\hat{Q}$:

$$\psi(x) = \sum_n a_n \phi_n(x) \ . \tag{17.17}$$

Therefore the expectation value of $\hat{Q}$ is

$$\begin{aligned} \left\langle \hat{Q} \right\rangle &= \int \psi^* \hat{Q} \psi \, \mathrm{d}x = \int \left(\sum_m a_m^* \phi_m^* \right) \hat{Q} \left(\sum_n a_n \phi_n \right) \mathrm{d}x \\ &= \sum_{m,n} a_m^* a_n \int \phi_m^* \hat{Q} \phi_n \, \mathrm{d}x \\ &= \sum_{m,n} a_m^* a_n q_n \int \phi_m^* \phi_n \, \mathrm{d}x = \sum_n |a_n|^2 q_n \ , \end{aligned} \tag{17.18}$$

where we have used the orthogonality relation of the eigenfunctions ϕ_n. For a single measurement, the probability of a result q_n is given by $|a_n|^2$.

Now we ask how measurement influences the system. Let the measurement have the result q_l. Immediately after the measurement, we assume it to be repeated. If the measurements are to make physical sense, we have to claim that the experimental result does not change: the second measurement also has to result in the value q_l. Since this value is assumed to result with certainty, the probabilities of the various results measured in the second experiment are

$$|a_n|^2 = \delta_{nl} \ .$$

Thus it follows that before the second measurement the system's wave function was ϕ_l. We may say: the wave function of the system in the beginning, given by ψ, has changed because of the measurement. The system's state after the measurement is given by ϕ_l:

$$\psi \xrightarrow[\text{by the measurement of } Q \text{ with the result } q_l]{} \phi_l \ .$$

That change of the wave function caused by a measurement is called *wave-function reduction*. The main problem of a measurement theory is to establish at what point in time this reduction takes place.

For instance, if a particle moves through one slit of a double-slit experiment, this is not a measurement of the particle's position, and thus the wave function is not reduced to an eigenfunction of the position operator unless we observe which of the two slits the particle moves through. Some physicists interpret this to mean that a quantum-mechanical measurement requires the presence of a human observer. Another interpretation proposes that the wave function is reduced when the experimental result is registered by an apparatus. On the other hand, such an apparatus has to be able to be described by a (naturally very complicated) wave function. The question then arises when will that wave function be reduced.

This problem may be circumvented by demanding macroscopic objects, for instance particle detectors, to follow the laws of classical mechanics exactly. This idea has the serious disadvantage that the boundaries between a classical and quantum-mechanical region are unclear. In addition, there are no experimental facts indicating the existence of such a boundary.

Now let us discuss the problem of wave-function reduction by considering the Stern–Gerlach experiment. We have an apparatus oriented in the z direction (abbreviated by SGz), and a beam of spin-$\frac{1}{2}$ particles splitting into two parts according to the two eigenvalues $\pm\frac{1}{2}\hbar$ of s_z. The corresponding eigenvectors are Z_+ and Z_-. Before the experiment, the wave function is given by

$$Z = \frac{(Z_+ + Z_-)}{\sqrt{2}} \ . \tag{17.19}$$

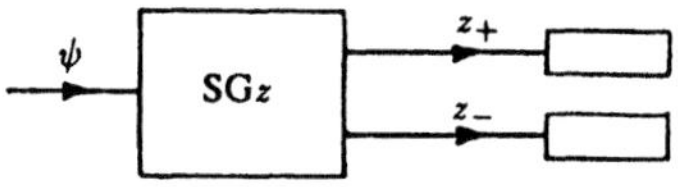

Fig. 17.6. A Stern–Gerlach experiment (schematic)

The experiment may be illustrated graphically:

The measurement and thus the reduction of the wave function, is not obtained, however, by just moving through a Stern–Gerlach magnet. To understand this, we look at the following experiment:

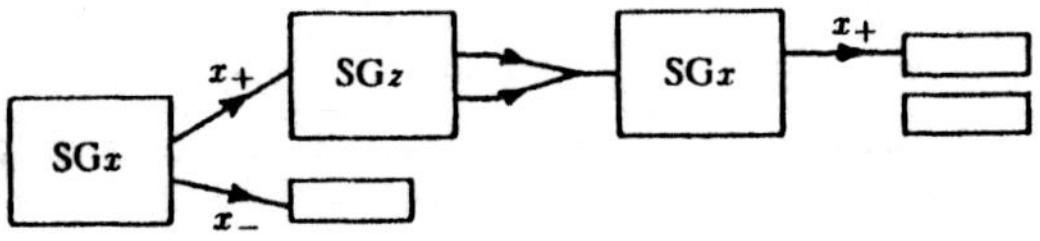

Fig. 17.7. Experimental set-up for multiple Stern–Gerlach measurements

The particle beam with $s_x = +\frac{1}{2}\hbar$ moves through an apparatus in the z direction, where it is split up. The particle beams are reunited so that it is impossible to say which path was taken by which particle. Thus no information on the value of s_z has been gained. If there is no phase difference between the two paths, the second measurement made in the x direction will yield the same result: the wave function was not reduced (filtered) to Z_+ or Z_-; it is still Z. This consideration shows us that for measuring a spin component a detector has to be present in the path the particle has chosen. This may be done by filtering out one of the two beams behind the SGz apparatus. This filtering out is then, so to speak, the detector just mentioned. But now a reduction of the wave function takes place, caused by the existence of this second detector.

To develop a complete quantum-mechanical measurement theory, it is necessary to treat the measuring apparatus quantum mechanically. Thus new problems arise. To clarify this point, let us look again at a Stern–Gerlach apparatus oriented in the z direction. Now, the passage of a particle is to be registered by a detector described by a wave function. The three possible states of the detector are χ_0 before registering the particle, χ_+ if $s_z = +\frac{1}{2}\hbar$ and χ_- if $s_z = -\frac{1}{2}\hbar$. First we consider the case of the incoming particle being in an eigenstate of $\hat{s}_z$. If the particle's spin is described by Z_+, the wave function of the total system, consisting of particle and detector, is given by

$$\psi_0 = Z_+\chi_0 \ . \tag{17.20}$$

After the particle has moved through the apparatus, the total wave function is

$$\psi_+ = Z_+\chi_+ \ . \tag{17.21}$$

The state of the incoming particle is Z_- and thus

$$\psi_0 = Z_-\chi_0 \ . \tag{17.22}$$

The state of the system after the experiment follows as

$$\psi_- = Z_-\chi_- \ . \tag{17.23}$$

Now we look at the case of the particle *not* being in an eigenstate of $\hat{s}_z$. For instance, before the experiment, its wave function is χ_+. The initial state of the system is then

$$\psi_0 = \chi_+\chi_0 = \frac{(Z_+ + Z_-)\chi_0}{\sqrt{2}} \ . \tag{17.24}$$

The time evolution of the total system is described by the time-dependent Schrödinger equation with the time-independent Hamiltonian. Thus it follows that each part of the sum on the right-hand side of (17.24) develops in time as described by (17.21) and (17.23). Therefore, after the measurement, the total wave function is

$$\psi = \frac{(Z_+\chi_+ + Z_-\chi_-)}{\sqrt{2}} \; . \tag{17.25}$$

Some of the problems of a measurement theory are encountered in (17.25). First we notice that the detector's state is not χ_0, χ_-, or χ_+, but a combination of χ_+ and χ_-. If the detector had a pointer capable of staying in three positions according to the three possible states of the detector, the pointer would be placed between two of these positions. Thus our experimental experience is contrary to the assumption that macroscopic objects are described by a wave function.

A second problem suggested by (17.25) is that it contradicts the postulate of the reduction of wave functions, claiming that the wave function is in one of the states $Z_+\chi_+$ or $Z_-\chi_-$ after the measurement.

A third problem arises in trying to solve the first two: if the system – particle and detector – is considered as one quantum-mechanical system, wave-function reduction should take place when the state of the total system is measured. Thus we should use a second apparatus to measure the state of the detector, and afterwards the wave function would be reduced to either $Z_-\chi_-$ or $Z_+\chi_+$. But that additional experimental apparatus is again part of a larger total system whose wave function is of the form of (17.25). This procedure could be repeated arbitrarily often. Obviously, there is no fixed point at which we could determine that the wave function is reduced.

17.6 Schrödinger's Cat

These apparently paradoxical results can be illustrated more vividly by a *Gedankenexperiment* conceived by Erwin Schrödinger. We assume that the result of the previously discussed quantum-mechanical measurement triggers a rifle pointing at a cat. The total "experimental setup" should be placed in a box to be opened later on to observe the cat's state (alive or dead). Making that observation, the result of the measurement can be deduced. In analogy to the previous discussion, the wave function of the box is χ_+ or χ_-, depending on whether the cat is alive or dead. The total wave function of the box containing cat, particle and detector is again given by (17.25). This, of course, would have the consequence that the state of the cat before opening the box would be neither alive nor dead. If we reject this absurd conclusion, we have to ask when the wave-function reduction takes place: When the particle enters the apparatus? When the cat dies? Or at some other time?

In the following, we present different attempts to formulate a quantum theory of measurement, trying to solve all the problems caused by wave-function reduction.

17.7 Subjective Theories

One theory solving all these problems was developed by E.P. Wigner. All we know of the physical world is the information reaching our brains through our senses and remaining in our mind. Therefore, Wigner postulated that wave-function reduction takes place when the information arrives at our brain. Thus particle, detector and cat remain in states described by wave functions of the form (17.25) until the box is opened by someone. Then the wave function is reduced.

Although this theory is consistent with all observed results, it is nevertheless unsatisfactory for various reasons. First, the theory assumes that the human mind is of a different nature than the physical material world, to which the brain also belongs. It is difficult to believe this assumption. It should be possible to describe the natural world with objective concepts, independent of our existence or our interaction with nature. Furthermore, the entire problem is shifted into an inaccessible region: if all physical knowledge exists only in the mind, and the mind is not an object of physical analysis, all of physics (natural science) loses its objective relevance. Finally, it is difficult to explain how minds of different persons reach the same conclusions concerning the results of physical experiments, if we do not admit the existence of an objective physical world.

Apart from these objections, the subjective theories do give an explanation of quantum-mechanical measurement, and some philosophers and natural scientists think that they offer the best explanation available at this time.

17.8 Classical Measurements

A completely different hypothesis is based on the postulate that classical mechanics may not be just a limit of quantum mechanics, but an independent theory, in which measured results are definitely fixed at all times. Detectors and cats should comply with physical laws different from those followed by particles belonging to the atomic and subatomic parts of the world. The reduction of the wave function takes place when particles interact with macroscopic objects. Since this interaction cannot be treated quantum mechanically, the problems discussed above do not occur.

This kind of hypothesis has the major disadvantage of introducing two different theories describing, in a complicated manner, the behaviour of material objects of different magnitude. A second problem is how to determine the boundary between macroscopic and microscopic objects, as can again be illustrated with the help of a Stern–Gerlach experiment. As shown in Chap. 1, this experiment is usually performed with atoms [and not with much smaller (microscopic) objects like electrons]. The possible paths of the atoms may be so far apart that they can be treated as classical objects. Furthermore, there is no reason why the experiment should not be performed with the spins of considerably more massive particles, e.g. uranium nuclei. Previously, we noted that the wave function stays unchanged if, after the splitting up, the paths of flight are reunited, so that there is no way to decide which path a particle has chosen. These classical objects should also be led by the guiding field. Furthermore, the beam of classical particles should interfere at a double slit, as discussed at the beginning of this chapter.

It is extraordinarily difficult to perform a Stern–Gerlach experiment in the manner just described. Both possible paths through the apparatus have to be of identical length (up to an accuracy of 10^{-6} m) to reconstruct the original spin wave function. A real test of the possible existence of a guiding field for classical particles has not yet been made. New experiments examining the tunnelling effect of electrical flow across superconducting connections may prove to be such a test. The theory of classical measurement can only be acceptable if the classical result of an experiment differs from the quantum-mechanical predictions.

17.9 The Copenhagen Interpretation

This measurement theory, which has been widely accepted for the longest period of time, was developed by Niels Bohr and his colleagues in Copenhagen. Their premise is the impossibility of separating the quantum-mechanical system from the measuring apparatus. Thus a spin-$\frac{1}{2}$ particle approaching an SGz apparatus has to be considered as a totally different system from a similar particle approaching an SGx apparatus. The problem of wave-function reduction does not occur, since the choice of wave functions (Z_+ and Z_- or χ_+ and χ_-) is determined by the details of the experimental setup. If we reunite both separated beams again as discussed above, we change the experimental setup and thus also change the corresponding wave functions. This aspect of the Copenhagen interpretation is supported by the experimental and theoretical results concerning systems of correlated particles mentioned above. As we have seen, particle pairs behave as a single system with properties that cannot be explained by the properties of the individual particles (see e.g. Chap. 15).

One further idea of the Copenhagen interpretation is that of *complementarity*. Some properties (e.g. position, momentum, the x and z components of angular momentum etc.) form complementary pairs. From this point of view, it follows naturally as a principal property of nature that every attempt to determine one variable yields an uncertainty in the complementary variable. An example is given by polarized light: the question of linearly polarized light being left- or right-handedly polarized is obviously senseless. The opinion of the Copenhagen school is that in quantum mechanics, the attempt to determine position and momentum at the same time is equally senseless.

The Copenhagen interpretation does not consider the nature of the measuring apparatus. Although it does solve the problem of wave-function reduction, it leaves unclear whether macroscopic objects, including measuring apparatuses have to be treated by wave functions, or if a totally different theory for that task is necessary.

17.10 Indelible Recording

Up to now, our discussion has shown that the difficulties of a quantum theory of measurement result from a contradiction between the time-dependent Schrödinger equation [yielding (17.25)] and the reduction postulate; perhaps the reduction postulate is not compatible with quantum mechanics (the Schrödinger equation). Now we will examine this contradiction more carefully, following Belinfante's[6] argumentation, which points out the contradiction and solves it.

We recall that after measurement the wave function of the total system – particle and detector – takes, according to quantum mechanics (the time-dependent Schrödinger equation), the form $\psi = (Z_+\chi_+ + Z_-\chi_-)/\sqrt{2}$ [see (17.25)], whereas, according to the reduction postulate, it should be $Z_+\chi_+$ or $Z_-\chi_-$. Clearly both predictions concern the probabilities of the possible results of a measurement. Therefore they must not be applied to *one* experiment with a single system; a *large number* of experiments has to be performed, either on the same system or on other identical systems, before predictions can be checked. Only if the system is already in an eigenstate of the measuring operator and quantum mechanics predicts a certain result, is a single measurement (a single system) sufficient. A multitude of identical experiments is called an *ensemble*.

Let us consider an ensemble of SGz measurements of particles with a wave function ψ_+. After measurement, the ensemble resides, because of the Schrödinger equation, in a so-called "pure state" with a wave function of the form (17.25), whereas, according to the reduction postulate, the ensemble should be in a "mixed state", i.e. each of the two eigenstates of $\hat{s}_z$ should be populated by one-half of the particles.

Let us consider how to decide whether the ensemble is in a pure or a mixed state. For this a second measurement by a person or another apparatus (e.g. looking at the counting rates of the detectors) is necessary. That measurement described by the operator $\hat{Q}$ should follow the reduction postulate (i.e. the counting rates can be read off unambiguously). If the system is in a pure state, the expectation value of $\hat{Q}$ follows from (17.25):

$$\langle \hat{Q} \rangle = \frac{1}{2}\int\limits_\tau (Z_+^*\chi_+^* + Z_-^*\chi_-^*)\hat{Q}(Z_+\chi_+ + Z_-\chi_-)\,d\tau \ , \tag{17.26}$$

where all variables necessary for the specification of particle and apparatus are contained in the volume element $d\tau$. Multiplication yields

$$\begin{aligned}\langle \hat{Q} \rangle = &\frac{1}{2}\int Z_+^*\chi_+^*\hat{Q}Z_+\chi_+\,d\tau + \frac{1}{2}\int Z_-^*\chi_-^*\hat{Q}Z_-\chi_-\,d\tau\\ &+\mathrm{Re}\left\{\int Z_+^*\chi_+\hat{Q}Z_-\chi_-\,d\tau\right\} \ .\end{aligned} \tag{17.27}$$

Here we have taken into account the Hermiticity of $\hat{Q}$.

[6] F.J. Belinfante: *Measurement and Time Reversal in Objective Quantum Theory* (Pergamon, Oxford 1978).

To calculate the properties of a mixed state, we have to consider that the expectation value of $\hat{Q}$ in a mixed state is equal to the average of the expectation values, which are calculated by separate measurements with the wave functions $Z_+\chi_+$ and $Z_-\chi_-$. Since the number of particles is the same in both states, it holds that

$$\langle\hat{Q}\rangle' = \frac{1}{2}\int Z_+^*\chi_+^*\hat{Q}Z_+\chi_+\,\mathrm{d}\tau + \frac{1}{2}\int Z_-^*\chi_-^*\hat{Q}Z_-\chi_-\,\mathrm{d}\tau \ . \tag{17.28}$$

A comparison of (17.27) and (17.28) shows both expectation values to be identical if

$$Q_{+-} = \int Z_+^*\chi_+^*\hat{Q}Z_-\chi_-\,\mathrm{d}\tau = 0 \ . \tag{17.29}$$

It can immediately be seen that a similar condition also guarantees that the probability distributions [not just the averages, as in (17.27) and (17.28)] of the possible results of the measurement of $\hat{Q}$ are independent of whether the system is in a pure or a mixed state. If we can show that expressions of the form Q_{+-} vanish for all physically possible operators $\hat{Q}$, then pure and mixed states are indistinguishable. Then the reduction postulate would be compatible with quantum mechanics; it would, in fact, be a consequence of it.

Now we want to consider the conditions under which the two states ψ_+ and ψ_- yield a vanishing integral (17.29). The quantity $|Q_{+-}|^2$ can be interpreted as being proportional to the probability of a transition between the states $Z_+\chi_+$ and $Z_-\chi_-$, caused by the action of the measurement operator $\hat{Q}$. If Q_{+-} vanishes, a transition between the states is impossible, meaning that the particle would have changed the state of the detector irreversibly; we could describe it as an *indelible recording of the event*. This is just the property we usually ascribe to a measuring apparatus: it registers the result until it is returned to its initial state by an external action.

So we see that the reduction of the wave function is induced by the measuring process itself, namely, by the recording of the result; it is *not* induced by human perception or the classical behaviour of macroscopical objects. Furthermore, we see that reduction is not absolutely necessary; after an indelible recording has been made, the total system may be described either by a pure form (17.25) or a mixed form consisting of reduced wave functions. Nevertheless, it is more useful to choose reduced wave functions, since the results of further measurements of the particle can be calculated without knowledge of the complicated details of the first apparatus. Reduction enables us to describe an isolated physical system without considering the other systems which have interacted irreversibly with it; we could do so, but this is unnecessary, since we know that a description with pure states yields the same results.

One objection to this argument is the fact that no process is totally irreversible: there is always a small but finite probability, for instance, of the detector changing its state. There are two possibilities: either the apparatus remains isolated, and the change takes place by chance, or the apparatus is manipulated by an external influence (e.g. by a person). In the first case, the erasure of the recording means that the measurement – and thus the reduction – has not taken place, as discussed above. In the second case, we observe that the assumption of a pure state can, in principle, only be made for a wave function describing everything that interacts with the apparatus. Although the experimental result was erased, the information on the detector's state caused an irreversible change in the interacting system (possibly including a human being). As an extreme case, the total universe may be considered one system. But if we assume that the probability of the reversal of an "irreversible" change is finite, we contradict ourselves, because in a universe occupying a state already occupied some time ago, we could not find out whether time has passed or not. Here we assume (contrary to subjective theories) that our minds are a part of the physical universe.

The principle of indelible recording combined with the ideas of the Copenhagen interpretation yields an objective and economical quantum theory of the measurement process. Nevertheless, this theory is not universally accepted. Some scientists think that the difficulty in determining when an indelible recording is made cannot be solved by the argument briefly described here.

17.11 The Splitting Universe

We end our survey of measurement theories with E. Everett's concepts based on the idea of an irreversible change in the universe. He assumes that the universe does not end in one of the various possible states as a result of a measurement, but that all possible results really take place. Accordingly, the universe must split up into a number of different noninteracting – and thus noncommunicating – universes. So if we observe an apparently random result of a quantum-mechanical experiment, all results will have taken place in a totally deterministic manner. We observe a special result in one branch of the splitting universe but other "versions" of our existence observe different results in different branches. Of course, the splitting of the universe is not restricted to the cases in which a measurement is taking place, but occurs for each quantum event. Thus the universe has to be thought of as continuously splitting.

The splitting-universe model treats the problem of wave-function reduction in a manner similar to that of the indelible-recording concept. Integrals like (17.29) vanish, because now the states $Z_+\chi_+$ and $Z_-\chi_-$ belong to different universes not interacting with each other a priori. In fact, many of the ideas in the last section were first developed in connection with the splitting universe and then used by Belinfante for his model.

The concept of a splitting universe is naturally uneconomical, and the idea of an infinite number of universes can never be proved, since individual universes do not interact. This model is thus accepted by only a small number of scientists, although its clearly deterministic character is attractive.

17.12 The Problem of Reality

A fundamental question of any theory of the physical world deals with the nature of that which really exists. This problem is particularly acute in quantum mechanics, but a complete discussion of this philosophical topic would exceed the range of this book. Therefore we restrict ourselves to a short introduction to the basic ideas involved.

For quantum mechanics, the wave function has no direct physical meaning, but is a theoretical construction that can be used to derive the probabilities of later events. On the other hand, we have seen in this chapter that the assumption of additional properties of quantum systems (namely, hidden variables), which are not contained in the wave function, lead to results contrary to experiment. If the wave function is not physical, and there are no hidden variables, what, in fact, *does* exist?

One possible answer to this question is again offered by *subjectivism*: all we know to exist is that which we perceive through our senses; therefore only they are real. Another answer is given by *positivism*, which claims that the question we are posing is senseless, since the existence or nonexistence of objects which we do not perceive cannot be verified. Assuming that we are part of a gigantic brain (computer) where all objects of the universe, their motion etc. exist as fabrications (as computer games), it would not be possible to distinguish between that brain and a real world.

The base of an objective theory of reality could be given by the concept of indelible recording. These recordings or the irreversible changes in the universe really exist; any real reversible change cannot be observed, and therefore statements concerning their existence are pointless. If we call these reversible changes unobservable, we are not necessarily referring to the influence of a human observer: such reversible "events" just have no influence on the later development of the universe. These unimportant phenomena we identify with the total universe regaining a state that already occurred during its evolution, or with the "everyday" example of a spin-$\frac{1}{2}$ particle travelling through one channel of a Stern–Gerlach experiment without having its path recorded. Thus quantum mechanics is a theoretical construction, enabling us to predict sequential, irreversible events in the universe. Although we speak of wave functions and particles, only their irreversible effects can be considered objectively existent.

Finally, we again stress the point that the purpose of this chapter was not to solve the conceptual problems of quantum mechanics, but to show that such problems exist, and that a variety of possible, often amusing answers exist as well. Quantum mechanics has been extraordinarily successful in predicting quantities such as energy spectra, transition probabilities, cross sections etc.; nevertheless, the consideration of problems concerning the philosophy of science can enrich our thinking and provide additional insight into our world and, perhaps, even ourselves.

Subject Index

Names given here in italic will also be found in the bibliographical notes.

Printing: Saladruck, Berlin
Binding: Lüderitz & Bauer, Berlin

Printed in the United States
43556LVS00005B/1-44